ALLE · ZEIT · WACH · 1842

G. Gross S. Jablonska H. Pfister H.E. Stegner
(Editors)

Genital Papillomavirus Infections

Modern Diagnosis and Treatment

With 177 Figures and 65 Tables

Springer-Verlag
Berlin Heidelberg New York
London Paris Tokyo
Hong Kong Barcelona

Editors

Prof. Dr. Gerd Gross
Hautklinik und -Poliklinik
Universitäts-Krankenhaus Eppendorf,
Martinistr. 52, 2000 Hamburg 20, FRG

Prof. Dr. Stefania Jablonska
Department of Dermatology,
Warsaw School of Medicine,
Koszykowa 82 a, 02-008 Warsaw, Poland

Prof. Dr. Herbert Pfister
Institut für klinische und molekulare Virologie,
Universität Erlangen-Nürnberg,
Loschgestr. 7, 8520 Erlangen, FRG

Prof. Dr. Hans-Egon Stegner
Frauenklinik,
Universitätskrankenhaus Eppendorf,
Martinistr. 52, 2000 Hamburg 20, FRG

ISBN-13:978-3-642-75725-9 e-ISBN-13:978-3-642-75723-5
DOI: 10.1007/978-3-642-75723-5

Library of Congress Cataloging-in-Publication Data.
Genital papillomavirus infections: modern diagnosis and treatment/G. Gross ... [et al.] (eds.)
Includes bibliographical references.
ISBN-13:978-3-642-75725-9
1. Papillomavirus diseases. 2. Generative organs – Infections. Gross, G. (Gerd) [DNLM: 1. Genital Neoplasms. Female. 2. Genital Neoplasms. Male. 3. Tumor Virus Infections. WP 145 G331] RC168.P15G46 1990
616.6'50194–dc20 DNLM/DLC for Library of Congress 90-10355

Softcover reprint of the hardcover 1st edition 1990

2127/3020-543210 – Printed on acid-free paper

Dedicated to Harald zur Hausen
whose pioneering studies contributed
so much to our understanding
of the role of HPV in genital cancer

Preface

The International Symposium on Genital Papillomavirus Infections was held in Hamburg on 3–5 February 1989 and is documented in this volume. The primary aim of the symposium was to contribute to the increasing body of information on the importance of papillomaviruses in diverse areas of research and clinical medicine. Special emphasis was placed on the diagnosis and treatment of papillomavirus disease of the genital organs.

This volume was designed to keep the clinician and the clinical laboratory scientist informed of the still growing field of papillomavirus infections of the lower genital tract. As such, it should appeal to medical and graduate students and research fellows as well as to primary investigators and clinical subspecialists interested in tumor viruses and especially in clinical and basic aspects of human papillomavirus (HPV). An attempt has been made to bring together data from clinical, epidemiological, pathological, immunological, molecular genetic, and virological research of HPV infections. Clinical features, modern diagnostic procedures, and treatment modalities available for anogenitally located HPV disease constitute the major part of this book. Nevertheless, sections on the basics of molecular biology and immunology are essential in helping to understand the unique role which papillomaviruses play both as sexually transmitted infectious agents and as possible etiologic factors in the origin and development of genital cancer. The data presented in this volume clearly indicate just how much gynecology, dermatology and venereology have profited from the great progress which has been made in molecular biology and biotechnology in the past decade.

Due to these advances a number of etiologically unknown diseases of the epithelia of the lower genital tract could be linked to the papillomaviruses. Today nucleic acid probes are available for the most frequently detectable genital HPV types which permit type-specific HPV diagnosis both in biopsy specimens and in cellular swabs. One has to take into account, however, that only in the last 15 years increasing attention has been directed to HPV (see Table). Because HPV could not be successfully propagated in tissue cultures, in contrast to the other members of the PAPOVA-virus group, they remained refractory to conventional virological studies from the time Ciuffo reported on the viral nature of human skin warts in 1907. Although the structure and the molecular weight of papillomavirus DNA was determined in 1965 by

Table. Milestones in genital papillomavirus research

Authors		References
Ciuffo (1907)	Initial demonstration of the viral nature of human skin warts	[4]
Shope and Hurst (1933)	Isolation and characterization of cottontail-rabbit (Shope) papillomavirus (CRPV)	[22]
Rous and Beard (1935) Rous and Kidd (1938) Rous and Friedewald (1944)	Development of rabbit skin carcinomas as a synergistic interaction between CRPV and chemical cocarcinogens	[19, 20, 21]
Crawford (1965) Klug and Finch (1965)	Initial determination of structure and molecular weight of papillomavirus DNA from human warts	[5, 13]
Almeida et al. (1969) Dunn and Ogilvie (1969)	Electron microscopic detection and characterization of viral particles in human genital warts	[1, 8]
Zur Hausen et al. (1974)	Initial hybridization studies with cRNA probes derived from a plantar wart: different virus types are responsible for common warts and for genital warts	[24]
Gissmann and zur Hausen (1976) Gissmann et al. (1977) Orth et al. (1977)	Establishment of the plurality of papillomavirus types	[9, 10, 18]
Meisels et al. (1976)	Association between papillomaviruses and cervical dysplasias by cytologic observation of Koilocytotic cells	[17]
Della Torre et al. (1978) Laverty et al. (1978)	Electron microscopic demonstration of papillomavirus particles in cervical dysplasia	[6, 14]
Gissmann and zur Hausen (1980)	Characterization of HPV 6 from a condyloma acuminatum	[11]
Law et al. (1979)	Introduction of gene cloning techniques and hybridization at nonstringent conditions into papillomavirus research	[15]
De Villiers et al. (1981)	Cloning of HPV 6 DNA in bacterial vectors	[23]
Gissmann et al. (1982)	Cloning and characterization of HPV 11 DNA from a laryngeal papilloma	[12]
Duerst et al. (1983)	Characterization of HPV 16 DNA and identification of this virus type in cervical carcinoma biopsies	[7]
Boshart et al. (1984) Lorincz et al. (1986) Beaudenon et al. (1986)	Characterization of HPV 18 DNA [3] and of two further HPV 16-related HPV types HPV 31 [16] and HPV 33 [2]	[2, 3, 16]
Zur Hausen (1986)	Concept of genital cancer as a result of a failing host cell control of persisting viral genes	[25]

Crawford, HPV genomes were not detected in genital warts until 1980 by Gissmann and zur Hausen.

In spite of the rapid progress in HPV detection using molecular virological techniques, the treatment of papillomavirus-related disorders is still non-specific. Selective HPV chemotherapeutics comparable to the nucleoside analogs which inhibit herpes simplex virus replication are not available. The main goal of the chapter on therapy was to review current approaches to the treatment of genital HPV diseases, such as podophyllin application, cryotherapy, and CO_2-laser surgery, while pointing out the potential and the preliminary experience with interferons. These glycoproteins, which exhibit antitumor, antiviral, and immunological properties, have been shown to possess high activity on papillomaviruses. Today, interferons provide a safe and effective treatment for HPV-related genital disorders. The contributions on interferons clearly demonstrate that their use has broadened the spectrum of the therapeutic armentarium in HPV disease.

The role of the host's immune system in the course of papillomavirus infections, which is not yet understood at all, is probably of great importance in the development and regression of lesions but possibly also in the origin of malignant conversion of induced tumors. Upcoming efforts in the development of vaccines, which probably will be of major importance in the prevention of sexually transmitted HPV diseases, are also outlined prospectively.

Some topics in the book are controversial and varying opinions may be represented here. These discussions, however, reflect the rapidly evolving nature of the field. If, at this point, the data are incomplete, they are, nonetheless, sufficiently persuasive to prompt physicians from the various clinical disciplines, virologists, immunologists, pathologists, and epidemiologists to ask questions that have not been asked before. Close cooperation between scientists and clinicians will further the progress in diagnosis, management, and etiology of genital HPV diseases. Hopefully this will result in a more efficient management and control of these viruses and the disorders associated with them. Better comprehension of genital HPV infections may also reveal completely the role of HPV in sexually transmitted disease and as the causal or casual factor in the genesis of cancer of the lower genital tract.

This symposium was initiated and organized by Hans-Egon Stegner of the Department of Gynecology and Obstetrics of the University Hospital Hamburg-Eppendorf and me. About 350 scientists from all parts of the world, especially from Europe, participated. This meeting would not have been possible without the hard work of a large number of people. In particular I would like to thank my colleagues from the scientific committee and the members of the local organizing committee. I wish to express my gratitude to all the authors for their outstanding contributions, and to all the participants of the meeting for the lively discussions. The success of the symposium and the production of this volume are due to the generous support of Hoffmann-La Roche Inc., Basel (Switzerland), and Grenzach-Wyhlen (Federal Republic of Germany), which is gratefully appreciated. I extend special thanks to Theodor Nasemann, previous Director of the Department of Dermatology, University

Hospital Hamburg-Eppendorf for promoting this meeting and to Springer-Verlag for the efficient secretarial organization, the help and continuous cooperation in assisting the publication of this book.

Hamburg, October 1990 Gerd Gross

References

1. Almeida JD, Oriel JD, Stannard LD (1969) Characterization of the virus found in human genital warts. Microbios 3:225–229
2. Beaudenon S, Kremsdorf D, Croissant O, Jablonska S, Wain-Hobson S, Orth G (1986) A novel type of human papillomavirus associated with genital neoplasias. Nature 321:246–249
3. Boshart M, Gissmann L, Ikenberg H, Kleinheinz A, Scheuerlen W, zur Hausen H (1984) A new type of papillomavirus DNA, its presence in genital cancer biopsies and in cell lines derived from cervical cancer. Embo J 3:1151–1157
4. Ciuffo G (1907) Innesto positivo con filtrato di verruca vulgare. G Ital Mal Venereol 48:12–17
5. Crawford LV (1965) A study of papilloma virus DNA. J Mol Biol 13:362–372
6. Della Torre G, Pilotti S, de Palo G, Rilke F (1978) Viral particles in cervical condylomatous lesions. Tumori 64:459–463
7. Duerst M, Gissmann L, Ikenberg H, zur Hausen H (1983) A papillomavirus DNA from a cervical carcinoma and its prevalence in cancer biopsy samples from different geographic regions. Proc Natl Acad Sci USA 80:3812–3815
8. Dunn AE, Ogilvie MM (1968) Intranuclear virus particles in human genital wart tissue: observation on the ultrastructure of epidermal layer. J Ultrastruct Res 22:282–295.
9. Gissmann L, zur Hausen H (1976) Human papillomaviruses: physical mapping and genetic heterogeneity. Proc Natl Acad Sci USA 73:1310–1313
10. Gissmann L, Pfister H, zur Hausen H (1977) Human papillomaviruses (HPV): characterization of four different isolates. Virology 76:569–580
11. Gissmann L, zur Hausen H (1980) Partial characterization of viral DNA from human genital warts (condylomata acuminata) Int J Cancer 25:605–609
12. Gissmann L, Diehl V, Schulz-Coulon H, zur Hausen H (1982) Molecular cloning and characterization of human papillomavirus DNA from a laryngeal papilloma. J Virol 44:393–400.
13. Klug A, Finch JT (1965) Structure of viruses of the papilloma-polyoma type. I. Human wart virus. J Mol Biol 11:403–423
14. Laverty CR, Booth N, Hills E, Cossart Y, Wills EJ (1978) Noncondylomatous wart virus infection of the postmenopausal cervix. Pathology 10:373–378

15. Law MF, Lancaster WD, Howley PM (1978) Conserved polynucleotide sequences among the genomes of papilloma viruses. J Virol 32:199–211
16. Lorincz AT, Lancaster WD, Temple GF (1986) Cloning and characterization of the DNA of a human papillomavirus from a woman with dysplasia of the uterine cervix. J Virol 58:225–229
17. Meisels A, Fortin R (1976) Condylomatous lesions of the cervix and vagina. I. Cytological patterns. Acta Cytol (Baltimore) 20:505–509
18. Orth G, Favre M, Croissant O (1977) Characterization of a new type of human papillomavirus that causes skin warts. J Virol 24:108–120
19. Rous P, Beard JW (1935) The progression to carcinoma of virus-induced rabbit papillomas (Shope). J Exp Med 62:523–548
20. Rous P, Kidd JG (1936) The carcinogenic effect of a papillomavirus on the tarred skin of rabbits. I. Description of the phenomenon. J Exp Med 67:399–422
21. Rous P, Friedewald WF (1944) The effect of chemical carcinogens on virus-induced rabbit carcinomas. J Exp Med 79:511–537
22. Shope RE, Hurst EW (1933) Infectious papillomatosis of rabbits, with a note on the histopathology. J Exp Med 58:607–624
23. De Villiers EM, Gissmann L, zur Hausen H (1981) Molecular cloning of viral DNA from human genital warts. J Virol 40: 932–935
24. Zur Hausen H, Meinhof W, Scheiber W, Bornkamm GW (1974) Attempts to detect virus specific DNA sequences in human tumors. I. Nucleic acid hybridizations with complementary RNA of human wart virus. Int J Cancer 13:650–656
25. Zur Hausen H (1986) Intracellular surveillance of persisting viral infections: human genital cancer resulting from a failing cellular control of papillomavirus-gene expression. Lancet 2:489–491

List of Contributors

J. Auböck
Universitätsklinik für Dermatologie und Venerologie, Anichstraße 35,
6020 Innsbruck, Austria

G. Bandieramonte
Division of Diagnostic Oncology and Outpatient Clinic,
Istituto Nazionale Tumori, Milan, Italy

R. Barrasso
Institut Pasteur and Colposcopy Clinic at Hôspital Pasteur,
24 rue du Docteur Roux, 75015 Paris, France

B. J. Bart
Department of Dermatology, Hennepin County Medical Center,
701 Park Ave South, Minneapolis, MN 55415, USA

K. R. Beutner
Department of Dermatology, University of California, 1516 Napa Street,
Vallejo, CA 94590, USA

O. Braun-Falco
Dermatologische Klinik und Poliklinik der Universität München,
Frauenlobstraße 9–11, 8000 München 2, FRG

E. W. Breitbart
Hautklinik und -Poliklinik, Universitäts-Krankenhaus Eppendorf,
Martinistraße 52, 2000 Hamburg 20, FRG

J. Brzoska
BIOFERON biochemische Substanzen GmbH & Co.,
Erwin-Rentschler-Straße 21, 7958 Laupheim, FRG

P. Buchmann
Departement Chirurgie, Klinik für Viszeralchirurgie,
Universitätsspital Zürich, Rämistraße 100, 8091 Zürich, Switzerland

F. Colla
Departement für Innere Medizin, Universitätsspital Zürich, Gloriastraße 31, 8091 Zürich, Switzerland

E. Dachów-Siwiec
Department of Dermatology, Warsaw School of Medicine, Koszykowa 82a, 02-008 Warsaw, Poland

J. A. DiPaolo
Laboratory of Biology, National Cancer Institute, Bethesda, MD 20892, USA

C. H. Evans
Laboratory of Biology, National Cancer Institute, Bethesda, MD 20892, USA

P. Fritsch
Universitätsklinik für Dermatologie und Venerologie, Anichstraße 35, 6020 Innsbruck, Austria

P. M. Furbert-Harris
Laboratory of Biology, National Cancer Institute, Bethesda, MD 20892, USA

R. Grob
Institut für Immunologie und Virologie, Universitätsspital Zürich, Gloriastraße 30, 8028 Zürich, Switzerland

G. Gross
Hautklinik und -Poliklinik, Universitäts-Krankenhaus Eppendorf, Martinistraße 52, 2000 Hamburg 20, FRG

M. Hagedorn
Hautklinik, Städtische Kliniken Darmstadt, Heidelberger Landstraße 379, 6100 Darmstadt 13, FRG

D. Haina
Gesellschaft für Strahlen- und Umweltforschung mbH, Abteilung für Angewandte Optik, 8042 Neuherberg, FRG

H. zur Hausen
Institut für Virusforschung, Deutsches Krebsforschungszentrum, Im Neuenheimer Feld 506, 6900 Heidelberg, FRG

S. Heinzl
Frauenklinik, Universitätskliniken, Kantonsspital Basel, Schanzenstraße 46, 4031 Basel, Switzerland

R. P. Henke
Institut für Pathologie, Universitäts-Krankenhaus Eppendorf,
Martinistraße 52, 2000 Hamburg 20, FRG

U. Hohenleutner
Dermatologische Klinik und Poliklinik der Universität München,
Frauenlobstraße 9–11, 8000 München 2, FRG

C. Huber
Abteilung Klinische Immunbiologie, Universitätsklinik für Innere Medizin,
Anichstraße 35, 6020 Innsbruck, Austria

H. Ikenberg
Abteilung Frauenheilkunde und Geburtshilfe II, Klinikum der
Albert-Ludwigs-Universität, Hugstetterstraße 55, 7800 Freiburg, FRG

S. Jablonska
Department of Dermatology, Warsaw School of Medicine, Koszykowa 82 A,
02-008 Warsaw, Poland

J. Kellokoski
Department of Oral Pathology and Radiology, Institute of Dentistry,
University of Kuopio, POB 6, 70211 Kuopio, Finland

M. von Knebel-Doeberitz
Institut für Virusforschung/ATV, Deutsches Krebsforschungszentrum,
Im Neuenheimer Feld 506, 6900 Heidelberg, FRG

R. Koronel
Division of Diagnostic Oncology and Outpatient Clinic,
Istituto Nazionale Tumori, Milan, Italy

G. von Krogh
Department of Dermatovenereology, Karolinska Hospital,
104 01 Stockholm, Sweden

M. Landthaler
Dermatologische Klinik und Poliklinik der Universität München,
Frauenlobstraße 9–11, 8000 München 2, FRG

T. Löning
Institut für Pathologie, Universitäts-Krankenhaus Eppendorf,
Martinistraße 52, 2000 Hamburg 20, FRG

S. Majewski
Department of Dermatology, Warsaw School of Medicine, Koszykowa 82 A,
02-008 Warsaw, Poland

R. Mäntyjärvi
Department of Clinical Microbiology, University of Kuopio,
70211 Kuopio, Finland

W. Mazurkiewicz
Institute of Venereology, Warsaw School of Medicine, Koszykowa 82a,
02-008 Warsaw, Poland

M. Meandzija
Geburtshilflich-gynäkologische Klinik und Poliklinik der Universitäts-Frauenklinik, Universität Bern, Schanzenstraße 1, 3012 Bern, Switzerland

K. Milde-Langosch
Institut für Pathologie, Universitäts-Krankenhaus Eppendorf,
Martinistraße 52, 2000 Hamburg 20, FRG

H. F. Nauth
Zytologisches Laboratorium, Alexanderstraße 20, 7000 Stuttgart 1, FRG

D. Niederwieser
Abteilung Klinische Immunbiologie, Universitätsklinik für Innere Medizin,
Anichstraße 35, 6020 Innsbruck, Austria

H.-J. Obert
BIOFERON biochemische Substanzen GmbH & Co.,
Erwin-Rentschler-Straße 21, 7958 Laupheim, FRG

J. D. Oriel
Department of Genito-Urinary Medicine, University College Hospital,
Gower Street, London WC1E 6AU, UK

G. De Palo
Division of Diagnostic Oncology and Outpatient Clinic,
Istituto Nazionale Tumori, Milan, Italy

H. Pfister
Institut für Klinische und Molekulare Virologie,
Universität Erlangen-Nürnberg, Loschgestraße 7, 8520 Erlangen, FRG

S. Pilotti
Division of Anatomic Pathology and Cytology, Istituto Nazionale Tumori,
Milan, Italy

F. Rilke
Division of Anatomic Pathology and Cytology, Istituto Nazionale Tumori,
Milan, Italy

A. Rivière
Institut für Pathologie, Universitäts-Krankenhaus Eppendorf,
Martinistraße 52, 2000 Hamburg 20, FRG

R. Rüdlinger
Dermatologische Klinik, Universitätsspital Zürich, Gloriastraße 31,
8091 Zürich, Switzerland

O. J. Rustad
Department of Dermatology, The University of Minnesota, Box 98,
Mayo Bldg., Minneapolis, MN 55455, USA

J. Saastamoinen
Department of Gynecology and Obstetrics, Kuopio University,
Central Hospital, Koljonniemenkatu 2, 70100 Kuopio, Finland

G. Schlunck
Frauenklinik, Klinikum der Universität Ulm, Prittwitzstraße 43,
7900 Ulm, FRG

S. Schmidt
Institut für Gerontologie, Heimerich-Straße 58, 8500 Nürnberg, FRG

A. Schneider
Frauenklinik, Klinikum der Universität Ulm, Prittwitzstraße 43,
7900 Ulm, FRG

A. Singer
Whittington Hospital, St. Mary's Wing, Highgate Hill,
London N19 5NF, UK

B. Stefanon
Division of Diagnostic Oncology and Outpatient Clinic,
Istituto Nazionale Tumori, Milan, Italy

H.-E. Stegner
Frauenklinik und -Poliklinik, Universitäts-Krankenhaus Eppendorf,
Martinistraße 52, 2000 Hamburg 20, FRG

R. Steiner
Departement für Frauenheilkunde, Frauenklinik, Universitätsspital Zürich,
8091 Zürich, Switzerland

K. Syrjänen
Department of Pathology, University of Kuopio; and Kuopio Cancer
Research Centre, University of Kuopio, POB 6, 70211 Kuopio, Finland

S. M. Syrjänen
Department of Oral Pathology and Radiology, Institute of Dentistry,
University of Kuopio, POB 6, 70211 Kuopio, Finland

K.-H. Tiedemann
Dermatologische Klinik, Universitätsklinikum Rudolf Virchow,
Augustenburger Platz 1, 1000 Berlin 65, FRG

J. C. Vance
Department of Dermatology, Hennepin County Medical Center,
701 Park Ave South, Minneapolis, MN 55415, USA

D. Wagner
Cytologisches Laboratorium, Burgunder Straße 1, 7800 Freiburg, FRG

C. D. Woodworth
Laboratory of Biology, National Cancer Institute, Bethesda, MD 20892, USA

Contents

Epidemiology

Molecular Biology

Diagnosis

Clinical Aspects

Therapy

Epidemiology

HPV in Genital Squamous Cell Tumors: Epidemiology and Clinical Synopsis *

K. J. Syrjänen

Recent Advances in Understanding an Ancient Disease

Condyloma acuminatum (venereal wart) caused by human papillomavirus (HPV) has been recognized as a disease entity since antiquity [19]. During the past 10 years, significant new data on the epidemiology, molecular biology, and biological behavior of HPV infections as well as on their close association with a variety of human squamous cell tumors have been accumulated [2, 6, 8, 9, 12, 16, 21, 25–28, 46–49]. HPV infections in the genital tract are a sexually transmitted disease (STD), and their reported rapid increase in most countries has been attributed to a dramatic changed in sexual habits during the past two decades, e.g., the early onset of sexual activity, frequent sexual partners, poor sexual hygiene, and inadequate preventive measures [3, 19, 20, 34].

Since 1976, it has been well recognized that HPV-induced lesions in the female genital tract (flat, inverted, and papillary condylomas) are frequently (from 50% to 80% in the different series reported) associated with cervical intraepithelial neoplasia (CIN), carcinoma in situ (CIS), and frankly invasive squamous cell carcinomas, even on light microscopic examination [17, 26, 33]. Using immunohistochemical (IP-PAP) techniques, viral structural protein expression can be demonstrated in CIN lesions, CIS, and less frequently in invasive carcinomas [31]. With the recently developed molecular biological techniques, more than 55 different HPV types have been recognized in less than 10 years [2, 9, 16, 21, 22, 25, 28]. Only some of the known HPV types seem to infect the genital tract, i.e., HPV 6, 11, 16, 18, 31, 33, 35, 39, 42, 43, 44, 45, 50, 51, 52, 53. So far, the DNA of few HPV types has been found integrated in the host cell DNA [6, 9, 13, 48, 49]. Based on these integration properties and their frequent detection in invasive carcinomas, HPV types have been classified as low risk types (HPV 6 and 11, 42) and high risk types (HPV 16 and 18, 39), a significant risk for the development of an invasive cancer being ascribed to infections by the latter [22, 23, 36, 37, 48, 49].

Such a potential for clinical progression has been established for cervical HPV infections in a prospective follow-up study conducted in our clinic since 1981, without therapeutic interventions [35–37]. During the 4-year mean follow-up available by now, a certain percentage of the HPV infections regress

* The original studies included in this review have been supported in part by a research grant from the Finnish Cancer Society, by PHS grant number 5 R01 CA 42010-03 awarded by the National Cancer Insitute, DHHS, a research grant from the Social Insurance Institution of Finland, as well as a research grant from the Finnish Cancer Institute

Genital Papillomavirus Infections
Eds.: Gross/Jablonska/Pfister/Stegner

spontaneously (35%), the majority seems to persist (50%), and most importantly, a substantial proportion (15%) makes clinical progress even into the stage of CIS, the immediate precursor of an invasive carcinoma [14, 37, 40]. When analysed using the life-table technique, clinical progression was associated significantly ($p<0.00001$) with the histological grade of the lesions in the first biopsy, the probability of progression varying significantly between the four groups (i.e., NCIN, CIN I, CIN II, and CIN III) [14].

Concomitantly, the data provided by the rapidly progressing molecular biological research have significantly contributed to our understanding of the molecular mechanisms involved in HPV replication and cell transformation by these viruses, as will be discussed later during this meeting [2, 8, 16, 21–23, 25, 30]. Man is the only host of HPV, keratinocyte being the target cell. No applicable in vitro cultivation system exists for papillomaviruses so far [2, 8, 11, 30]. Papillomaviruses are capable of inducing cell transformation in well-defined experimental systems both in vitro and in vivo. Data on the proteins involved in regulation of the viral life cycle are emerging [22, 23]. The recent discovery of HPV DNA in other squamous cell carcinomas (e.g., oral, skin, anal, laryngeal, esophageal, and bronchial) [4, 29, 30, 32, 39] has led to suggestions that certain HPV types might represent at least contributing (if not the single most important) etiological agents in human squamous cell carcinogenesis [29, 48, 49].

The above data have been subjected to a growing number of specialized review articles [2, 12, 16, 21, 22, 25, 27, 28, 46, 49]. The purpose of the present communication is to briefly review the epidemiology of genital HPV infections and to underline the importance of their proper clinical appreciation as potentially precancerous lesions in both sexes, deserving careful examination, treatment, and follow-up to prevent progression to malignancy. The possibilities for detecting the risk groups as well as the approaches needed for an effective prevention are discussed briefly as well.

Natural History of Genital HPV Infections

To elucidate the natural history (i.e., the clinical behavior without treatment) of HPV infections, a prospective follow-up study on women infected with this virus was started by our clinic in 1981. During the past 7 years, more than 500 women have been invited to participate in the study, and by now their cervical HPV lesions have been followed-up for a mean of 4 years (M+SD, 50+20 months), using colposcopy, PAP smears, and punch biopsies, but without any kind of therapy [35–37, 40]. Undoubtedly, the nationwide mass-screening program for detection of cervical precancer lesions continued in Finland since the early 1960s has contributed to the willingness of the women to participate in such a long-term follow-up program.

Using a wide panel of techniques applied to cytological smears and cervical punch biopsy specimens, including light microscopy, immunohistochemistry, monoclonal antibodies, and recombinant DNA technology (i.e., HPV typing with different methods), a substantial amount of new data has been gleaned during the past few years on the biological behavior of genital HPV infections (reviewed in [28] and [29]). According to our experience, the factors associated with the behavior of genital HPV infections and the link between this virus and genital squamous cell cancer can be summarized as follows [7, 27–29, 34–37, 46, 47]:

1. HPV infection in the genital tract constitutes a sexually transmitted disease, and is associated with the risk factors known to predispose women to cervical carcinoma.
2. HPV involvement in benign, precancer, and malignant genital lesions is evidenced by morphological, immunohistochemical and DNA hybridization techniques.
3. The natural history of cervical HPV lesions is equivalent to that of CIN, i.e., they are potentially progressive to carcinoma in situ if left untreated.
4. Genital HPV infections with a long latency most likely exist in both sexes.
5. The malignant potential of HPV lesions seems to depend on the virus type and the physical state of its DNA, i.e., whether or not integrated.
6. The malignant transformation most probably requires synergistic actions between HPV and other carcinogens or infectious agents.
7. The immunological defence mechanisms of the host are likely to modify the course of HPV infections (the efficacy in man remains to be established).

Therapeutic Considerations of Genital HPV Infections

Prompted by the rapidly accumulated evidence on cervical HPV infections as potentially precancerous lesions, an urgent need has developed to critically evaluate the efficiency of the therapeutic measures used against HPV infections. The literature is increasing on the therapeutic approaches used to eradicate genital HPV infections, as lately summarized by von Krogh and Rylander [43]. Currently in our project an extensive series of women is being recruited to determine which is the most appropriate set of therapeutic measures against these infections. The women invited into this treatment group are randomized according to four commonly used means of treatment; conization, cryotherapy, carbon dioxide laser, and interferon [45]. Such a set-up enables comparisons between the treated and nontreated women followed up, concerning the clinical course of their HPV infections. This, in turn, will hopefully help in clarifying the issue on the most effective treatment modalities against this infection. It is to be emphasized that the only way to critically evaluate the treatment practice of HPV infections must be based on a long-term follow-up of the treated patients. Equally important is to relate the treatment results with the natural history of the disease, i.e., the rate of spontaneous regression, obtained from a series of identically followed up nontreated patients [37, 40, 45].

Preliminary results based on analysis of a series of 119 women with HPV infections of the uterine cervix and/or vagina suggest that the efficacy of cryotherapy and CO_2-laser vaporization is very similar [45]. Thus, after a mean follow-up of 14 months (SD 6 months) posttreatment, the cure rates of laser vaporization and cryotherapy were practically identical (>90%), being significantly higher than the spontaneous regression rate ($p<0.001$). This suggests that treatment by either cryotherapy or CO_2-laser vaporization significantly changes the natural history of genital HPV infections. More patients and longer follow-up are still needed, however, to fully establish the efficacy of the different treatment modalities available for gynecological HPV infections [45]. Furthermore, practically nothing is known about the effects of male treatment on the clinical course of HPV infections in the female sexual partners [10, 20]. Our tentative experience with interferon treatment of HPV lesions in both sexes will be summarized separately.

Clinical Significance of Genital HPV Infections

It can be argued that in countries like Finland, where the age-adjusted incidence rates of cervical cancer have dramatically declined since the 1950s (from $137/10^6$ to $60/10^6$ in 1980, attributable to the nationwide mass-screening program in effect since early 1960s), carcinoma of the uterine cervix no longer represents such a major clinical problem as it used to. The same applies to the other Scandinavian countries, except Norway. However, WHO estimates that the annual worldwide incidence of cervical carcinoma is about 450000 cases, and that other squamous cell malignancies of the genital tract account for approximately 150000 additional cases [1, 5, 42]. Even with adequate medical intervention, some 45% to 50% of these patients eventually die of their disease.

Epidemiological evidence strongly suggests that cervical carcinomas derive from a sexually transmitted disease [1, 5, 24, 42, 46]. Thus, many of the risk factors for cervical cancer are similar to those found to predispose women to HPV infections. These include the onset of sexual relations at an early age, sexual promiscuity, lower sosioeconomic status, and multiple episodes of other venereal infections [19, 24, 36]. The association of different genital carcinomas (in both sexes) with HPV is now very convincing, the evidence having been discussed elsewhere [28, 49].

Figures from different clinics demonstrate that the prevalence and incidence of genital HPV infections have dramatically increased in most countries since the 1970s [3, 19]. HPV infections currently pose a serious therapeutic problem, following the establishement of their relationship to genital tract malignancies [10, 20, 43]. Despite the vast number of studies completed during the past few years, no reliable data have been presented on the true prevalence and incidence rates of HPV infections among unselected populations. To assess these figures, a mass-screening program was inaugurated by our group in 1985–1986, which focused on an unselected cohort of 22-year-old women in Kuopio province [41]. Thus, in 2 successive years, the same cohort of women was screened using routine PAP smears. This age group was selected because previous evidence from both retrospectively and prospectively collected series indicated that the peak age of HPV-infected women falls between 20 and 24 years [17, 26].

In 1985, 2013 women were invited of which 1289 attended. One year later, 1768 women of those 2013 were again invited to participate and the number of screenees at the second round was 1069. The prevalence of HPV infection among the 22-year-old women was about 3% at the beginning of the follow-up and about 7% 1 year later. The crude annual incidence was 7.0% (Table 1). The prevalence figures for those attending both rounds of screening and for those attending only once indicate that the nonattenders have an approximately 25% higher risk of HPV infection than the attenders. Adjusting for this source of bias would increase the incidence estimate to 8% [41] (Table 2). Based on these figures, and those derived from a random sample of 2084 (out of 28861) routine PAP smears examined in our laboratory, the lifetime risk of cervical HPV infections was estimated. According to the estimates for the lifetime risk, one-half of the women having a sexually active life would experience at least one HPV infection within 10 years. Up to 79% of the Finnish females would contract at least one HPV infection during the ages

Table 1. Prevalence of HPV infection in 1985–1986 among women born in 1963 and resident in the Kuopio University Central Hospital area

Attendance	Number		Prevalence (%)
	Total	Positive	
Prevalence at 1st round			
All	1289	35	2.7
Attenders of 2nd round	735	19	2.5
Nonattenders of 2nd round	514	16	3.1
Prevalence at 2nd round			
All	1069	76	7.1
Attenders of 1st round	735	51	6.9
Nonattenders of 1st round	334	25	7.5

Table 2. Annual incidence of HPV infection 1985–1986 among women born in 1963 and resident in the Kuopio University Central Hospital area

	Woman-years at risk	Number of new HPV-positive cases	Incidence (%)
Crude	716	50	7.0
Adjusted for nonresponse			8.0

of 20 to 79 years [41]. It should be appreciated, however, that this 79% lifetime risk of contracting HPV infection or even the observed 15% clinical progression rate for HPV infections [14, 40] by no means signify an identical risk for development of cervical cancer (i.e., $0.79 \times 0.15 = 11\%$). This indicates that there are factors regulating the development of an invasive carcinoma from a CIS lesion which are poorly understood at the moment.

It should be pointed out that these figures do not include any estimates on the occurrence of genital HPV infections in males, because no reliable data for the basis of such estimates are currently available [10, 20]. It is also self-evident that HPV infections will reach a clinical importance of an entirely different magnitude, in the event that this virus should prove to be involved in the development of other squamous cell carcinomas, e.g., oral, esophageal, laryngeal, and bronchial. Circumstantial evidence to suggest that this might be the case is available by now, and has recently reviewed [2, 25, 29, 30, 32, 39, 46, 47].

Early Detection and Recognition of the Risk Groups

During the past 10 years in which the association of HPV infections with cervical cancer has appreciated, the treatment of these lesions has varied among the different clinics. In most clinics, however, cervical HPV infections are treated like classical CIN lesions, e.g., by conization or more recently by laser evaporation [43, 45]. This is feasible because no means has been available for distinguishing between HPV lesions at increased risk for progression into cancer from those that will regress spontaneously. As long as clinicians do not have any means to predict the clinical outcome of the disease, overtreatment of some HPV infections (those eventually regressing) is inevitable. This is in spite of the fact that a cone treatment administered to these young women not infrequently leads to insufficiency of the cervical channel, resulting in failure to conceive or making these women vulnerable to spontaneous abortions [43]. In the long run, this might have widespread consequencies in countries with low birth rates in general.

One of the central aims of our follow-up project, from its very beginning, has been to develop measures that could distinguish HPV-infected women at increased risk for progression from those eventually regressing without treatment [36]. Based on our experience from the long-term prospective follow-up and aided by the modern DNA techniques, the outlook is reasonably good that this will be achieved in the future [29, 37]. There seems to be a higher risk for clinical progression in infections caused by HPV 16, as compared with any other HPV types analysed (HPV 6, 11, 18, 31 and 33) [14, 37, 40]. More follow-up data are needed, however, on the clinical behavior of the newly recognized types HPV 31, 33, 35, and 39, which have been found in invasive genital carcinomas as well. The other important prognostic determinants seem to be the histological grade of the lesion at the time of diagnosis and to some extent, the grade of cellular atypia in the PAP smear, as has recently been discussed [14, 37, 40].

Undoubtedly, because routine PAP smear is the only feasible means to conduct mass screening of genital HPV infections in women, this will remain the technique on which early detection of this disease will be based also in the future [17, 33]. In practice, it may be possible to detect the risk patients by techniques applicable to PAP smears or cervical punch biopsies, using one of the DNA-hybridization techniques. At the moment, a most suitable technique for routine use seems to be the in situ DNA hybridization, which can be applied directly on PAP smears and punch biopsies. The major advantage of the latter is the possibility to complete the grading of CIN in the same biopsy, which is another important prognostic predictor [14, 28, 37]. So far, however, the limiting factor preventing the adoption of in situ hybridization into routine use has been the radioactive HPV DNA probes used [11]. However, techniques are being currently developed where the radioactive label in the probes is to be replaced by nonradioactive markers such as biotin [38]. When the problems related to the slightly inferior sensitivity of these new techniques are solved these methods will most certainly become the first applicable in the large-scale screening of genotype-specific HPV infections [38]. As emphasized previously recognition of the lesions at risk for malignant transformation will have major implications in therapeutic considerations of genital HPV infections [43, 45].

Outlook for Effective Prevention

HPV infection has been shown to constitute a sexually transmitted infectious disease, the epidemiology and biological behavior of which have been elucidated to some detail only recently [19, 28, 34, 36, 37, 48]. Males harboring the infection are suspected to transmit the virus to their female sexual partners, but only fragmentary data are available on the major epidemiological parameters, such as length of the incubation period, efficacy of transmission, virus reservoirs outside the genital tract, and of course modes of transmission other than sexual [10, 19, 20, 24]. This is because of the lack of well-controlled partner studies, where both sexual partners had been examined, treated, and prospectively followed-up [15, 18]. This in turn is due to the fact that the lesions are asymptomatic, and only a small fraction of HPV infections in the male genital tract manifest themselves, the vast majority being subclinical and undetectable by the naked eye [7, 10]. The proper examination of males necessitates use of colposcopy and application of 5% acetic acid on the entire genital area, as will be repeatedly emphasized throughout this volume [10, 20, 44].

To elucidate the role of the male as potential carrier, the systematic examination of the male sexual partners of HPV-infected women was started in the Kuopio project in 1986. By combining colposcopy, biopsy, and HPV DNA typing, the main aim is to assess the role of the male as a reservoir of latent or subclinical HPV infections and their nonintended transmission to the female sexual partners [7]. This also permits screening of the possible risk males, because HPV typing can be applied to examination of the biopsy specimens or scrape samples from the male genital lesions.

In our approach, the male sexual partners of the women included in the treatment group are invited to participate. This enables us to evaluate the efficacy of the different treatment modalities on transmission, reinfection, and eventual cure of HPV infections in both partners. Because of the fact that all the male partners of the treatment group women cannot be traced, two groups will arise spontaneously those women whose partners are examined and treated, and those whose partners are not. Such a design permits critical evaluation of the true effects of male treatment on the clinical course of female HPV infections. Except for providing important data on the epidemiology of this infection, this information will hopefully help in designing the optimal schedule of examination and treatment for both sexes.

Ultimately, if such an effective treatment schedule can be established, these lines could lead to effective intervention in the sexual transmission of HPV, and could hopefully offer a possibility to prevent the spread of the disease. These expectations are based on the concept that early detection of HPV infection and screening of the risk groups in both sexes should offer the best grounds for early treatment, and thereby forms a sound basis for prevention of this increasingly common sexually transmitted disease. In this respect, we should not forget the advantages offered by vaccination, even if the realistic prospects for development of an effective vaccine seem to lie only in the distant future.

References

1. Brinton LA (1986) Current epidemiological studies – emerging hypotheses. Banbury Rep 21:17–28
2. Broker TR, Botchan M (1986) Papillomaviruses: retrospectives and prospectives. Cancer Cells 4:17–36
3. Chuang T-Y, Perry HO, Kurland LT, Ilstrup DM (1984) Condyloma acuminatum in Rochester, Minn, 1950–1978. I. Epidemiology and clinical features. Arch Dermatol 120:469–475
4. de Villiers E-M, Schneider A, Gross G, zur Hausen H (1986) Analysis of benign and malignant urogenital tumors for human papillomavirus infection by labelling cellular DNA. Med Microbiol Immunol 174:281–286
5. Doll R (1986) Implications of epidemiological evidence for future progress. Banbury Rep 21:321–332
6. Dürst M, Gissmann L, Ikenberg H, zur Hausen H (1983) A papillomavirus DNA from a cervical carcinoma and its prevalence in cancer biopsy samples from different geographic regions. Proc Natl Acad Sci USA 80:3812–3814
7. Ferenczy A, Mitao M, Nagai N, Silverstein SJ, Crum CP (1985) Latent papillomavirus and recurring genital warts. N Engl J Med 313:784–788
8. Giri I, Danos O (1986) Papillomavirus genomes: from sequence data to biological properties. Trends Genet 2:227–232
9. Gissmann L, Boshart M, Dürst M, Ikenberg H, Wagner D, zur Hausen H (1984) Presence of human papillomavirus in genital tumors. J Invest Dermatol 83:26s–28s
10. Gross G (1987) Lesions of the male and female external genitalia associated with human papillomaviruses. In: Syrjänen KJ, Gissmann L, Koss L (eds) Papillomaviruses and human disease. Springer, Berlin Heidelberg New York, pp 197–234

11. Gupta JW, Gendelman HE, Naghashfar Z, Gupta P, Rosenshein N, Sawada E, Woodruff JD, Shah KV (1985) Specific identification of human papillomavirus type in cervical smears and paraffin sections by in situ hybridization with radioactive probes: a preliminary communication. Int J Gynecol Pathol 4:211–218
12. Howley PM (1983) The molecular biology of papillomavirus transformation. Am J Pathol 113:414–421
13. Ikenberg H, Gissmann L, Gross G, Grussendorf-Conen E-I, zur Hausen H (1983) Human papillomavirus type 16-related DNA in genital Bowenś disease and bowenoid papulosis. Int J Cancer 32:563–565
14. Kataja V, Syrjänen K, Mäntyjärvi R, Väyrynen M, Syrjänen S, Saarikoski S, Parkkinen S, Yliskoski M, Salonen JT, Castrén O (1989) Prospective follow-up of cervical HPV infections: life-table analysis of histopathological, cytological and colposcopic data. Eur J Epidemiol 5:1–7
15. Levine RU, Crum CP, Herman E, Silvers D, Ferenczy A, Richart RM (1984) Cervical papillomavirus infection and intraepithelial neoplasia: a study of male sexual partners. Obstet Gynecol 64:16–20
16. McCance DJ (1986) Human papillomaviruses and cancer. Biochem Biophys Acta 823:195–205
17. Meisels A, Fortin R, Roy M (1976) Condylomatous lesions of cervix and vagina. I. Cytologic patterns. Acta Cytol 20:505–509
18. Obalek S, Jablonska S, Beaudenon S, Walczak L, Orth G (1986) Bowenoid papulosis of the male and female genitalia: risk of cervical neoplasia. J Am Acad Dermatol 14:433–444
19. Oriel JD (1981) Genital warts. Sex Transm Dis 8:326–329
20. Oriel JD (1987) Genital and anal papillomavirus infections in human males. In: Syrjänen KJ, Gissmann L, Koss L (eds) Papillomaviruses and human disease. Springer, Berlin Heidelberg New York, pp 182–196
21. Pfister H (1984) Biology and biochemistry of papillomaviruses. Rev Physiol Biochem Pharmacol 99:112–181
22. Pfister H, Fuchs PG (1987) Papillomaviruses: particles, genome organisation and proteins. In: Syrjänen KJ, Gissmann L, Koss L (eds) Papillomaviruses and human disease. Springer, Berlin Heidelberg New York, pp 1–18
23. Schwarz E (1987) Transcription of papillomavirus genomes. In: Syrjänen KJ, Gissmann L, Koss L (eds) Papillomaviruses and human disease. Springer, Berlin Heidelberg New York, pp 444–466
24. Singer A, Reid BL, Coppleson M (1976) A hypothesis: the role of a high-risk male in the etiology of cervical carcinoma. Am J Obstet Gynecol 126:110–115
25. Smith KT, Campo MS (1985) Papillomaviruses and their involvement in oncogenesis. Biomed Pharmacother 39:405–414
26. Syrjänen KJ (1983) Human papillomavirus (HPV) lesions in association with cervical dysplasias and neoplasias. Obstet Gynecol 62:617–622
27. Syrjänen KJ (1984) Current concepts on human papillomavirus (HPV) infections in the genital tract and their relationship to intraepithelial neoplasia and squamous cell carcinoma. Obstet Gynecol Surv 39:252–265
28. Syrjänen KJ (1986) Human papillomavirus (HPV) infections of the female genital tract and their associations with intraepithelial neoplasia and squamous cell carcinoma. Pathol Ann 21:53–89
29. Syrjänen KJ (1987) Papillomaviruses and cancer. In: Syrjänen KJ, Gissmann L, Koss L (eds) Papillomaviruses and human disease. Springer, Berlin Heidelberg New York, pp 468–503
30. Syrjänen SM (1987) Human papillomavirus infections in the oral cavity. In: Syrjänen KJ, Gissmann L, Koss L (eds) Papillomaviruses and human disease. Springer, Berlin Heidelberg New York, pp 104–137
31. Syrjänen KJ, Pyrhönen S (1982) Immunoperoxidase demonstration of human papilloma virus (HPV) in dysplastic lesions of the uterine cervix. Arch Gynecol 233:53–61

32. Syrjänen K, Syrjänen S (1987) Human papillomavirus in bronchial squamous cell carcinomas. Lancet I:168–169
33. Syrjänen KJ, Heinonen UM, Kauraniemi T (1981) Cytological evidence of the association of condylomatous lesions with the dysplastic and neoplastic changes in uterine cervix. Acta Cytol 25:17–22
34. Syrjänen K, Väyrynen M, Castren O, Yliskoski M, Mäntyjärvi R, Pyrhönen S, Saarikoski S (1984) Sexual behaviour of the females with human papillomavirus (HPV) lesions in the uterine cervix. Br J Vener Dis 60:243–248
35. Syrjänen K, de Villiers E-M, Väyrynen M, Mäntyjärvi R, Parkkinen S, Saarikoski S, Castrén O (1985) Cervical papillomavirus infection progressing to invasive cancer in less than three years. Lancet I:510–511
36. Syrjänen K, Väyrynen M, Saarikoski S, Mäntyjärvi R, Parkkinen S, Hippeläinen M, Castrén O (1985) Natural history of cervical human papillomavirus (HPV) infections based on prospective follow-up. Br J Obstet Gynaecol 92:1086–1092
37. Syrjänen K, Mäntyjärvi R, Parkkinen S, Väyrynen M, Saarikoski S, Syrjänen S, Castrén O (1986) Prospective follow-up in assessment of the biological behaviour of cervical HPV-associated dysplastic lesions. Banbury Rep 21:167–177
38. Syrjänen S, Partanen P, Syrjänen K (1987) Comparison of the in situ DNA hybridization protocols using ^{35}S-labeled probes and biotin-labeled probes in detection of HPV DNA sequences. Cancer Cells 5:329–336
39. Syrjänen S, Syrjänen K, Mäntyjärvi R, Collan Y, Kärjä J (1987) Human papillomavirus (HPV) DNA sequences in squamous cell carcinomas of the larynx demonstrated by in situ DNA hybridization. J Otorhinolaryngol Relat Spec 49:175–186
40. Syrjänen K, Mäntyjärvi R, Saarikoski S, Väyrynen M, Syrjänen S, Parkkinen S, Yliskoski M, Saastamoinen J, Castren O (1988) Factors associated with progression of cervical human papillomavirus (HPV) infections into carcinoma in situ during a long-term prospective follow-up. Br J Obstet Gynaecol 95:1096–1102
41. Syrjänen K, Hakama M, Saarikoski S, Väyrynen M, Yliskoski M, Syrjänen S, Kataja V, Castrén O (1990) Prevalence, incidence and estimated life-time risk of cervical human papillomavirus (HPV) infections in nonselected Finnish female population. Sex Transm Dis (in press)
42. Vessey MP (1986) Epidemiology of cervical cancer: role of hormonal factors, cigarette smoking and occupation. Banbury Rep 21:29–43
43. von Krogh G, Rylander E (1987) Clinical management of human papilloma virus infections. In: Syrjänen KJ, Gissmann L, Koss L (eds) Papillomaviruses and human disease. Springer, Berlin Heidelberg New York, pp 296–333
44. Väyrynen M (1986) Natural history of cervical HPV infections. Colposcopic assessment and biological considerations. Publ Univ Kuopio Orig Rep 2:1–83
45. Yliskoski M, Saarikoski S, Syrjänen K, Syrjänen S, Castrén O (1990) Cryotherapy and CO_2-laser vaporization in the treatment of cervical and vaginal human papillomavirus (HPV) infections. Acta Obstet Gynecol Scand (in press)
46. zur Hausen H (1977) Human papillomaviruses and their possible role in squamous cell carcinomas. In: Current topics in microbiology and immunology, vol 78. Springer, Berlin Heidelberg New York, pp 1–30
47. zur Hausen H (1982) Human genital cancer: synergism between two virus infections or synergism between a virus infection and initiating events? Lancet II:1370–1372
48. zur Hausen H (1986) Intracellular surveillance of persisting viral infections. Human genital cancer results from deficient cellular control of papillomavirus gene expression. Lancet II:489–491
49. zur Hausen H, Gissmann L, Schlehofer JR (1984) Viruses in the etiology of human genital cancer. Prog Med Virol 30:170–186

Papillomavirus Infection in the Male: Results of a Peniscopic Study

G. De Palo, R. Koronel, B. Stefanon, S. Pilotti, G. Bandieramonte, and F. Rilke

Introduction

Human papillomavirus (HPV) infection is considered a sexually transmitted disease (STD). Therefore the male sexual partner should play a compulsory role in the epidemiology of HPV infection. Nevertheless, whereas a large number of contributions have appeared in recent years on HPV infection in females, the frequency of HPV infection in male partners of women with HPV infection or HPV-associated cervical intraepithelial neoplasia (CIN) or vulvar intraepithelial neoplasia (VIN), the sites and the morphologic patterns of HPV infection in males are unclear.

The aim of this study was to determine the frequency, morphology, and sites of HPV infection in male sexual partners of females with HPV infection or HPV-associated CIN or VIN.

Materials and Methods

Study population

All male partners of women with HPV infection of the lower genital tract or with HPV-associated CIN or VIN seen in the Colposcopic Clinic of the Division of Diagnostic Oncology were advised to undergo a medical examination in the same Division.

After recording the case history on previous treatments for condylomata acuminata, gonorrhea, syphilis and herpes, number of partners and sexual habits with special reference to the use of condoms, a careful clinical examination of the external genitalia and a colposcopic examination of the penis were performed. The latter examination was performed in three phases: a direct examination under the colposcope without acetic acid solution; thereafter the penis and the scrotum were covered with gauzes soaked in 5% acetic acid solution for 3–5 min; and finally an observation with the colposcope at different magnifications of the penile shaft, foreskin, glans, collum glandis penis and urethral meatus. In the presence of visible lesions by colposcope guided microbiopsies under colposcope without local anesthesia were performed. Photographs were taken in selected cases.

From March to December 1987, 196 male sexual partners of women with HPV infection of the lower genital tract or HPV-associated CIN or VIN were examined.

Genital Papillomavirus Infections
Eds.: Gross/Jablonska/Pfister/Stegner

Table 1. Frequency of peniscope positivity for HPV infection in 196 males

Female histology	Peniscopy		Aspect of Lesions			
	Positive	Negative	Florid	White patch	Macula	Fingering
HPV-infection						
Cervix ± vagina ± vulva	40	25	1	7	30[a]	2
Vagina ± vulva	53	36	3[b]	10	34	6
HPV-associated CIN II-III	29	5	1[b]	2	25	1
HPV-associated CIN II + VIN II–III	7	–	–	–	7	–
VIN III	1	–	–	–	1	–
Total	130 (66.3%)	66	5 (3.9%)	19 (14.6%)	97 (74.6%)	9 (6.9%)

[a] One case had contemporaneous presence of macular and fingering lesions
[b] One case had contemporaneous presence of florid and macular lesions

The diagnosis in females were performed by colposcopy and histology: 154 had HPV infection of the lower genital tract, 34 had HPV-associated CIN grade II or III, 7 had CIN III plus VIN II–III, and 1 had VIN grade III (Table 1).

The median age of the male patients was 35 years (range 18–66 years). All patients reported monogamous relationships and all patients were asymptomatic. Six patients had been treated elsewhere with electrodiathermy or cryosurgery for condylomata acuminata; these patients have been considered elsewhere as cured. No patient had had other venereal diseases. Eleven patients had circumcision at a young age for phimosis. Only three males always began intercourse with a condom.

Males positive for HPV infection were treated with CO_2-laser vaporization. All males positive and treated, such as patients negative at the first control were followed every 4–6 months.

Peniscopic Terminology

The peniscopic aspects of HPV infection in the male have been classified into four types, as follows:

Florid lesion. This type of lesion has the aspect of an exophytic proliferation, with an irregular surface, made up of rough asperities. Evident to the naked eye it may be single or multiple. The color is that of the subject's skin or mucosa. The penile shaft, foreskin, and glans may be equally affected. On the glans and internal face of the foreskin it may have a cauliflower appearance. After 5% acetic acid solution it appears snow-white (Fig. 1).

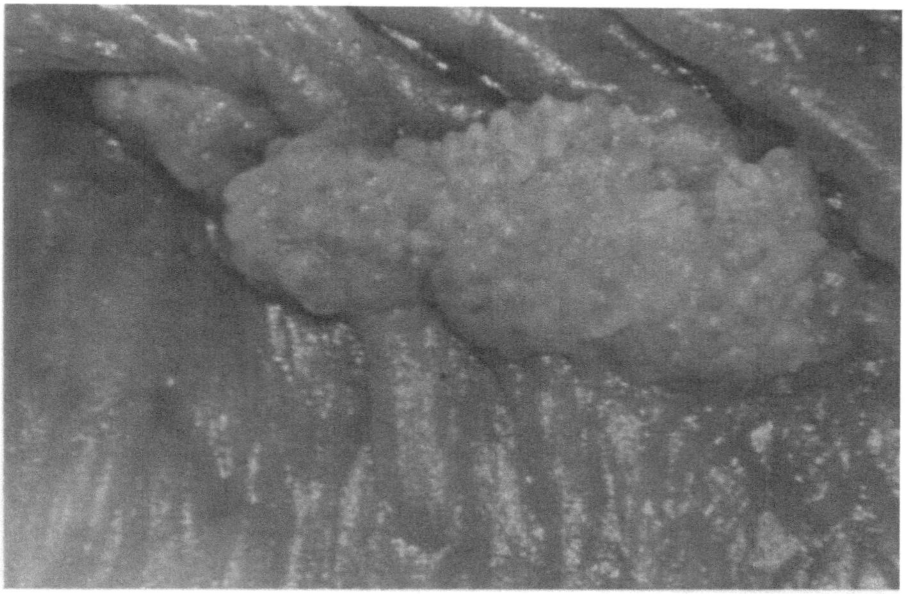

Fig. 1. Florid HPV infection

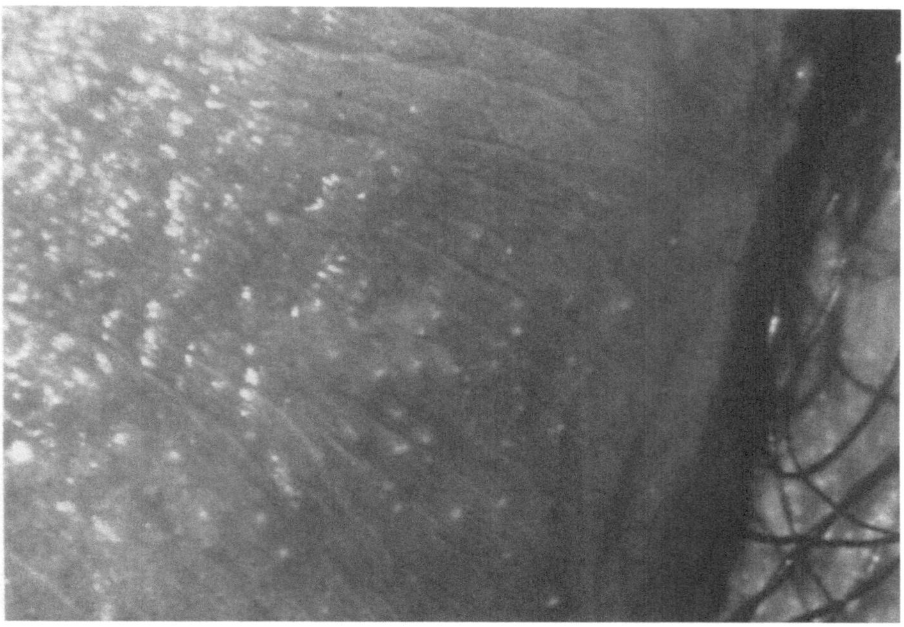

Fig. 2. White patch HPV infection

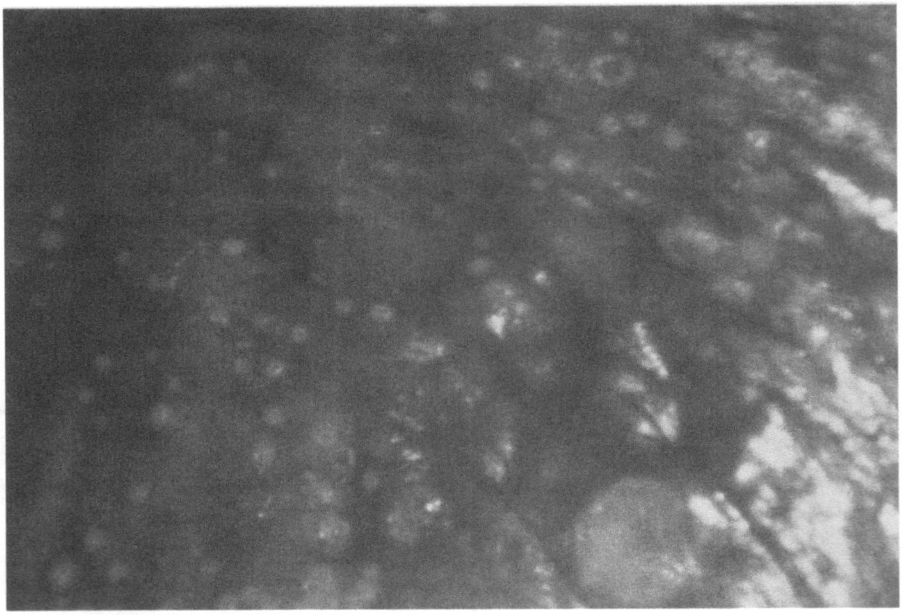

Fig. 3. Macular HPV infection

White patch. It is the presence of circumscribed, small, slightly raised microproliferations with a strong acetoreactivity, usually multiple, and localized in the external-internal surface of the foreskin and the penile shaft. White patch condylomatosis appears only after long exposure to a 5% acetic acid solution (Fig. 2), whereas without acetic acid these lesions may be clinically identifiable as small asperities with a yellow color when localized in the penile shaft.

Macula. It is characterized by the presence of well-demarcated, translucent, flat or slightly elevated, multiple (sometimes more than 10), round, small (1–3 mm or more) and aceto-white areas with regular borders; their surface is smooth or umbilicate. These lesions appear only after prolonged exposure to a 5% acetic acid solution. Their color is snow-white or ice-white. At high magnification punctation may be sometimes observed. The glans and internal surface of the foreskin are the sites of macules (Fig. 3).

Fingering. This is a particular aspect consisting of a smooth aceto-white area on the apex of those acuminate structures distributed in rows circumferentially around the coronal sulcus that are of angiofibromatous nature known as "hirsutoid papillomas of the coronal margin of the glans penis" [19] or "pearly penile papules" [1]. It is a superimposed HPV infection on these curious wart-like conditions (Fig. 4).

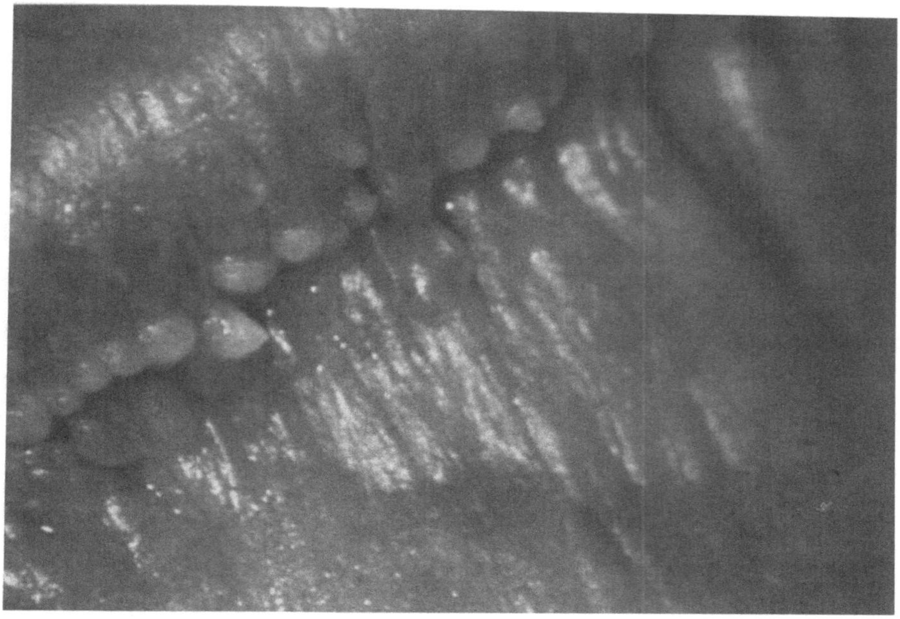

Fig. 4. Fingering HPV infection

Histologic Diagnostic Criteria

The minimal requirement for the histologic diagnosis of HPV infection of the penile epithelium was the presence of small foci of unequivocal koilocytosis with bi- or multinucleation and/or dyskeratosis. Acanthosis and parakeratosis alone were not accepted as sufficient evidence. The histologic criteria for the diagnosis of intraepithelial neoplasia associated with HPV infection have been previously described [13]

Results

The histologic diagnosis of female patients and the peniscopic findings in positive males are reported in Table 1. Of 196 patients, HPV-related lesions were found in 130. Of these, 5 had florid condylomatosis visible to the naked eye, 2 on the penile shaft and 3 on the foreskin, and 125 had lesions visible only with peniscopy. Of 3 patients with florid condylomatosis visible on the foreskin, 2 had the synchronous presence of macular lesions. Five patients refused biopsy. Of 125 biopsed patients on the more suggestive area, 101 were positive for HPV infection. Only 1 patient had penile intraepithelial neoplasia (PIN)

Table 2. Correlation between peniscopy and histology

Female histology	Males positive by histology	Peniscopy			
		Florid	White patch	Macula	Fingering
HPV-infecfion					
Cervix ± vagina ± vulva	32/38[a]	1	4	26	1
vagina ± vulva	39/51[a]	3[b]	7	24	5
HPV-associated CIN II–III	23/28[c]	1[b]	–	21	1
HPV-associated CIN III + VIN II–III	7/7	–	–	7	–
HPV-associated VIN III	0/1	–	–	–	–
Total	101/125 (81%) 101/191 (52.9%)	5	11	78	7

[a] Two patients refused biopsy
[b] One case had contemporaneous presence of florid and macular lesions
[c] One patient refused biopsy

Table 3. Site and type of HPV infection in 101 histologically positive males

Female histology	Glans		Glans + internal face of foreskin	Internal face of foreskin			Shaft or external face of foreskin	
	Macula	Finger-ing	Macula	Florid	White patch	Macula	Florid	White patch
HPV-infection								
Cervix ± vagina ± vulva	–	1[a]	3	1	1	23	–	3
Vagina ± vulva	–	5	2	1[b]	5	22	2	2
HPV-associated CIN II–III ± VIN II–III	1	1	7	1[b]	–	20	–	–
Total	1	7	12	3	6	65	2	5

[a] Contemporaneous presence of macular lesions
[b] Contemporaneous presence of florid and macular lesions

grade II. The incidence of penile HPV infection histologically documented was 53% (Table 2).

Males who used condoms for intercourse were negative. In 4 of 6 patients previously treated for condylomata acuminata and considered cured, peniscopy showed evidence of macular lesions.

The preferential site of HPV infection in the male was the internal surface of the foreskin where it was observed in 86/101 (85%). Infrequent sites were the glans and the shaft alone (8% and 7% respectively). The most common type was the macular lesion, which accounted for 78/101 (77%) cases (Table 3).

The correlation with the type of HPV infection in females showed that of 41 male partners of women with HPV-associated CIN or VIN (one patient refused biopsy), 30 were positive (73%) and in one of these PIN grade II was diagnosed. In the group of partners of women with HPV infection alone (154 cases, 4 of whom refused biopsy) 71 were positive (47.3%). The aspect of HPV infection was macular in 28/30 (93.3%) in the first group and in 50/71 (70%) in the second group.

Discussion

HPV infection is a sexually transmitted disease, although transmission by fomites is occasionally possible [10]. It is clear, therefore that male sexual partners should play a role in the epidemiology of HPV infection and may be at risk for venereal transmission of the HPV. Nevertheless, most male sexual partners of women with HPV infection have no evident HPV infection.

Clinically detectable lesions of male genitalia are condyloma acuminatum [11] and bowenoid papulosis [18]. Subclinical papilloma virus infection has been described by many authors [2, 3, 9, 14–16].

In our series of 196 patients examined, 53% were histologically positive for HPV infection, which is inferior to that reported by other authors [2, 9, 15, 16]. Nevertheless, it should be remembered that of our 130 positive cases florid condylomatosis visible to the naked eye was determined in only 3.8% versus percentages of 17% [16], 28.5% [9], and 33% [2] reported in other series, and that in our series strict criteria of histologic diagnosis were adopted, although the presence of HPV DNA in about 45% of clinically suspected lesions without histologic characteristics of HPV infection has been reported in other series [3].

As in cervical and vulvar HPV lesions, the morphologic aspect of HPV infection in the male is not unique, and the natural history of these lesions is unknown. It is not clear whether the different forms are an expression of their evolution from subclinical to florid condylomatosis or whether they are an expression of different types of HPV.

Peniscopy under long exposure to a 5% acetic acid solution is today the most significant tool to evaluate the male for HPV infection. It reveals subclinical HPV infection as macular and white patch lesions in other wises undiagnosed, as demonstrated by the fact that in this series of 5 cases with florid condylomatosis visible to the naked eye, 2 had contemporaneous macular lesions, and of 6 patients previously treated and judged cured, 4 had persistent HPV infection in the form of macules. Nevertheless the aceto reactivity seems not to be specific for HPV infection. An aceto reactivity of genital mucosa may be observed in the presence of a specific inflammation and also in the presence of healing epithelium after electrodiathermy or cryosurgery treatments, as reported in other series [3]. In the first cases it is possible to observe before acetic acid a diffuse mild erythema of the glans and foreskin.

Since the percentage of positivity on the penis is inferior to that expected, the urethra, the prostate and the seminal vesicles have been suspected to be the reservoir of the HPV. DNA sequences of HPV 5 and 2 were found in the semen of patients with chronic HPV infection [12]. Although HPV 5 and 2 are not associated with lesions of the genital tract, the presence of viral genome indicates that sexual transmission of HPV may occur by this route [12].

Condylomata acuminata or exophitic papillomas in the urethra have been reported by many authors [4, 7, 8, 11]. Whereas condylomata acuminata of the urethra are always symptomatic with symptoms similar to malignancy, flat condylomata could be asymptomatic. They could be responsible for venereal transmission and could spontaneously regress or persist without transformation into papillomatous symptomatic forms. At this time the frequency of flat condylomatous lesions of the male urethra is unknown. Their presence might be ascertained with intraurethral endoscopy, urinary cytology, and intraurethral cytology. Inconsistent results have been reported with urinary cytology [9] and urethral cytology [16, 17]; in the recent series of Cecchini et al. [6] urethral cytology by cytobrush was positive for HPV infection in 43% of partners of women with HPV infection in the absence of visible lesions by peniscopy.

Few data have been reported on the correlation between female and male HPV infection, as regards the type of lesions and the correlation with intraepithelial neoplasia. In the series of Barrasso et al. [3] lesions showing histologic features of condyloma were found in 41% of partners of women with cervical flat condyloma and in only 5% of partners of women with CIN; in contrast lesions with histologic features of PIN were found in 33% of partners of women with CIN and only in 1% of partners of women with cervical flat condyloma. From our series emerges a difference in percentage of positivity for HPV infection among men whose female partners had HPV-associated CIN or VIN, or HPV infection alone of the lower genital tract. The first group showed histologic positivity for HPV infection in 73% and the second group in 47% of cases. The aspect of lesions was macular in 93% in the first group and in 70% in the second one. Nevertheless from our series it seems that HPV is not prone to cause atypia in the male. These data, that may be explained on the basis of the different epithelium of the penis in comparison with that of the cervix, are in accordance with the fact that an epidemic of PIN in association with HPV infection has not been reported in countries where HPV infection in females is frequent [5].

In conclusion, although the logical reservoir for the transmission of HPV infection would be the male, the frequency of HPV infection in the external genitalia of male sexual partners of women with HPV infection or HPV-associated CIN or VIN is largely inferior to what would be expected. Other sites such as the prostate, urethra, and seminal vesicles should be investigated using other techniques to adequately screen HPV infection in the male. Screening and treatment of HPV infection in men is useful in the prevention of infection and reinfection in women, and in view of the possible risk in the men themselves.

References

1. Ackerman BA, Kornberg R (1973) Pearly penile papules. Acral angiofibromas. Arch Dermatol 108:673–675
2. Barrasso R, Guillemotonia A, Catalan F, Coupez F, Siboulet A (1986) Lesions genitales masculines a papillomavirus. Interet de la colposcopie. Ann Dermatol Venereol 113:787–795
3. Barrasso R, de Brux J, Croissant O, Orth G (1987) High prevalence of papillomavirus-associated penile intraepithelial neoplasia in sexual partners of women with cervical intraepithelial neoplasia. N Engl J Med 317:916–923
4. Bissada NK, Cole AT, Fried FA (1974) Extensive condylomas acuminata of the entire male urethra and the bladder. J Urol 112:201–203
5. Boon ME, Schneider A, Hogewoning CJA, Van der Kwast TH, Bolhuis P, Kok LP (1988) Penile studies and heterosexual partners. Peniscopy, cytology, histology, and immunohistochemistry. Cancer 61:1652–1659
6. Cecchini S, Cipparrone I, Confortini M, Scuderi A, Meini L, Piazzesi G (1988) Urethral cytology of cytobrush specimens. A new technique for detecting subclinical human papillomavirus infection in men. Acta Cytol 32:314–317
7. Dean P, Lancaster WD, Chun B, Jenson B (1983) Human papilloma virus structural antigens in squamous papillomas of the male urethra. J Urol 129:873–875
8. Debenedictis TJ, Marmar JL, Praiss DE (1977) Intraurethral condylomas acuminata: management and review of the literature. J Urol 118:767–769
9. Levine RR, Crum CP, Herman E, Silvers D, Ferenczy A, Richart RM (1984) Cervical papillomavirus infection and intraepithelial neoplasia: a study of male sexual partners. Obstet Gynecol 64:16–20
10. McCance DJ, Campion MJ, Baram A, Singer A (1986) Risk of transmission of human papillomavirus by vaginal specula. Lancet 2:816–817
11. Oriel JD (1971) Natural history of genital warts. Br J Vener Dis 47:1–13
12. Ostrow RS, Zachow KR, Niimura M, Okagaki T, Muller S, Bender M, Faras AJ (1986) Detection of papillomavirus DNA in human semen. Science 231:731–733
13. Pilotti S, Rilke F, De Palo G, Della Torre G, Alasio L (1981) Condylomata of the uterine cervix and koilocytosis of cervical intraepithelial neoplasia. J Clin Pathol 34:532–541
14. Rosemberg SK (1985) Subclinical papilloma viral infection of male genitalia. Urology 26:554–557
15. Sand PK, Bowen LW, Blischke SU, Ostergard DR (1986) Evaluation of male consorts of women with genital human papilloma virus infection. Obstet Gynecol 68:679–681

16. Sedlacek TV, Cunnane M, Carpiniello V (1986) Colposcopy in the diagnosis of penile condyloma. Am J Obstet Gynecol 154:494–496
17. Stefanon B, Pilotti S, Koronel R, Alasio L, Rilke F, De Palo G (1987) Papilloma virus infection (PVI) in sexual partners of women affected by PVI of the lower genital tract associated or not to intraepithelial neoplasia. 6th World Congress Pathology and Colposcopy, San Paulo, May 11–14 (abstr MT-09)
18. Wade TR, Kopf AW, Ackerman AB (1978) Bowenoid papulosis of the penis. Cancer 42:1890–1903
19. Winer JH, Winer LH (1955) Hirsutoid papillomas of coronal margins of glans penis. J Urol 74:375–378

Male HPV-Associated Lesions: Epidemiology and Diagnostic Criteria

R. Barrasso and G. Gross

Epidemiology

Genital warts are benign proliferations associated with human papillomavirus (HPV) types 6 and 11 and are assumed to be sexually transmitted. Recently, epidemiologic data have been accumulated on the sexual transmission of other genital papillomaviruses, now considered to be potentially oncogenic. Cervical condyloma or intraepithelial neoplasia have been detected in 76% of female partners of men with genital warts [1]. Moreover, cervical intraepithelial neoplasia has been found in partners of men with bowenoid papulosis [2] or HPV-16 associated atypical lesions of external genitalia, and the coexistence of cervical intraepithelial neoplasia and vulvar bowenoid papulosis has been reported.

With the introduction of the colposcope [3] and the acetic acid test [4] for examination of male genitalia, 53%–80% of male partners of women with cervical flat condyloma or intraepithelial neoplasia have been shown to present genital lesions mostly displaying histologic features of HPV infection [3–5]. More than half of these lesions are subclinical since they appear only after the application of acetic acid [6]. Most interestingly, subclinical lesions may show features of intraepithelial neoplasia [6] and may contain potentially oncogenic HPVs [7–9].

Histologic features of lesions detected in regular partners have been shown to correlate, in that about one-third of male partners of women with cervical intraepithelial neoplasia present penile intraepithelial neoplasia [7], and 9% of partners of women with grade III cervical intraepithelial neoplasia present grade III penile intraepithelial neoplasia [9], while about 40% of partners of women with cervical flat condyloma show penile condyloma, but only 2% show intraepithelial neoplasia [7]. This strongly supports the sexual transmission and the oncogenic potential of HPVs and further suggests that different HPV types have different specific cytopathic effects.

When the presence of HPV infection has been virologically studied in women with disease and in their regular partners, one-third [10] to one-half [11] of couples have been shown to harbor the same HPV type. In our series (G. Orth, R. Barrasso, B. Huynh et al., in preparation), regular partners with typical condylomata acuminata of external genitalia are mostly infected by HPV type 6 or 11. About 50% of partners of women with cervical intraepithelial neoplasia present genital lesions. In 32% of cases, as we have shown [7], histology demonstrates penile intraepithelial neoplasia, and virology detects the same potentially oncogenic HPV type (60% of cases) or dif-

Genital Papillomavirus Infections
Eds.: Gross/Jablonska/Pfister/Stegner

ferent potentially oncogenic HPV types in specimens from partners, but never HPV 6, 11, or 42. The same figures are evidenced in regular partners with lesions of external genitalia showing features of vulvar and penile intraepithelial neoplasia. However, in about 20% of partners of women with intraepithelial neoplasia, the histology of the male lesions shows features of condyloma, papilloma, or only minimal histologic changes. In such cases, different HPV types are found in specimens from partners. Thus, out of 100 regular male partners of women with HPV 16-associated cervical intraepithelial neoplasia, we may expect to find about 20 individuals presenting HPV 16-associated penile intraepithelial neoplasia. This confirms, in our opinion, the existence of the male reservoir of sexually transmissible, potentially oncogenic HPVs and the need for screening and treatment of male lesions.

Lesions are often numerous on external genitalia. Moreover, infection may be multifocal in women (cervix, vagina, vulva). Double infection may be detected in single specimens or in specimens from different genital sites. The rate of double infection is higher in specimens from scrapings than in specimens from biopsies, thus suggesting that the sampling technique could account for some of the discrepancies when only limited biopsy sampling is performed for viral studies. Prolonged incubation times, acquired immunity, and early regression of lesions could also explain some of the discrepancies, thus underlying our lack of knowledge of the biology of HPV infection. These factors render the study of the sexual transmission of HPV difficult.

Studies on the use of urethral cytology to screen men has produced unsatisfactory results [3, 12]. Viral DNA has been identified in penile scrapings from men with colposcopically and histologically normal tissue [13, 14], thus stressing the probably high prevalence of latent infection and, again, our poor knowledge of the mechanisms of its reactivation.

The finding of DNA of potentially oncogenic HPVs in clinical and subclinical lesions from male genitalia and their prevalence in regular partners of women with HPV-associated disease suggests the existence of a male reservoir of these viruses, underlying the need for the identification and treatment of these lesions. For this, colposcopy is currently considered the most precise technique to identify HPV-associated lesions of male genitalia.

Diagnosis

Condylomata acuminata are exophytic protuberances with lobated or irregular surface, pink-reddish or white-grayish depending mainly on the location. If numerous, they become confluent. They are prevalent on the inner aspect of the prepuce [15]. Extension of warts to the proximal urethra has rarely been reported [16]. From a practical point of view, this means that systematic urethroscopy is not needed in men with meatal warts. Condylomata extending to the surrounding areas and the anal condylomas all have characteristics of typical condylomata acuminata. Condylomata acuminata must be dis-

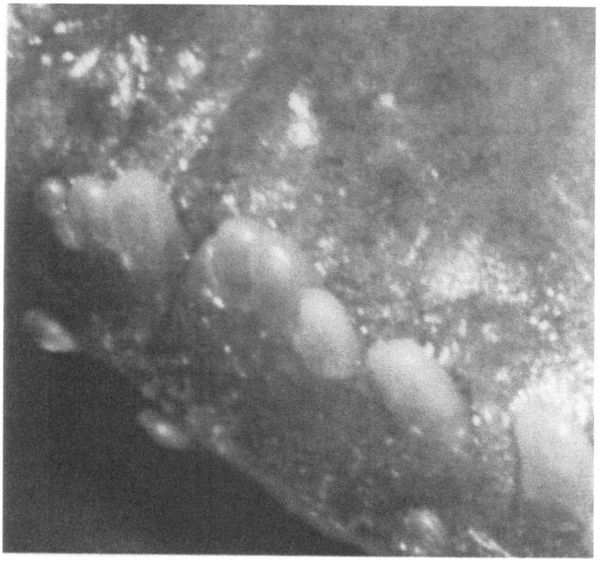

Fig. 1. Physiologic pearly papules of the coronal sulcus. HPV DNA research on specimens from such lesions is always negative

tinguished from pearly papules [17], which appear as parallel rows of discrete acuminate structures distributed circumferentially around the coronal sulcus. Colposcopic magnification helps in their distinction from condylomata acuminata since their surface is mostly smooth, and they do not show the typical vascular pattern presented by mucosal condylomata (Fig. 1).

Histologically, condylomata acuminata present an epidermal proliferation with variously pronounced hyper- and parakeratosis and, generally, koilocytosis. However, after months of persistence, lesions may not show koilocytosis and thus resemble papillomas. HPV DNA types 6 and 11 are detected in most condylomata acuminata [18–20].

Some genital papillomas have been found to be associated with HPV infection. These proliferative lesions have a more papillomatous surface and are larger than condylomata acuminata. Usually pedunculated, they are darker than the surrounding skin. They are preferentially located on the shaft. Some lesions closely resemble skin warts. Histology shows pronounced epidermal proliferation and hyperkeratosis. Koilocytosis is usually absent. Pigmented papillomas mostly contain HPV 6 or 11.

Cutaneous wart-like lesions (Fig. 2) localized on genital and perianal skin may contain cutaneous HPV types, as yet uncharacterized HPVs, or rarely HPV 6. These lesions must be considered infectious, since they may be found in regular partners of women with vulvar condylomas or papillomas or with cervical flat condyloma.

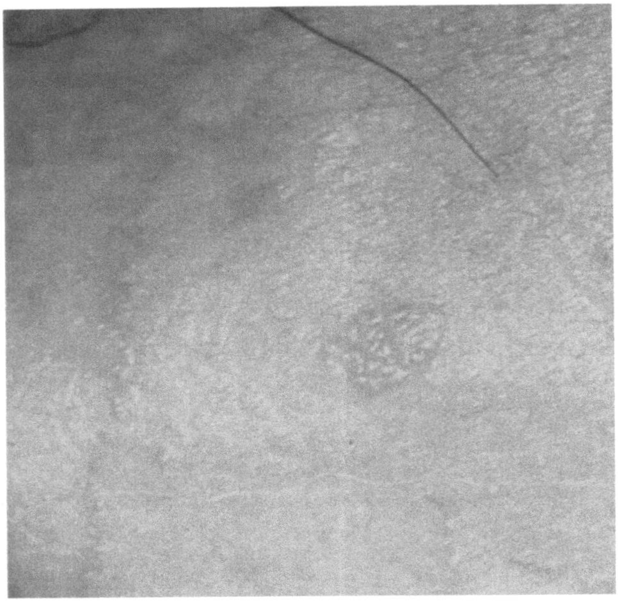

Fig. 2. Cutaneous wart-like genital lesions. Histology showed features of papilloma, without koilocytosis. Virologic studies detected the DNA of an as yet uncharacterized HPV type

Papules are defined as clearly outlined lesions, variably elevated but not pedunculated, with round or dome-shaped, slightly hyperkeratotic or smooth surface. The term bowenoid papulosis was introduced to describe some of them, mostly reddish or brownish, showing histologic features of Bowen's atypia. Correlative colposcopic, histologic, and virologic studies have allowed the characterization of several types of papular lesions [10, 21], which can be distinguished by their localization and by their color. Nonpigmented papules are frequently evidenced on the penile shaft, alone or together with condylomata acuminata. Some of them represent early condylomata, but others grow maintaining a papular morphology. Pigmented papules on the penile shaft and other cutaneous genital areas are easily identifiable since they are mostly brownish or even blackish. When located on mucosa, papules may be translucent, reddish, or leukoplakia-like. They mostly show a clear vascular punctuation on the top [7], allowing their distinction from small pearly papules. Often small, they may be undetected without the aid of the colposcope. The application of 5% acetic acid produces a strong, well-demarcated reaction which usually preserves the punctate vessels.

Histologically, nonpigmented and translucent papules generally show features of condyloma. Epidermal proliferation is discrete and often endophytic. On the other hand, the histology of pigmented, red, or leukoplasic papules shows features of intraepithelial neoplasia, i.e., epidermal proliferation with

a

b

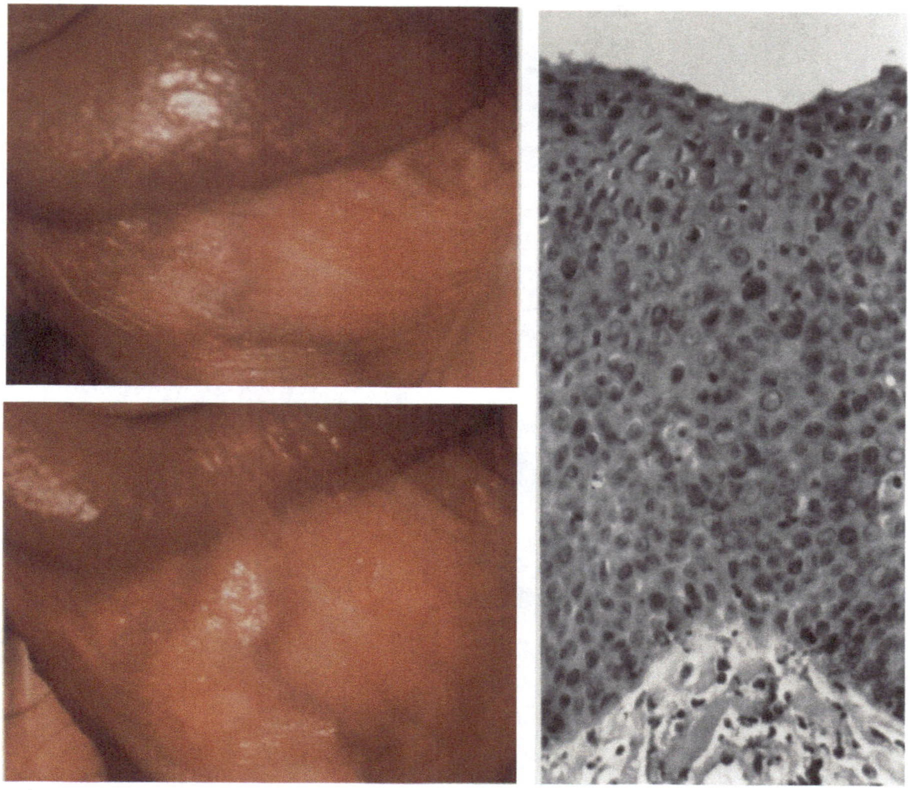

c

Fig. 3 a–c. Erythematous macules. **a** Red areas before acetic acid. **b** Marked whitening after the application of acetic acid. **c** Histology: intraepithelial neoplasia. Virologic studies detected HPV 16

numerous, partially abnormal mitotic figures and atypical pleomorphic cells with hyperchromatic and clumped nuclei, and pronounced dyskeratosis. Virologic studies confirm the morphologic distinction into two categories [21]; nonpigmented and translucent papules mostly contain the same HPVs as evidenced in genital condylomas. In cases of pigmented, red, or leukoplasic papules there is a clear prevalence of HPV 16 [2, 10, 21–23]. The other potentially oncogenic HPV types may also be detected [24]. For the clinician the identification, among papules, of the potentially atypical lesions is thus of great importance since a diagnostic biopsy is recommended and the treatment guidelines will not be the same.

As previously described, the application of acetic acid on penile epithelium may produce the appearance of white areas. Among areas whitening after the acetic acid test on penile epithelium, a distinction must be made among macules and nonspecific acetowhitening since the latter is not associated with HPV infection. Macules are defined [7] as well-demarcated flat areas showing

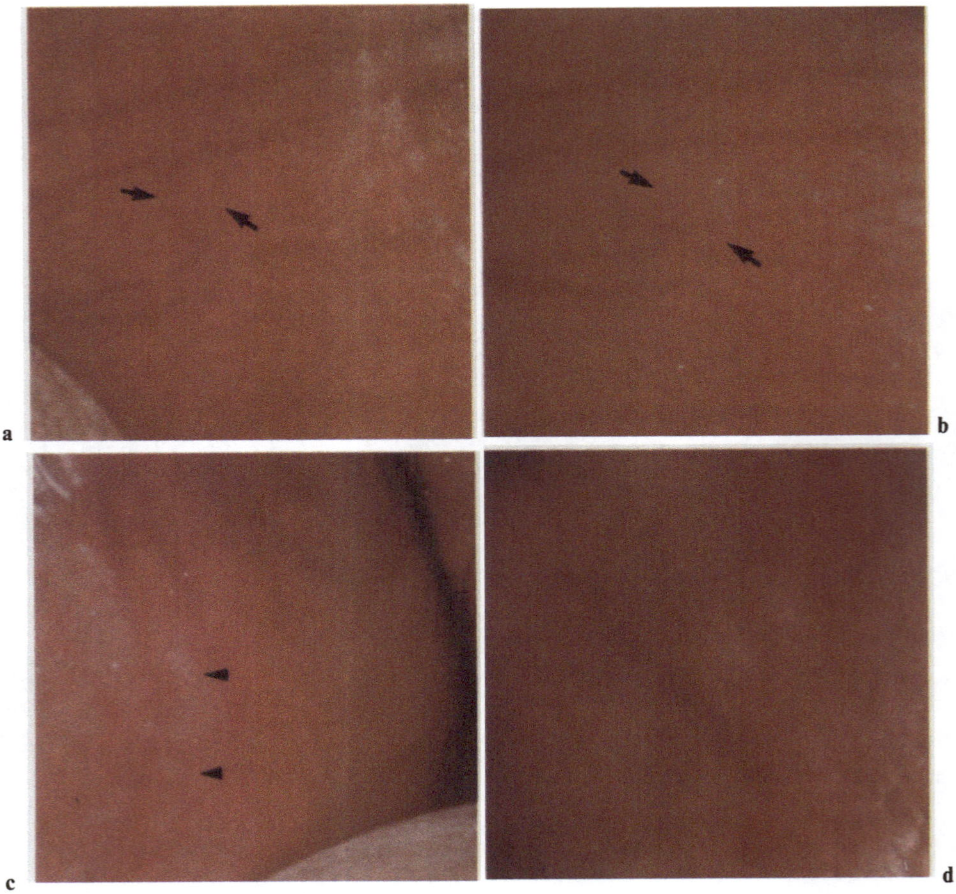

Fig. 4 a–f. Macules. **a** Epidermal appearance before application of acetic acid. **b** The same lesion after application of acetic acid. **c, d** Penile macules detected only after application of acetic acid in two other patients. **e, f** Histology of lesions shown in c and d, respectively: papillomatosis, marked acanthosis, parakeratosis, absence of clear koilocytosis. Virologic studies detected HPV 42 DNA in lesions from the three patients

a white reaction after acetic acid application and containing colposcopically detectable capillary loops (Figs. 3, 4). They correspond to pink or reddish areas with capillary loops detected by the colposcope or to areas of epithelial surfaces that had been normal on colposcopy before the test [6]. When macules are defined on the basis of these morphologic criteria, almost all of them contain HPV DNA and show histologic features of HPV infection. Macules are usually found on the prepuce or gland or, occasionally, in the urethral meatus. They are rareley detected on the penile skin. Macules may be the only feature, or they may be associated with condylomata acuminata or

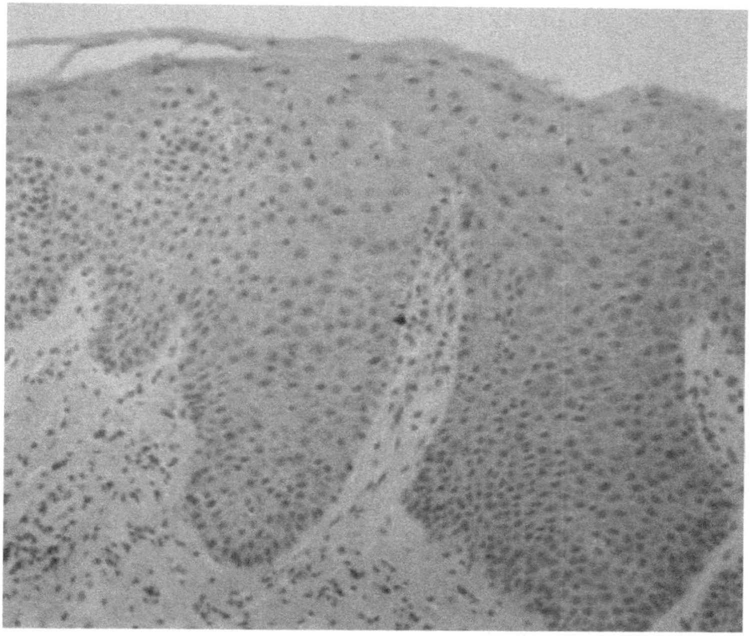

e

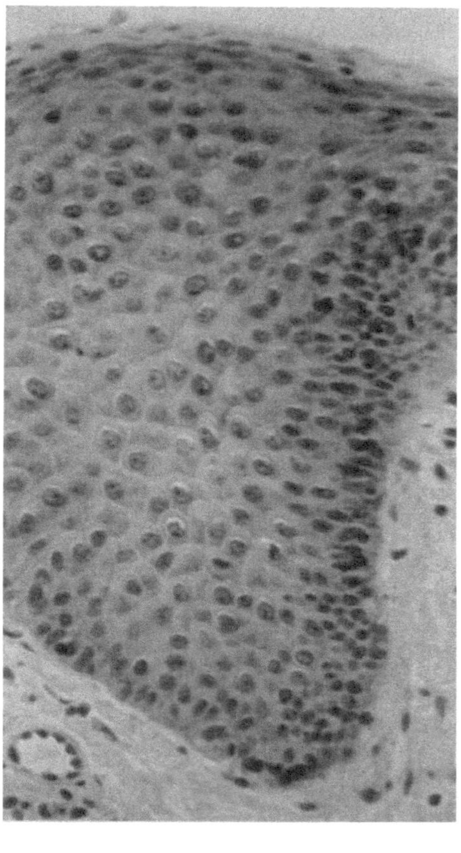

f

papules. They may even occur around clinically identifiable lesions, thus better showing the extent of viral infection. Moreover, their appearance at a site where previous condylomas have been treated and healed is suggestive of the persistence of HPV infection.

The colpohistologic correlative study of macules has allowed a further distinction [21]. Areas whitening on a previously erythematous epithelium almost exclusively show histologic features of intraepithelial neoplasia (Fig. 3). For therapeutic purposes, therefore, examination before acetic acid application is most relevant since erythematous macules must be treated as atypical lesions, as well as red or pigmented papules. Virologic studies confirm these observations since erythematous macules usually contain potentially oncogenic HPVs [21].

As for macules whitening on a previously normal epithelium, the colposcopic feature (Fig. 4) can be described as a small (few millimeters in diameter), well-delimited, white area. Before the application of acetic acid the penile mucosa looks clinically normal. The high magnification of the colposcope, however, may disclose a tiny white halo or an epithelial area devoid of and surrounded by the normal penile vessels. A well-defined white area appears 2–3 min after application of acetic acid. Its surface is rough but never shows asperities or micropapillae. Often, the compactness of the white reaction may give the impression of the existence of a slightly elevated area even in the presence of macules. Capillary vessels are very well identified all over the area.

Histology always shows papillomatosis, marked acanthosis, and a discrete parakeratosis. Koilocytosis is discrete, mostly focal, or absent. Nuclear atypia and atypical mitotic figures are almost always absent. The term endophytic papilloma has been proposed [21] for such lesions. HPV type 42 is by far the most prevalent type encountered (more than 70% of positive specimens), and as yet uncharacterized HPV types are associated with most of the remaining specimens [7, 21]. Papillomavirus types 6 and 11 and potentially oncogenic HPVs are virtually never associated with these macules. These data show that this lesion exhibits specific morphologic and histologic features, and that it is associated mostly with infection by a specific HPV type; condylomata acuminata have also been associated with infection by HPV 6 and 11.

In situ hybridization is positive in more than two-thirds of such lesions (O. Croissant, personal communication), even in the absence of koilocytosis, suggesting that they may be infectious, and that they should be treated.

We stress that penile macules must be distinguished from nonspecific acetowhitening, mostly devoid of capillary loops (Fig. 5). These features are not associated with HPV infection but usually have an irritative or inflammatory origin. Such areas generally show histologic inflammatory features and/or very mild degrees of acanthosis and parakeratosis, without koilocytosis or papillomatosis. HPV DNA is virtually never detected in specimens from such areas. From the pathologist's point of view, it must therefore be stressed that parakeratosis alone does not represent a sufficiently specific feature of HPV infection.

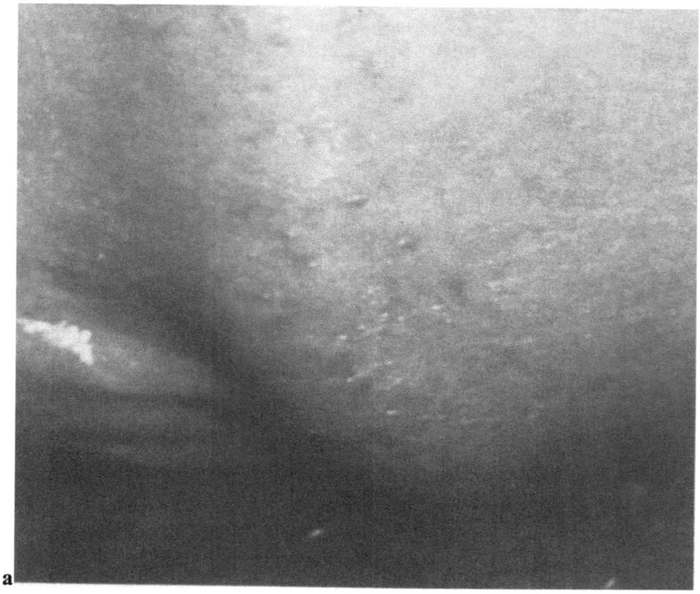
a

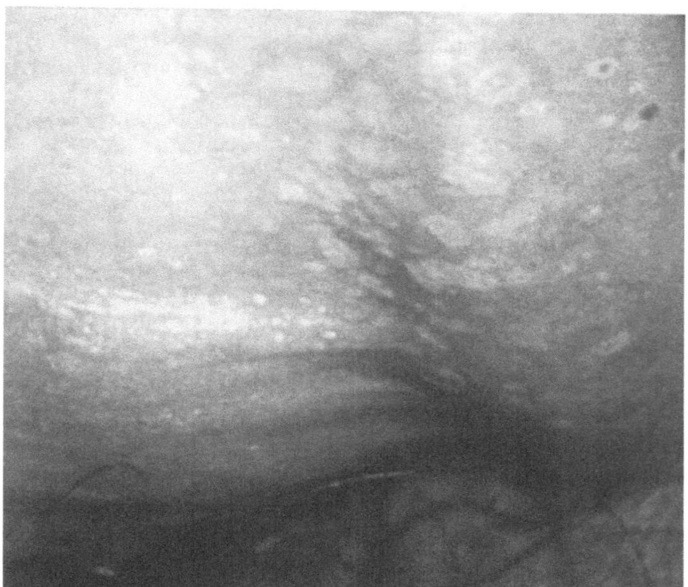
b

Fig. 5 a, b. Nonspecific acetowhitening of penile epithelium. **a** Erythematous areas before acetic acid. **b** Whitening after acetic acid. Virologic studies did not detect HPV DNA sequences. The colposcopic examination after treatment for *Candida albicans* was normal

From a clinical point of view, we stress that more than 80% of white reactions of penile epithelium are not specific. Thus, in the absence of a clear vascular punctuation in such areas at the colposcopic examination or when white areas are detected without a colposcope, local treatment is recommended and a diagnostic biopsy or treatment for warts should be performed only in case of persistence of the white areas.

In conclusion, colposcopy, histology, and virology allow the identification of HPV-associated subclinical lesions of male genitalia, avoiding overdiagnosis of the frequently encountered inflammatory reactions of penile epithelium.

Epidemiologic evidence accumulated so far suggests that these lesions must be considered a reservoir of sexually transmissible virus and must therefore be treated for the prevention of cervical cancer.

Acknowledgements. We are indebted to Gerard Orth for all the virologic studies reported in this chapter, for personal communications, and for his advice and encouragement during our research. We are very grateful to Odile Croissant for personal communications and fruitful discussions.

References

1. Campion MJ, Singer A, Clarkson PK et al. (1985) Increased risk of cervical neoplasia in consorts of men with penile condylomata acuminata. Lancet 1:943
2. Obalek S, Jablonska S, Beaudenon S et al. (1986) Bowenoid papulosis of the male and female genitalia: risk of cervical neoplasia. J Am Acad Dermatol 14:433
3. Levine RU, Crum CP, Herman E et al. (1984) Cervical papillomavirus infection and intraepithelial neoplasia: a study of the male sexual partner. Obstet Gynecol 64:16
4. Sedlacek TV, Cunnane M, Carpiniello V (1986) Colposcopy in the diagnosis of penile condyloma. Am J Obstet Gynecol 154:494
5. Sand PK, Bowen LW, Blischke SO et al. (1986) Evaluation of male consorts of women with genital human papillomavirus infection. Obstet Gynecol 68:679
6. Barrasso R, Guillemotonia A, Catalan F et al. (1986) Lésions génitales masculines à papillomavirus: intérêt de la colposcopie. Ann Dermatol Venereol 113:787
7. Barrasso R, de Brux J, Croissant O et al. (1987) High prevalence of papillomavirus-associated penile intraepithelial neoplasia in sexual partners of women with cervical intraepithelial neoplasia. N Engl J Med 317:916
8. Syrjänen SM, von Krogh G, Syrjänen KJ (1987) Detection of human papillomavirus DNA in anogenital condylomata in men using in situ DNA hybridization applied to paraffine sections. Genitourin Med 63:32
9. Campion MJ, McCance DJ, Mitchell HS et al. (1988) Subclinical penile human papillomavirus infection and dysplasia in consorts of women with cervical neoplasia. Genitourin Med 64:90
10. Gross G, Ikenberg H, de Villiers EM et al. (1986) Bowenoid papulosis: a venereally transmitted disease as reservoir for HPV 16. Banbury Rep 21:149
11. Schneider A, Sawada E, Gissmann L et al. (1987) Human papillomaviruses in women with a history of abnormal Papanicolaou smears and in their male partners. Obstet Gynecol 69:554
12. Nahhas WA, Marshall RL, Ponziani J et al. (1986) Evaluation of urinary cytology of male sexual partners of women with cervical intraepithelial neoplasia and human papilloma virus infection. Gynecol Oncol 24:279

13. Grussendorf-Conen El, de Villiers EM, Gissmann L (1986) Human papillomavirus genomes in penile smears of healthy men. Lancet 2:1092
14. Rosemberg SK, Reid R, Greenberg M et al. (1988) Sexually transmitted papillomaviral infection in the male. II. The urethral reservoir. Urology 32:47
15. Von Krogh G, Syrjänen SM, Syrjänen KJ (1988) Advantage of human papillomaviruses typing in the clinical evaluation of genitoanal warts. Experience with in situ deoxyribonucleic acid hybridization technique applied on paraffine sections. J Am Acad Dermatol 18:495
16. Gartman E (1956) Intraurethral verruca acuminata in men. J Urol 75:717
17. Johnson BL, Baxter DL (1964) Pearly penile papules. Arch Dermatol 90:166
18. De Villiers EM, Gissmann L, zur Hausen H (1981) Molecular cloning of viral DNA from human genital warts. J Virol 40:932
19. Gissmann L, Diehl V, Schultz-Coulon HJ et al. (1982) Molecular cloning and characterisation of human papillomavirus DNA derived from a laryngeal papilloma. J Virol 44:393
20. Gissmann L, Wolnik L, Ikenberg H et al. (1983) Human papillomaviruses types 6 and 11 DNA sequences in genital and laryngeal papillomas and in some cervical cancers. Proc Natl Acad Sci USA 80:560
21. Barrasso R, Jablonska-S (1989) The clinical, colposcopic and histologic spectrum of male HPV-associated genital lesions. In: Winkler B, Richart RM (eds) Human papillomaviruses. Elsevier, New York. Clin Pract Gynecol 2:73
22. Ikenberg H, Gissmann L, Gross G et al. (1983) Human papillomavirus type 16-related DNA in genital Bowen's disease and in bowenoid papulosis. Int J Cancer 32:563
23. Obalek S, Jablonska S, Orth G (1985) HPV-associated intraepithelial neoplasia of external genitalia. Clin Dermatol 3:104
24. Beaudenon S, Kremsdorf D, Croissant O et al. (1986) A novel type of human papillomavirus associated with genital neoplasias. Nature 321:246

Molecular Biology

Molecular Biology of Genital HPV Infections

H. Pfister

Introduction

Human papillomaviruses (HPV) affect squamous epithelia of the skin and mucosa (Pfister 1984). A subgroup of HPV types shows a strong preference for the anogenital region and appears associated with genital warts and flat lesions, nowadays referred to as intraepithelial neoplasias to denote their potential to develop into squamous cell carcinomas (zur Hausen 1989). Experimental transmission and follow-ups of naturally occurring transmissions between sexual partners as well as in vitro transformation studies leave little doubt that HPVs are the cause of the benign and precancerous lesions they are associated with. Several HPV types (e.g., HPV 6 and 11) are mainly detected in genital warts, flat condylomas, and low grade intraepithelial neoplasias whereas others (e.g., HPV 16 and 18) are frequently associated with severe intraepithelial neoplasias and invasive cancer. This led to the concept of two groups of HPV, which differ in their carcinogenic potential, but the role of HPV in tumor progression and maintenance of the malignant phenotype is still poorly understood.

This review tries to summarize recent data on the molecular biology of papillomavirus infections, which may become relevant to the design of new diagnostic and therapeutic approaches and may help to better understand activities of the virus in malignant conversion.

Physical Properties of Papillomaviruses

The virion particles have a cubic protein shell, which is composed of one major structural protein with a molecular mass of about 55000 and a minor component with an average molecular mass of 70000 (Pfister and Fuchs 1987). They contain a double-stranded circular DNA genome of roughly 8000 base-pairs. Due to the lack of a cell culture system for in vitro propagation of papillomaviruses it has usually not been possible to obtain enough particles to raise specific antisera for serologic typing. As a consequence, the viruses have been typed by comparing their DNA sequences, which were directly cloned from biopsy material by ligation to appropriate vector DNAs and amplified in bacteria to obtain sufficient amounts for biochemical characterization and for use as diagnostic reagents. The relationship of different HPV DNAs is estimated from their cross-hybridization, i.e., from the ability of viral DNA single strands to form stable hybrid double strands under controlled ex-

Genital Papillomavirus Infections
Eds.: Gross/Jablonska/Pfister/Stegner

Table 1. Human papillomavirustypes grouped according to DNA sequence homology. Types that preferentially infect the genital tract are marked by an asterisk

A	B[a]	C		D[a]		E[a]	F	G	H[a]	I	J	K	L	M	N	O	P	Q	R	S	T
1	2	4	5	9	24	6*	7	16*	18*	26	30*	33*	34*	35*	39*	41	43*	48	50	54*	56*
	27		8	15		11*	40*	31*	45*	51*	53*	52*									
	57*		12	17								58*									
	29		14	37		13			32												
			19	38		44*			42*												
	3		20			55*															
	10		21																		
	28		22																		
			23																		
			25																		
			36																		
			46																		
			47																		
			49																		

[a] Subgroups can be differentiated within groups B, D, E, and H, which comprise rather closely related viruses

perimental conditions. The genomes of many HPV types do not cross-hybridize at all when tested under so-called stringent conditions, which allow less than 20% mismatches in base-pairing between heterologous DNA strands. It is therefore obvious that a tumor may appear false negative for HPV DNA if tested by nucleic acid hybridization with an unrelated HPV DNA as probe. Viruses showing less than 50% cross-hybridization when assayed under stringent conditions are considered as different types. According to this standard, 58 types of HPV have now been catagorized (Table 1), and new types are being recognized on a regular basis (de Villiers 1989). The viruses can be grouped according to DNA homology, which correlates by and large with the pathogenic properties although there are a number of exceptions to this rule.

All HPV genomes show sufficient homology to cross-hybridize under so-called relaxed conditions, which allow double-strand formation in spite of extensive mismatches. This was and is widely used to screen tumors for the presence of unknown HPV types.

The genome organization of all papillomaviruses sequenced to date is very uniform (Fig. 1). All putative protein-coding sequences (open reading frames, ORFs) are located on one DNA strand and occupy similar positions relative to each other. The ORFs are classified as early (E) or late (L) to denote their expression in the viral replication cycle. Early ORFs encode proteins involved in transcription control, viral DNA replication, and oncogenic transformation. The late ORFs encode structural proteins. The E and L ORFs form two blocks, which are separated by noncoding sequences, which are referred to as long control regions because they contain signal sequences controlling viral

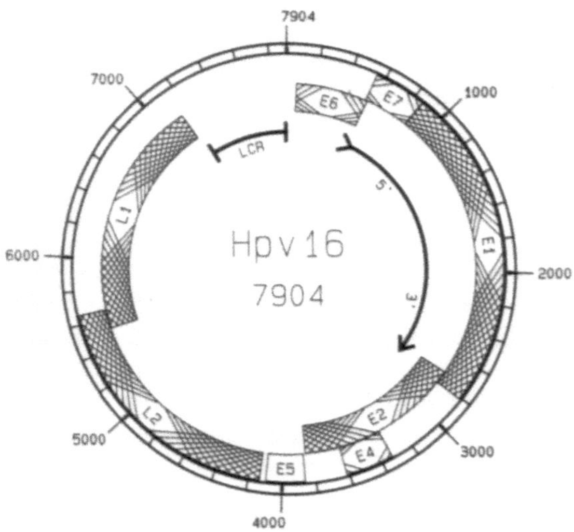

Fig. 1. Genome organization of HPV 16 deduced from the DNA sequence (Howley and Schlegel 1988). The positions of the long control region (*LCR*) and the major open reading frames (*E1* to *E7*, *L1*, and *L2*) are shown. There is no E3 in HPV 16. The nucleotide numbers are noted. *Arrow* indicates the 5′–3′ polarity of transcription

replication and gene expression. The genomes are usually highly homologous within reading frames E1, E2, and L1. This explains the close relationship of the major structural proteins, which are encoded by L1 and share group-specific antigenic epitopes. These are recognized by cross-reactive antisera raised against detergent-disrupted virus particles. Such sera are also used as diagnostic reagents to detect virus capsid proteins in tumors without the experimenter having to know the exact virus type.

Biology of HPV Infection

Papillomaviruses have a specific tropism for squamous epithelial cells and are assumed to infect primarily cells from the basal layer. This infection is abortive and may remain latent over months and probably years (Pfister 1984). Nothing is known about the physical state of the viral DNA and the virus-specific gene expression in this phase. The induction of a clinically visible lesion is probably triggered by the activation of additional virus genes although there is presently again no information available about the underlying mechanisms and the genes involved. In vitro studies, which will be discussed below, suggest that virus encoded proteins lead to increased cell proliferation and abnormal, delayed differentiation. Both effects result in a thickening of the epidermis and mainly of the spinous layer. As the keratinocytes migrate upward through the epidermis, they undergo a program of differentiation and gradually become more permissive for papillomaviruses (Pfister 1984). Vegetative viral DNA synthesis can be detected by in situ hybridization techniques in cells of the stratum spinosum and stratum granulosum, and viral capsid protein production and virus assembly occur in the upper stratum spinosum and in the stratum granulosum. Viral gene expression in undifferentiated keratinocytes appears to be restricted at the level of transcription initiation and termination (Baker and Howley 1987), but the molecular basis for the differentiation linked control is not yet known.

Viral Gene Functions

The functions of viral genes were initially analysed with bovine papillomavirus (BPV) 1, which is exceptional in inducing fibropapillomas in vivo and being easily able to transform fibroblasts in vitro. Later studies indicated, however, that many results can be transferred to HPVs. Virus encoded proteins are involved in the control of viral DNA replication and gene expression as well as in cell transformation and represent potential targets for therapeutic intervention with HPV infection.

Gene Transcription

The transcription pattern of papillomaviruses is very complex. In BPV1 there are at least six promoters for early transcripts and one additional promoter for

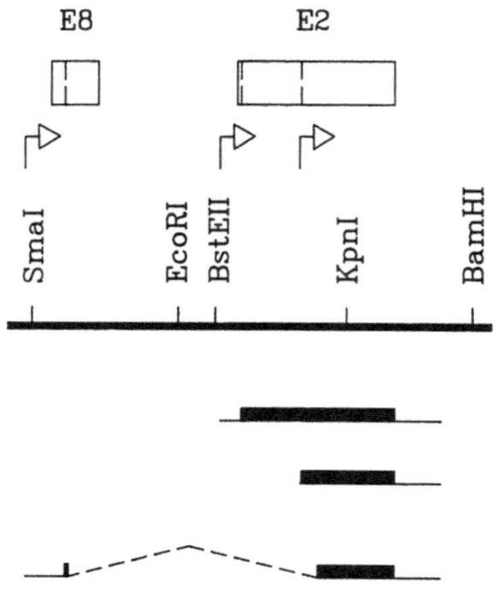

Fig. 2. Map location and coding potential (*black bars*) of colinear E2 RNAs of BPV 1. The *solid lines* indicate nontranslated regions; the *slanted, dashed line* represents an intervening sequence removed by splicing. *Top,* Positions of open reading frames E8 and E2, three relevant translation start codons (*vertical, dashed lines*), and three different promoters (*bent arrows*). *Middle,* Part of the restriction enzyme cleavage map of BPV1 (see Fig. 3). Modified from Baker (1990)

late mRNAs (Baker 1990). Three separate promoters for example give rise to different, partially overlapping transcripts of ORF E2 (Fig. 2). The RNAs encode proteins with opposite functions in transcription regulation (see below). Diverse splice events may link different ORFs, which leads to fusion proteins regarding individual reading frames (Fig. 2). One should therefore note that ORFs E1–E8 do not usually correspond to only one protein.

Replication Control

The 3′ part of the BPV1 ORF E1 encodes a function to ensure the extrachromosomal replication of viral DNA (Lusky and Botchan 1985). At least three proteins seem to be involved in the control of a stable copy number of the BPV1 DNA during latent infection (Fig. 3). An E6 and an E6/E7 fusion protein guarantee high copy number maintenance whereas a protein from the 5′ end of ORF E1 seems to prevent undue run-away replication (Berg et al. 1986a, b; Lusky and Botchan 1986a; Roberts and Weintraub 1986).

Transcription Control

The transcriptional activity of BPV1 is affected by the DNA binding E2 proteins. The full length E2 acts as activator of transcription whereas both the carboxy terminal fragment of E2 and an E8/E2 fusion protein repress transcription (Spalholz et al. 1985; Lambert et al. 1987; Choe et al. 1989). This was confirmed for a number of HPVs including HPV16 (Cripe et al. 1987; Hirochika et al. 1987; Phelps and Howley 1987). The three proteins are able to bind via their common carboxy terminal part to the palindromic sequence ACCGN$_4$CGGT, which appears several times in the long control regions of papillomaviruses and occasionally within the early region close to additional promoters (Androphy et al. 1987b; Haugen et al. 1987; Moskaluk and Bastia

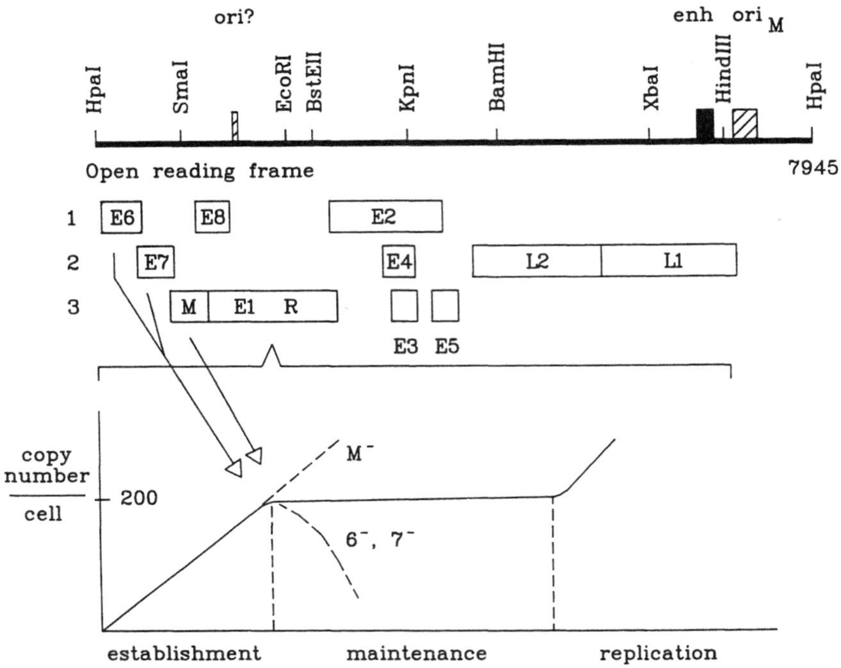

Fig. 3. Functions involved in the DNA replication of BPV1. *Top,* Physical map of BPV1 and positions of two sequences, which support plasmid maintenance (*hatched areas*). One of them appears to function as origin of maintenance DNA replication during latent infection (ori_M). The meaning of the other (*ori?*) is unknown. The position of a transcription enhancer element (*enh*), which is essential for the activity of ori_M is indicated by a *black bar* (Lusky and Botchan 1986b). *Middle,* Position of BPV1 open reading frames. E1 encodes a function R essential for extrachromosomal replication and a function M for down-modulation of the BPV1 DNA copy number. *Bottom,* Copy number of BPV1 DNA during establishment of infection, maintenance, and vegetative replication. The effects of E1M, E6, and E7 are illustrated by the behavior of mutants at the transition from establishment to maintenance (see also text)

1987; Spalholz et al. 1987; Hermonat et al. 1988; McBride et al. 1988). It is assumed that the amino terminal of the full length E2 protein interacts with essential cellular transcription factors to enhance transcriptional activity whereas binding alone appears to be inhibitory (Giri and Yaniv 1988). BPV1 E1 mutant transformed cells reveal higher steady-state concentrations of viral RNA than wild-type transformed cells, which is partially due to an altered E2 regulatory circuit (Lambert and Howley 1988; Schiller et al. 1989). The mechanism how E1 functions influence E2 expression is not yet known.

HPV 16 E7 was shown to encode another transcriptional activator, which can replace the regulating adenovirus E1a protein and shares aminoacid sequence homology with E1a (Phelps et al. 1988). An HPV 18 DNA sequence, which stimulates transcription is dependent upon the viral E6 gene product for function (Gius et al. 1988).

The long control regions of papillomaviruses contain a number of transcriptional enhancer sequences. These are by definition able to stimulate viral transcription relatively independent of their location and orientation to the promoters. One of them, the E2 responsive element, depends on activation by binding of the E2 activator protein (Hirochika et al. 1987; Phelps and Howley 1987; Spalholz et al. 1987; Hawley-Nelson et al. 1988). Additional constitutive enhancers were identified in HPV 11, 16, and 18 (Cripe et al. 1987; Swift et al. 1987; Gius et al. 1988; Steinberg et al. 1989). Of special interest is a glucocorticosteroid-reactive element found in the long control regions of genital HPVs, which is activated by dexamethasone with resulting increase in transcription (Gloss et al. 1987). This raised speculations about the influence of hormones on viral gene expression and indirectly on oncogenic activity (Pater et al. 1988). Although glucocorticoids and estrogens or progesterones act via similar DNA consensus sequences, one should be aware that the physiological activities of the hormones of interest are quite different.

Transformation

In the case of BPV1, E6, E7, and E5 revealed transforming activity when tested with mouse fibroblasts (DiMaio and Neary 1990). E5 appears particularly conserved among those papillomaviruses that induce fibropapillomas in vivo but is not even detectable in several HPVs.

Some HPVs are also capable of converting immortalized mouse fibroblasts to the tumorigenic state. Oncogenic transformation of NIH3T3 and Rat 1 cells has been achieved by using only the E6/E7 ORFs of HPV 18 (Bedell et al. 1987) and HPV 16 E7 ORF sufficed to transform Rat 3Y1 cells (Kanda et al. 1988). Using a specific Rat cell line as assay system to distinguish two groups of immortalizing oncogenes, a functional similarity could be demonstrated between HPV 16 E7, SV40 large T, and adenovirus E1a (Vousden and Jat 1989). The *myc* oncogene, which is also known as immortalizing gene, could not replace the HPV 16 E7 function.

More relevant to human tumorigenesis is the finding that HPV 16, 18, 31, or 33 can immortalize primary cultures of human foreskin keratinocytes or human cervical epithelia cells (Dürst et al. 1987b; Pirisi et al. 1987; Kaur and McDougall 1988; Woodworth et al. 1989). Keratinocytes electroporated with HPV 6, 11, 16, or 18 DNA all exhibited an increased cellular proliferation but only HPV 16 or 18 induced an altered differentiation behavior so that cells gave rise to proliferating, poorly stratified colonies when grown in the presence of serum and calcium (Schlegel et al. 1988). Transfection of keratinocytes with DNA of HPV 16, 18, 31, or 33, followed by in vitro culture under conditions, which allow stratification and differentiation, led to histological pictures reminiscent of cervical intraepithelial neoplasias of grade 1–3 (McCance et al. 1988). Established keratinocyte cell lines contain integrated viral DNA, are usually aneuploid but not tumorigenic in nude mice.

A genetic dissection showed that HPV 16 E6 and E7 proteins cooperate to immortalize keratinocytes withE7 apparently playing the major role in this process (Hawley-Nelson et al. 1989). In this respect it is of particular interest that the HPV 16 E7 protein forms a complex with the retinoblastoma gene product (Dyson et al. 1989). The retinoblastoma gene is regarded as a tumor suppressor gene and the observed protein interaction could disrupt normal control mechanisms.

HPV Infection and Malignant Conversion

The exact role of HPV in malignant conversion is not yet clear. The persistence of HPV DNA in genital cancers and the continued expression of the transforming proteins E6 and E7 (Smotkin and Wettstein 1986; Androphy et al. 1987a; Seedorf et al. 1987) suggest that viral functions are basically required for carcinoma development and for the maintenance of the malignant state. This is supported by the notion that specific HPV types imply a particular risk of malignant conversion, which is based on the prevalence of specific HPV DNAs in carcinomas and on the progression of precancerous lesions associated with different HPVs (zur Hausen 1989). Although HPV 6, 11, 16, and 18 are similarily prevalent in the normal population, HPV 16 clearly prevails in squamous cell carinomas of the cervix and HPV 18 is most frequently detected in adenocarcinomas (Tase et al. 1988). In prospective studies, CIN lesions harboring HPV 16 and 18 have been found to have the highest potential of progression to more severe lesions. A few differences in the molecular biology of HPV types with assumed higher or lower oncogenic potential became obvious but the meaning for pathogenic differences is not yet clear.

Oncogenic and less oncogenic genital HPVs generate the E7 mRNA by different mechanisms (Smotkin et al. 1989). It is transcribed from a separate promoter in the case of HPV 6 and HPV 11. In HPV 16 and 18, both E6 and E7 mRNAs are transcribed from the same promoter but processed in a different way (Fig. 4). E7 protein appears to be translated after the removal of an intron within ORF E6. The different strategies could affect the relative expression of E6 and E7 under various conditions and thus influence the oncogenic activity of the viruses. The splice event in HPV 16 and 18 furthermore leads to the translation of a modified E6 protein, called E6*, which shares the amino terminal part with E6 but has a short specific carboxy terminal encoded by the second exon. The use of an alternative splice acceptor site even leads to a second E6* protein in the case of HPV 16. The biological activity of the E6* proteins is not yet known.

In contrast to HPV 6, 11, 33, or 52 (Lancaster et al. 1983; Beaudenon et al. 1986; Yajima et al. 1988), HPV 16 and 18 reveal a higher tendency to integrate into the host cell genome (zur Hausen 1989). A significant percentage of HPV 16 positive cervical cancers, which give no evidence for integration (probably more than 30% of the cases), suggests, however, that integration is not absolutely required for tumor progression although it may play an important role if it happens (Fuchs et al. 1989). Integration may occur already in mild dysplasias (Lehn et al. 1988) and most or all of the viral DNA appears in-

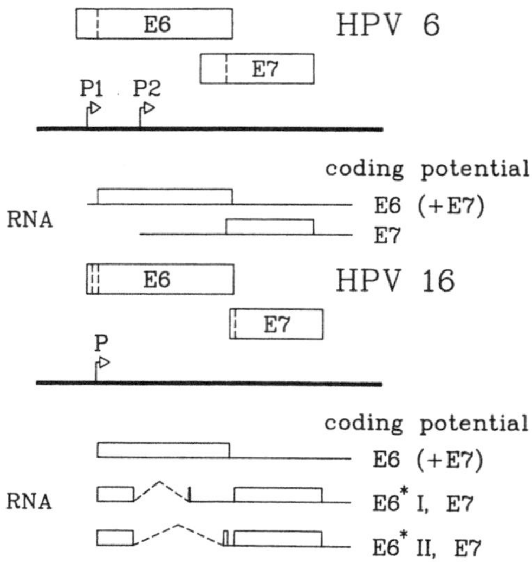

Fig. 4. Map location and coding potential (*open bars*) of colinear E6–E7 RNAs of HPV 6 (*top*) and HPV 16 (*bottom*). *Solid lines* indicate nontranslated regions; *slanted, dashed lines* intervening sequences removed by splicing. *Bent arrows* represent promoters. *Vertical dashed lines* within open reading frames show the positions of relevant translation start codons. Modified from Smotkin et al. (1989)

tegrated in many cancers. Once established the association between viral and cellular DNA is very stable as exemplified by DNA from lymph node metastasis of cervical cancers usually showing the same virus- and integration-specific restriction enzyme cleavage pattern as DNA from the primary tumor (Fuchs et al. 1989).

The insertion of viral DNA into cellular sequences allows special mechanisms of carcinogenesis in that viral transcription control signals could activate cellular genes and vice versa. The disruption of the circular viral genome in the course of integration usually implies additional qualitative and quantitative changes in viral gene expression.

Integration of the HPV 16 or HPV 18 DNA in cervical carcinomas does not appear to be specific with regard to the host chromosomes. In two cervical cancer derived cell lines integration has been mapped to chromosome 8 in the vicinity of the *myc* gene, which then turned out to be more actively transcribed than in other cell lines (Dürst et al. 1987a) but this is obviously no general mechanism for tumor progression.

With respect to the viral genome, integration occurs in a specific manner with an almost regular disruption of the E1 and E2 ORFs (zur Hausen 1989). Both the lack of E1 and particularly the lack of the E2 repressor will result in the deregulated expression of the transforming genes E6 and E7.

The disruption of the circular viral genome furthermore removes 3′ located splice acceptor sites and the viral early transcription termination signals. Cellular sequences 3′ of the integration site are used instead of those leading to viral cellular fusion transcripts. In view of the known influence of 3′ sequences on the stability of mRNA, it seems possible that the cellular sequences account for the increased half-life of viral transcripts in malignant cell lines (Kleiner et al. 1986). This implies elevated steady-state levels of viral mRNAs and consequently higher concentrations of transforming proteins.

Consequences on viral gene expression may also be expected from duplications, insertions, and deletions within the long control regions of extrachromosomally persisting HPV 6 or 16 DNA from vulval carcinomas (Rando et al. 1986; Kennedy et al. 1987). A large insertion in plasmid DNA of HPV 16 from a cervical cancer has not been mapped (Fuchs et al. 1989) but could fulfill a similar task.

Summary

The anogenital skin and mucosa can be infected by at least 22 different HPV types. They reveal genus specific capsid antigens and their DNAs cross-hybridize under relaxed conditions. Hybridization under stringent conditions requires type-specific probes for detection. The viruses spend about 50% of their genetic information to encode proteins regulating viral replication and transcription to enable efficient persistence. Roughly 1000 base-pairs are non-coding but contain *cis*-active, partially protein-binding control sequences. At least two genes (E6 and E7) are involved in oncogenic transformation. HPV types seem to differ in their carcinogenic potential. Malignant conversion may be preceded by integration or mutations of HPV DNA.

References

Androphy EJ, Hubbert NL, Schiller JT, Lowy DR (1987a) Identification of the HPV-16 E6 protein from transformed mouse cells and human cervical carcinoma cell lines. EMBO J 6:989–992

Androphy EJ, Lowy DR, Schiller JT (1987b) Bovine papillomavirus E2 *trans*-activating gene product binds to specific sites in papillomavirus DNA. Nature 325:70–73

Baker CC (1990) Bovine papillomavirus type 1 transcription. In: Pfister H (ed) Papillomaviruses and human cancer. CRC Press, Boca Raton, pp 91–112

Baker CC, Howley PM (1987) Differential promoter utilization by the bovine papillomavirus in transformed cells and productively infected wart tissues. EMBO J 6:1027–1035

Beaudenon S, Kremsdorf D, Croissant O, Jablonska S, Wain-Hobson S, Orth G (1986) A novel type of human papillomavirus associated with genital neoplasias. Nature 321:246–249

Bedell MA, Jones KH, Laimins LA (1987) The E6–E7 region of human papillomavirus type 18 is sufficient for transformation of NIH 3T3 and Rat-1 cells. J Virol 61:3635–3640

Berg LJ, Lusky M, Stenlund A, Botchan M (1986a) Repression of bovine papillomavirus replication is mediated by a virally encoded *trans*-acting factor. Cell 46:753–762

Berg LJ, Singh K, Botchan M (1986b) Complementation of a bovine papillomavirus low-copy-number mutant: evidence for a temporal requirement of the complementing gene. Mol Cell Biol 6:859–869

Choe J, Vaillancourt P, Stenlund A, Botchan M (1989) Bovine papillomavirus type 1 encodes two forms of a transcriptional repressor: structural and functional analysis of new viral cDNAs. J Virol 63:1743–1755

Cripe TP, Haugen TH, Turk JP, Tabatabai F, Schmid PG, Dürst M, Gissmann L, Roman A, Turek LP (1987) Transcriptional regulation of the human papillomavirus-16 E6–E7 promoter by a keratinocyte-dependent enhancer, and by viral E2 *trans*-activator and repressor gene products: implications for cervical carcinogenesis. EMBO J 6:3745–3753

De Villiers EM (1989) Heterogeneity of the human papillomavirus group. J Virol 63:4898–4903

DiMaio D, Neary K (1990) The genetics of bovine papillomavirus type 1. In: Pfister H (ed) Papillomaviruses and human cancer. CRC Press, Boca Raton, pp 113–144

Dürst M, Croce CM, Gissmann L, Schwarz E, Huebner K (1987a) Papillomavirus sequences integrate near cellular oncogenes in some cervical carcinomas. Proc Natl Acad Sci USA 84:1070–1074

Dürst M, Dzarlieva-Petrusevska RT, Boukamp P, Fusenig NE, Gissmann L (1987b) Molecular and cytogenetic analysis of immortalized human primary keratinocytes obtained after transfection with human papillomavirus type 16 DNA. Oncogene 1:251–256

Dyson N, Howley PM, Muenger K, Harlow E (1989) The human papillomavirus-16 E7 oncoprotein is able to bind to the retinoblastoma gene product. Science 243:934–936

Fuchs PG, Girardi F, Pfister H (1989) Human papillomavirus 16 DNA in cervical cancers and in lymph nodes of cervical cancer patients: a diagnostic marker for early metastases? Int J Cancer 43:41–44

Giri I, Yaniv M (1988) Structural and mutational analysis of E2 *trans*-activating proteins of papillomaviruses reveals three distinct functional domains. EMBO J 7:2823–2829

Gius D, Grossman S, Bedell MA, Laimins LA (1988) Inducible and constitutive enhancer domains in the noncoding region of human papillomavirus type 18. J Virol 62:665–672

Gloss B, Bernard HU, Seedorf K, Klock G (1987) The upstream regulatory region of the human papillomavirus-16 contains an E2 protein-independent enhancer which is specific for cervical carcinoma cells and regulated by glucocorticoid hormones. EMBO J 6:3735–3743

Haugen TJ, Cripe TP, Ginder GD, Karin M, Turek LP (1987) *Trans*-activation of an upstream early gene promoter of bovine papillomavirus 1 by a product of the viral E2 gene. EMBO J 6:145–152

Hawley-Nelson P, Androphy EJ, Lowy DR, Schiller JT (1988) The specific DNA recognition sequence of the bovine papillomavirus E2 protein is an E2-dependent enhancer. EMBO J 7:525–531

Hawley-Nelson P, Vousden KH, Hubbert NL, Lowy DR, Schiller JT (1989) HPV 16 E6 and E7 proteins cooperate to immortalize primary human foreskin keratinocytes. EMBO J 8:3905–3910

Hermonat PL, Spalholz BA, Howley PM (1988) The bovine papillomavirus P_{2443} promoter is E2 *trans*-responsive: evidence for E2 autoregulation. EMBO J 7:2815–2822

Hirochika H, Broker TR, Chow LT (1987) Enhancers and *trans*-acting E2 transcriptional factors of papillomaviruses. J Virol 61:2599–2606

Howley PM, Schlegel R (1988) The human papillomaviruses. Am J Med 85:155–158

Kanda T, Furuno A, Yoshiike K (1988) Human papillomavirus type 16 open reading frame E7 encodes a transforming gene for Rat 3Y1 cells. J Virol 62:610–613

Kaur P, McDougall JK (1988) Characterization of primary human keratinocytes transformed by human papillomavirus type 18. J Virol 62:1917–1924

Kennedy IM, Simpson S, MacNab JCM, Clements JB (1987) Human papillomavirus type 16 DNA from a vulvar carcinoma in situ is present as head-to-tail dimeric episomes with a deletion in the non-coding region. J Gen Virol 68:451–462

Kleiner E, Dietrich W, Pfister H (1986) Differential regulation of papillomavirus early gene expression in transformed fibroblasts and carcinoma cell lines. EMBO J 6:1945–1950

Lambert PF, Howley PM (1988) Bovine papillomavirus type 1 E1 replication-defective mutants are altered in their transcriptional regulation. J Virol 62:4009–4015
Lambert PF, Spalholz BA, Howley PM (1987) A transcriptional repressor encoded by BPV-1 shares a common carboxy terminaldomain with the E2 transactivator. Cell 50:69–78
Lancaster WD, Kurman RJ, Sanz LE, Perry S, Jenson AB (1983) Human papillomavirus: detection of viral DNA sequences and evidence for molecular heterogeneity in metaplasias and dysplasias of the uterine cervix. Intervirology 20:202–212
Lehn H, Villa LL, Marziona F, Hilgarth M, Hillemans HG, Sauer G (1988) Physical state and biological activity of human papillomavirus genomes in precancerous lesions of the female genital tract. J Gen Virol 69:187–196
Lusky M, Botchan M (1985) Genetic analysis of bovine papilloma-virus type 1 *trans*-acting replication factors. J Virol 53:955–965
Lusky M, Botchan M (1986a) A bovine papillomavirus type 1-encoded modulator function is dispensable for transient viral replication but is required for establishment of the stable plasmid state. J Virol 60:729–742
Lusky M, Botchan MR (1986b) Transient replication of bovine papilloma virus type 1 plasmids: *cis* and *trans* requirements. Proc Natl Acad Sci USA 83:3609–3613
McBride AA, Schlegel R, Howley PM (1988) The carboxy-terminal domain shared by the bovine papillomavirus E2 transactivator and repressor proteins contains a specific DNA binding activity. EMBO J 7:533–539
McCance DJ, Kopan R, Fuchs E, Laimins LA (1988) Human papilloma-virus type 16 alters human epithelial cell differentiation in vitro. Proc Natl Acad Sci USA 85:7169–7173
Moskaluk C, Bastia D (1987) The E2 "gene" of bovine papilloma-virus encodes an enhancer binding protein. Proc Natl Acad Sci USA 84:1215–1218
Pater MM, Hughes GA, Hyslop DE, Nakshatri H, Pater A (1988) Glucocorticoid-dependent oncogenic transformation by type 16 but not type 11 human papillomavirus DNA. Nature 335:832–835
Pfister H (1984) Biology and biochemistry of papillomaviruses. Rev Physiol Biochem Pharmacol 99:111–181
Pfister H, Fuchs PG (1987) Papillomaviruses: particles, genome organization, and proteins. In: Syrjänen K, Gissmann L, Koss LG (eds) Papillomaviruses and human disease. Springer, Berlin Heidelberg New York, pp 1–18
Phelps WC, Howley PM (1987) Transcriptional *trans*-activation by the human papillomavirus type 16 E2 gene product. J Virol 61:1630–1638
Phelps WC, Yee CL, Muenger K, Howley PM (1988) The human papillomavirus type 16 E7 gene encodes transactivation and transformation functions similar to those of adenovirus E1A. Cell 53:539–547
Pirisi L, Yasumoto S, Feller M, Doniger J, DiPaolo JA (1987) Transformation of human fibroblasts and keratinocytes with human papillomavirus type 16 DNA. J Virol 61:1061–1066
Rando RF, Groff DE, Chirikjian JG, Lancaster WD (1986) Isolation and characterization of a novel human papillomavirus type 6 DNA from an invasive vulvar carcinoma. J Virol 57:353–356
Roberts JM, Weintraub H (1986) Negative control of DNA replication in composite SV40-bovine papilloma virus plasmids. Cell 46:741–752
Schiller JT, Kleiner E, Androphy EJ, Lowy DR, Pfister H (1989) Identification of bovine papillomavirus E1 mutants with increased transforming and transcriptional activity. J Virol 63:1775–1782
Schlegel R, Phelps WC, Zhang Y-L, Barbosa M (1988) Quantitative keratinocyte assay detects two biological activities of human papillomavirus DNA and identifies viral types associated with cervical carcinoma. EMBO J 7:3181–3187
Seedorf K, Krammer G, Dürst M, Suhai S, Röwekamp W (1987) Identification of early proteins of the human papillomavirus type 16 (HPV 16) and type 18 (HPV 18) in cervical carcinoma cells. EMBO J 6:139–144

Smotkin D, Wettstein FO (1986) Transcription of human papilloma-virus type 16 early genes in a cervical cancer and a cancer-derived cell line and identification of the E7 protein. Proc Natl Acad Sci USA 83:4680–4684
Smotkin D, Prokoph H, Wettstein FO (1989) Oncogenic and nononcogenic human genital papillomaviruses generate the E7 mRNA by different mechanisms. J Virol 63:1441–1447
Spalholz BA, Yang Y-C, Howley PM (1985) Transactivation of a bovine papillomavirus transcriptional regulatory element by the E2 gene product. Cell 42:183–191
Spalholz BA, Lambert PF, Yee CL, Howley PM (1987) Bovine papillomavirus transcriptional regulation: localization of the E2-responsive elements of the long control region. J Virol 61:2128–2137
Steinberg BM, Auborn KJ, Brandsma JL, Taichman LB (1989) Tissue site-specific enhancer function of the upstream regulatory region of human papillomavirus type 11 in cultured keratinocytes. J Virol 63:957–960
Swift FV, Bhat K, Younghusband HB, Hamada H (1987) Characterization of a cell type-specific enhancer found in the human papillomavirus type 18 genome. EMBO J 6:1339–1344
Tase T, Sato S, Wada Y, Yajima A, Okagaki T (1988) Prevalence of human papillomavirus type 18 DNA in adenocarcinoma and adenosquamous carcinoma of the uterine cervix occurring in Japan. Tohoku J Exp Med 156:47–53
Vousden KH, Jat PS (1989) Functional similarity between HPV 16 E7, SV40 large T and adenovirus E1a proteins. Oncogene 4:153–158
Woodworth CD, Doniger J, DiPaolo JA (1989) Immortalization of human foreskin keratinocytes by various human papillomavirus DNAs corresponds to their association with cervical carcinoma. J Virol 63:159–164
Yajima J, Noda T, de Villiers E-M, Yajima A, Yamamoto K, Noda K, Ito Y (1988) Isolation of a new type of human papillomavirus (HPV 52b) with a transforming activity from cervical cancer tissue. Cancer Res 48:7164–7172
zur Hausen H (1989) Papillomaviruses as carcinomaviruses. In: Klein G (ed) Advances in viral oncology, vol 8. Raven, New York, pp 1–26

Biological Significance of Human Papillomavirus Early Gene Expression in Human Cervical Carcinoma Cells

M. von Knebel Doeberitz and H. zur Hausen

Introduction

For about 150 years infectious agents have been implicated as etiological factors for cervical cancer in women (Rigoni-Stern 1842; zur Hausen 1977b). A venereal transmission of these agents has been proposed since sexual promiscuity, early onset of sexual activity, and poor hygienic conditions emerged as potential risk factors. Genital infection with the herpes simplex virus (HSV) type 2 was suspected as a potential risk factor. Patients suffering from invasive cervical carcinoma were in a significantly higher percentage found to be HSV-seropositive than the respective control groups (Rawls et al. 1977; Melnick and Adam 1978). However, the virus itself or its genetic material has not regularly been found in cancer biopsy specimens (zur Hausen 1982) and a prospective study recently performed by Vonka and colleagues could not confirm HSV 2 infection as a potential risk factor for subsequent development of cervical cancer (Vonka et al. 1984).

The first reports on a potential involvement of human papillomavirus infections in carcinogenesis were published 1974 (zur Hausen et al. 1974). The identification of DNA sequences of specific types of human papillomaviruses, i.e., HPV 16 (Dürst et al. 1983) and HPV 18 (Boshart et al. 1984) in cervical carcinoma specimens stimulated extensive epidemiological investigations on a possible association of specific HPV types with cervical cancer and its respective precursor lesions. As seen from recent studies, about 95% of cervical cancer biopsy specimens contain DNA sequences of human papillomaviruses. The genotypes HPV 16 and 18 were identified in about 70%–80% of cervical cancer biopsy specimens, HPV 31, 33, 35, and 39 are found in about 15%, and only in approximately 5% not yet known HPV DNA can be detected (for review see Gissmann et al. 1987).

However, infections with these particular HPV types are ubiquitous. In 10%–30% of cervical smears of healthy women HPV infection could be demonstrated (de Villiers et al. 1987). More recent studies performed with the sensitive polymerase chain reaction (PCR) technique identified HPV 16 sequences in genital epithelial cells of more than 50% of male patients attending a dermatology clinic (Ch. von Knebel Doeberitz and L. Gissmann, unpublished results). Thus, clinically inapparent HPV infections are very widespread and only a minor percentage of infected people develop cervical cancer. In addition, the latency period between primary infection and the development of cancer has been estimated to be in the order of 20–50 years (zur Hausen 1986). Furthermore, cervical cancer is monoclonal in origin (zur

Genital Papillomavirus Infections
Eds.: Gross/Jablonska/Pfister/Stegner

Hausen 1986), thus only one out of many thousand infected cells undergoes malignant transformation. These findings clearly demonstrate that HPV infection per se is not sufficient to cause cancer. Other factors acting on the viral and cellular level are absolutely mandatory for the development of invasive disease (zur Hausen 1986).

Malignant cells arise within nonmalignant precursor lesions. These are subdivided according to their degree of histological atypia into three different stages referred to as cervical intraepithelial neoplasia (CIN) I to III. HPV 16 or 18 DNA can be detected also in lesions with mild atypia (CIN I). They usually progress to more advanced ones (CIN II or III) and finally to invasive cancer. Thus, it is likely that several factors act together in leading an HPV-infected keratinocyte to a precancerous and finally to the fully malignant state. Each factor is thereby contributing its part to escape control of growth.

Factors concerning the virus or its genetic information have to be separated from those concerning structural alterations within the host cell. Studies with somatic cell hybrids on cellular factors have revealed that they act in a recessive fashion, since they can be supplemented by normal nonmalignant cells (Stanbridge 1982; Stanbridge et al. 1976; Saxon et al. 1986). A model how cellular factors may interact with papillomaviruses in cervical epithelial cells has been recently proposed (zur Hausen 1986).

Molecular studies on the HPV DNA in cervical carcinoma cells either grown in vivo or in vitro as cell lines revealed some characteristic features (Fig. 1). The viral DNA is integrated into the host cell genome. Apparently no specific location within the host cell chromosomes is preferred. The viral genome however is disrupted in the E1 or E2 open reading frame (orf) leading to functional inactivation of at least the E2 orf. The noncoding region (ncr) and the orfs E6 and E7 are consistently preserved, while other viral genes, especially those encoding late viral functions (L1 and L2), may be deleted in the integrated state. The E6 and E7 orfs are regularly transcribed into mRNA and proteins derived therefrom were identified in cervical cancer cells (Schwarz et al. 1985; Smotkin and Weltstein 1986; Shirasava et al. 1987; Baker et al. 1987; Seedorf et al. 1987; Androphy et al. 1987; Oltersdorf et al. 1987). These data strongly suggest that continuous expression of the E6 and E7 orfs is somehow involved in growth control of malignant cervical cancer cells.

Further support for an oncogenic potential of HPVs was derived from experiments, demonstrating that transfection of the DNA of HPV 16, 18, 31, 33, and 35 is capable of immortalizing human foreskin keratinocytes (Dürst et al. 1987; Pirisi et al. 1987; Woodwarth et al. 1989). However, HPV-immortalized keratinocytes are not tumorigenic in nude mice and it is not yet clear, whether continuous expression of the E6 and E7 orfs is sufficient to induce the malignant state of human keratinocytes.

Within the scope of research for a causal strategy for treatment of invasive cervical cancer we are interested in the biological functions of the E6 and E7 orfs in malignant cervical cancer cells. Our interest is focused on the question whether the specific modulation of the expression level of these two genes has an influence on the phenotype and the growth pattern of human cervical can-

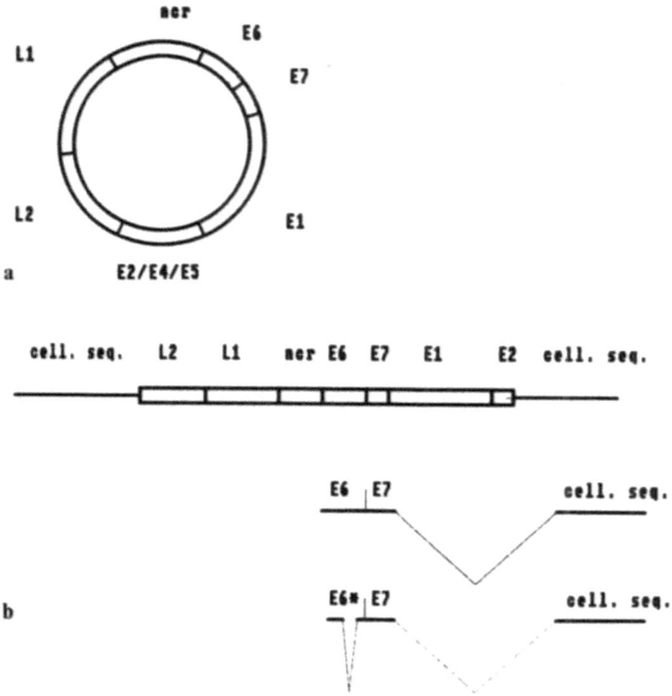

Fig. 1 a, b. Integration and transcription pattern of human papillomaviruses in human cervical carcinoma cells. The double-stranded circular DNA of the virus is disrupted in the E1 or E2 open reading frame (orf). The noncoding region (*ncr*) and the E6 and E7 orf are consistently preserved. Other parts of the viral DNA can be deleted in the integrated state. Two different types of viral cellular fusion transcripts are found: **a** transcripts containing the complete E6 and E7 sequences and **b** a spliced form in which an intron in the E6 orf is spliced out leading to the E6* transcript. The cellular flanking sequences do not contain major orfs and most likely provide stabilizing functions without coding for proteins (Schneider-Gädicke and Schwarz 1986; Le and Defendi 1988)

cer cells. If suppression of the E6 and E7 gene expression leads to decreased cell growth or modification of the malignant potential of cervical cancer cells, an approach to inhibit the genetic or biochemical activities of these genes might provide the basis for a new and causal therapeutical concept.

Results

In studies to be reported here the effect of specific modulation of HPV E6 and E7 gene expression on the phenotype of HPV 18 positive cervical carcinoma cell lines (Boshart et al. 1984) was analysed.

The cell lines used for our experiments are listed in Table 1. In experiments performed on C4-1 cells the antisense RNA technique was applied to alter

Table 1. HPV-18-positive cervical cancer cell lines

Cell line	Origin	Reference
C4-I	Squamous cell carcinoma, both cell lines derived from the same	Auersperg and Hawryluk (1962)
C4-II	biopsy specimen	Auersperg (1969)
HeLa	Adenocarcinoma of the cervix	Gey et al. (1952); Jones et al. (1971)
SW 756	Squamous cell carcinoma	Freedman et al. 1982

HPV 18 E6/E7 gene expression specifically without interfering with other genes (von Knebel Doeberitz et al. 1988).

Specific inhibition of gene expression of eukaryotic cells with antisense RNA was first described by Izant and Weintraub (1985). If in a given cell a RNA sequence is expressed that is complementary to a particular mRNA (therefore called antisense RNA) both RNA sequences form double-stranded hybrid molecules by base pairing (Fig. 2). These double-stranded RNA molecules are rapidly degraded in the nucleus and they can not be processed correctly and are not transported into the cytoplasm. However, even if they are in the cytoplasm they prevent binding of ribosomes to the mRNA and therefore interfere with protein translation. The effect is that synthesis of a specific protein encoded by the mRNA is inhibited (for review see Green et al. 1986).

HPV 18 E6 and E7 antisense RNA was introduced in C4-1 cervical carcinoma cells with a hormone inducible eukaryotic expression vector (Fig. 3). In this construction we used a control region that only permits RNA synthesis from adjacent DNA sequences in the presence of glucocorticoid hormones. Glucocorticoid responsive promoter and enhancer sequences were derived from the mouse mammary tumor virus long terminal repeat (MMTV-LTR; Fasel et al. 1982). HPV 18 E6 and E7 sequences derived from a cDNA clone (Schneider-Gädicke and Schwarz 1986) were cloned in an inverted orientation into that expression vector. Cells harboring the resulting vector pM6neo express complementary antisense RNA to the HPV 18 E6/E7 mRNA only in the presence of glucocorticoid hormones. Additionally the construction contained the pSV2neo derived geneticin resistance gene (neo) allowing selection of transfected cells in tissue culture (Southern and Berg 1982).

This vector was transferred into C4-1 cells and expression of antisense RNA upon dexamethasone treatment of cells harboring the vector was confirmed by Northern blot analysis as is here shown for three different clones (Fig. 4).

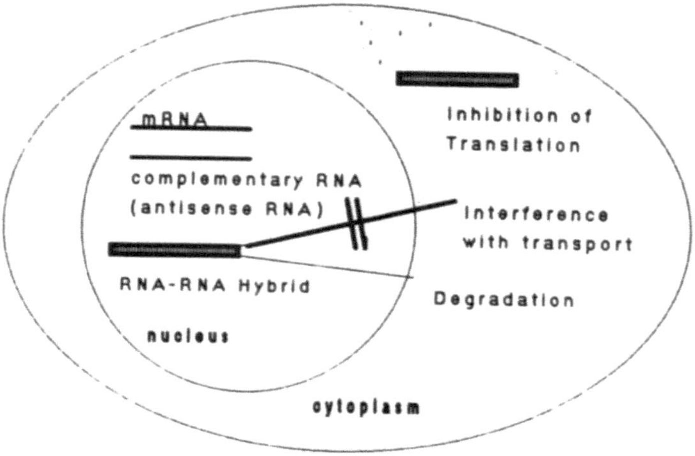

Fig. 2. Inhibition of gene expression by antisense RNA. RNA hybrid molecules formed between a mRNA sequence and a complementary "antisense" RNA lead to interference with protein synthesis by the shown mechanisms

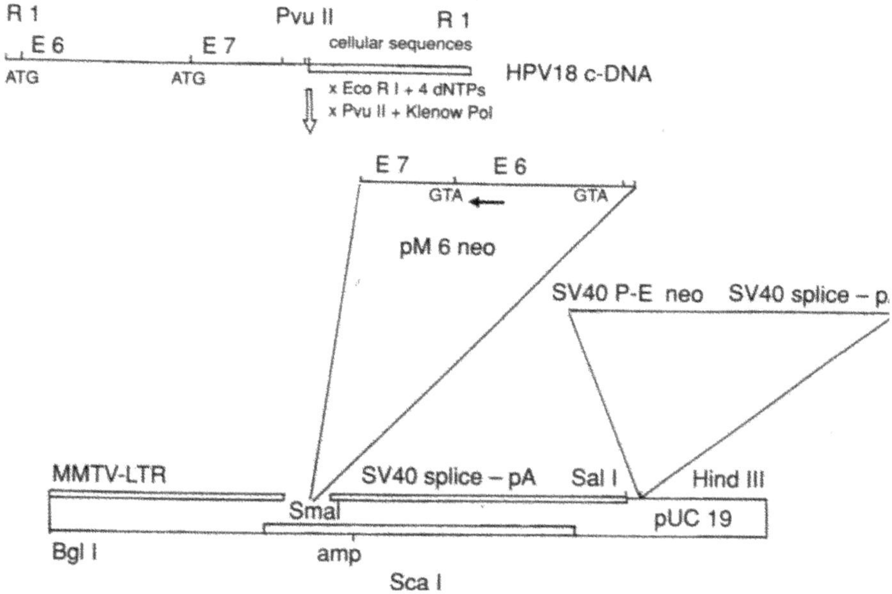

Fig. 3. Construction of a glucocorticoid inducible eukaryotic expression vector. The dexamethasone responsive enhancer and promoter element was derived from the MMTV-LTR (Fasel et al. 1982). It controls transcription of HPV 18 E6 and E7 sequences derived from a cDNA clone 7–23 (Schneider-Gädicke and Schwarz 1986) and cloned in an inverted orientation downstream to the MMTV-LTR. Cellular flanking sequences were removed at the *Pvu II* site. SV40 polyadenylation and splice signals provided for stability of HPV 18 antisense transcripts were derived from the pSV2neo plasmid. The dominant geneticin resistance gene was also derived from the pSV2neo plasmid

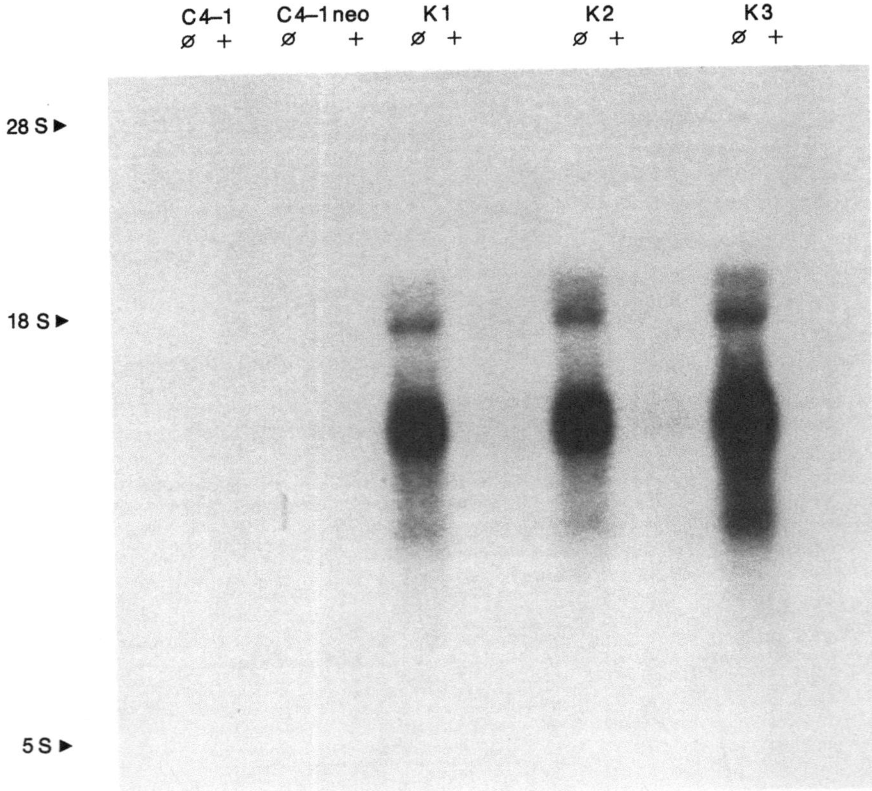

Fig. 4. Northern blot analysis of cytoplasmic RNA for the presence of HPV 18 E6 and E7 antisense transcripts. RNA was separated in a 1% denaturing formaldehyde gel and blotted onto a Gene Screen Plus filter (NEN, Boston, USA). The filter was hybridized with a ^{32}P-labelled strand-specific probe for the antisense RNA. Antisense transcripts can only be detected in C4-1 subclones harboring the pM6neo plasmid and cultured in the presence of dexamethasone

Three antisense RNA expressing clones were compared with parental C4-1 cells and C4-1 cells harboring only the geneticin resistance gene (C4-1 neo). Both cell populations did not express HPV 18 E6 and E7 antisense sequences upon glucocorticoid treatment.

The HPV 18 E7 protein synthesis was reduced in C4-1 cells expressing HPV 18 E6/E7 antisense RNA (Fig. 5) compared with dexamethasone treated controls. However, hormone treatment itself of C4-1 cells without antisense RNA significantly increased the E7 gene expression compared with untreated controls. Thus, these clones differ in the expression level of HPV early genes under dexamethasone treatment. Antisense clones express lower levels of HPV 18 E7 protein compared with normal C4-1 cells grown under the same conditions. The inhibition of protein synthesis was restricted to the HPV early gene products, since various other proteins (e.g., the receptor for the

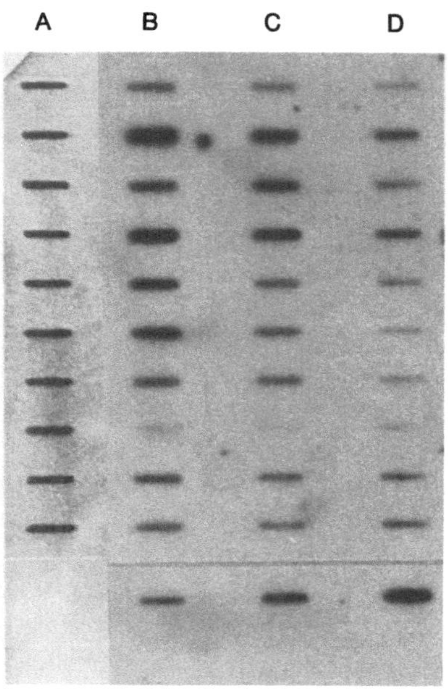

Fig. 5. HPV 18 E7 protein expression in C4-1 cells and the antisense clones either treated or not with dexamethasone. Nitrocellulose bound cytoplasmic protein extracts were exposed to an HPV 18 E7 antiserum (Seedorf et al. 1987). *Lane A* shows a Coommassie brilliant blue staining of a nitrocellulose filter with 20 μg of protein per slot to control that indeed comparable amounts of protein were loaded in each slot. *Lanes B, C,* and *D* show an X-ray exposure of a filter with 40, 20, and 10 μg of protein extract respectively incubated with the E7 antiserum and subsequently with ^{125}J-labelled protein A. As a quantitative standard 2, 4, and 8 ng of bacterially expressed E7-fusion protein were loaded on the same filter. Dexamethasone treatment of C4-1 and C4-1neo cells led to increased E7 protein synthesis. In the presence of antisense RNA (K1 dex, K2 dex, and K3 dex) dexamethasone did not increase E7 synthesis. Compared with dexamethasone treated C4-1 cells (C4-1 dex and C4-1neo dex) reduced levels of the E7 protein were found

epidermal growth factor, MHC class I and II antigenes, β_2-microglobulin, transferrin receptor, and others) were not expressed at different levels in C4-1 cells or antisense subclones.

Next we analysed the biological properties of the C4-1 cells expressing HPV 18 early gene products at different levels. Within 10 days in tissue culture C4-1 cells with a higher rate of HPV E6/E7 gene expression formed larger colonies derived from a single cell compared with C4-1 cells with antisense RNA grown under the same conditions and therefore lower levels of HPV E6/E7 gene expression (Fig. 6).

The morphology of antisense RNA expressing cells with reduced HPV early gene expression was rounded up, maybe reflecting some aspects of cellular differentiation. Dexamethasone treatment of C4-1 cells without antisense RNA leads to larger colonies in tissue culture. These cells expressed higher levels of the HPV early genes.

The anchorage independent growth of cells in soft agar is a good parameter correlating with the tumorigenicity of cancer cells. C4-1 cells without antisense RNA formed much larger colonies in the presence of dexamethasone in soft agar. Although grown under the same conditions, antisense RNA expressing cells did not form larger colonies in soft agar. Thus, cell growth of C4-1 cervical cancer cells is correlated with the expression level of HPV 18 E6 and E7 genes. Enhanced gene expression induced by dexamethasone leads to increased cell growth. Inhibition of viral gene expression by antisense RNA is correlated with decreased cell growth.

Steroid hormones, in particular in the form of oral contraceptives were suspected to be potential risk factors for the development of invasive cervical cancer (Brinton et al. 1986; Beral et al. 1988). As discussed above, glucocorticoids stimulate the proliferation rate of C4-1 cervical cancer cells by increasing HPV 18 E6 and E7 gene expression. Glucocorticoids and progesterones bind to an identical target sequence on the DNA, thereby regulating the expression of genes. Thus, responsiveness to glucocorticoids or progesterones of a given gene is dependent on the presence of the respective receptor in a cell, but does not differ at the sequence level of the gene that is to be regulated by these hormones (von der Ahe et al. 1985; Beato 1989). Therefore, it was of interest to investigate whether these steroid hormones have a systematic effect on the HPV early gene expression in cervical cancer cells. In our hands progesterones, estrogens, and androgens had no influence on viral gene expression in the HPV 18 positive cell lines C4-1, C4-2, HeLa, and SW 756. However, this may be due to lack of functional receptors in these cells.

Dexamethasone as synthetic glucocorticoid had no effect on Hela cells but stimulated the HPV early gene expression in C4-1 and C4-2 cells (Fig. 7). Both cell lines were established from the same biopsy specimen (Table 1). However, surprisingly dexamethasone treatment strongly reduced the HPV 18 gene expression in SW 756 cells. Thus, in cell lines derived from three different tumors dexamethasone elicits three different types of response concerning the HPV early gene expression. Factors specific for each tumor cell, therefore, strongly influence glucocorticoid mediated regulation of integrated viral genes.

In line with the association of the cellular proliferation rate and HPV early gene expression observed in C4-1 cells, the proliferation rate of other HPV 18 positive cells treated with dexamethasone reflected the level of HPV 18 E6 and E7 gene expression. In Hela cells, where dexamethasone had no influence on the viral gene expression, no significant alteration of the proliferation rate was observed. In C4-2 cells the hormone induced the E6 and E7 expression and led to an increased proliferation rate. In contrast, in SW 756 cells dexamethasone treatment reduced the HPV early gene expression and led to significant inhibition of cellular proliferation (Fig. 8). Expression of other proteins suspected to be involved in growth regulation (e.g., the receptor for the epidermal growth factor) was uniformly upregulated in all cell lines by dexamethasone.

A systematic effect of steroid hormones on the HPV 18 early gene expression in different cell lines can be excluded. However, the level of HPV E6 and E7 gene expression is in all cell lines tested directly correlated with growth

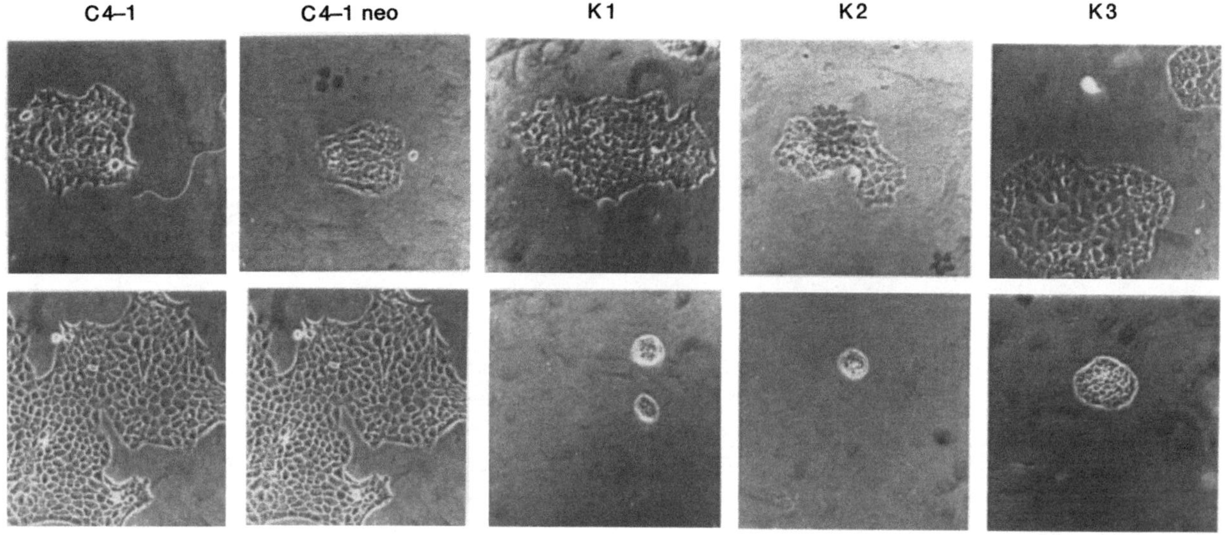

Fig. 6. Morphology of C4-1 cells with varying levels of HPV 18 early gene expression. C4-1 and C4-1neo cells grown in the presence of dexamethasone and therefore expressing higher levels of the HPV 18 early gene products (see Fig. 5) formed larger colonies derived from single cells within 10 days. C4-1 cells expressing HPV 18 antisense RNA upon hormone treatment and therefore expressing reduced levels of the viral early genes formed smaller colonies with an altered morphology

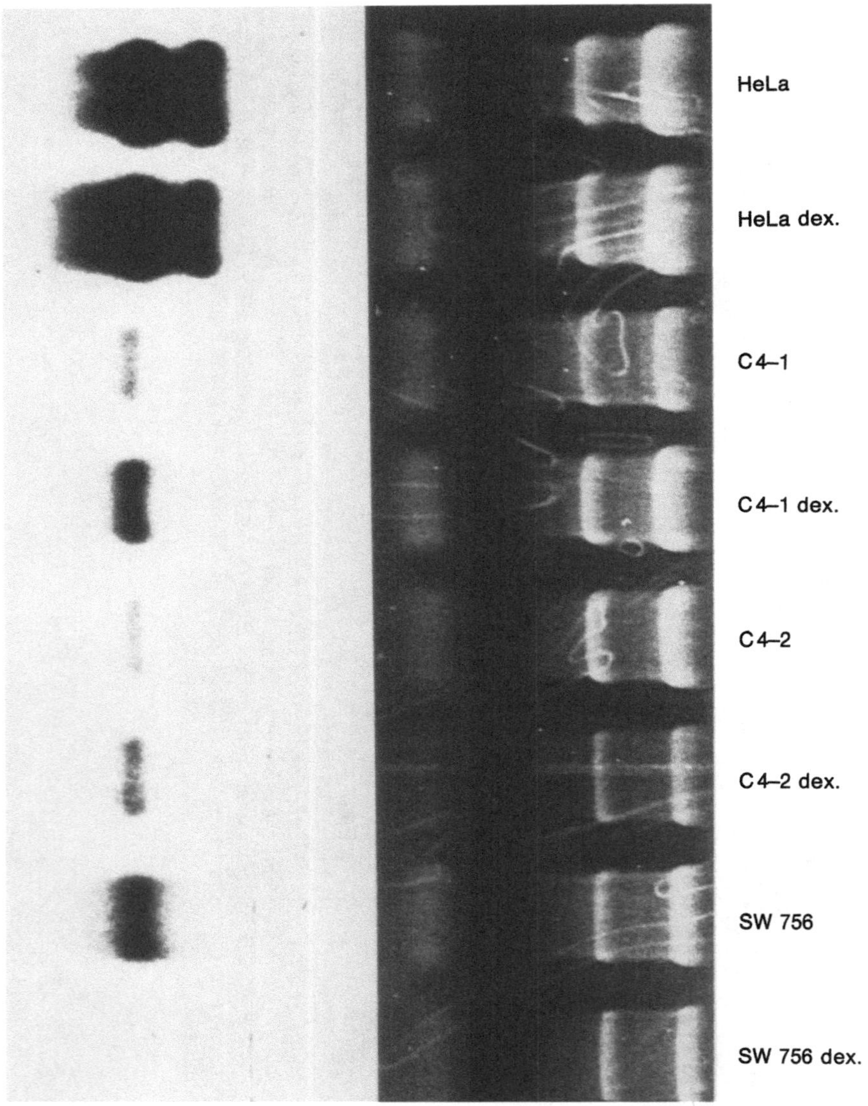

Fig. 7. Influence of dexamethasone on the steady state level of the HPV 18 E6 and E7 mRNA in four HPV-18-positive cervical carcinoma cell lines: 10 μg of cytoplasmic RNA of hormone-treated or untreated cells were separated electrophoretically in a 1% nondenaturing MOPS gel and blotted on a gene screen plus filter (NEN, Boston, USA). The blot was hybridized with a ^{32}P-labelled probe for the HPV 18 E6 and E7 sequences. Dexamethasone had no influence on the HPV 18 E6 and E7 mRNA steady state level in HeLa cells. In C4-1 and C4-2 cells E6 and E7 expression was increased and in SW 756 cells E6 and E7 mRNA was significantly reduced by hormone treatment

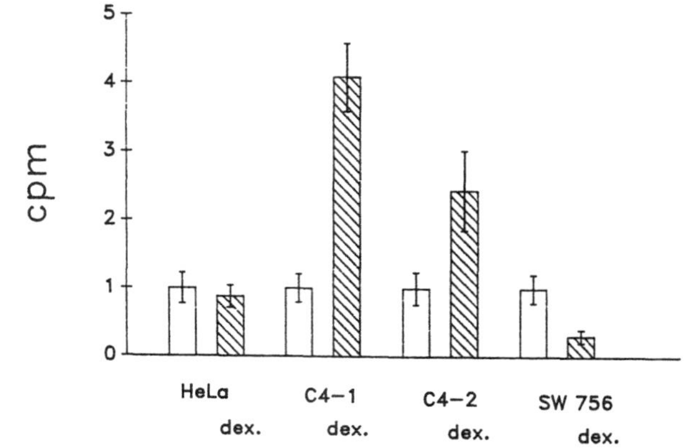

Fig. 8. Relative ^{3}H-thymidine incorporation rate of cervical cancer all lines cultured with or without dexamethasone. Cells grown either in the presence or absence of dexamethasone for 1 week were pulsed with ^{3}H-thymidine. The incorporation of radiolabelled thymidine reflecting the DNA-synthesis rate and therefore also the proliferation rate of the cells was determined as described (von Knebel Doeberitz et al. 1988). Each value was determined 24-fold and the standard deviation is given in the figure. In line with the level of HPV 18 early gene expression (Fig. 7) the proliferation rate of HeLa cells was not significantly affected by dexamethasone. In C4-1 and C4-2 cells the hormone stimulated the proliferation rate and in SW 756 cells glucocorticoid treatment significantly reduced the cellular proliferation rate

properties. Dexamethasone treatment of SW 756 cells led to a strong reduction of the viral gene expression. This is accompanied by significant phenotypical changes of the cells (Fig. 9). Hormone-treated SW 756 cells resembled somewhat the phenotype of fibroblasts, while untreated cells clearly show an epithelial morphology.

Growth in soft agar elucidated strong differences of the biological properties of hormone-treated or untreated SW 756 cells. Significant differences of the size of soft agar colonies were observed. In the presence of dexamethasone only a marginal proliferation of SW 756 cells was observed and only one-fifth of the seeded cells formed detectable colonies. The untreated control cells expressing HPV 18 E6 and E7 mRNA, however, proliferated well in soft agar and formed much larger colonies with a cloning efficiency of about 100% in the same course of time.

Are these dexamethasone-induced phenotypical changes in SW 756 cells indeed linked to reduced HPV 18 E6/E7 gene expression? To approach this question we introduced a hormone-inducible eukaryotic expression vector that permits HPV expression in the presence of dexamethasone. The inducible eukaryotic expression vector mentioned earlier was therefore reconstructed, now expressing the HPV 18 E6/E7 sequences under control of the dexamethasone inducible MMTV-LTR in the correct coding or sense orientation. This vector was transferred into SW 756 cells. Cells harboring this construction and expressing the E6 and E7 sequences also in the presence of

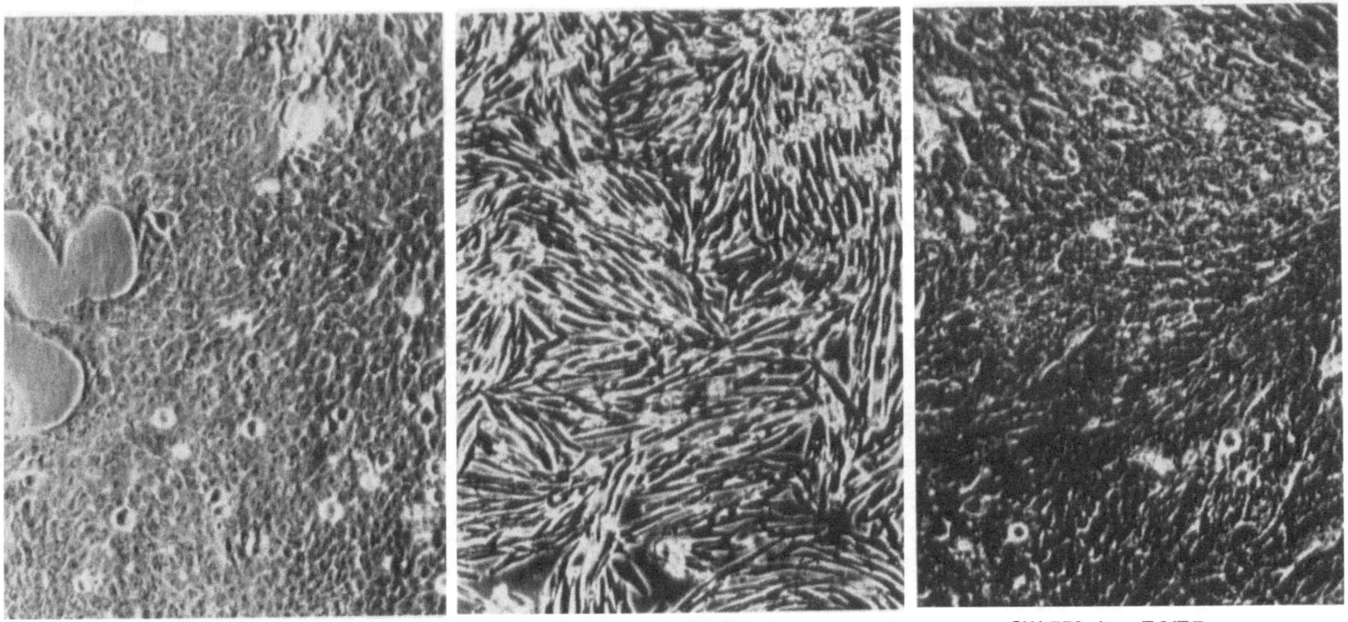

SW 756 | SW 765 dex. E6/E7 neg. | SW 756 dex. E6/E7 pos.

Fig. 9. Morphology of SW 756 cells with varying rates of HPV 18 E6 and E7 gene expression. Dexamethasone treatment of SW 756 cells leads to significantly reduced viral gene expression and severe changes of the morphology in tissue culture (SW 756 dex E6/E7 neg). Upon hormone treatment for 1 week the cells can not grow densely. They die before reaching a confluent state. Reintroduction of HPV 18 E6 and E7 gene expression with a glucocorticoid inducible eukaryotic expression vector apparently allows the cells to grow confluently even in the presence of the hormone as is here demonstrated for a subclone of the SW 756 cell line expressing HPV early genes also in the presence of dexamethasone (SW 756 dex E6/E7 pos) dexamethasone were compared with those not expressing the viral genes in the presence of the hormone.

SW 756 cells expressing the E6/E7 sequences under dexamethasone again revealed some morphological characteristics like the parental cell line without dexamethasone treatment. Most prominently this is the ability to form dense cell layers. In contrast SW 756 cells without expression of the E6/E7 genes in the presence of dexamethasone were not able to grow to confluency (Fig. 9). If the cells reached the confluent state a part of them detached and died. Dexamethasone-treated SW 756 cells harboring a hormone-inducible E6 and E7 expression vector formed dense cell layers. In these cells the E6 and E7 genes were expressed even in the presence of the hormone. In the confluent state they resembeled very much the morphology of untreated SW 756 cells (Fig. 9). Therefore, the effects of dexamethasone on at least some biological properties of SW 756 cells correlate directly with the expression of papillomavirus early genes.

Conclusions

Experiments described here reveal a tight association of the biological characteristics of HPV 18 positive cervical cancer cells with the rate of viral E6 and E7 gene expression. In C4-1 and C4-2 cells the rate of HPV 18 early gene expression is enhanced by glucocorticoid hormones. This is accompanied by an increased proliferation rate of the cells. Specific inhibition of the increased viral gene expression with antisense RNA abolishes the proliferation increase in these cells.

In the Hela cell line, dexamethasone has no effect either on viral gene expression or on the proliferation rate of the cells.

In SW 756 cells, however, dexamethasone leads to a strong reduction of viral gene expression which is accompanied by a severe alteration of their growth characteristics and presumably converting them to a nonmalignant phenotype. Reintroduction of the E6 and E7 genes under control of a hormone-inducible expression vector apparently reverses at least some of the glucocorticoid hormone-induced effects on SW 756 cells.

Taken together, these data support the concept that continuous expression of the viral E6 and E7 genes is an essential prerequisite for the malignant phenotype of cervical cancer cells. The development of specific strategies to inhibit the expression or the biochemical function of the viral E6 and E7 genes might provide the basis for a new therapeutic strategy or prevention of this widespread type of human cancer.

References

Androphy EJ, Hubbert NL, Schiller JT, Lowy DR (1987) Identification of the HPV 16 E6 protein from transformed mouse cells and human cervical carcinoma cell lines. EMBO J 6:989–992

Auersperg N (1969) Histiogenetic behavior of tumors. I. Morphologic variation in vitro and in vivo of two related human carcinoma cell lines. JNCI 43:151–173

Auersperg N, Hawryluk AP (1962) Chromosome observations on three epithelial cell structures derived from carcinomas of the human cervix. JNCI 28:605–627

Baker CC, Phleps WC, Lindgren V, Braun MJ, Gonda MA, Howley PM (1987) Structural and transcriptional analysis of human papillomavirus type 16 sequences in cervical carcinoma cell lines. J Virol 61:962–971

Beato M (1989) Gene regulation by steroid hormones. Cell 56:335–344

Beral V, Hannaford P, Kay C (1988) Oral contraceptive use and malignancies of the genital tract. Lancet 10:1331–1334

Boshart M, Gissmann L, Ikenberg H, Kleinheinz A, Scheurlen W, zur Hausen H (1984) A new type of papillomavirus DNA its presence in genital cancer and in cell lines derived from cervical cancer. EMBO J 3:1151–1157

Brinton LA, Huggins GR, Lehman HF, Mallin K, Savitz DA, Trapido E, Rosenthal J, Hoover R (1986) Long term use of oral contraceptives and risk of invasive cervical cancer. Int J Cancer 38:339–344

de Villiers EM, Schneider A, Miklaw H, Papendick U, Wagner D, Wesch H, Wahrendorf J, zur Hausen H (1987) Human papillomavirus infections in women with and without abnormal cervical cytology. Lancet 26:703–706

Dürst M, Dzarlieva-Petrusevska RT, Boukamp P, Fusenig N, Gissmann L (1987) Molecular and cytogenetic analysis of immortalized human primary keratinocytes obtained after transfection with human papillomavirus type 16 DNA. Oncogene 1:251–256

Dürst M, Gissmann L, Ikenberg H, zur Hausen H (1983) A papillomavirus DNA from a cervical carcinoma and its prevalence in cancer biopsy samples from different geographic regions. Proc Natl Acad Sci USA 80:3812–3815

Fasel N, Pearson K, Buetti E, Diggelmann H (1982) The region of mouse mammary tumor virus DNA containing the long terminal repeat includes a long coding sequence and signals for hormonally regulated transcription. EMBO J 1:3–7

Freedman RS, Bowen JM, Leibovitz A, Pathak S, Siciliano MJ, Gallager HS, Giovanella BC (1982) Characterization of a cell line (SW 756) derived from a human squamous cell carcinoma of the uterine cervix. In Vitro 18:719–726

Gey GO, Coffman WD, Kubiciek MT (1952) Tissue culture studies of the proliferative capacity of cervical carcinoma and normal epithelium. Cancer Res 12:264–265

Gissmann L, Dürst M, Oltersdorf T, von Knebel Doeberitz M (1987) Human papillomaviruses and cervical cancer. Cancer Cells 5:275–280

Green PJ, Pines O, Inouye M (1986) The role of antisense RNA in gene regulation. Annu Rev Biochem 55:569–597

Izant IG, Weintraub H (1985) Constitutive and conditional suppression of exogenous and endogenous genes by antisense RNA. Science 229:345–352

Jones HW, Mc Kusick VA, Harper PS, Wuu KD (1971) The HeLa cells and a reappraisal of its origin. Obstet Gynecol 38:945–949

Le J-Y, Defendi V (1988) A viral cellular junction fragment from a human papillomavirus type 16 positive tumor is competent in transformation of NIH 3T3 cells. J Virol 62:4420–4426

Melnick JL, Adam E (1978) Epidemiological approaches to determining whether herpesvirus is the etiological agent of cervical cancer. Prog Exp Tumor Res 21:49–69

Oltersdorf T, Seedorf K, Röwekamp W, Gissmann L (1987) Identification of human papillomavirus type 16 E7 protein by monoclonal antibodies. J Gen Virol 68:2933–2938

Pirisi L, Yasumoto S, Feller M, Doninger JK, Di Paolo JA (1987) Transformation of human fibroblasts and keratinocytes with human papillomavirus type 16 DNA. J Virol 61:1061–1066

Rawls WE, Bachetti S, Graham FL (1977) Relationship of herpes simplex viruses to human malignancies. Current topics in microbiology and immunology, vol 77. Springer, Berlin Heidelberg New York, pp 71–94

Rigoni-Stern D (1842) Fatti statisici relativi alle malatie cancerose. G Servre Prog Pathol Terap 2:507–517
Saxon PJ, Srivatsan ES, Stanbridge EJ (1986) Introduction of human chromosome 11 via microcell transfer controls tumorgenic expression of HeLa cells. EMBO J 5:3461–3466
Schwarz E, Freese UK, Gissmann L, Mayer W, Roggenbuck B, zur Hausen H (1985) Structure and transcription of human papillomavirus sequences in cervical carcinoma cells. Nature 314:111–114
Schneider-Gädicke A, Schwarz E (1986) Different human cervical carcinoma cell lines show similar transcription patterns of human papillomavirus type 18 early genes. EMBO J 5:2285–2292
Seedorf K, Oltersdorf T, Krämmer G, Röwekamp W (1987) Identification of early proteins of the human papillomaviruses type 16 and 18 in cervical carcinoma cells. EMBO J 6:139–144
Shirasava H, Tomita Y, Sekiya S, Takamizawa H, Simizu B (1987) Integration and transcription of human papillomavirus type 16 and 18 sequences in cell lines derived from cervical carcinomas. J Gen Virol 68:583–591
Smotkin D, Wettstein F (1986) Transcription of human papillomavirus type 16 early genes in a cervical cancer and a cancer derived cell line and identification of the E7 protein. Proc Natl Acad Sci USA 83:4680–4684
Southern PJ, Berg P (1982) Transformation of mammalian cells to antibiotic resistance with a bacterial gene under control of the SV 40 early region promotor. J Mol Appl Genet 1:327–341
Stanbridge EJ (1976) Suppression of malignancy in human cells. Nature 260:17–20
Stanbridge EJ, Der CJ, Doersen CJ (1982) Human cell hybrids: Analysis of transformation and tumorgenicity. Science 215:252–259
von der Ahe D, Janich S, Scheidereit C, Renkawitz R, Schütz G, Beato M (1985) Glucocorticoid and progesterone receptors bind to the same sites in two hormonally regulated promotors. Nature 313:706–709
von Knebel Doeberitz M, Oltersdorf T, Schwarz E, Gissmann L (1988) Correlation of modified human papillomavirus early gene expression with altered growth properties in C4-1 cervical carcinoma cells. Cancer Res 48:3780–3786
Vonka V, Kanka J, Jelinek J, Subrt I, Suchanek A, Havrankova A, Vachal M, Hirsch I, Domorazkova E, Zavadova H, Richterova V, Naprstkova J, Dvorakova V, Svoboda B (1984) Prospective study on the relationship between cervical neoplasia and herpes simplex type-2 virus. Int J Cancer 33:49–60
Woodwarth CD, Doniger J, Di Paolo JA (1989) Immortalization of human foreskin keratinocytes by various human papillomavirus DNAs corresponds to their association with cervical carcinoma. J Virol 63:159–164
zur Hausen H (1977 a) Cell virus gene balance hypothesis of carcinogenesis. Behring Inst Mitt 61:23–30
zur Hausen H (1977 b) Human papillomaviruses and their possible role in squamous cell carcinomas. In: Current topics in microbiology and immunology, vol 78. Springer, Berlin Heidelberg New York, pp 1–30
zur Hausen H (1982) Human genital cancer: synergism between two virus infections or synergism between a virus infection and initiating events? Lancet 2:1370–1373
zur Hausen H (1986) Intracellular surveillance of persisting viral infections: human genital cancer results from defficient cellular control of human papillomavirus gene expression. Lancet 2:489–491
zur Hausen H, Meinhof W, Scheiber W, Bornkamm GW (1974) Attempts to detect virus specific DNA sequences in human tumors: I. Nucleic acid hybridization with complementary RNA of human wart virus. Int J Cancer 13:650–656

Diagnosis

Detection and Typing of Genital Papillomaviruses: Nucleic Acid Hybridization and Type-Specific Antigen

A. Schneider and G. Schlunck

Introduction

Human papillomaviruses (HPV) are classified on the basis of the relatedness of their nucleic acids (Coggin and zur Hausen 1979). Different hybridization techniques allow the discrimination of 23 genital HPV types, of which HPV 6, 11, 16, and 18 are most prevalent (zur Hausen and Schneider 1987). Serologic tests had been of no value for the taxonomic discrimination of genital HPVs since the genital lesions produce only a limited amount of viral antigen with a concentration too low for detection by immunization tests. Now that the complete DNA sequences of several genital HPV types are known, type-specific viral proteins have recently become available. The viral antigens are synthesized in bacterial expression vectors, and antisera are produced by immunization of animals (Seedorf et al. 1987).

Demonstration of HPV by Nucleic Acid Hybridization

Basic Principles of Hybridization

The basic principle of hybridization is the duplex formation between single nucleic acid strands (either DNA or RNA) of different origins. For HPV diagnosis usually cloned viral DNA or RNA is used to detect HPV DNA in infected cells. Since the viral DNA is double-stranded, the DNA strands must be separated to produce accessible targets for the probe. This dissociation process of the two nucleic acid strands is called denaturation or melting (Fig. 1). The stability of the duplex is due to hydrogen bonds between the complementary basepairs. Both strands separate at a specific melting temperature (T_m) when thermal mobility exceeds the binding force of the hydrogen bonds. Besides temperature, other factors such as concentration of formamide and salt ions, length of the nucleic acid strands, and guanosine-cytosine content of the nucleic acid influence the stability of the molecule (McConaughy et al. 1969). Additionally, the stability of a probe-target complex is affected by the homology between the corresponding nucleotides of the hybridizing strands. A high degree of nucleotide matching results in a high stability of the nucleic acid complex. The stability of the nucleic acid complex is reflected by the conditions of stringency used in the hybridization experiment (Fig. 2). Under high-stringency conditions the temperature of the hybridization is near the T_m

Genital Papillomavirus Infections
Eds.: Gross/Jablonska/Pfister/Stegner

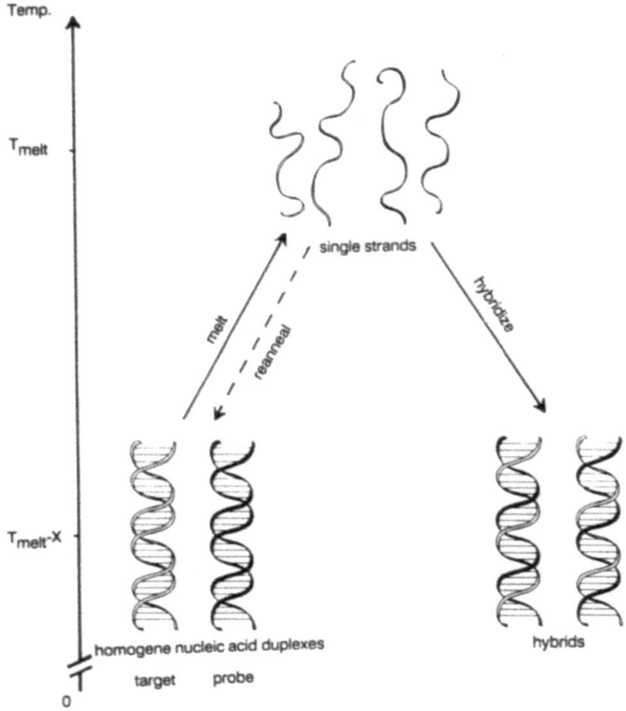

Fig. 1. Principles of hybridization. DNA as well as RNA build helical secondary structures. This conformation is stabilized by hydrogen bonds which are formed between complementary bases. Increasing temperature leads to a characteristic point ("melting" temperature, T_m) at which the thermal mobility of each nucleic acid strand exceeds the binding force of the hydrogen bonds between the corresponding basepairs. Thus, both strands separate ("denaturation" or "melting" of the nucleic acid complex). Increase in the formamide concentration or decrease in the salt concentration of the hybridization solution also has a destabilizing effect on the nucleic acid complex. Subsequently lowering of temperature leads to a reverse effect, and reannealing of the separated nucleic acid strands occurs. If reannealing occurs between nucleic acids originating from different molecules, this is called hybridization. *Above,* single strands of nucleic acids separate at T_m. *The two white strands* symbolize the target (e.g., HPV DNA derived from infected cells), whereas *the two black strands* symbolize the probe (e.g., HPV plasmid DNA). *Below,* the possible results after decreasing of the temperature are shown. The homogeneous strands reanneal and form duplexes identical to the ones before denaturing (*left*). Alternatively, hybrids are build between the nonhomogeneous strands (*right*), and the probe hybridizes with the target. Since the probe is labeled, this newly formed complex can be detected. When the nucleotide sequences of probe and target are identical, the likelihood for reannealing or hybridization is equal

of the nucleic acid complex (18° C below T_m), whereas under low-stringency conditions the difference between the temperatures is 40° C. The degree of stringency of the hybridization experiment is chosen to control the quality of the hybrids. In a highly stringent experiment the two strands of a hybrid must be strongly attached to each other to withstand unfavorable environmental

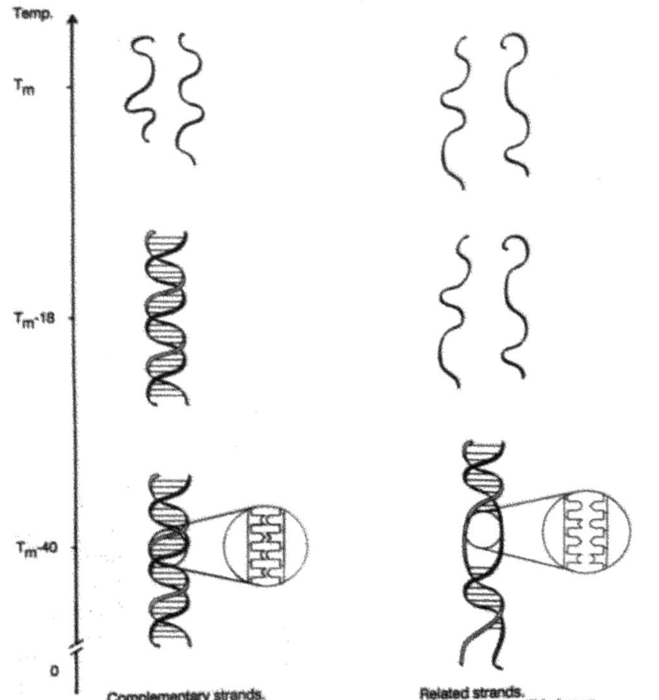

Fig. 2. Principles of stringency. The pairing of the nucleic acid strands depends on the conditions of stringency present in the environment and the homology between the corresponding nucleotides. Thus, at T_m complete and incomplete homologous strands are separated. By lowering the temperature to T_m $-18°$ C (stringent conditions) only homologous strands hybridize. Decreasing the temperature to T_m $-40°$ C allows also nonhomologous strands to partly hybridize. *Above,* strands of probe (*black*) and target (*white*). The opposing strands have homologous (*left*) and partly homologous (*right*) nucleotides. Thus, the *left probe* represents plasmid DNA of an HPV type identical to the target, whereas the *right probe* stands for plasmid DNA of another HPV type. Under stringent conditions (*centre*) only the homologous HPV DNAs hybridize (*left*) whereas the related HPV DNAs (*right*) remain separated. Nonstringent conditions with further decrease in the temperature (*below*) lead also to hybridization of the strands from different HPV types (*right*). The magnification shows that in certain areas of the nonhomologous strands no pairing occurs due to mismatches (*right*), whereas in the homologous strands all bases are paired (*left*)

conditions such as high temperature. On the other hand, conditions of low stringency allow even weak hybrids to remain annealed. Hence, under high-stringency conditions HPV types identical to the HPV probe used are identified whereas with low-stringency conditions related or unknown HPV types can be detected.

To visualize the hybrid complex formed between cloned nucleic acid probe and target DNA, the probe must be labeled. Labeling can be done either by isotopes or chemically. The most common method for labeling of double-

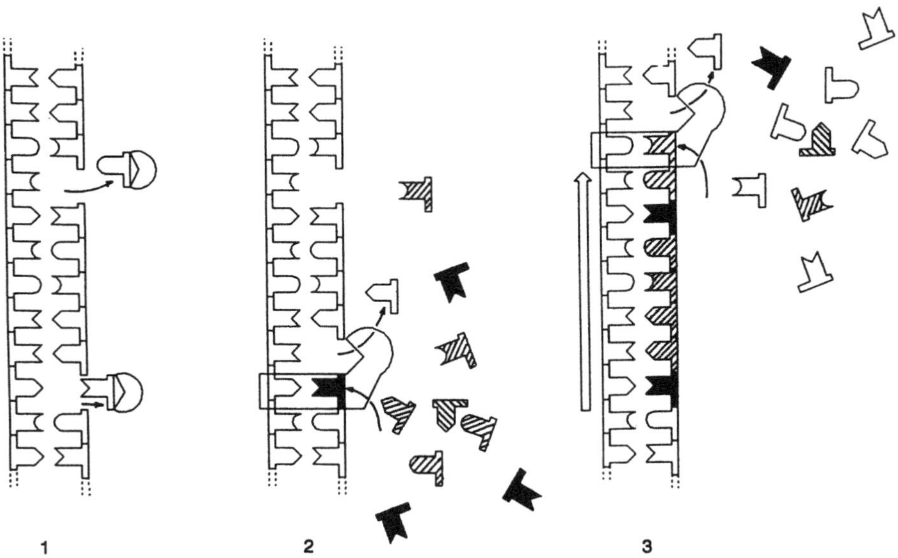

Fig. 3. Principles of nick translation of double-stranded DNA. In the first step of the reaction DNAse produces "nicks" in both strands of the nucleic acid (*1*). DNA polymerase I starts to "repair" the nicked strand: the 3′-5′ exonuclease activity of the DNA polymerase I excises small groups of nucleotides adjacent to the "nick" and creates "gaps" (*2*); simultaneously the polymerase fills the nucleotide gaps with nucleotides of which one or several types are labeled (*2, 3*). Since only part of the labeled nucleotides are incorporated, the labeled probe must be separated from the unincorporated nucleotides by gel filtration after nick translation. This prevents background labeling during the hybridization experiment. ◌, DNAse; ◌, DNA polymerase I; ◌, intrinsic nucleotides; extrinsic nucleotides: ◌, nonlabeled; ◌, labeled

stranded DNA is nick translation (Fig. 3; Rigby et al. 1977). Oligomeric primer extension is also used for labeling of DNA. Randomly synthesized oligonucleotides, usually 6 bases long (hexamers), anneal to the denatured single DNA strands (Maniatis et al. 1982). A DNA polymerase (e.g., Klenow polymerase) binds to the hexamer "primer" and uses the opposite strand as a template while adding nucleotides which are partly labeled to the primer. Single-stranded nucleic acids can be labeled using a T4 polynucleotide kinase which transfers the gamma-phosphate of ATP to the 5′-hydroxyl end (end-labeling; Maniatis et al. 1982). Alternatively, single-stranded RNA can be produced by transcription of a viral DNA template which is cloned in a vector plasmid downstream of a promoter sequence. The promoter can be recognized by a RNA polymerase (e.g., SP6 polymerase; Cox et al. 1984). Asymmetric RNA probes produced by in vitro transcription yield the highest sensitivity due to a high specific activity and due to a lack of self-reannealing of the nucleic acid (Cox et al. 1984).

Southern Blot Hybridization

Southern blot hybridization is used for the detection of HPV DNA in cellular DNA specimens (Southern 1975). Tissues or cells are homogenized and proteins enzymatically digested (Fig. 4). Cellular DNA is extracted using phenol and chloroform. Between 10 and 12 μg cellular DNA is digested with a restriction endonuclease into DNA fragments of different sizes, and the frag-

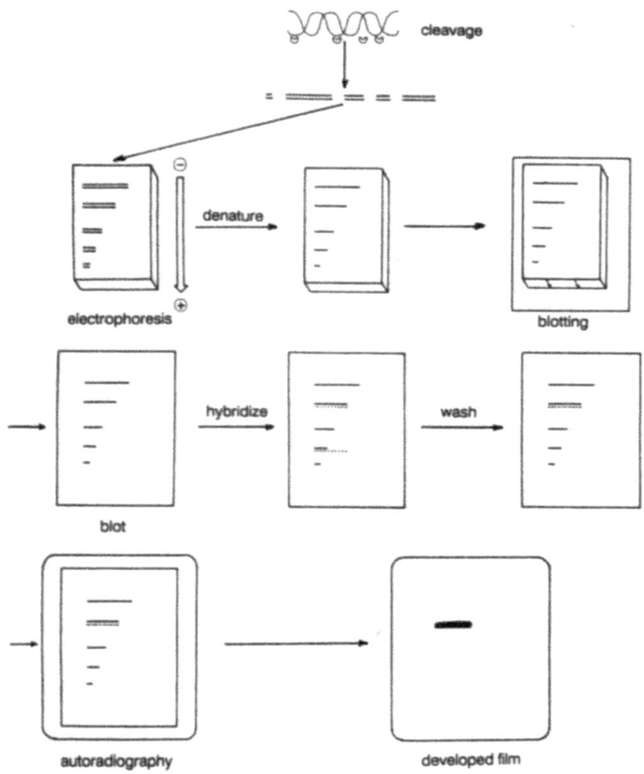

Fig. 4. Principles of Southern blot hybridization. Cellular DNA which has been extracted from tissue or cells is digested by a restriction enzyme (*above*). Thus, DNA fragments of defined length are produced. Given that HPV DNAs contain cutting sites for the enzyme, the digestion creates a characteristic pattern of DNA fragments which is different for each HPV type and for each restriction enzyme. The fragments are separated by agarose gel electrophoresis according to their length. Since long fragments migrate more slowly than short fragments, they remain nearer the starting point (negative electrode). The various DNA bands are stained by ethidiumbromide and visualized with UV light. The DNA is denatured by alkali treatment and transferred onto a membrane (nylon, nitrocellulose). Thus, the DNA is bound to a support and accessible for hybridization, which is done with labeled single-stranded probes (*dotted line*). Washing removes nonspecifically bound probes. Depending on the conditions of stringency used during hybridization it can be controlled if identical (high-stringency) or related (low-stringency) HPV types are identified. The labeled probe-target complex can be visualized by autoradiography if isotope-labeled probes are used. To visualize chemically labeled probes the filter is counterstained. ♡, Restriction enzyme

ments are separated by gel electrophoresis. Thus, DNA fragments of varying length are created and separated. The fragment pattern which results from the digestion of the HPV DNA contained in the bulk of cellular DNA is specific for each HPV type. The various DNA fragments are stained with a fluorescent dye and visualized by ultraviolet illumination. The DNA is denatured and transferred onto a nylon membrane or nitrocellulose filter. After prehybridization to block non-specific binding sites, the filter is hybridized with the cloned labeled HPV DNA or RNA. Thus, the labeled nucleic acid can anneal to the single-stranded viral target DNA which is bound to the filter. Defined concentrations of various HPV DNAs are included on the filter and serve as controls. The membrane is hybridized for 2–48 h under specific conditions of stringency. Then the membrane is washed and either exposed for autoradiography (when isotope labeling is used) or counterstained (when chemical probes are used).

As an alternative to the method described above, instead of the cloned HPV DNA the cellular DNA can be labeled ("reverse hybridization"). The labeled cellular DNA is hybridized with various HPV DNAs which have been blotted onto a membrane (Gissmann and Schwarz 1985). In contrast to Southern blot hybridization, only 0.1–1 μg cellular DNA is needed with this technique. Additionally, the presence of a large spectrum of different HPVs can be tested in one experiment, whereas with Southern blot usually only one specific HPV type can be identified when stringent conditions are applied. On the other hand, the sensitivity of Southern blot hybridization is 10–100 times higher than that of reverse hybridization (0.1 HPV genome equivalents per cell are detected compared to about 10 HPV genome equivalents per cell).

Dot (or Slot) Blot Hybridization

Compared to Southern blot, dot blot hybridization omits two steps, the digestion of cellular DNA and the separation of the different DNA fragments by gel electrophoresis. After extraction the DNA is denatured using heat and alkaline treatment. The nucleic acids are either filtrated on a membrane with a manifold or dotted directly onto the filter (Kafatos et al. 1979; Thomas 1980). Varying amounts of cloned viral nucleic acid are included in the experiment as controls. Hybridization and washing is performed under conditions identical to those of Southern blot. After autoradiography or counterstaining typical hybridization signals are seen (Fig. 5). Since the cellular DNA can be concentrated on a small filter area, only small amounts are necessary (0.1–1 μg). This has the disadvantage that due to high cellular DNA concentration background signals may mimic a positive result. Therefore, hybridization can only be performed under stringent conditions. Since a large number of samples can be examined in one experiment, this technique (modified slot blot procedure) became commercially available (ViraPap, ViraType). In this procedure a simple DNA extraction method and ^{32}P-labeled RNA probes are used. After hybridization a ribonuclease treatment is included to reduce background signals. With one kit 50 specimens can be examined.

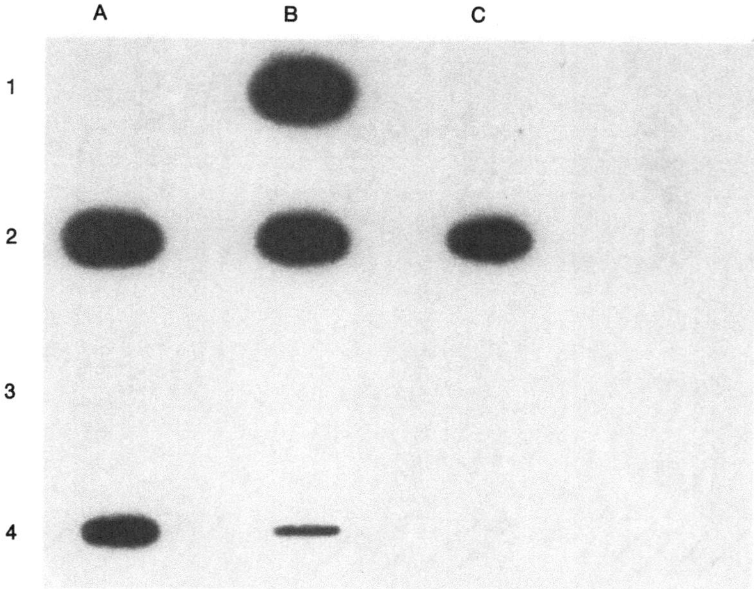

Fig. 5. Example of result of slot blot hybridization. Autoradiogram of a nylon membrane with 12 DNA samples derived from paraffin-embedded, formalin-fixed cervical biopsies. The DNA samples had been submitted to 40 cycles of DNA amplification (PCR) using primers from the E6 region of HPV 16. The PCR product was identified by hybridization under high-stringency conditions with a ^{32}P-end-labeled 20-oligonucleotide probe, which was complementary to the central portion of the 163 basepair target in the E6 region. Autoradiography was done for 4 days. *Row 1*, plasmid DNA of HPV 11 (*A*), 16 (*B*), and 18 (*C*); 10 ng prior to PCR. *Row 2*, different amounts of CaSki cells, which contain approximately 600 HPV 16 copies per cell; 10^4 (*A*), 10^3 (*B*), and 10^2 (*C*) cells per slot. *Rows 3, 4*, DNA from cervical biopsies. Samples *4A* and *4B* are strongly positive, sample *4C* weakly positive for HPV 16

Sandwich Hybridization

Nucleic acid sandwich hybridization is a rapid method which allows testing of large numbers of clinical samples (Parkkinen et al. 1986; Parkkinen 1988). Two nonoverlapping fragments of the viral genome are selected (Fig. 6). One fragment ("capture") is immobilized on a matrix such as nitrocellulose. The other fragment is labeled and serves as "probe." During hybridization the sample which contains the "target" DNA is mixed with the two fragments. Annealing with the capture and the probe occurs at separate parts of the target DNA. After washing the result, i.e. the amount of labeled hybrid which has been bound by the capture-target complex is measured by a liquid scintillation counter. Thus, the whole procedure can be completed in less than 24 h. A total of $1–5 \times 10^5$ HPV DNA molecules can be detected with this method.

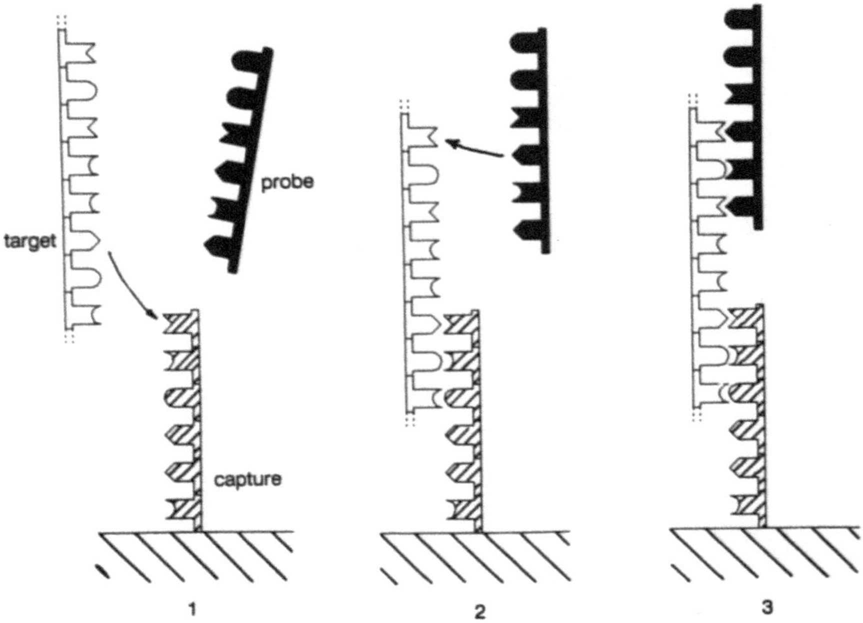

Fig. 6. Principles of the sandwich hybridization technique. The sandwich hybridization uses two adjacent, nonoverlapping restriction fragments of the HPV genome. One fragment (capture) is immobilized onto a solid support such as nitrocellulose or nylon (*1*). The second fragment (probe) is labeled radioactively or chemically (*1*). The target nucleic acid in solution is allowed to form hybrids with the capture (*2*) and the probe (*3*). target; capture; probe

Filter In Situ Hybridization

In filter in situ hybridization extraction of cellular DNA can be omitted (Wagner et al. 1984). However, this method can only be applied to cellular samples and not to biopsies. Swabs are taken from different areas of the lower genital tract and the cells are suspended in phosphate-buffered saline. The cells are filtered on nitrocellulose membranes using a manifold. Cell lysis and DNA denaturation of the cell-covered filters is done by alkaline treatment. After neutralization the filters are baked and stored. Hybridization is performed under conditions of high stringency. After washing, the specimens are exposed for autoradiography and the results are read after an interval of 1–7 days (Fig. 7). Filters are interpreted as positive when clear spots can be distinguished from background signals. In lesions where cells contain only low copy numbers, specific signals cannot be differentiated from background. It is estimated that 10000 HPV DNA copies per cell are necessary to produce a specific dot on the autoradiogram (Gissmann et al. 1986).

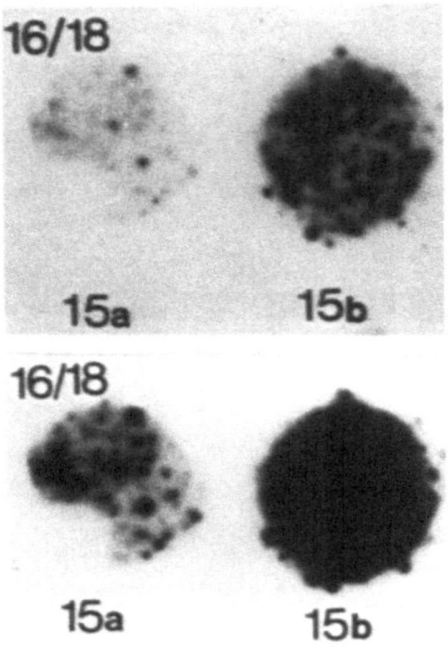

Fig. 7. Example of result of filter in situ hybridization. Filters with cells from a grade II cervical intraepithelial neoplasia (CIN II) were hybridized with a mixture of ^{32}P-labeled HPV 16 and 18 probes. Autoradiography was performed for 24 h (*upper row*) and 5 days (*lower row*). The different amount of signals is due to variation in the concentration of infected cells on the filter (1.2×10^5 cells on *15a* and 2.8×10^5 on *15b*)

In Situ Hybridization

With this technique the distribution of viral nucleic acids can be allocated in tissues and cells. The fixed or frozen material is placed on glass slides. If fixatives are used which cross-link cellular proteins (e.g., formaldehyde), pretreatment must include proteinase digestion. Isotope or chemically labeled nucleic acids are used as probes, and hybridization is performed under stringent conditions. Mainly asymmetric RNA probes are used for hybridization due to their higher sensitivity, and RNA digestion is included in the washing process (Stoler and Broker 1986). The labeled probes are finally visualized by dipping in nuclear track emulsion and by autoradiography when isotope labeling is used (Fig. 8). If biotinylated probes are used, chemical reactions such as the peroxidase-antiperoxidase or streptavidin complexes take place. Known HPV-positive and HPV-negative control sections are included in each experiment. Major problems of in situ hybridization are loss of sections, deterioriation of morphology, and variation of sensitivity. Different approaches have been recommended for increasing adherence of sections to the slides using poly-D lysine, poly-L lysine, Elmer's glue, gelatine, and siliconization (Nagai et al. 1987). For fixation formaldehyde, glutaraldehyde, paraformaldehyde, Bouin's and Carnoy's fixatives have been used. Only the latter fixative does not need pretreatment with proteinase K and thus yields the best morphologic preservation of the tissue. Generally, isotope-labeled probes yield higher sensitivity than chemically labeled probes (Table 1). However, the latter have the great advantage that the labeled nucleic acids can be kept for a long time and cause no hazard to the technician.

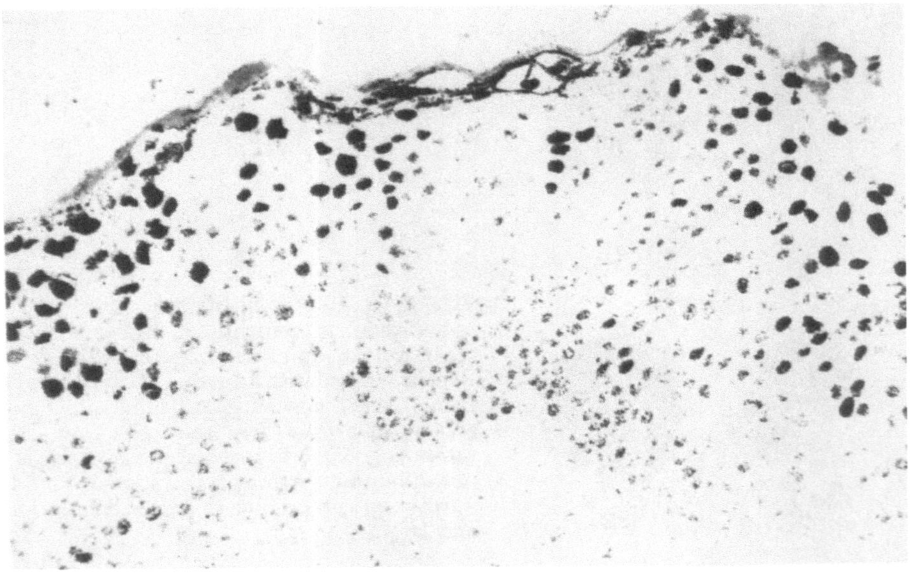

a

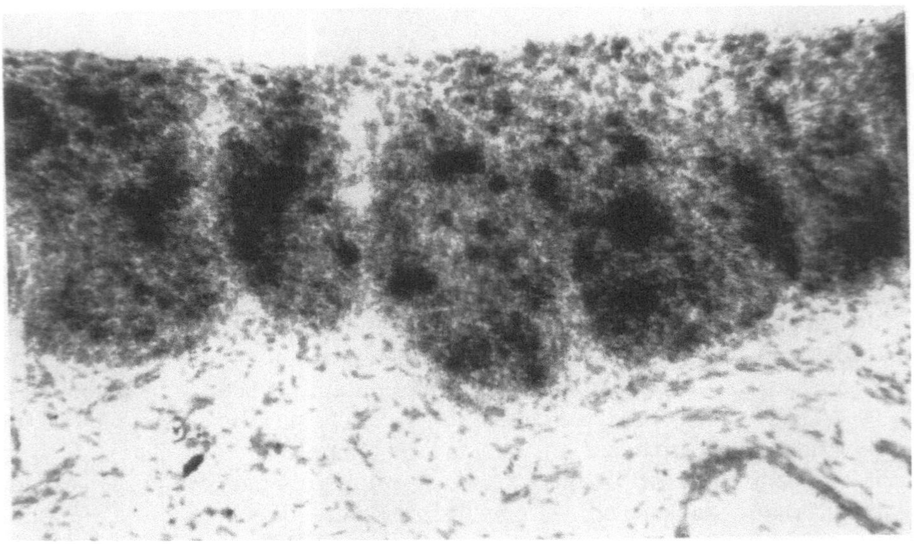

b

Fig. 8 a, b. Examples of results of in situ hybridization (photomicrographs). *Above,* a CIN I positive for HPV 18; *below,* a CIN II positive for HPV 31. The cone biopsies were fixed in Carnoy's fixative and paraffin-embedded. Asymmetric RNA was used for hybridization labeled with ^{3}H (*above*) or ^{35}S (*below*). The distribution of grains is more diffuse for the ^{35}S-labeled probe due to the longer path length of the isotope. Autoradiography was performed for 45 days (*above*) and 5 days (*below*). (× 180)

Table 1. Sensitivity of different in situ hybridization techniques

Reference	Labeling	
	^{35}S	Biotin
Burns et al. (1987)		10
Crum et al. (1988)	50	800
Schneider et al. (1987a)	20–50	
Syrjänen et al. (1988)	10–50	1–2
Walboomers et al. (1988)		20

Numbers indicate the lowest amount of HPV copies detectable

HPV DNA Amplification

With the polymerase chain reaction (PCR) single genes are amplified in vitro (Saiki et al. 1985). A sequence of between 100 and 1000 basepairs is selected from the known target DNA to be amplified (Fig. 9). For each end of the complementary DNA strands oligonucleotides of about 20 bases length are identified and synthesized ("primers"). The PCR consists basically of three steps: denaturation, primer annealing, and primer extension. The target DNA is denatured at 95° C, which is followed by annealing of the primer oligomers at a temperature between 37° C and 55° C. The primers anneal with the complementary nucleotide sequence on the separated strands. DNA polymerase extends the primers using the single-strand DNA as template at 72° C, which results in two daughter strands. Thus, the target DNA is amplified exponentially with the number of cycles. Since heat-stable DNA polymerase became available (*Taq* polymerase) the polymerase must be added only once at the beginning of the experiment. The amplified DNA can be submitted to Southern blot, dot blot (Fig. 5), or sandwich hybridization under identical conditions as described above. With the PCR technique minute amounts of HPV DNA can be detected, which makes it the standard for sensitivity of HPV DNA identification (Shibata et al. 1988).

Application of Different Hybridization Techniques

The method of hybridization to be chosen depends on the type, status, and size of the clinical sample. In addition, expected number of HPV DNA molecules in the sample and the practicability of the method are important factors. DNA extracted from tissues or cells can be examined by all techniques. Only PCR, sandwich, in situ and filter in situ hybridization allow the examination of cells without extraction. All methods can be used on fresh or frozen material. Since DNA can be extracted from fixed material, all techni-

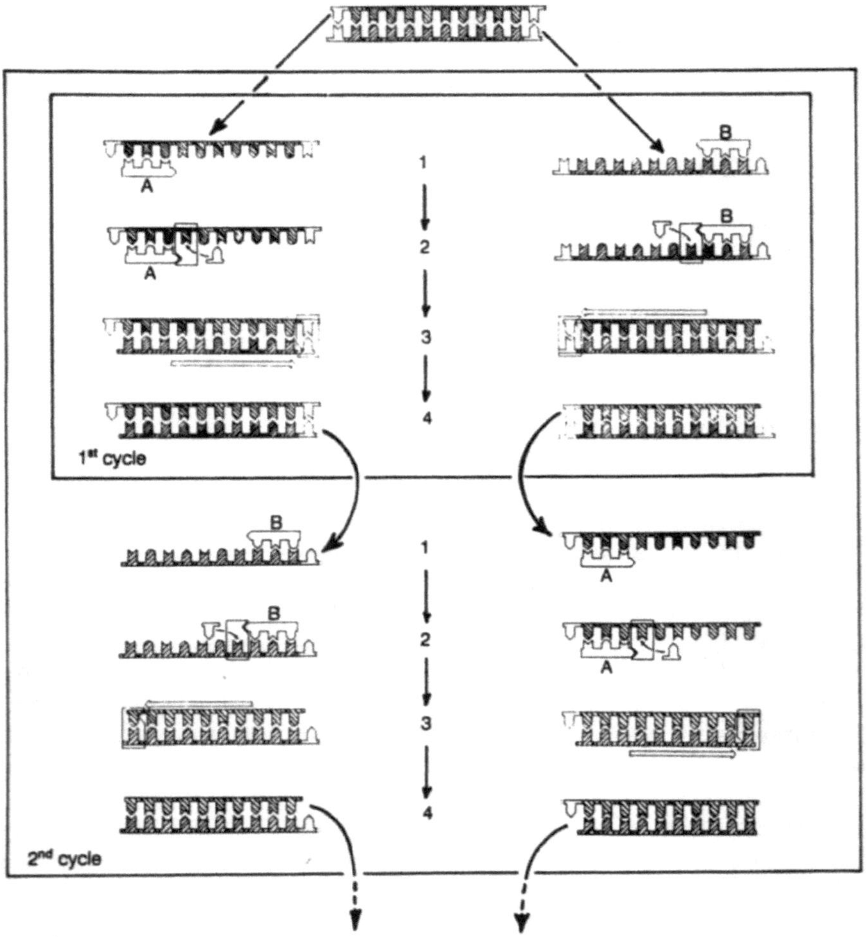

Fig. 9. Principles of polymerase chain reaction (*PCR*). A specific region of a HPV sequence (*hatched area*) is selected for amplification (usually between 100 and 1000 bases long). Oligonucleotides of about 20 bases length are synthesized, which are complementary to the opposing ends of the selected target DNA strands (*primers A* and *B*). *Small box*, the first cycle of the PCR. The unamplified DNA is denatured (denaturation at 92° C–95° C) and incubated with *primers A* and *B* which recognize their complementary DNA sequence on the separated strands (annealing at 45° C–65° C; *1*). The primers are extended by adding heat-stable *Taq DNA polymerase* to the assay (primer extension at 73° C; *2*). The single-strand DNA is used as template, and two daughter strands are produced (*3, 4*). In the second cycle (*large box*) the DNA is again denatured (*1*), the primers anneal to their specific sites (*2*), and the polymerase starts another extension reaction (*3, 4*). Thus four new daughter strands are synthesized. After each cycle the products serve as templates in the subsequent cycles. Thus, under ideal conditions the DNA fragment is amplified exponentially ($n = 2^i$, where i is the number of cycles). Under normal conditions a DNA fragment can be amplified up to 10^5-fold. Primers: 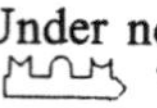 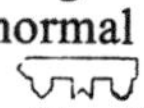 . For simplification the 20 bases long primers are shown to be only 3 bases long. Taq DNA polymerase

ques with the exception of filter in situ hybridization can be used with such material. However, the fixation process affects the status of the DNA. Especially fixation with Bouin's fixative leads to degradation and DNA loss during the extraction process, whereas Carnoy's fixative preserves high molecular weight DNA (unpublished data). The amount of DNA which can be extracted from tissues or swabs depends on the cellularity of the specimen. If more than 500 000 cells are available, about 10 μg cellular DNA can be recovered, which is sufficient for proper performance of all techniques. If the yield of DNA is less than 10 μg but more than 1 μg, this amount is sufficient for sandwich and dot blot hybridization. DNA amplification is the only suitable method if less than 1 μg target DNA is available.

The value of different tests can be compared by estimating their sensitivity and specificity. For evaluation of the sensitivity of various HPV hybridization techniques two definitions must be distinguished. The sensitivity of a method can be defined as the number of HPV DNA molecules detectable (molecular-biological definition) or as the percentage of true positive cases in all patients with disease (clinical definition). Since PCR can detect as few as 10 HPV DNA molecules in a specimen, it is considered the golden standard for HPV detection with respect to molecular-biological sensitivity. However, it is not clear if the presence of a few HPV DNA molecules in the lower genital tract should be considered a disease. Latent genital HPV infections have such a high prevalence that their detection may be of no clinical importance and may not be considered pathologic (Gissmann et al. 1989). Therefore, the clinical sensitivity of the different methods with respect to HPV disease cannot be estimated at the moment. Southern blot hybridization detects between 0.01 and 0.1 HPV genomes per cell if a total of 500 000 cells are available. This means a total of 5000–50 000 HPV molecules. The HPV DNA detection rate of sandwich and dot blot hybridization is almost as high as that of Southern blot. In situ hybridization and filter in situ hybridization have the lowest sensitivity.

The molecular-biological definition of specificity refers to the reliability of distinction between the various HPV types, whereas the clinical definition concerns the percentage of true negatives in all cases without disease. In Southern blot electrophoretic separation of the various DNA fragments produces a typical restriction pattern of the viral DNA and thus allows proper identification of each HPV type. Additionally, electrophoresis separates viral target DNA and host genome. Fewer background signals and no false positive hybridization reactions occur. Thus, Southern blot hybridization seems the golden standard for molecular biological and clinical specificity since false-positive results are very unlikely.

The main problem of PCR is unvoluntary contamination of specimens with plasmid DNA or interspecimen contamination. Even the exchange of only a few HPV molecules renders a positive result. Therefore, the specificity of PCR must be controlled in each experiment with a number of positive and negative controls. However, interspecimen contamination remains a constant problem of this method.

Dot blot hybridization may also yield false-positive results if the stringency conditions are not controlled rigidly. In filter in situ hybridization especially mucus and detritus may cause background signals which can be misinterpreted as specific. Background is also a consistent problem with in situ hybridization. Especially low copy number infections such as in invasive cancer may be misdiagnosed as HPV positive when in fact high background level is observed.

The reproducibility of the different techniques has been investigated systematically in a recent study for Southern blot hybridization (Brandsma et al. 1989). The rate of agreement for the discrimination of HPV-positive and HPV-negative samples was between 67% and 97%. Regarding HPV type specificity, an agreement rate between 50% and 92% was observed. The validity of the different hybridization techniques is usually measured in comparison with Southern blot hybridization. For in situ hybridization a complete agreement could be found for all condylomata acuminata positive for HPV 6 (Beckmann et al. 1985). The analysis of tissues with cervical intraepithelial neoplasia positive for HPV 16 detected HPV with in situ hybridization in 10 out of 15 cases previously found to be positive by Southern blot (Crum et al. 1986). The slot dot blot hybridization technique (ViraPap, ViraType) was compared with Southern blot independently. Regarding HPV detection an intermediate agreement (κ value of 0.58) was found (Mifflin et al. 1988). Identical HPV types were identified in all cases positive by both methods. Thus, multiple factors must be taken into account when HPV results of different studies are compared (Table 2). Analyzing the prevalence rate of HPV infection in asymptomatic women, the numbers vary between 1.3% and 29%. Besides the fact that the hybridization techniques used vary considerably in their

Table 2. HPV DNA detection rates in cervical smears or biopsies from cytologically negative women detected by different hybridization techniques

Reference		HPV-positive	Method
Burk et al. (1986)	$n=$ 29	29%	SB
De Villiers et al. (1987)	$n=$ 8755	9%	FISH
Martinez et al. (1988)	$n=$ 68	3%	SB
McCance et al. (1985)	$n=$ 17	18%	SB
Pratili et al. (1986)	$n=$ 311	1.3%	DB
Schneider et al. (1987b)	$n=$ 92	28.3%[a]	SB
	$n=$ 96	12.5%[b]	SB
Wagner et al. (1984)	$n=$ 36	11%	FISH
Wickenden et al. (1985)	$n=$ 19	11%	DB
Wickenden et al. (1987)	$n=$ 215	19.5%	DB

DB, Dot blot hybridization; FISH, filter in situ hybridization; SB, Southern blot hybridization

[a] Pregnant women

[b] Nonpregnant women

sensitivity, other factors such as age distribution, smoking, and sexual habits are different for various populations.

Labor and cost of the different hybridization techniques correlate directly with the sensitivity and specificity of the test. Thus, Southern blot hybridization and PCR are less practicable than screening tests such as dot blot, sandwich, or filter in situ hybridization. However, automation of DNA diagnostic techniques is advancing rapidly and may result in more cost-effective procedures (Landegren et al. 1988).

Detection of HPV by Type-Specific Proteins

Specific regions of genomes of the genital HPV types 6, 11, 16, and 18 have been expressed in procaryotic expression vectors (Table 3). Using these vectors, proteins specific for the different open reading frames (see chapter by H. Pfister, pp 37) were produced and could detect type-specific antibodies in human sera. Comparing the immune response in human sera to fusion

Table 3. Type-specific antigens and antibodies of genital HPV types

Reference	HPV type	ORF	AB
Banks et al. (1987a,b)	6	*L1, L2*	*mc, pc*
	16	*L1, L2*	
	18	*E6*	
Bernard et al. (1987)	18	E7	
Brown et al. (1988)	11	*E4*	*pc*
Browne et al. (1988)	16	*L1*	*pc*
Firzlaff et al. (1988)	6	*L1, L2*	*pc*
	16	*L1, L2*	
Jenison et al. (1988)	6b	E2, E4, E6, E7, L1, L2	
	16	E1, E2, E4, E6, E7, L1, L2	
Jochmus-Kudielka et al. (1989)	16	E4	
	16	E7	
Li et al. (1987, 1988)	6	L1	*pc*
		E2	
Seedorf et al. (1987)	16	E4, E6, E7, L1	
	18	E1, E6, E7	
Tomita et al. (1987a,b)	6b	L1, *L2*	*pc*
	16	L1	

ORF, Open reading frame of the HPV genome; ORFs against which antibodies have been generated are marked in italics (*L* or *E*); AB, antibody; *mc*, monoclonal; *pc*, polyclonal proteins from different open reading frames the reactivity varies considerably. When fusion proteins of the early (E) and late (L) open reading frames of HPV 16 were used, the most frequent reactivity was seen against the late HPV gene products (Galloway and Jenison 1990). In HPV 16 infection seroreactivity against the E4 proteins seems to be a marker for recent infection and virus replication, whereas reactivity against E7 may be indicative for virus persistence and cancer risk (Jochmus-Kudielka et al. 1989). Monoclonal and polyclonal antibodies were generated against these HPV type-specific proteins (Table 3). Sensitivity and specificity of these new diagnostic tools still await further evaluation.

References

Banks L, Spence P, Androphy E, Hubbert N, Matlashewski G, Murray A, Crawford L (1987a) Identification of human papillomavirus type 18 E6 polypeptide in cells derived from human cervical carcinomas. J Gen Virol 68:1351–1359

Banks L, Matlashewski G, Pim D, Churcher M, Roberts C, Crawford L (1987b) Expression of human papillomavirus type 6 and type 16 capsid proteins in bacteria and their antigenic characterization. J Gen Virol 68:3081–3089

Beckmann AM, Meyerson D, Daling JR, Kiviat NB, Fenoglio CM, McDougall JK (1985) Detection and localisation of human papillomavirus DNA in human genital condylomas by in situ hybridization with biotinylated probes. J Med Virol 16:265–273

Bernard HU, Oltersdorf T, Seedorf K (1987) Expression of the human papillomavirus type 18 E7 gene by a cassette-vector system for the transcription and translation of open reading frames in eukaryotic cells. EMBO J 6:133–138

Brandsma J, Burk R, Lancaster W, Pfister H, Schiffman MH (1989) Inter-laboratory variation as an explanation for varying prevalence estimates of human papillomavirus infection. Int J Cancer 43:260–262

Brown DR, Chin MT, Strike DG (1988) Identification of human papillomavirus type 11 E4 gene products in human tissue implants from athymic mice. Virology 165:262–267

Browne HM, Churcher MJ, Stanley MA, Smith GL, Minson AC (1988) Analysis of the L1 gene product of human papillomavirus type 16 by expression in a vaccinia virus recombinant. J Gen Virol 69:1263–1273

Burk RD, Kadish AS, Calderin S, Romney SL (1986) Human papillomavirus infection of the cervix detected by cervicovaginal lavage and molecular hybridization: correlation with biopsy results and Papanicolaou smear. Am J Obstet Gynecol 154:982–989

Burns J, Graham AK, Frank C, Fleming KA, Evans MF, McGee JOD (1987) Detection of low-copy human papilloma virus DNA and m-RNA in routine paraffin sections of cervix by non-isotopic in situ hybridization. J Clin Pathol 40:858–864

Coggin R, zur Hausen H (1979) Workshop on papillomaviruses and cancer. Cancer Res 39:545–546

Cox KH, de Leon DV, Angerer LM, Angerer RC (1984) Detection of mRNAs in sea urchin embryos by in situ hybridization using asymmetric RNA probes. Dev Biol 101:485–502

Crum CP, Nadai N, Levine RU, Silverstein S (1986) In situ hybridization analysis of HPV 16 DNA sequences in early cervical neoplasia. Am J Pathol 123:174–182

Crum CP, Nuovo G, Friedman D, Silverstein SJ (1988) A comparison of biotin and isotope-labelled ribonucleic acid probes for in situ detection of HPV-16 ribonucleic acid in genital precancers. Lab Invest 58:354–359

De Villiers E-M, Wagner D, Schneider A, Wesch H, Miklaw H, Wahrendorf J, Papendick U, zur Hausen H (1987) Human papillomavirus infections in women without and with abnormal cervical cytology. Lancet 1:703–705

Firzlaff JM, Kiviat NB, Beckmann AM, Jenison SA, Galloway DA (1988) Detection of human papillomavirus capsid antigens in various squamous epithelial lesions using antibodies directed against the L1 and L2 open reading frames. Virology 164:467–477

Galloway DA, Jenison SA (1990) Characterization of the humoral immune response to genital papillomaviruses. Mol Biol Med 7:59–72

Gissmann L, Schwarz E (1985) Cloning of papillomavirus DNA. In: Becker Y (ed) Recombinant DNA research and virus. Nijhoff, Boston, pp 173–197

Gissmann L, Forbes B, Pawlita M, Schneider A (1986) Filter in situ hybridization – a sensitive method to detect papillomavirus DNA single cell. In: Lerman LS (ed) DNA probes, applications in genetic and infectious disease and cancer. Cold Spring Harbor Laboratory, Cold Spring Harbor, pp 157–162

Gissmann L, Kirchhoff T, von Knebel-Döberitz C, Jochmus-Kudielka I, Meinhardt G, Schneider A (1989) Analysis of repeated cervical swabs from Pap-negative women for the presence of HPV 16 DNA. J Cell Biochem [Suppl] 13C:171

Jenison SA, Firzlaff JM, Langenberg A, Galloway DA (1988) Identification of immunoreactive antigens of human papillomavirus type 6b by using *Escherichia coli*-expressed fusion proteins. J Virol 62:2115–2123

Jochmus-Kudielka I, Schneider A, Braun R, Kimmig R, Koldovsky U, Schneweis KE, Seedorf K, Gissmann L (1989) Antibodies against the human papillomavirus type 16 early proteins in human sera: correlation of anti-E7 reactivity with cervical cancer. J Natl Cancer Inst 81:1698–1704

Kafatos FC, Jones CW, Estratiadis A (1979) Determination of nucleic acid sequence homologies and relative concentration by a dot blot hybridization procedure. Nucleic Acids Res 7:1541–1552

Landegren U, Kaiser R, Caskey CT, Hood L (1988) DNA diagnostics – molecular techniques and automation. Science 242:229–237

Li CH, Shah KV, Seth A, Gilden RV (1987) Identification of the human papillomavirus 6b L1 open reading frame protein in condylomas and corresponding antibodies in human sera. J Virol 61:2684–2690

Li CH, Gilden RV, Showalter SD, Shah K (1988) Identification of the human papillomavirus E2 protein in genital tract tissues. J Virol 62:606–609

Maniatis T, Fritsch EF, Sambrook J (1982) Molecular cloning. A laboratory manual. Cold Spring Harbor Laboratory, Cold Spring Harbor

Martinez J, Smith R, Framer M, Resau J, Alger L, Daniel R, Gupta J, Shah K, Naghashfar Z (1988) High prevalence of genital tract papillomavirus infection in female adolescents. Paediatrics 82:604–608

McCance DJ, Campion MJ, Clarkson PK, Chesters PM, Jenkins D, Singer A (1985) Prevalence of human papillomavirus type 16 DNA sequences in cervical intraepithelial neoplasia and invasive carcinoma of the cervix. Br J Obstet Gynaecol 92:1101–1105

McConaughy BR, Laird CD, McCarthy BJ (1969) Nucleic acid reassociation in formamide. Biochemistry 8:3289–3295

Mifflin TE, Bruns DE, Savory J (1988) Detection of papillomavirus DNA. Clin Chem 34:1359

Nagai N, Nuovo G, Friedman D, Crum CP (1987) Detection of papillomavirus nucleic acids in genital precancers with the in situ hybridization technique. Int J Gynecol Pathol 6:366–379

Oltersdorf T, Seedorf K, Röwekamp W, Gissmann L (1987) Identification of human papillomavirus type 16 E7 protein by monoclonal antibodies. J Gen Virol 68:2933–2938

Parkkinen S (1988) Nucleic acid sandwich hybridization in detection of HPV 16 DNA: technique and its clinical application. J Virol Methods 19:69–77

Parkkinen S, Mäntyjärvi R, Syrjänen K, Ranki M (1986) Detection of human papillomavirus DNA by the nucleic acid sandwich hybridization method from cervical scraping. J Med Virol 20:279–288

Pratili MA, LeDoussal V, Harvey P, Laval C, Bertrand F, Jibard N, Croissant O, Orth G (1986) Human papillomaviruses in the epithelial cells of the cervix uteri: frequency of types 16 and 18. Preliminary results of a clinical, cytologic and viral study. J Gynecol Obstet Biol Reprod (Paris) 15:45–50

Rigby PJW, Dieckmann M, Rhodes C, Berg P (1977) Labelling deoxyribonucleic acid to high specific activity in vitro by nick translation with DNA polymerase I. J Mol Biol 113–237–252

Saiki RK, Scharf S, Faloona F, Mullis KB, Horn GT, Erlich HA, Arnheim N (1985) Enzymatic amplification of β-globin genomic sequences and restriction site analysis for diagnosis of sickle cell anemia. Science 230:1350–1354

Schneider A, Oltersdorf T, Schneider V, Gissmann L (1987a) Distribution pattern of human papillomavirus 16 genome in cervical neoplasia by molecular in situ hybridization of tissue sections. Int J Cancer 39:717–721

Schneider A, Hotz M, Gissmann L (1987b) Prevalence of genital HPV-infections in pregnant women. Int J Cancer 40:198–201

Seedorf K, Oltersdorf T, Krämmer G, Röwekamp W (1987) Identification of early proteins of the human papilloma viruses type 16 (HPV 16) and type 18 (HPV 18) in cervical carcinoma cells. EMBO J 6:139–144

Shibata DK, Arnheim N, Martin WJ (1988) Detection of human papilloma virus in paraffin-embedded tissue using the polymerase chain reaction. J Exp Med 167:225–230

Southern EM (1975) Detection of specific sequences among DNA fragments separated by gel electrophoresis. J Mol Biol 98:503–517

Stoler M, Broker TR (1986) In situ hybridization detection of human papillomavirus DNAs and messenger RNAs in genital condylomas and a cervical carcinoma. Hum Pathol 17:1250–1258

Syrjänen S, Partanen P, Mäntyjärvi R, Syrjänen K (1988) Sensitivity of in situ hybridization techniques using biotin- and 35 S-labelled human papilloma virus (HPV) DNA probes. J Virol Methods 19:225–238

Thomas PS (1980) Hybridization of denatured RNA and small DNA fragments transferred to nitrocellulose. Proc Natl Acad Sci USA 77:5201–5205

Tomita Y, Shirasawa H, Sekine H, Simizu B (1987a) Expression of the human papillomavirus type 6b L2 open reading frame in *Escherichia coli:* L2-β-galactosidase fusion proteins and their antigenic properties. Virology 158:8–14

Tomita Y, Shirasawa H, Simizu B (1987b) Expression of human papillomavirus types 6b and 16 L1 open reading frames in *Escherichia coli:* detection of a 56,000-Dalton polypeptide containing genus-specific (common) antigens. J Virol 61:2389–2394

Wagner D, Ikenberg H, Boehm N, Gissmann L (1984) Identification of human papillomavirus in cervical swabs by deoxyribonucleic acid in situ hybridization. Obstet Gynecol 64:767–772

Walboomers JM, Melchers WJ, Mullink H, Meijer CJ, Struyk A, Quint WG, van der Noordaa J, ter Schegget J (1988) Sensitivity of in situ detection with biotinylated probes of human papilloma virus type 16 DNA in frozen tissue sections of squamous cell carcinomas of the cervix. Am J Pathol 131:587–594

Wickenden C, Steele A, Malcolm AD, Coleman DV (1985) Screening for wart virus infection in normal and abnormal cervices by DNA hybridization of cervical scrapes. Lancet 1:65–67

Wickenden C, Malcolm AD, Byrne M, Smith C, Anderson MC, Coleman DV (1987) Prevalence of HPV DNA and viral copy numbers in cervical scrapes from women with normal and abnormal cervices. J Pathol 153:127–135

Zur Hausen H, Schneider A (1987) The role of papillomaviruses in human ano-genital cancer. In: Howley P, Salzmann NP (eds) The Papovaviridae, the papillomaviruses. Plenum, New York, pp 245–263

Human Papillomavirus DNA in Invasive Genital Carcinomas

H. Ikenberg

Introduction

Cancer of the cervix has the epidemiologic characteristics of a sexually transmitted disease (Rotkin 1973). During the past decade the human papillomaviruses (HPV) have emerged as the most probable venereal agent in the pathogenesis of this disease.

The identification of specific HPV types from the majority of premalignant and malignant cervical lesions suggests a causative role for these viruses. The lack of adequate in vitro systems blocked papillomavirus research for a long time. By the arrival of molecular cloning techniques papillomaviruses could be established as a heterogeneic group (Gissmann and zur Hausen 1976; Gissmann et al. 1977).

Meanwhile 60 individual HPV types of numerous extragenital and genital benign and malignant diseases have been isolated. Twenty of these were detected in premalignant and malignant genital lesions (de Villiers 1989). Only 10 different genotypes are predominantly identified in malignant genital tumors and two isolates, HPV 16 and 18, account for about 90% of HPV-positive genital carcinomas tested so far. In this chapter the presence of HPV DNA in genital carcinomas will be reviewed.

Cervical Carcinoma

Primary Invasive Carcinoma

Prevalence of HPV

Numerous studies have been conducted since the cloning of HPV 16 in 1983, most of them applying Southern blot hybridization. There is a broad range of HPV positivity, from 18% in a small series of Fukushima et al. (1985) to 89% in a larger study of Lorincz et al. (1987 b) reaching 90% and 100%, respectively, in a group tested by polymerase chain reaction (PCR; Xiao et al. 1988; Tidy et al. 1989 b). By far most of the studies reported a HPV prevalence rate in cervical carcinoma of 60%–80% (reviews: Pfister 1987; zur Hausen 1987, 1989 a, b; zur Hausen and Schneider 1987). We analyzed 178 primary invasive cervical carcinomas by Southern blot hybridization under stringent and non-stringent conditions for the prevalence of HPV DNA; 70.2% of the tumors contained HPV DNA.

Genital Papillomavirus Infections
Eds.: Gross/Jablonska/Pfister/Stegner

A similar percentage of HPV positivity was found in different geographical regions of the world with high (Panama; Prakash et al. 1985) and with low incidence (Israel; Mitrani-Rosenbaum et al. 1988) of cervical cancer. In conservative rural areas with a low incidence of venereal diseases, high (Xiao et al. 1988; Ji et al. 1990) and low (Fukushima et al. 1985; King et al. 1989) rates of HPV positivity in cervical carcinoma were observed.

The association of HPV and cervical carcinoma does not seem to be a newly established phenomenon. Collins et al. (1988) detected HPV DNA in 80% of tumors dating back to 1932 using in situ hybridization of histological slides while Shibata et al. (1988) found HPV DNA in six cervical carcinomas of the late 1940s by PCR on the tissue section.

HPV 16 is the most prevalent virus type in all studies published. It was found in 57.3% of the 178 tumors we analyzed. This rate is similar to results reported by other groups (Dürst et al. 1983; Boshart et al. 1984; reviews: Pfister 1987; zur Hausen 1987, 1989 a, b; zur Hausen and Schneider 1987). The prevalence of HPV 18 (2.8%) was in line with former reports from Europe (Boshart et al. 1984; Fuchs et al. 1987) and in contrast to the higher detection rates for this type in the USA, only partially due to a higher percentage of adenocarcinomas in those studies (Lorincz et al. 1987 b). Of the lesions in our study 9.5% hybridized only under nonstringent conditions with the probes used, indicating the presence of HPV-related sequences (HPV X) in the tumors. This rate is lower than reported by Lorincz et al. (1987 b) and Prakash et al. (1985), but in line with some other previous reports (Dürst et al. 1983; Boshart et al. 1984). Many studies only used stringent hybridization conditions and therefore could not evaluate the prevalence of HPV-related sequences.

Other HPV types seem to be only rarely present in cervical carcinoma. In our series the DNA of HPV 11 was detected only once in a tumor which additionally harbored HPV-related sequences. HPV 31, 34, and 35 DNA could not be disclosed in any tumor. HPV 6 was identified in one of 20 cervical carcinomas (Gissmann et al. 1984).

HPV 31 cloned from a CIN I was detected in 4 of 62 tumors (6.5%; Lorincz et al. 1986), HPV 33 which originated from a cervical carcinoma in one of 53 cases (1.9%; Beaudenon et al. 1986). HPV 35 which was isolated from an endocervical adenocarcinoma was identified in one of 50 cervical squamous carcinomas and in one of 6 cervical adenocarcinomas (Lorincz et al. 1987 a). HPV 39, isolated from a penile bowenoid papule was found in 1 of 120 (3.3%) invasive cervical carcinomas (Beaudenon et al. 1987). HPV 45 isolated from a CIN I lesion was identified in one of 52 cervical carcinomas (Naghashfar et al. 1987). HPV 51 was cloned from a cervical condyloma. It could be found in one of five cervical carcinomas (Nuovo et al. 1988).

Correlation with Histology

Invasive cervical carcinoma is a heterogeneous disease showing wide variations in histologic and clinical characteristics. Prognostic importance has been attributed to clinical stage, lymph node involvement, cell type, depth of invasion, and lymphatic permeation. The clinical course of cervical cancer varies greatly. Some patients have unusually rapid progression of clinically limited disease while others have relatively benign clinical outcome with a large primary tumor burden. A possible variant of cervical cancer with rapid progression, perhaps by bypassing cervical intraepithelial neoplasia (CIN) stages, has been postulated (Berkeley et al. 1980; Bain and Crocker 1983).

Only few reports dealing with the prevalence of HPV in such lesions have described the detailed histopathologic and clinical features of the tumors analyzed (Meanwell et al. 1987; Barnes et al. 1988; Wilcynski et al. 1988; King et al. 1989; Walker et al. 1989).

Adenocarcinomas and adenosquamous carcinomas of the cervix more often contain HPV 18 than squamous tumors. Smotkin et al. (1986) detected HPV 18 in four of five adenosquamous carcinomas. Walker et al. (1989) found 8 of 12 and King et al. (1989) 15 of 27 adenocarcinomas positive for HPV 18. By in situ hybridization on formalin-fixed, paraffin-embedded tissue Tase et al. (1988) found HPV 18 in all but one case of 17 of 40 (42.5%) HPV-positive adenocarcinomas and half of 16 of 44 (36.4%) HPV-positive adenosquamous carcinomas. In our series adenocarcinomas tended less often to contain HPV 16 and more often HPV 18 or HPV-related sequences. The association of HPV (mainly HPV 18) with cervical adenocarcinoma is underlined by the presence of HPV DNA and RNA in the majority of preinvasive glandular lesions of the cervix (Farnsworth et al. 1989; Okagaki et al. 1989; Tase et al. 1989).

We did not observe any difference in HPV types between keratinizing and nonkeratinizing or large cell and small cell carcinomas. In contrast Walker et al. (1989) found a strong association of HPV 16 with keratinization. Here 30 of 33 (91%) of the large cell keratinizing and only 9 of 29 (31%) of the large cell nonkeratinizing cervical carcinomas contained HPV 16. Two studies revealed a significant association of HPV 18-positive tumors with anaplasia (grade III of differentiation; Barnes et al. 1988; Walker et al. 1989) while one study found no such relationship (King et al. 1989).

Correlation with Clinical Course

The HPV status showed no correlation with the clinical stage of the cervical carcinoma in the 178 patients we studied. No influence of the HPV status on parametrial or vaginal involvement, cervical invasion, lymph-vascular invasion, and lymph node metastasis was noted in 79 surgically examined patients in our group and 69 patients analyzed by Walker et al. (1989).

While King et al. (1989) detected no relationship between HPV type and the presence of nodal metastases, in the smaller study of Barnes et al. (1989) HPV 18-positive patients had nodal metastasis almost twice as often as the HPV 16-positive women.

In remarkable contrast to most of these findings, women with HPV 18-containing tumors had a significantly higher recurrence rate according to a recent study (Walker et al. 1989). The tendency of HPV 18-containing tumors to recur at a higher rate was independent of clinical stage, histologic type, and tumor grade. Of 29 women with stage Ib, grade III disease, 57% with HPV 18 had recurrence at 20-month follow-up, compared with 25% of patients with HPV 16 and 10% with no HPV. The overall mortality for patients with HPV 18-positive tumors was also higher but did not reach statistical significance.

In our study with a mean follow-up of 24 months for 178 patients, no significant differences in survival and recurrence rate were observed between HPV 16-positive, HPV X-positive and HPV-negative tumors. The number of HPV 18-positive lesions (5) was to small for statistical analysis.

Meanwell et al. (1987) found no difference in regard to survival between 47HPV 16-positive and HPV-negative patients after 20.5 months.

A similar survival time for 85 HPV 16-positive, HPV 18-positive or HPV-negative patients with cervical carcinoma was reported by King et al. (1989) in a retrospective study by in situ hybridization of paraffin-embedded formalin-fixed tissue.

The hypothesis that HPV 18-related cancers are more aggressive and progress more rapidly from CIN to invasion without showing the classically described premalignant changes is suggested by the results of two studies. Walker et al. (1989) found a high incidence (44%) of normal Papanicolaou smears during the 3 years preceding the diagnosis of cervical cancer in patients with HPV 18-positive tumors, compared to only 17% of women with HPV 16-positive cancers. Kurman et al. (1988) identified HPV 18 in a much lower percentage of CIN (3%) than in invasive cancers (22%). The presence and expression of HPV 18 DNA in the majority of permanent cervical carcinoma cell lines tested, among them the HeLa line which has been in culture for nearly 40 years (Boshart et al. 1984; Pater and Pater 1988) indicates a special oncogenic potential of this virus. Patients with HPV 18-positive cervical carcinoma are about 10 years younger than women with HPV 16-positive or HPV-negative lesions (Barnes et al. 1988; Walker et al. 1989). This observation, which was not confirmed by King et al. (1989), again, might reflect a potential for more rapid progression.

HPV Copy Number

HPV copy number per cell has been correlated with clinical and histopathological data of cervical tumors only in few studies. In one small group reported by Scholl et al. (1985) the poorly differentiated lesions had higher copy numbers of HPV 16 DNA. Meanwell et al. (1987) found no correlation with the clinical course of the disease. In our patients no significant relationship between copy number and clinical stage, histologic type, tumor grade, and clinical outcome existed. Only a weak trend was noted towards higher copy number in stage III/IV tumors and in less differentiated lesions.

Physical State of the Viral DNA

The physical state of the HPV DNA seems to be correlated with the progression of the malignant state. In condylomatous lesions and CIN I the DNA is mostly episomal although early integration events may occur (Lehn et al. 1988; Schneider-Maunoury et al. 1987). In higher grade CIN integration events occur more frequently and in the majority of invasive cervical carcinomas at least some of the viral DNA is integrated (Lehn et al. 1988). Cervical cell lines contain only integrated HPV DNA (Pater and Pater 1985).

While there is no common site of integration on the host chromosome in different tumors (Dürst et al. 1987), a remarkable consistency of the disruption of the viral genome was revealed. The integration interrupts the E1 and E2 open reading frame in almost all cases studied (Schwarz et al. 1985; Pater and Pater 1985). This might interfere with normal regulation of other early viral gene expressions by E2 encoded proteins (Cripe et al. 1987).

Matsukura et al. (1989) tested 34 HPV 16-positive cervical carcinomas. Some 70% showed only episomal viral DNA, and 30% contained integrated HPV 16 DNA. No correlation between the form of the DNA and the clinical stage of the tumors was noted.

Choo et al. (1987) found in 37.5% of HPV 16-positive cervical carcinomas only integrated sequences, in 37.5% only episomal DNA was detected, and 25% contained the viral DNA in both episomal and integrated form.

Fuchs et al. (1989) had identical findings in 14 HPV 16-positive primary cervical cancers. Thirteen of 16 involved lymph nodes contained HPV 16. Integrated viral DNA showed the same pattern in primary tumors and in metastases. Interestingly, the level of episomal HPV DNA was considerably lower in some nodes. The prognostic significance of these findings remains to be established.

Recurrences

Only few recurrences of cervical carcinomas have been analyzed for HPV DNA up to now. Seven of eight vaginal or cervical recurrences tested by us contained HPV DNA (five of them HPV 16, two HPV-related sequences). In two cases where the corresponding primary lesions were available for analysis, differences were revealed between primary lesion and recurrence. One vaginal recurrence of a HPV-negative cervical carcinoma contained HPV 16 DNA, while the corresponding primary lesion to another HPV-negative vaginal recurrence harbored HPV-related sequences.

Conclusions

The regular presence of HPV DNA in cervical cancer, but also a consistent pattern of DNA integration (Dürst et al. 1985) and transcription (Schwarz et al. 1985) in cervical cancer cells, has led to the suggestion that HPV infection is necessary, although alone not sufficient to cause cervical cancer (zur Hausen 1986). HPV DNA could be detected in cell lines derived from cervical cancer (Pater and Pater 1985), in lymph node (Lancaster et al. 1986; Fuchs et al.

1989) and distant metastases (Stremlau et al. 1985; Walboomers et al. 1987; Wilcynski et al. 1988) of cervical carcinoma. HPV 11 DNA can induce dysplasias in heterografted cervical tissue (Kreider et al. 1985). Finally HPV 16 (Dürst et al. 1987) and HPV 18 (Kaur and McDougall 1988) can immortalize human keratinocytes. HPV 16 turns initiated cells tumorigenic (Yasumoto et al. 1986). All these data argue in favor of an active role of HPV in genital carcinogenesis.

It is apparent that the HPV infection alone is not sufficient for cancer induction. HPV infections are ubiquitous. At least 10% of sexually active women with clinically normal cervix are HPV positive (Wagner et al. 1984; de Villiers et al. 1987). Up to 30% of pregnant women are positive (Schneider et al. 1987b). Comparing these prevalence rates with the incidence of cervical cancer it becomes clear that only a very low percentage of HPV-infected women will develop invasive cancer. Therefore additional factors must be active in carcinogenesis.

Until recently Southern blotting (Southern 1975; Maniatis et al. 1982) was regarded as the most sensitive method of detecting viral genomes in cellular tissue (Gissmann et al. 1987). Using this method HPV DNA was demonstrated in 18%–90% of invasive cervical carcinomas (zur Hausen 1987; zur Hausen and Schneider 1987). Considerable variation exists between studies from different laboratories and geographical regions as to the absolute HPV prevalence rate and the relative prevalence of the individual types. Generally the most commonly identified HPVs in cervical cancer are type 16 (which in nearly all studies is the most common type with an overall incidence of 35%–60%), HPV 18, and HPV 31 (Dürst et al. 1983; Boshart et al. 1984; reviews: Pfister 1987; zur Hausen 1987; zur Hausen and Schneider 1987). Some of this variation may be due to differences in the geographical distribution of the viruses. Differences in the hybridization conditions used (stringency, specific radioactivity, exposure time) are an important source of variation. A considerable degree of variation among laboratories in the use of the same detection methods was revealed by Brandsma et al. (1989). An additional role could be attributed to the histological types of cervical carcinomas analyzed in the different series.

Recently the newly developed polymerase chain reaction (PCR), which amplifies target sequences, was used to investigate the prevalence of HPV DNA in cervical carcinoma. By this technique HPV 16 was detected in 100% of cervical carcinoma tissue, in 80% of scrapes from sexually active women with normal cervices, but not in control specimens (Tidy et al. 1989a; Young et al. 1989). In contrast to Southern blotting PCR does not allow the copy number and the physical state of the viral sequences of interest to be analyzed. The need for extensive controls in the use of the PCR is illustrated by the retraction of the publication of HPV subtype 16b which was probably due to a contamination (Tidy and Farrell 1989).

Summarizing the numerous reports cited there exists a substantial body of epidemiological evidence linking specific HPV types to cervical cancer. HPV 18 seems to be related to more aggressive forms of the disease. But the comparison of nucleic acid hybridization data from different laboratories is difficult, and carefully controlled long-term epidemiologic analyses are still lacking. This may be partially due to the fact that the HPV types most frequently found in cervical cancer (HPV 16 and HPV 18) have only been known for 7 years. The practical and methodic problems with nucleic acid hybridization techniques already mentioned may also have contributed to this deficiency. Certainly a larger number of cases have to be analyzed with a longer follow-up time. An elegant alternative to circumvent the apparent organizational and time problems of a prospective study might be the analysis of routinely processed paraffin-embedded formalin-fixed tissue specimens which has been made possible by the means of the PCR. Shibata et al. (1988) applying this method detected HPV genomes in 40-year-old material. Until now seroepidemiology made only minimal contributions to our knowledge of the association of papillomaviruses and cancer. A recent report showing different rates of antibody response to specific HPV 16-encoded proteins between cervical cancer patients and controls (Jochmus-Kudielka et al. 1989) will probably intensify seroepidemiological research on this field.

Vulvar Carcinoma

Introduction

Cancer of the vulva accounts for approximately 5% of all cancers of the female genital tract. The incidence of invasive disease is highest in the eighth decade of life. More than 85% of invasive vulvar carcinomas are well-differentiated, keratinizing squamous cell lesions (Morley 1986). The incidence of double primary tumors is about 10%, most of these lesions being cervical carcinomas (Choo 1980).

Multifocality is quite common in women with vulvar condylomas and vulvar intraepithelial neoplasia (VIN) (Wilkinson et al. 1981). Over 75% of patients with invasive carcinoma, however, have unifocal disease (Morley 1986). Less is known about predisposing factors for vulvar carcinoma (Stegner 1986). While approximately 50% of invasive carcinomas of the vulva arise in an area of chronic vulvar dystrophy only 2%–4% of patients with chronic vulvar disease will finally experience a malignancy in this area.

The hypothesis that vulvar carcinoma develops from HPV-induced dysplasia was evoked by several observations. Of the patients with preinvasive vulvar carcinoma 20%–42% have or have had preexisting condyloma acuminatum (Husseinzahdeh et al. 1989). Up to 81% of patients with cervical HPV infection also had vulvar disease (Spitzer et al. 1989). Sixty percent of vulvar intraepithelial neoplasia (VIN) lesions contain HPV DNA (Gupta et al. 1987). Malignant transformation of vulvar condyloma is a rare but repeatedly reported event (Daling et al. 1984; Husseinzadeh et al. 1989).

HPV Prevalence

Squamous Carcinoma

HPV DNA sequences could be found in 50% of 108 primary invasive squamous cell carcinomas of the vulva tested so far in 11 different studies. HPV 16 was by far the most prevalent type, being present in 30% of the lesions (Green et al. 1982; Dürst et al. 1983; Bergeron et al. 1986; Gross et al. 1986; MacNab et al. 1986; de Villiers et al. 1986; Sutton et al. 1987; Buscema et al. 1988; Carson et al. 1988; Ikenberg et al. 1988; Venuti and Marcante 1989).

In 10% of the tumors HPV 6 or 11 was identified. This percentage is lower than in verrucous carcinoma of the vulva but significantly higher than in cervical carcinoma.

Six percent of the tumors contained HPV-related sequences which only hybridized under conditions of low stringency. HPV 10 was detected in two vulvar carcinomas (Green et al. 1982). Venuti and Marcante (1989) found HPV 18 in 12 of 15 invasive carcinomas of the vulva using dot blot hybridization. In 1 of 13 vulvar carcinomas analyzed for this type, HPV 35 was identified (Lorincz et al. 1987 a).

Most reports did not indicate the copy number of viral DNA per cell. Sutton et al. (1987), Ikenberg et al. (1988), and Venuti and Marcante (1989) found between 1 and 100 copies HPV DNA per cellular genome.

The grade of differentiation and keratinization of the tumors and the clinical tumor stage and nodal status of the lesions was reported in two studies (Sutton et al. 1987; Ikenberg et al. 1988). Here no relationship between the above-mentioned variables and the HPV status of the vulvar carcinomas was noticed. No correlation between clinical course of the disease and HPV status of the primary tumors was observed. The mean age of HPV-positive and -negative patients showed no significant differences.

HPV DNA was identified in inguinal lymph node metastases in vulvar carcinoma patients. For details see "Metastatic Disease."

Twelve samples from tumor-free abdominal skin in HPV-typed vulvar carcinoma patients considered as corresponding normal tissue specimens were negative for HPV DNA in our data (Ikenberg et al. 1988), while MacNab et al. (1986) reported the detection of HPV DNA in the normal tissue of three patients with vulvar carcinomas.

The majority of the studies cited applied Southern blotting mainly under stringent and nonstringent conditions, while Sutton et al. (1987) used dot blot hybridization and de Villiers et al. (1986) hybridization with labeled cellular DNA. It is interesting to note that in vulvar squamous carcinoma the same range of papillomavirus types is identified as in lesions of the genital mucosa. With the exception of HPV 10 in two cases (Green et al. 1982), none of the HPV types which are present in extragenital squamous cell carcinoma could be disclosed in vulvar carcinoma. This may be partially due to a small range of labeled HPV probes in several studies, and some of the HPV-related sequences detected in vulvar lesions might represent papillomaviruses of extragenital types. But obviously the HPV type is not only determined by the tissue type of a target cell. A major role can be contributed to a field effect which may be exerted by hormonal influences or by continuous virus spread from infected cells into neighboring tissues.

Adenosquamous Carcinoma

Adenosquamous carcinoma of the vulva is regarded as a variant of squamous cell carcinoma (Lever 1947) containing a mucin-producing glandular component (Underwood et al. 1978), it has a poorer prognosis than squamous carcinoma. Only 1 of 16 adenosquamous carcinomas of the vulva contained HPV-related sequences according to a study by Carson et al. (1988). This underlines the possibility that this disease is a separate histomorphologic and clinical entity.

Condylomatous Carcinoma

Condylomatous carcinoma is an invasive squamous cell carcinoma integrated with a condyloma acuminatum. The lesion is often multifocal and tends to occur in immunocompromised patients. In a study by Downey et al. (1988) HPV 6 and HPV 16 were identified in two of nine cases, respectively, while HPV 2 and HPV related sequences were found in one lesion. This pattern resembles the situation in squamous carcinoma of the vulva.

Buschke-Löwenstein Tumor

Buschke-Löwenstein tumor is a cauliflowerlike neoplasm of male and female external genitalia, intermediate between a condyloma and a carcinoma.

While the lesions show a locally destructive growth pattern with downgrowth into the subcutaneous fat, deeper invasion, local lymph node involvement, and distant metastases are absent. In rare cases a transformation into a metastasizing carcinoma was observed (zur Hausen 1977). Morphologically Buschke-Löwenstein tumors are similar to verrucous carcinoma (Ackerman 1948), the oral florid papillomatosis. In a verrucous carcinoma of the larynx (Brandsma et al. 1986) and of the oral mucosa (Adler-Storthz et al. 1986) HPV 16 and HPV 2 were detected.

Southern blotting consistently shows the presence of HPV 6 or HPV 11 DNA in Buschke-Löwenstein tumors (Gissmann et al. 1982; Zachow et al. 1982; Lehn et al. 1984; Mathieu et al. 1985; Boshart and zur Hausen 1986; de Villiers et al. 1986; Jablonska et al. 1987). In one case the presence of HPV 33 was reported (Jablonska et al. 1987). A metastasizing recurrency of a Buschke-Löwenstein tumor exhibited HPV 6 DNA in relapse and metastasis (Jablonska et al. 1987). These data underline the close association of Buschke-Löwenstein tumors and condylomata acuminata, most of which contain HPV 6 or HPV 11 DNA.

Recurrences

All seven recurrent vulvar malignancies examined in one study (Ikenberg et al. 1988), including five squamous carcinomas, one fibrosarcoma, and one zylindroma contained no HPV DNA, while Sutton et al. (1987) detected HPV 16 DNA in one of two squamous carcinoma recurrences.

Vaginal Carcinoma

Introduction

Primary invasive carcinoma of the vagina is among the rarest malignancies of the female genital tract. It accounts for 1%–2% of all gynecological cancers. The mean age of patients with vaginal carcinoma is 60–63 years (Rutledge 1967; Pride et al. 1979; Podczaski and Herbst 1986). The majority of vaginal malignancies are metastases (Gompel and Silverberg 1977). These lesions may be the first manifestations of an occult neoplasm often arising from endometrial or cervical carcinomas. Squamous cell carcinomas make up 90% of primary vaginal carcinomas. Adenocarcinoma counts for 5% of all vaginal malignancies.

There is only scanty information about risk factors for carcinomas of the vagina. The prevalence of these tumors as of the vaginal intraepithelial neoplasms (VAIN) is much lower than that of cervical carcinoma although the squamous epithelium of the vagina is continuous with that of the cervix. This could be explained by the absence of metaplastic epithelium, the putative target for initiating agents, in the vagina. This is in line with the finding that young women with a history of intrauterine exposure to diethylstilbestrol (DES) who show an increased incidence of VAIN and clear cell adenocarcinoma of the vagina have also higher rates of metaplastic epithelial changes in the vagina (Robboy et al. 1984). However, in few cases the development of carcinoma in the vagina from a preexisting condyloma in young women has been reported (Beck 1984).

Squamous neoplasia of the female genital tract often is multifocal. Women with CIN and invasive carcinoma of the cervix are at risk to develop secondary neoplasias in the remaining squamous epithelium of the vulva and the vagina (Choo 1980; Woodruff 1981). Up to 60% of vaginal carcinomas occur in women with a history of cervical cancer or CIN (Podczaski and Herbst 1986; Gallup et al. 1987; Manetta et al. 1988).

The HPV infection is a potential agent for multifocal neoplasia in the genital tract. It has been shown that simultaneous or successive infection of various sites of the anogenital area with human papillomaviruses is a common event (McCance et al. 1985; Schneider et al. 1987a).

A high percentage of preinvasive neoplastic lesions of the vagina harbors HPV DNA (Schneider et al. 1987a). A history of cervical neoplasia was associated with a significantly higher rate of HPV infection than found in patients with benign disease in the past. In 5% of these HPV-positive patients HPV 16-positive VIN could be diagnosed in a study of Schneider et al. (1987a).

Prevalence of HPV

Due to the low incidence of carcinoma of the vagina only a limited number of these tumors have been tested for the prevalence of HPV DNA.

In one recent study (Ikenberg et al. 1990) HPV DNA was identified by Southern blotting in 10 of 18 patients with vaginal carcinoma. HPV 16 was the most prevalent type. It could be identified in 5 of 15 primary squamous cell carcinomas, in one primary adenocarcinoma, in one vulvar recurrence, and in two lymph node metastases of an HPV-negative primary tumor.

HPV-related sequences were detected in one primary squamous carcinoma and in an inguinal metastasis of another squamous carcinoma where the primary tumor was not available for testing.

Only 3 of 18 patients in this study had a history of malignant cervical disease. The HPV copy number per cell ranged from 0.5 to 50. Neither HPV type nor the amount of viral DNA correlated with clinical stage or histological grade of differentiation or keratinization.

Mitrani-Rosenbaum et al. (1988), using Southern blotting, identified HPV 16 in one of three vaginal carcinomas tested. One case analyzed by labeling cellular DNA was found negative (de Villiers et al. 1986). In a retrospective study Ostrow et al. (1988) detected HPV DNA in 3 of 14 carcinomas of the vagina by in situ hybridization on formalin-fixed paraffin-embedded tissues. Two of the lesions contained HPV-related sequences and one HPV 16. No data about history and clinical stage were given. A possible explanation for the lower prevalence of HPV in this study is the reduced sensitivity of the in situ hybridization technique compared with the Southern blot method.

HPV 6 was identified by Southern blotting in two cases of verrucous carcinoma of the vagina by Okagaki et al. (1984). Verrucous carcinoma (Ackerman 1948) is a rapidly growing tumor with little nuclear atypia which lacks invasion and metastases. No uniform definition of these lesions is given by different authors (Kraus and Perez-Mesa 1966, Okagaki et al. 1984). The presence of HPV 6, which is by far the most prevalent type in condylomata acuminata, supports the notion that verrucous carcinoma of the vagina is closely related to condyloma acuminatum.

The hybridization experiments in carcinomas of vagina and cervix give similar results with differences in detail. In the vagina the number of HPV-positive tumors is lower and a smaller range of types is found. This might be due to the rather low number of cases tested to date. There could also be other papillomavirus types present in vaginal neoplasms, too far related with the probes to be detectable even under sensitive conditions by cross hybridization. No significant differences existed in the mean age of HPV-positive and -negative patients.

HPV DNA was also detected in inguinal lymph nodes of patients with carcinoma of the vagina. For details see "Metastatic Disease."

Six tumor-free specimens from cervix and myometrium in HPV-typed vaginal carcinoma patients as corresponding normal tissue samples were HPV negative (Ikenberg et al. 1990).

The results of the cited studies indicate a key role for human papillomaviruses also in carcinoma of the vagina. The enormous difference in the incidence of cervical and vaginal cancer underlines the importance of additional factors in genital carcinogenesis. In the vagina this may be the lack of the metaplastic squamous epithelium of the cervical transformation zone as locus minoris resistentiae to HPV infection and extrinsic physical or chemical factors. Additionally intrinsic components such as the individual genetic and immunologic constitution which modifies controlling host cell genes certainly play an important role in the process of malignant conversion (zur Hausen 1989b).

Endometrial Carcinoma

Introduction

Carcinoma of the endometrium has become the most frequent neoplasm of the female genital tract. Its peak incidence is in the seventh decade. Premenopausal women are rarely involved. Most endometrial carcinomas are adenocarcinomas (Kaiser and Pfleiderer 1989). The mechanisms of pathogenesis remain unclear. In sharp contrast to cervical carcinoma epidemiological studies exhibited no evidence for any role of a transmissible agent in the etiology of this tumor. Already in 1842 Rigoni-Stern found uterine cancer to be virtually nonexistent in nuns.

While an association between CIN and adenocarcinoma in situ of the endocervix is established, no such correlation between CIN and endometrial hyperplasia or carcinoma has been reported (Weisbrot et al. 1972).

Continuous estrogen stimulation without compensative gestagen activity is being intensively discussed as a predisposing condition for endometrial carcinoma. Endometrial hyperplasias and well-differentiated carcinomas are associated with long-standing estrogenic stimulation unopposed by progesterone (Siiteri 1981; Ferenczy et al. 1983). The precise role of estrogen in endometrial carcinogenesis is by no means clear. Possibly estrogens act as promoters in the growth of hyperplastic or neoplastic cells. Clinical experience suggests there are two types of endometrial carcinoma: highly differentiated estrogen and progesterone receptor-positive tumors found in corpulent women or after estrogen intake which have a good prognosis and early metastasizing less-differentiated tumors which are generally receptor negative. The first step in the genesis of endometrial carcinomas is the development of cystic glandular hyperplasia, a strong proliferation of endometrial glands and stroma, but not a truly precancerous lesion. Then adenomatous hyperplasia may develop with its excessive increase of glandular cells and the parallel decrease of stroma. Up to 20% of these lesions progress to cancer in 1–10 years (Kaiser and Pfleiderer 1989). In a study with [^{3}H] in situ hybridization HPV DNA was not present in 16 cases of microglandular endocervical hyperplasia, while 6% of 36 cases of endocervical glandular dysplasia and 67% of 21 endocervical adenocarcinomata in situ contained HPV 6, 16, or 18 (Okagaki et al. 1989).

Prevalence of HPV

Adenocarcinomas of the endocervix contain HPV DNA sequences in about half of the tested specimens. In contrast to cervical squamous cell carcinoma mainly HPV 18 is identified in these lesions (Smotkin et al. 1986; Tase et al. 1988; Wilczynski et al. 1988; King et al. 1989). The HeLa cell line which is derived from a cervical adenocarcinoma also harbors HPV 18 DNA (Boshart et al. 1984). Finally, a glucocorticoid- and progesterone-responsive element in the HPV 16 enhancer region has been described (Gloss et al. 1987).

The findings cited above and the close anatomic and histologic relationship of the endometrial mucosa with the endocervical epithelium makes the endometrium a possible target for HPV infection. The prevalence of HPV DNA in normal, precancerous, and carcinomatous endometrium has been analyzed only by a few studies.

MacNab et al. (1986) using Southern blotting found HPV 16 DNA in two endometrial carcinomas and in four of seven specimens of normal myometrium (five of these patients had HPV 16-positive cervical carcinoma). Applying labeled cellular DNA de Villiers et al. (1986) did not detect HPV DNA in five endometrial cancers while 2 of 11 normal endometria contained HPV 16 genomes. In our own data we disclosed HPV 16 DNA (one copy/cell) and HPV-related sequences each in 1 case of 50 endometrial carcinomas of different histological subtypes and clinical stages by Southern bloting under nonstringent and stringent conditions.

Bergeron et al. (1988) failed to identify HPV DNA by Southern bloting in any of 7 normal, 5 hyperplastic (2 of them with atypia) and 16 neoplastic endometria. These data suggest that HPV infection of the endometrium is a rare event when the cervix is free of HPV-related disease. In the studies of H. Ikenberg (in preparation) and Bergeron et al. (1988) special attention was paid to avoid contamination of the endometrial material by, possibly, HPV DNA-containing cervical tissue. Most of the specimens were taken after hysterectomy and tissue excised as far away from the isthmus as possible. The endometrial specimens in the study of de Villiers et al. (1986) were taken by biopsy. At least 10% of women with negative cervical cytology, colposcopy, and histology carry HPV DNA (de Villiers et al. 1987). Both patients with HPV 16-positive endometrial tissue had had carcinoma in situ of the cervix. Five of seven of the patients in the study of MacNab et al. (1986) had HPV 16-positive cervical carcinoma. Adenocarcinoma of endocervical origin mainly contains HPV 18 (Smotkin et al. 1986; King et al. 1989). Although an infection with HPV types completely unrelated to those probed for cannot yet be ruled out, a role of HPV in the development of endometrial neoplasia seems quite unlikely.

Ovarian Carcinoma

Introduction

Ovarian carcinoma ranks third in incidence among genital cancers with 14–16 of 100000 women per year in western Europe and the USA, but it causes about half of the deaths due to genital carcinomas (Kaiser and Pfleiderer 1989). Women with a low number of pregnancies and no intake of oral contraceptives have a higher risk of ovarian cancer. The incidence of ovarian carcinoma increases significantly with the age, having its peak in the eighth decade. No further definite risk factors are known.

Prevalence of HPV

There is one study reporting the detection of HPV 6 in 10 of 12 ovarian cancers by in situ hybridization (Kaufmann et al. 1987). De Villiers et al. found no HPV DNA in seven carcinomas of the ovaries by labeling cellular DNA (1986). We analyzed 20 ovarian carcinomas of different histological types and clinical stages for the prevalence of HPV DNA by Southern blotting. In none of these tumors were HPV sequences detected. In a further recent study no HPV DNA was identified in 15 epithelial ovarian carcinomas by Southern blot and PCR (Leake et al. 1989). On the basis of these results there appears to be no association between HPV and ovarian neoplasia.

Penile Carcinoma

Introduction

Penile cancer is a rare disease in most parts of the world. In some regions of South America, Africa, and South East Asia, however, the incidence of this neoplasia is more than ten times higher.

For example, in Recife, Brazil, an incidence rate of 6.8 per 100000 inhabitants was reported (Waterhouse et al. 1976). Generally the relative frequencies of cervical and penile cancer in the same areas are strictly correlated with each other. This points to common etiological factors (Boon et al. 1989). Penile premalignant lesions contain the same HPV types at similar frequencies as dysplasias of the lower female genital tract (Gross et al. 1986). In more than

50% of male partners of HPV-positive women human papillomavirus sequences can be identified (Gross et al. 1986; Schneider et al. 1987c).

Prevalence of HPV

Dürst et al. (1983) first described the presence of HPV 16 DNA in one of four penile carcinomas. Boshart et al. (1984) reported HPV 16 in two and HPV 18 in one of ten tumors. HPV 18 was identified in 39% of 17 penile carcinomas from Brazil (Villa and Lopes 1986) by Southern blotting.

The striking parallels in the HPV status of premalignant and malignant cervical and penile lesions argue in favor of a causal role for these viruses. The lower incidence of invasive penile neoplasms may be due to the absence of a target at the penis for initiating agents comparable with the cervical transformation zone.

Anal Cancer

Introduction

Anal cancer is a rare tumor in the general population (0.6/100000) Young et al. 1981). Most of the lesions are squamous neoplasms. In homosexual men the incidence of the disease is much higher (Daling et al. 1982). Equally, a history of anogenital condyloma acuminatum is strongly associated with the occurrence of squamous anal cancer in both men and women (Daling et al. 1987). Holly et al. (1989) reported an increase in the relative risk of anal cancer of 12.6 for homosexual men with genital warts and of 4.4 for heterosexual males with the same history. An increased risk was also observed for patients with anal fistulas and heavy cigarette smoking. HPV 11 and HPV 16 DNA has been identified in anal intraepithelial neoplasia (G. Gross and H. Ikenberg, unpublished data).

Prevalence of HPV

HPV 16-related DNA was detected in one anal carcinoma by Scheurlen et al. (1986) and in a squamous carcinoma (SCC) of the anus arising in a condyloma acuminatum by Hill and Coghill (1986). Further single cases were reported by Beckman et al. (1985) and Wells et al. (1987).

Palmer et al. (1987) found HPV 16 in six of ten anal SCCs but not in ten controls using the Southern blot method. Interesting results were derived from a case-control study of 126 patients with malignant anal lesions applying in situ hybridization (Beckman et al. 1989). HPV 6 was found in 9% of 47 invasive tumors and in 17% of 23 CIS lesions. HPV 16 was contained in 11% of the invasive and in 23% of the CIS lesions, while 2% of the invasive tumors and 13% of the CIS harbored HPV-related sequences. Forty-two anal transitional cell carcinomas, 14 anal adenocarcinomas, and 110 colorectal adenocarcinomas were HPV negative.

One third of the anal cancer patients with detectable HPV DNA were infected with HPV 6. This rate of HPV 6-positive tumors is much higher than that for the cervix or vulva. In some HPV 6 genomes isolated from vulvar carcinomas and Buschke-Löwenstein tumors alterations in the noncoding region of the viral genome compared to the prototype HPV 6b were revealed (Boshart et al. 1986; Rando et al. 1986; Kasher and Roman 1988). By an altered control of early gene transcription this could augment the oncogenic potential of the virus. Another explanation for the observed association of HPV 6 with anal cancer is a coinfection with other HPV types not detectable under the relatively insensitive conditions of in situ hybridization.

In conclusion these data implicate a role for HPV in the genesis of anal squamous cancer.

Metastatic Disease

Regional Lymph Node Metastases

Lymphogenic spread of cervical carcinoma is mainly to pelvic and paraaortic nodes. Carcinoma of the vagina metastazises to pelvic or inguinal lymph nodes depending on the site of the primary lesion, while vulvar carcinoma preferentially involves inguinal nodes.

Cervical Carcinoma

HPV DNA also persists in lymph node metastases of these cancers. Lancaster et al. (1986) first described the detection of HPV 16 DNA or HPV 16 related sequences in pelvic and paraaortic lymph node metastases of six cervical carcinomas (among them one adenocarcinoma). The viral sequences had the same restriction pattern and copy number as the primary lesions. Fuchs et al. (1989) found HPV 16 DNA in 10 of 11 histologically fully involved lymph node metastases of 9 HPV 16-positive cervical cancers. Three of five micrometastases contained also HPV 16. Viral sequences in the metastases from cancers with integrated HPV DNA shared the same cleavage pattern and the same copy number as those in the primary tumors. This stability is in line with the unvarying HPV DNA integration pattern in cervical cancer cell lines cultured over decades (Schwarz et al. 1985). In remarkable contrast to this, two nodes from cancers with predominantly extrachromosomal HPV 16 DNA harbored only much smaller amounts of viral DNA than the primary lesions, and in one case no more HPV DNA was detectable. This indicates the possibility that plasmids are lost during metastatic tumor spread. Defects in autonomous viral DNA replication could also account for the absence of HPV plasmids in established cell lines. We made similar observations in our own data. All six histologically proven lymph node metastases of HPV-positive cervical carcinomas contained the same HPV type. Only minimal changes in copy number and restriction pattern occurred (Ikenberg et al. 1986).

Vulvar Carcinoma

One tumor invaded inguinal lymph node of a HPV 16-positive vulvar carcinoma also contained HPV 16 DNA, while two lymph node metastases of a HPV-negative tumor were free of HPV DNA (Ikenberg et al. 1988). Contrary to this, in another study one lymph node of an HPV-negative primary tumor contained HPV 6/11 and a nodal metastasis of a HPV 16/18-positive primary lesion was positive for HPV 6/11 (Sutton et al. 1987).

Vaginal Carcinoma

HPV DNA was also detected in inguinal lymph nodes of patients with carcinoma of the vagina (Ikenberg et al. 1990). In one histologically positive node of a patient with a HPV 16-positive recurrence the same type with unvaried restriction pattern and copy number was found. HPV-related sequences were disclosed in an inguinal node where the corresponding primary tumor could not be tested. Two histologically positive nodes of the same patient with HPV-negative primary tumor contained HPV 16. This parallels the observation of Sutton et al. (1987) who identified HPV 6/11 DNA in a nodal metastasis of a HPV-negative vulvar carcinoma.

Histologically Negative Lymph Nodes

A particularly interesting finding was the detection of HPV 16 DNA in numerous histologically negative lymph nodes in patients with HPV 16-positive cervical carcinomas (Lancaster et al. 1986; Ikenberg et al. 1986; Fuchs et al. 1989). One of seven lymph nodes without histological signs of malignant invasion also contained HPV 16 DNA in the study of Lancaster et al. (1986). HPV 16 DNA was detected in 18 of 59 tested histologically tumor-free nodes in 9 patients by Fuchs et al. (1989). While 13 nodes with HPV DNA were obtained from 3 patients without any proof of metastasis, 5 HPV-positive nodes were derived from 4 women with metastasis to other pelvic nodes. Only in two patients all nodes tested were histologically and virologically negative. The HPV-positive histologically tumor-free lymph nodes contained only small amounts of integrated viral DNA with the same cleavage pattern as the primary tumor. No data were given about the clinical follow-up of these patients. Ikenberg et al. (1986) reported the presence of HPV 16 DNA in 3 of 17 histologically tumor-free pelvic lymph nodes of 15 patients with HPV-positive cervical carcinoma. After a follow-up of 2 and 4 years, respectively, no recurrence was observed.

Micrometastasis which has not been detected is not likely a reason for these findings because the larger part of the specimens tested was examined by histological serial sections, and HPV DNA was deteced in a high number of nodes. A HPV infection of histologically normal tissue as it is frequently diagnosed in normal cervical epithelium (Wagner et al. 1984; de Villiers et al. 1987; Meanwell et al. 1987) is unknown in the lymphatic system of healthy women or even in the blood of cervical cancer patients (H. Ikenberg, unpublished data). While the HPV sequences prevalent in normal tissue are generally extrachromosomal, the viral sequences in histologically negative lymph nodes were almost exclusively integrated and showed the same cleavage pattern as the HPV DNA in the corresponding primary tumor. The finger print character of integration patterns makes it most likely that these sequences originate from the primary tumors. Up to now it cannot be decided whether the viral sequences indicate early metastasis or residual DNA of tumor cells already destroyed by immune cells. In situ hybridization on lymph node sections might solve this question.

Distant Metastases

HPV DNA was also detected in several distant metastases of cervical car cinomas. We identified HPV 16 and HPV-related sequences in six of nine metastases to the intestine and to supraclavicular lymph nodes by Southern blotting. Only in one of the three cases where the corresponding primary tumor was available for analysis did the metastasis contain less HPV DNA than the primary lesion. Additionally, we found identical HPV 16 DNA in primary tumor, liver, and muscle metastases of a cervical carcinoma where the tissue was taken 24 h post mortem at autopsy (H. Ikenberg, in preparation).

Walboomers et al. (1987) identified HPV 16 in high copy number in a pelvic metastasis (or recurrent disease) of a patient with endocervical adenocarcinoma using in situ hybridization. The HPV DNA was exclusively located in the nuclei of the malignant epithelial cells and not in the stroma.

Wilcynski et al. (1988) found HPV 16 or 18 in four of five distant metastases of cervical carcinomas. The primary tumors of these patients were not available for analysis. HPV 16 DNA was detected in cells from peritoneal washings of two patients with invasive cervical cancer who had the identical HPV type found in the primary tumor or in cervicovaginal cells (Anderson et al. 1987). Manias et al. (1989) reported the presence of identical HPV 11 DNA in a primary perianal squamous cell carcinoma and its metastases to lung and liver in a renal transplant recipient.

The presence of HPV DNA in primary and metastatic tumors has also been reported from extragenital carcinomas in humans. Approximately 25% of patients with epidermodysplasia verruciformis (EV) a rare autosomal recessive disease characterized by varying degrees of decreased cell-mediated immunity and increased susceptibility to HPV infection undergo malignant conversion of HPV associated cutaneous lesions. Of the numerous HPV types associated with this disease only HPV 5 and HPV 8 were identified in primary and metastatic squamous cancers arising in EV patients (Ostrow et al. 1982). HPV 11 was found in a primary lung carcinoma and metastatic lymph node and hepatic lesions of one patient (Byrne et al. 1987). Finally Shibata et al. (1989) detected HPV 16 DNA in five of seven fine-needle aspirations of metastatic cervical carcinomas. In two cases in which the primary tumor was available for comparison the HPV DNA content of primary tumor and metastasis was identical.

All the cited findings strongly implicate HPV in the pathogenesis of genital and extragenital cancers.

References

Ackerman LV (1948) Verrucous carcinoma of the oral cavity. Surgery 23:670–678

Adler-Storthz N, Newland JR, Tessin BA, Yendall WA, Shillitoe EJ (1986) HPV 2 DNA in oral verrucous carcinoma. International Workshop on Papillomaviruses, September 1986, Cold Spring Harbor

Anderson LL, Ritter DB, Kadish AS, Runowicz CD, Burk RD (1987) Detection of human papillomavirus type 16 DNA in peritoneal washings from patients with cervical carcinoma. J Infect Dis 155:1349

Bain RW, Crocker AW (1983) Rapid onset of cervical cancer in an upper socioeconomic group. Am J Obstet Gynecol 146:366–371

Barnes W, Delgado G, Kurman RJ, Petrilli ES, Smith DM, Ahmed S, Lorincz AT, Temple GF, Jenson AB, Lancaster WD (1988) Possible prognostic significance of human papillomavirus type in cervical cancer. Gynecol Oncol 29:267–273

Beaudenon S, Kremsdorf D, Croissant O, Jablonska S, Wain-Hobson S, Orth G (1986) A novel type of human papillomavirus associated with genital neoplasias. Nature 321:246–249

Beaudenon S, Kremsdorf D, Obalek S, Jablonska S, Pehan-Armaudet G, Croissant O, Orth G (1987) Plurality of genital human papillomaviruses. Characterization of two new types with distinct biological properties. Virology 161:374–384

Beck I (1984) Vaginal carcinoma arising in vaginal condylomata. Case report. Br J Obstet Gynecol 91:503–505

Beckman AM, Daling JR, McDougall JK (1985) Human papillomavirus in anogenital carcinomas. J Cell Biochem [Suppl. 9c] 68

Beckman AM, Daling JR, Sherman KJ, Maden C, Miller BA, Coates RJ, Kiviat NB, Myerson D, Weiss NS, Hislop TG, Beagrie M, McDougall JK (1989) Human papillomavirus infection and anal cancer. Int J Cancer 43:1042–1049

Bergeron C, Naghashfar Z, Shah K, Fu Y, Ferenczy A (1986) Human papillomavirus type 16 in intraepithelial neoplasia (bowenoid papulosis) and coexistant invasive carcinoma of the vulva. International Workshop on Papillomaviruses, September 1986, Cold Spring Harbor

Bergeron C, Shah K, Daniel R, Ferency A (1988) Search for human papillomaviruses in normal, hyperplastic, and neoplastic endometria. Obstet Gynecol 72:383–387

Berkeley AS, LiVolsi VA, Schwartz PE (1980) Advanced squamous cell carcinoma of the cervix with recent normal Papanicolaou tests. Lancet II:375–376

Boon ME, Susanti I, Tasche MJ, Kok LP (1989) Human papillomavirus (HPV)-associated male and female genital carcinomas in a Hindu population. The male as vector and victim. Cancer 64:559–565

Boshart M, Gissmann L, Ikenberg H, Kleinhenz A, Scheurlen H, zur Hausen H (1984) A new type of papillomavirus-DNA, its presence in genital cancer biopsies and in cell lines derived from cervical cancer. EMBO J 3:1151–1157

Boshart M, zur Hausen H (1986) Human papillomaviruses in Buschke-Löwenstein tumors: physical state of the DNA and identification of a tandem duplication in the noncoding region of a human papillomavirus 6 subtype. J Virol 58:963–966

Brandsma JL, Steinberg BM, Abramson AL, Winkler B (1986) Presence of human papillomavirus type 16 related sequences in verrucous carcinoma of the larynx. Cancer Res 46:2185–2188

Brandsma J, Burk RD, Lancaster WD, Pfister H, Schiffman MH (1989) Interlaboratory variation as an explanation for varying prevalence estimates of human papillomavirus infection. Int J Cancer 43:260–262

Buscema J, Naghashfar Z, Sawada E, Daniel R, Woodruff JD, Shah K (1988) The predominance of human papillomavirus type 16 in vulvar neoplasia. Obstet Gynecol 71:601–606

Byrne J, Tsao M, Fraser R, Howley P (1987) Human papillomavirus 11 DNA in a patient with chronic laryngotracheobronchial papillomatosis and metastatic squamous cell carcinoma of the lung. N Engl J Med 317:873–878

Carson LF, Twiggs LB, Okagaki T, Clark BA, Ostrow RS, Faras AJ (1988) Human papillomavirus DNA in adenosquamous carcinoma and squamous cell carcinoma of the vulva. Obstet Gynecol 72:63–67

Choo CY (1980) Double primary epidermoid carcinoma of the vulva and the cervix. Gynecol Oncol 9:324–333

Choo KB, Pan CC, Liu MS, Ng HT, Chen CP, Lee YN, Chao CF, Mang CL, Yeh MY, Han SH (1987) Presence of episomal and integrated human papillomavirus DNA sequences in cervical carcinomas. J Med Virol 21:101–107

Collins JE, Jenkins D, McCance DJ (1988) Detection of human papillomavirus DNA sequences by in situ DNA-DNA hybridization in cervical intraepithelial neoplasia and invasive carcinoma: a retrospective study. J Clin Pathol 41:289–295

Cripe TP, Haugen TH, Turk JP, Tabatabai F, Schmid PG, Dürst M, Gissmann L, Roman A, Turek LP (1987) Transcriptional regulation of the human papillomavirus-16– E6-E7 promoter by a keratinocyte-dependent enhancer and by viral E2 transactivator and repressor gene products: implication for cervical carcinogenesis. EMBO J 6:3745–3754

Daling JR, Weiss NS, Klopfenstein LL, Cochran LE, Chow WH, Daifuku R (1982) Correlates of homosexual behavior and the incidence of anal cancer. JAMA 247:1988–1990

Daling JR, Chu J, Weiss NS, Emel L, Tamini HK (1984) The association of condylomata acuminata and squamous carcinoma of the vulva. Br J Cancer 50:533–535

Daling JR, Weiss NS, Hislop TG, Maden C, Coates RJ, Sherman KJ, Ashley RL, Beagrie M, Ryan JA, Corey L (1987) Sexual practices, sexually transmitted diseases, and the incidence of anal cancer. N Engl J Med 317:973–977

de Villiers EM (1989) Heterogeneity of the human papillomavirus group. J Virol 63:4898–4903

de Villiers EM, Schneider A, Gross G, zur Hausen H (1986) Analysis of benign and malignant urogenital tumors for human papilloma virus infection by labeling cellular DNA. Med Microbiol Immunol (Berlin) 174:281–284

de Villiers EM, Schneider A, Miklaw H, Papendick U, Wagner D, Wesch H, Wahrendorf J, zur Hausen H (1987) Human papillomavirus infections in women with and without abnormal cervical cytology. Lancet II:703–706

Downey GO, Okagaki T, Ostrow RS, Clark BA, Twiggs LB, Faras AJ (1988) Condylomatous carcinoma of the vulva with special reference to human papillomavirus DNA. Obstet Gynecol 72:68–73

Dürst M, Gissmann H, Ikenberg H, zur Hausen H (1983) A papillomavirus DNA from a cervical carcinoma and its prevalence in cancer biopsy samples from different geographic regions. Proc Natl Acad Sci USA 80:3812–3815

Dürst M, Kleinhenz A, Hotz M, Gissman L (1985) The physical state of human papillomavirus type 16 DNA in benign and malignant genital tumors. J Gen Virol 66:1515–1522

Dürst M, Dzoulieva-Petrusevska RT, Boukamp P, Fusenig NE, Gissmann L (1987) Molecular and cytogenetic analysis of immortalized human primary keratinocytes obtained after transfection with human papillomavirus type 16 DNA. Oncogene 1:251–256

Farnsworth A, Laverty C, Stoler MH (1989) Human papillomavirus messenger RNA expression in adenocarcinoma in situ of the uterine cervix. Int J Gynecol Pathol 8:321–330

Ferenczy A, Gelfand MM, Tzipris F (1983) The cytodynamics of endometrial hyperplasia and carcinoma. A review. Ann Pathol 3:189–202

Fuchs PG, Girardi F, Pfister H (1987) Papillomavirus infections in cervical tumors of Austrian patients. In: Steinberg BM, Brandsma JL, Taichman LB (eds) Papillomaviruses. Cold Spring Harbor Laboratory, Cold Spring Harbor, pp 297–300 (Cancer cells, vol 5)

Fuchs PG, Girardi F, Pfister H (1989) Human papillomavirus 16 DNA in cervical cancers and in lymph nodes of cervical cancer patients: a diagnostic marker for early metastases? Int J Cancer 43:41–44

Fukushima M, Okagaki T, Twiggs LB, Clark BA, Zachow KR, Ostrow RS, Faras AJ (1985) Histological types of carcinoma of the uterine cervix and the detectability of human papillomaviruses DNA. Cancer Res 45:3252–3255

Gallup TG, Talledo OC, Shah KJ, Hayes C (1987) Invasive squamous cell carcinoma of the vagina: a 14-year study. Obstet Gynecol 72:782–785

Gissmann L, zur Hausen H (1976) Human papillomaviruses: physical mapping and genetic heterogeneity. Proc Natl Acad Sci USA 73:1310–1313

Gissmann L, Pfister H, zur Hausen H (1977) Human papilloma viruses (HPV): characterization of four different isolates. Virology 76:569–580

Gissmann L, de Villiers EM, zur Hausen H (1982) Analysis of human genital warts (condylomata acuminata) and other genital tumors for human papillomavirus type 6 DNA. Int J Cancer 29:143–146

Gissmann L, Boshart M, Dürst M, Ikenberg H, Wagner D, zur Hausen H (1984) Presence of human papillomavirus in genital tumors. J Invest Dermatol 83:26S–28S

Gissmann L, Dürst M, Oltersdorf T, von Knebel-Döberitz M (1987) Human papillomaviruses and cervical cancer. In: Steinberg BM, Brandsma JL, Taichman LB (eds) Papillomaviruses. Cold Spring Harbor Laboratory, Cold Spring Harbor, pp 275–280 (Cancer cells, vol 5)

Gloss B, Bernard HU, Seedorf K, Klock K (1987) The upstream regulatory region of human papilloma-virus-16 contains an E2 protein-independent enhancer which is specific for cervical carcinoma cells and regulated by glucocorticoid hormones. EMBO J 6:3735–3743

Gompel D, Silverberg SC (1977) Pathology in gynecology and obstetrics. Lippincott, Philadelphia

Green M, Brackmann KH, Sanders PR, Loewenstein PM, Freel JH, Eisinger M, Susitlyk SA (1982) Isolation of a human papillomavirus from a patient with epidermodysplasia verruciformis. Presence of related viral DNA genomes in human urogenital tumors. Proc Natl Acad Sci USA 79:4437–4441

Gross G, Ikenberg H, de Villiers EM, Schneider A, Wagner D, Gissmann L (1986) Bowenoid papulosis: a venerally transmissible disease as reservoir for HPV 16. In: Peto R, zur Hausen H (eds) Viral etiology of cervical cancer. Cold Spring Harbor Laboratory, Cold Spring Harbor, pp 149–165 (Banbury Report 21)

Gupta J, Pilotti S, Shah KV, de Paolo G, Rilke F (1987) Human papillomavirus associated early vulvar neoplasia investigated by in situ hybridization. Am J Surg Pathol 11/6:430–434

Hill S, Coghill S (1986) Human papillomavirus in squamous carcinoma of anus. Lancet II:1333

Holly EA, Whittemore AS, Aston DA, Ahn DK, Nickoloff BJ, Kristiansen JJ (1989) Anal cancer incidence: genital warts, anal fissure or fistula, hemorrhoids, and smoking. J Natl Cancer Inst 81:1726–1731

Husseinzadeh N, Newman NJ, Wessler TA (1989) Vulvar intraepithelial neoplasia: a clinicopathological study of carcinoma in situ of the vulva. Gynecol Oncol 33:157–163

Ikenberg H, Kleine W, Pfleiderer A (1986) HPV DNA in primary tumors and lymph node metastases of genital carcinomas: correlation of HPV prevalence with clinical stage and histology. Fifth International Papillomavirus Workshop, September 1986, Cold Spring Harbor

Ikenberg H, Schwörer D, Pfleiderer A (1988) Nachweis von humaner Papillom-Virus (HPV)-DNA in Vulvakarzinomen. Geburtsh Frauenheilkd 48:776–780

Ikenberg H, Runge M, Göppinger A, Pfleiderer A (1990) Human papillomavirus (HPV) DNA in invasive carcinoma of the vagina. Obstet Gynecol (in press)

Jablonska S, Kawashima M, Obalek S, Szymanczyk J, Orth G (1987) Human papillomavirus-related cutaneous benign lesions and skin malignancies. In: Steinberg BM, Brandsma JL, Taichman LB (eds) Papillomaviruses. Cold Spring Harbor Laboratory, Cold Spring Harbor, pp 309–317 (Cancer cells, vol 5)

Ji HX, Syrjänen S, Syrjänen K, Wu AR, Chang FJ (1990) In situ hybridization analysis of HPV DNA in cervical precancer and cervical cancer from China. Arch Gynecol Obstet 247:21–29

Jochmus-Kudielka J, Schneider A, Braun R, Kimmig R, Koldovsky U, Schneeweis KE, Seedorf K, Gissmann L (1989) Antibodies against the human papillomavirus type 16 early proteins in human sera: correlation of Anti-E7 reactivity with cervical cancer. J Natl Cancer Inst 81(22):1698–1704

Kaiser R, Pfleiderer A (1989) Lehrbuch der Gynäkologie, 15th edn. Thieme, Stuttgart

Kasher MS, Roman A (1988) Characterization of human papillomavirus 6b isolated from an invasive squamous carcinoma of the vulva. Virology 165:225–233

Kaufman RH, Bernstein J, Gordon AN, Adam E, Kaplan AL, Adler-Stohrtz K (1987) Detection of human papillomavirus DNA in advanced epithelial ovarian carcinoma. Gynecol Oncol 27:340–349

Kaur P, Mc Dougall JK (1988) Characterization of primary human keratinocytes transformed by human papillomavirus type 18. J Virol 62:1917–1924

King LA, Tase T, Twiggs LB, Okagaki T, Savage JE, Adcock LL, Prem KA, Carson LF (1989) Prognostic significance of the presence of human papillomavirus DNA in patients with invasive carcinoma of the cervix. Cancer 63:897–900

Kraus FT, Perez-Mesa C (1966) Verrucous carcinoma: clinical and pathological study of 105 cases involving oral cavity, larynx and genitalia. Cancer 19:26–38

Kreider JW, Howett MK, Wolfe SA, Bartlett GL, Zaino RJ, Sedlacek TV, Mortel R (1985) Morphological transformation in vivo of human uterine cervix with papillomavirus from condylomata acuminata. Nature 317:639–641

Kurman RJ, Schiffman MH, Lancaster WD, Reid R, Jenson AB, Temple GF, Lorincz AT (1988) Analysis of individual human papillomavirus types in cervical neoplasia: a possible role for HPV 18 in rapid progression. Am J Obstet Gynecol 159:293–296

Lancaster WD, Castellano C, Santos C, Delgado G, Kurman RJ, Jenson AB (1986) Human papillomavirus DNA in cervical carcinoma from primary and metastatic sites. Am J Obstet Gynecol 154:115–118

Leake J, Woodruff JD, Searle C, Daniel R, Shah KV, Currie JL (1989) Human papillomavirus and epithelial ovarian neoplasia. Gynecol Oncol 34:268–273

Lehn H, Ernst TM, Sauer G (1984) Transcription of episomal papillomavirus DNA in human condylomata acuminata and Buschke-Löwenstein tumors. J Gen Virol 65:2003–2010

Lehn H, Villa LL, Marziona F, Hilgarth M, Hillemanns HG, Sauer G (1988) Physical state and biological activity of human papillomavirus genomes in precancerous lesions of the female genital tract. J Gen Virol 69:187–196

Lever WT (1947) Adenoacanthoma of the sweat glands: carcinoma of sweat glands with glandular and epidermal elements. Arch Dermatol Syphilol 56:151

Lorincz AT, Lancaster WD, Temple GF (1986) Cloning and characterization of the DNA of a new human papillomavirus from a woman with dysplasia of the uterine cervix. J Virol 58:225–229

Lorincz AT, Quinn AP, Lancaster WD, Temple GF (1987a) A new type of papillomavirus associated with cancer of the uterine cervix. Virology 159:187–190

Lorincz AT, Temple GF, Kurman RJ, Jenson AB, Lancaster WD (1987b) Oncogenic association of specific human papillomavirus types with cervical neoplasia. JNCI 79:671–677

MacNab JCM, Walkinshaw SA, Cordiner JW, Clements JB (1986) Human papillomavirus in clinically and histologically normal tissue of patients with genital cancer. N Engl J Med 315:1052–1058

Manetta A, Pinto JL, Larson JE, Stevens CW Jr, Pinto PS, Podszaski ES (1988) Primary invasive carcinoma of the vagina. Obstet Gynecol 72:77–81

Manias DA, Ostrow RS, Mc Glennen RC, Estensen DA, Faras AJ (1989) Characterization of integrated human papillomavirus type DNA in primary and metastatic tumors from a renal transplant recipient. Cancer Res 49:2514–2519

Maniatis T, Fritsch EF, Sambrook J (1982) Molecular cloning – a laboratory manual. Cold Spring Harbor Laboratory, New York

Mathieu A, Avril M-F, Duvillard P, George M, Orth G, Riou G, Prade M, Gerbaulet A, Wolff J-P, Michel G (1985) Tumeur de Buschke-Löwenstein. Trois localisations vulvaires. Association à l'HPV 6 dans un cas. Ann Dermatol Venereol 112:745–746

Matsukura T, Koi S, Sugase M (1989) Both episomal and integrated forms of human papillomavirus type 16 are involved in invasive cervical cancers. Virology 172:63–72

McCance D, Clarkson PK, Dyson JL, Walker PG, Singer A (1985) Human papillomavirus type 6 and 16 in multifocal intraepithelial neoplasias of the female lower genital tract. Br J Obstet Gynecol 92:1093–1100

Meanwell CA, Cox MF, Blackledge G, Maitland NJ (1987) HPV 16 DNA in normal and malignant cervical epithelium: implication for the etiology and behaviour of cervical neoplasia. Lancet I:703–707

Mitrani-Rosenbaum S, Gal D, Friedman M, Kitron N, Tsvieli R, Mordel N, Anteby SO (1988) Papillomaviruses in lesions of the lower genital tract in Israeli patients. Eur J Cancer Clin Oncol 24:725–731

Morley G (1986) Cancer of the vulva. In: Knapp RC, Berkowitz RS (eds) Gynecologic oncology. Macmillan, New York, pp 377–398

Naghashfar ZS, Rosenshein NB, Lorincz AT, Buscema J, Shah KV (1987) Characterization of human papillomavirus type 45, a new type 18-related virus of the genital tract. J Gen Virol 68:3073–3079

Nuovo GJ, Crum CP, de Villiers E-M, Levine RV, Silvesten SJ (1988) Isolation of a novel human papillomavirus (type 51) from a cervical condyloma. J Virol 62:1452–1455

Okagaki T, Clark B, Zachow KR, Twiggs LB, Ostrow RS, Pass F, Faras AJ (1984) Presence of human papillomavirus in verrucous carcinoma (Ackermann) of the vagina. Arch Pathol Lab Med 108:567–570

Okagaki T, Tase T, Twiggs LB, Carson LT (1989) Histogenesis of cervical adenocarcinoma with reference to human papillomavirus 18 as a carcinogen. J Reprod Med 34:639–644

Ostrow RS, Bender M, Niimura M, Sehi T, Kawashima M, Pass F, Faras AJ (1982) Human papillomavirus DNA in cutaneous primary and metastasized squamous cell carcinomas from patients with epidermodysplasia verruciformis. Proc Natl Acad Sci USA 79:1634–1638

Ostrow RS, Manias DA, Clark BA, Fukushima M, Okagaki T, Twiggs LB, Faras AJ (1988) The analysis of carcinomas of the vagina for human papillomavirus DNA. Gynecol Oncol 29:267–273

Palmer JG, Shepherd NA, Jass JR, Crawford LV, Northover JM (1987) Human papillomavirus type 16 DNA in anal squamous carcinoma. Lancet II:42

Pater MM, Pater A (1985) Human papilloma virus types 16 and 18 sequences in carcinoma cell lines of the cervix. Virology 145:313–318

Pater MM, Pater A (1988) Expression of human papillomavirus types 16 and 18 DNA in cervical carcinoma cell lines. J Med Virol 26:185–195

Pfister H (1987) Papillomaviruses and genital cancer. Adv Cancer Res 48:113–147

Podcaski E, Herbst AL (1986) Cancer of the vagina and fallopian tube. In: Knapp RC, Berkowitz RS (eds) Gynecologic oncology. Macmillan, New York, pp 399–424
Prakash SS, Reeves WC, Sisson GR, Brenes M, Gody J, Bacchetti S, de Britton RC, Rawls WE (1985) Herpes simplex virus type 2 and human papillomavirus type 16 in cervicitis, dysplasia and invasive cervical carcinoma. Int J Cancer 35:51–57
Pride GL, Schultz AE, Chuprevish TW, Buchlet DA (1979) Primary invasive squamous carcinoma of the vagina. Obstet Gynecol 53:218–225
Rando RF, Sedlacek TV, Hunt J, Jenson AB, Kurman RJ, Lancaster WD (1986) Verrucous carcinoma of the vulva associated with an unusual type 6 human papillomavirus. Obstet Gynecol 67:705–715
Rigoni-Stern DA (1842) Fatti statistici relativi alle mallattie cancerose che servirono di base alle poche cose delta dal dotte. G Progr Pathol Therap Ser 2:507–517
Robboy SJ, Koller KL, O'Brien P (1984) Increased incidence of cervical and vaginal dysplasia in 3.980 diethylstilbestrol-exposed young women. JAMA 252:2979–2983
Rotkin ID (1973) A comparison review of key epidemiological studies in cervical cancer related to current searches for transmissible agents. Cancer Res 33:1353–1367
Rutledge F (1967) Cancer of the vagina. Am J Obstet Gynecol 97:635–655
Scheurlen W, Stremlau A, Gissmann L, Hohn D, Zenner HP, zur Hausen H (1986) Rearranged HPV 16 molecules in an anal carcinoma and in a laryngeal carcinoma. Int J Cancer 38:671–676
Schneider A, de Villiers E-M, Schneider V (1987a) Multifocal squamous neoplasia of the female genital tract. Significance of human papillomavirus infection in the vagina after hysterectomy. Obstet Gynecol 70:294–298
Schneider A, Hotz M, Gissmann L (1987b) Increased prevalence of human papillomaviruses in the lower genital tract of pregnant women. Int J Cancer 40:198–201
Schneider A, Sawada E, Gissmann L, Shah K (1987c) Human papillomavirus in women with a history of abnormal Papanicolaou smears and in their male partners. Obstet Gynecol 69:554–562
Schneider-Maunoury S, Croissant O, Orth G (1987) Integration of human papillomavirus type 16 DNA sequences: a possible early event in the progression of genital tumors. J Virol 61:3295–3298
Scholl SM, Kingsley Pillers EM, Robinson RE, Farell PJ (1985) Prevalence of human papillomavirus type 16 DNA in cervical carcinoma samples in East Anglia. Int J Cancer 35:215–218
Schwarz E, Freese L, Gissmann L, Mayer W, Roggenbuck A, Stremlau A, zur Hausen H (1985) Structure and transcription of human papillomavirus sequences in cervical carcinoma cells. Nature 314:111–114
Shibata DK, Arnheim N, Martin WJ (1988) Detection of human papillomavirus in paraffin-embedded tissue using the polymerase-chain-reaction. J Exp Med 167(1):225–230
Shibata D, Cosgrove M, Arnheim N, Martin WJ, Martin SE (1989) Detection of human papillomavirus DNA in fineneedle aspirations of metastatic squamous-cell carcinoma of the uterine cervix using the polymerase chain reaction. Diagn Cytopathol 5:40–43
Siiteri PK (1981) Extraglandular oestrogen fraction and serum binding of oestradiol: relationship to cancer. J Endocrinol 89:119P–129P
Smotkin D, Berek JS, Fu YS, Hacker NF, Major FJ, Wettstein FO (1986) Human papillomavirus deoxyribonucleic acid in adenocarcinoma and adenosquamous carcinoma of the uterine cervix. Obstet Gynecol 68:241–244
Southern EM (1975) Detection of specific sequences among DNA fragments separated by gel electrophoresis. J Mol Biol 98:503–517
Spitzer M, Krumholz BA, Seltzer VL (1989) The multicentric nature of disease related to human papillomavirus infection of the female lower genital tract. Obstet Gynecol 79:303–307
Stegner HE (1986) Das Karzinom der Vulva. In: Zander J, Baltzer J (eds) Erkrankungen der Vulva. Urban & Schwarzenberg, München, pp 80–86

Stremlau A, Gissmann L, Ikenberg H, Stark M, Bannasch P, zur Hausen H (1985) Human papillomavirus type 16 related DNA in an anaplastic carcinoma of the lung. Cancer 55:1737–1740

Sutton GP, Stehman FB, Ehrlich CE, Roman A (1987) Human papillomavirus deoxyribonucleic acid in lesions of the female genital tract: evidence for type 6/11 in squamous carcinoma of the vulva. Obstet Gynecol 70:564–568

Tase T, Okagaki T, Clark BA, Manias DA, Ostrow RS, Twiggs LB, Faras AJ (1988) Human papillomavirus types and localization in adenocarcinoma and adenosquamous carcinoma of the uterine cervix: a study by in situ DNA hybridization. Cancer Res 48:993–998

Tase T, Okagaki T, Clark BA, Twiggs LB, Ostrow RS, Faras AJ (1989) Human papillomavirus DNA in glandular dysplasia and microglandular hyperplasia: presumed precursors of adenocarcinoma of the uterine cervix. Obstet Gynecol 73:1005–1008

Tidy J, Farrell PJ (1989) Retraction: human papillomavirus subtype 16b. Lancet II:1539

Tidy AJ, Parry GC, Ward P, Coleman DV, Peto J, Malcolm AD, Farrell PJ (1989) High rate of human papillomavirus type 16 infection in cytologically normal cervices. Lancet II:434

Tidy AJ, Vousden KH, Farrell PJ (1989) Relation between infection with a subtype of HPV 16 and cervical neoplasia. Lancet II:1225–1227

Underwood JW, Adcock LL, Okagaki T (1978) Adenosquamous carcinoma of skin appendages (adenoid squamous cell carcinoma, pseudoglandular squamous cell carcinoma, adenoacanthoma of the sweat glands of Lever) of vulva: a clinical and ultrastructural study. Cancer 42:1851–1858

Venuti A, Marcante ML (1989) Presence of human papillomavirus type 18 DNA in vulvar carcinomas and its integration into the cell genome. J Gen Virol 70:1587–1592

Villa LL, Lopes A (1986) Human papillomavirus DNA sequences in penile carcinomas in Brazil. Int J Cancer 37:853–855

Wagner D, Ikenberg H, Böhm N, Gissmann L (1984) Type specific identification of human papillomavirus in cervical smears by DNA in situ hybridization. Obstet Gynecol 62:767–772

Walboomers JM, Fokke HE, Polak M, Volkers H, Honthoff HJ, Barents J, van der Noordaa J, ter Schegget J (1987) In situ localization of human papillomavirus type 16 DNA in a metastasis of an endocervical adenocarcinoma. Intervirology 27:81–85

Walker J, Bloss JD, Liao SY, Berman M, Bergen S, Wilcynski SP (1989) Human papillomavirus genotype as a prognostic indicator in carcinoma of the uterine cervix. Obstet Gynecol 74:781–785

Waterhouse J, Muir SC, Correa P, Powell J (1976) Cancer incidence in five continents, vol. III. IARC Scientific publication no. 15. IARC, Lyon

Weisbrot IM, Stabinsky D, Davis DM (1972) Adenocarcinoma in situ of the uterine cervix. Cancer 29:1179–1187

Wells M, Griffith S, Lewis F, Dixon MF, Bird CC (1987) Identification of human papillomavirus in paraffin sections of anal condylomas and squamous carcinomas by in situ DNA hybridisation. J Pathol 151:A64

Wilcynski SP, Bergen S, Walker J, Liao S, Pearlman LF (1988) Human papillomaviruses and cervical cancer: analysis of histopathologic features associated with different viral types. Hum Pathol 19:697–704

Wilkinson EJ, Friedrich EG, Eduard G, Fu YS (1981) Multicentric nature of vulvar carcinoma in situ. Obstet Gynecol 58:69–75

Woodruff JD (1981) Carcinoma in situ of the vagina. Clin Obstet Gynecol 24:485–490

Xiao X, Cao M, Miller TR, Cao Z, Yen TSB (1988) Papillomavirus DNA in cervical carcinoma specimens from central China. Lancet II:902

Yasumoto S, Burkhardt AL, Doniger J, Di Paolo JA (1986) Human papillomavirus type 16 DNA-induced malignant transformation of NIH3T3 cells. J Virol 57:572–577

Young JL, Percy CL, Asire AJ (eds) (1981) Surveillance epidemiology and end results. Cancer incidence and mortality in the United States, 1973–77. NCI Monograph, NCI, Bethesda, p 57

Young LS, Bevan IS, Johnson MA, Blomfield PI, Bromidge T, Maitland NJ, Woodman CBJ (1989) The polymerase chain reaction: a new epidemiological tool for investigating cervical human papillomavirus infection. Br Med J 298:14–18

Zachow KR, Ostrow RS, Bender M, Watts S, Okagaki T, Pass F, Faras AJ (1982) Detection of human papillomavirus DNA in anogenital neoplasias. Nature 300:771–773

zur Hausen H (1977) Human papillomavirus and their possible role in squamous cell carcinomas. Curr Top Microbiol Immunol 78:1–30

zur Hausen H (1986) Intracullar surveillance of persisting vival infections. Human genital cancer results from deficient control of papillomavirus gene expression. Lancet II:489–491

zur Hausen H (1987) Papillomaviruses in human cancer. Cancer 59:1692–1696

zur Hausen H (1989a) Papillomaviruses as carcinomaviruses. In: Klein G (ed) Advances in viral oncology, vol 8, pp 1–27

zur Hausen H (1989b) Papillomaviruses in anogenital cancers as a model to understand the role of viruses in human cancers. Cancer Res 49:4677–4681

zur Hausen H, Schneider A (1987) The role of papillomaviruses in human anogenital cancer. In: Salzman NP, Howley PM (eds) The papovaviridiae, vol 2. Plenum, New York, pp 245–263

Clinical Aspects

HPV Infection and Precancer in Gynaecology – Diagnosis and Therapeutic Aspects

H.-E. Stegner

Both prevalence and incidence of genital HPV infections have significantly increased in most countries during the past 20 years (Powell 1978; Oriel 1981). The infection may cause a broad spectrum of characteristic dysplastic and neoplastic changes of the target tissue. Specific PV types have predilections for particular anatomical sites and tissues, affecting predominantly the different squamous epithelia of the anogenital area, the vagina, and both the original and metaplastic squamous epithelium of the uterine cervix. In metaplastic epithelium the proliferating cells susceptible to the infectious agent are superficially located. Thus, the immature transformation zone is the preferred target of HPV and other exogenous noxa, whereas in original squamous epithelium the compartment of mitotic active basal cells is protected by several rows of differentiating cells, and the susceptible cells can be reached only through epithelial defects and ulcerations.

Several features aid in determining that the genital tract is infected with a papillomavirus. These include the gross morphology, colposcopic appearance and histopathology of the epithelial lesions, characteristic cytopathic effects in the cervical smear, occurrence of intranuclear viral antigen in tissue sections and smears, and demonstration of viral particles by electron microscopy. In many cases the diagnosis of HPV infection can be made clinically by gross inspection. This holds true for the classical papillary condylomas, but at times the manifestations may be invisible to the naked eye, thus calling for supplementary diagnostic tools such as colposcopy to define and to locate the lesion. Application of 3% acetic acid or 2% toluidine blue may be helpful to visualize occult (subclinical) flat condylomas, dysplastic areas or micropapillary projections (spikes). Koilocytosis and dyskeratosis are the most striking cytological criteria. Based on these classical parameters, cytological identification of HPV infection can be made in approximately 20% of cases. Using additional cytopathological features such as multinucleation, clear cytoplasm and others, the diagnostic accuracy may reach more than 80%.

Approximately 1.5% of routinely taken smears from the cervix uteri reveal cytomorphological signs of papillomavirus infection (Reid et al. 1980; Meisels et al. 1982; Fu et al. 1983). The highest frequency is found in the 21- to 25-year age-group. In 10%–12% of prostitutes and teenagers with sexually transmitted diseases cytopathic effects indicative of HPV infection are found. Histologically, koilocytosis and other cytopathic effects can be recognized in a majority of cases of so-called mild dysplasia of the cervix and frequently coexist with precancerous dysplasia and carcinoma in situ; such association was observed in various proportions of cases but some morphological

Genital Papillomavirus Infections
Eds.: Gross/Jablonska/Pfister/Stegner

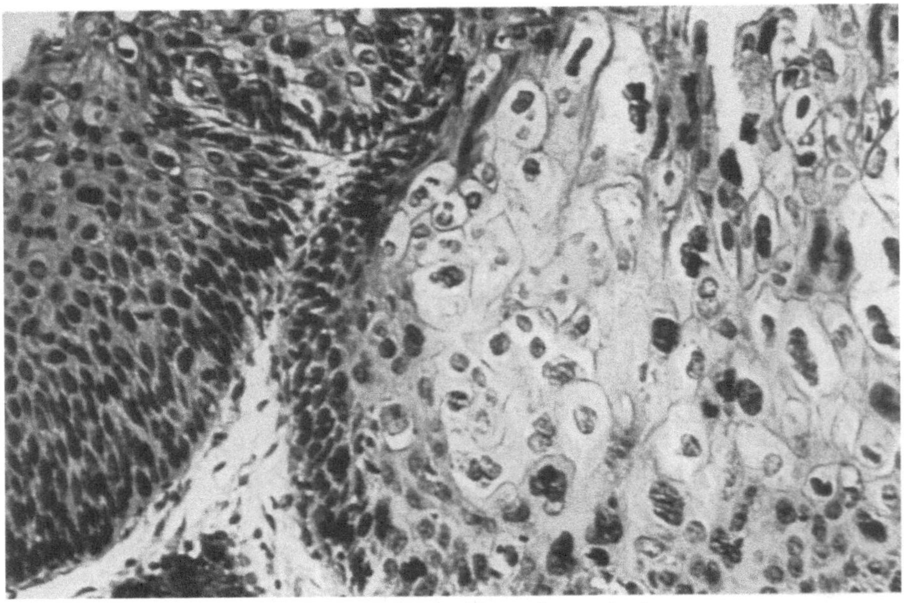

Fig. 1. Condylomatous dysplasia associated with carcinoma in situ

evidence of HPV infection may be found in virtually every instance of CIN or invasive squamous cancer (Fig. 1).

Using an immunocytochemical technique, HPV antigen has been found in more than 80% of condylomatous dysplasia and condyloma-associated CIN of the cervix (Woodruff et al. 1980; Morin et al. 1981; Kurman et al. 1982, 1983; Guillet et al. 1983; Syrjänen 1983; Warhol et al. 1984). The rate of electron-microscopical demonstration of viral particles depends on the mode of selection of the material. In general, the detection rate is low in a random sampling of tissue ranging from 25% to 30% but may reach 100% if focused on preselected koilocytotic areas (Sato et al. 1987). So-called perichromatin granules may be mistaken for viral inclusions. Complete virions have not been found in high grades of squamous dysplasia and infiltrating cancer (Della Torre et al. 1978; Hills and Laverty 1979; Morin and Meisels 1980; Ferenczy et al. 1981).

The identification of the specific viral type infecting cervical epithelial cells, however, requires special methods such as DNA hybridization with DNA probes of viral genomes. The Southern-blot hybridization technique using DNA extracted from fresh-frozen tissue has been the standard method for the identification of HPV-DNA from human lesions during the past few years. The direct intracellular demonstration of type-specific viral DNA is possible by in situ hybridization of smears or routinely processed paraffin sections using radioisotope (^{32}P, ^{35}S, ^{125}I, ^{3}H)-labelled or nonradioactive biotin-labelled viral DNA probes. In situ hybridization permits not only examination of fresh tissue samples but also retrospective diagnosis of specimens which are collected and stored over several years. Based on this technique, a large number of research groups have recently contributed to the present knowledge on the rate of HPV infections in defined female populations and to the topographical mapping of the distribution of infected cells within the lower genital tract. HPV-DNA is regularly found in condylomatous lesions, in cervical precancer and in a high proportion of carcinomas – both squamous and glandular – of the lower genital tract (Yoshikawa et al. 1985; de Villiers et al. 1986; Ikenberg et al. 1987; Tase et al. 1989).

In recent years, special attention has been paid to the question of the prevalence of HPV infection in nonselected populations and asymptomatic women. As evidenced by studies of healthy women, the subclinical persistence of HPV now exceeds 20% when calculated irrespective of the HPV type (Wagner et al. 1984; Mc Cance et al. 1985; Wickenden et al. 1985; Burk et al. 1986; Pratili et al. 1986; Schneider et al. 1986). The infection rate reaches a maximum in the fifth decade of life in women with normal smears, whereas HPV infection coincidental with abnormal cytological or colposcopical findings culminates in the age-group of 21- to 30-year-old women. The detection rate depends strongly on the sensitivity of the method applied. The introduction of highly sensitive methods such as the polymerase chain reaction (PCR) will certainly increase the proportion of positive cases. In up to 80% of the male partners of women with HPV-associated lesions clinical or subclinical manifestations of virogenic disease such as penile condylomas, hyperplastic epithelium or bowenoid papulosis can be found (Gross et al. 1985 b; Schneider 1987). This means that HPV infection is one of the most frequent sexually transmitted diseases of the lower genital tract.

Papillomaviruses are classified according to DNA sequence homology. Out of a spectrum of nearly 60 heterologous HPV types HPV 6, 11, 16, 18, 31, 33, 35, 39, 41, 42, 51, 52 and others were found in relatively high frequencies and varying proportions in squamous dysplasia and invasive cervical cancer. Some of them, such as types 6 and 11, and 16 and 31, are closely related. Regarding the continuum of histopathological lesions from mild dysplasia to higher grades of dysplasia leading to carcinoma in situ and finally to frankly invasive cancer, some correlations could be found between HPV-DNA and the degree of atypia in different squamous lesions. HPV-DNA was identified in nearly all squamous epithelia showing koilocytotic or dysplastic changes, but types 6 and 11 are found predominantly in flat or classical condylomatous lesions of the genital tract and have been reported rarely in carcinomas, while types 16, 18 and 31 are more preferentially associated with precancerous dysplasias and invasive cancer, thus indicating their higher oncogenic potential. Moreover, HPV 16, 18, 33 and 35 were shown to be integrated into the genome of squamous cells, whereas other papillomaviruses tested so far apparently persist in an episomal state within the tumours (Boshart et al. 1984; Dürst et al. 1985; Schwarz et al. 1985; Seedorf et al. 1985; Cole and Streek 1986; Fukushima et al. 1987; Lorincz et al. 1987). Thus, evidence is accumulating that infections by HPV 16 and 18 could be causative events for malignant transformation by changing the transcription pattern and interfering with specific control mechanisms in squamous epithelial cells.

What is the clinical significance of these virological findings? Can the HPV-DNA pattern be regarded as a better predictor of the malignant potential of a questionable epithelial lesion than the histological degree of cellular atypia? Should therapeutic strategies be changed and based on virological findings?

Premalignant neoplasia is synonymous with the criteria of aneuploidy, which histologically include abnormal mitosis and marked nuclear atypia in all layers of the epithelium. DNA cytophotometry has proven to be a powerful predictor of long-term behaviour in various grades of CIN. Diploid and polyploid lesions usually regress or occasionally persist, whereas aneuploid lesions persist or progress in the majority of cases. Simple condylomatous dysplasia caused by HPV 6 and 11 generally shows diploid or polyploid DNA histograms. Lesions harbouring HPV 16 and 18 DNA correspond to aneuploid patterns in a high proportion, but HPV 16 has also been found in epithelia of bland or conventional warts associated with other HPVs (Fig. 2a–e). These data suggest that a minority of mild dysplasias with normoploid DNA distribution contain "high-risk" virus types that may be premalignant. Moreover, normal squamous epithelium might be infected by HPV without showing any morphological sign of infection (Ferenczy et al. 1985; MacNar et al. 1986). Consequently, the presence of the virus is per se not equivalent to neoplastic events, and the conditions leading to the induction of dysplastic and precancerous changes are still incompletely understood.

If the assumption is true that HPV 16 and other types are integral to the development of malignant neoplasia, then a reevaluation of the current preventive and therapeutic strategies is mandatory. A definite answer will not be available until the molecular mechanisms of carcinogenesis by HPV are clarified. Current knowledge offers no convincing arguments for changing the present therapeutic strategy, which is based on classical histological criteria.

Any patient with an HPV infection of the cervix is a patient at risk. Women whose cervices have been infected by HPV have a 15–40 times greater risk of developing squamous cell cancer of the uterus than do noninfected women. Due to the heterogeneity of types, subtypes and recombinations involved in all grades of CIN, HPV-DNA typing cannot replace the classical morphological parameters for selecting patients for adequate treatment. Adequateness means therapy adjusted to the location, extension and severity of the lesion, using nonmutilating methods such as cryosurgery, electrocauterization and laser, provided that the preconditions for these procedures are fulfilled completely (Figs. 3, 4). HPV-induced dysplastic lesions are multifocal by nature. Precise determination of the localization of the lesions is therefore a precondition for adequate treatment. The size and distribution of the precursor lesions can reliably be determined by colposcopy. The most important aspect of the colposcopical examination is that it is an effective technique for dividing patients into two groups, those with invasive cancer and those who have CIN and/or vaginal or vulval precursor lesions. Here it is important to

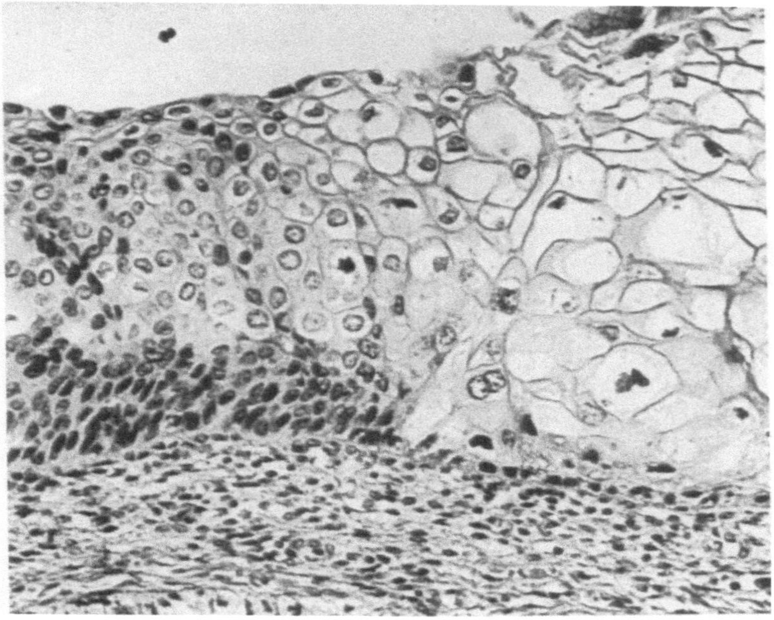

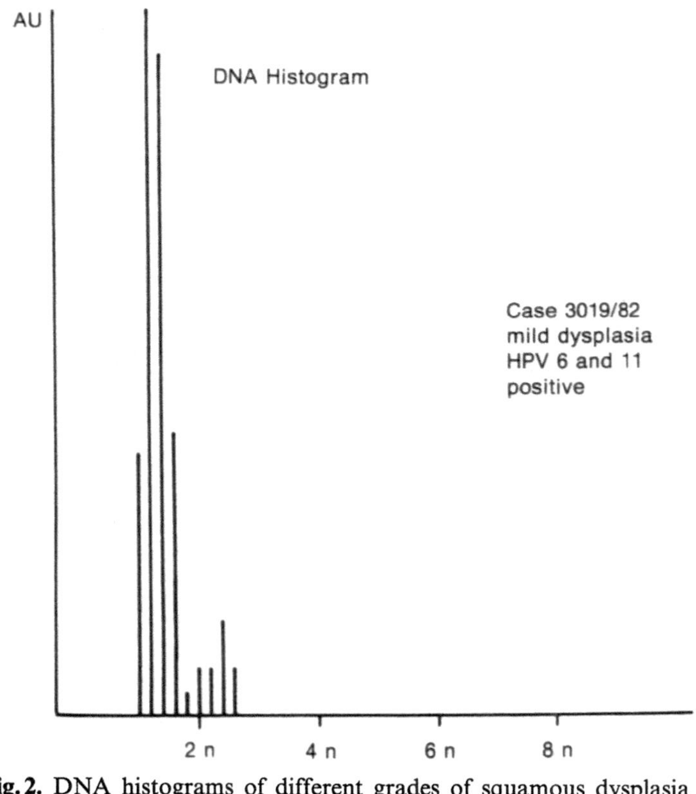

Fig. 2. DNA histograms of different grades of squamous dysplasia as correlated to HPV typing. **a** Mild dysplasia; **b** moderate dysplasia; **c** severe dysplasia

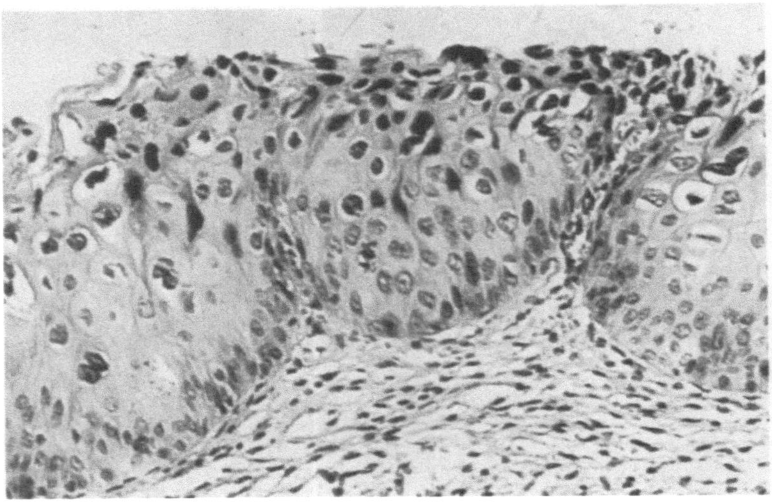

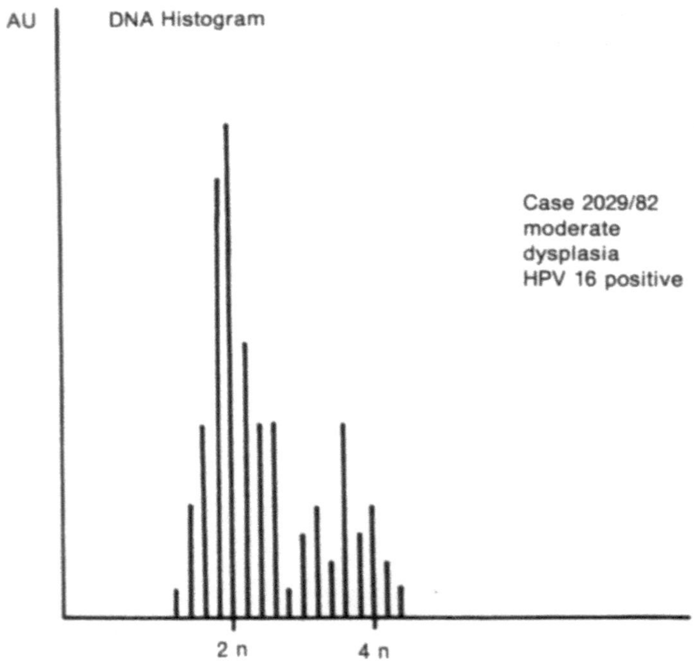

Fig. 2b

emphasize that colposcopy is not itself a diagnostic tool and that no therapeutic decisions should be based on colposcopy alone; however, colposcopy can localize the areas of greatest abnormality which require histological sampling. Application of surface-destroying treatment procedures such as hot cautery, cryosurgery and laser vaporization is possible only if invasive cancer is definitely ruled out (Singer and Jordan 1978; Heinzl 1981; Burke 1982;

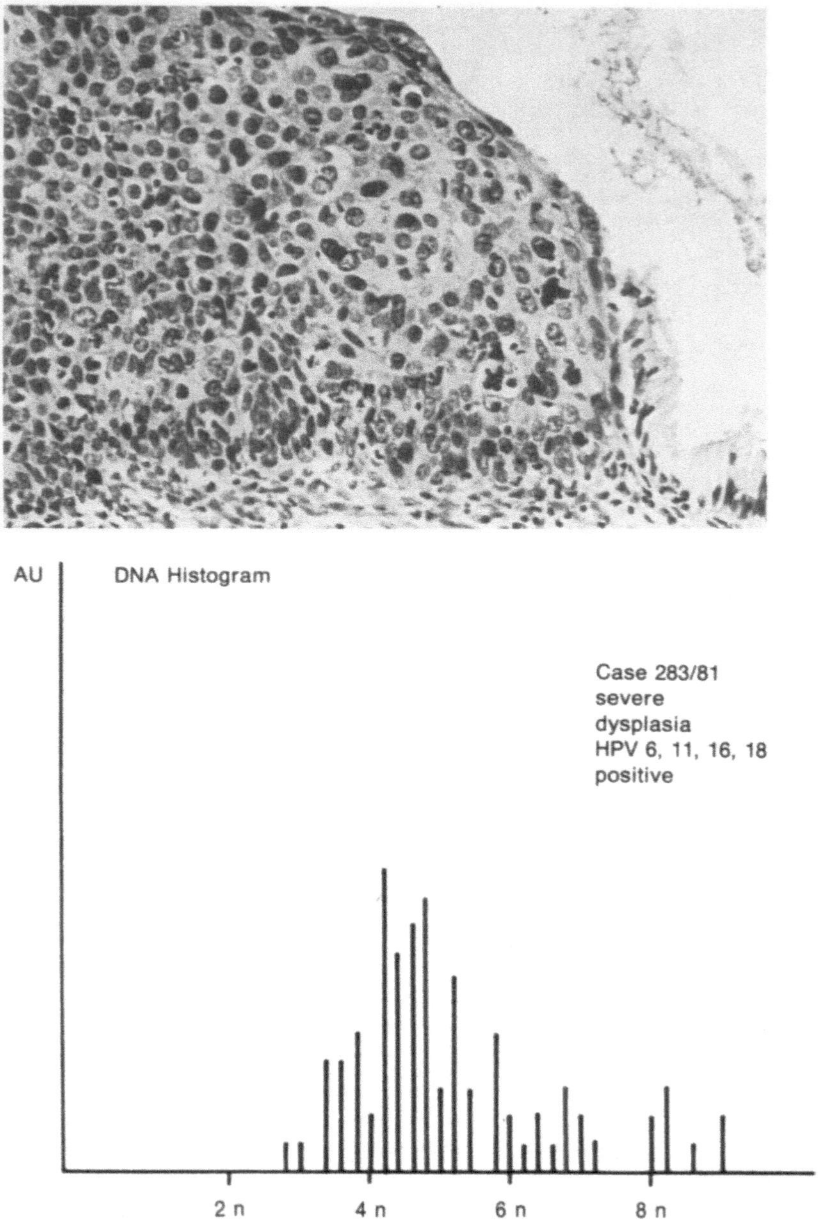

Fig. 2c

Crum and Richart 1980; Heinzl 1988; Hilgarth et al. 1988). This also holds true for HPV-associated epithelial lesions of the external genitalia. A vulvar lesion frequently misinterpreted as intraepithelial neoplasia, and therefore treated by more or less extensive and radical surgery, is the so-called bowenoid papulosis, a clinically benign variety of HPV 16-induced lesions predominantly found in young women and characterized by multifocal flat-

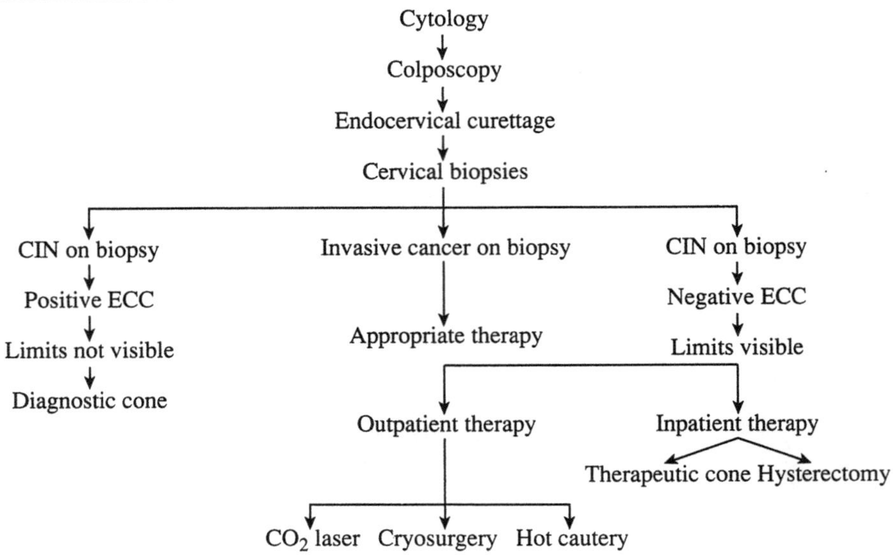

Fig. 3. Protocol for evaluation of patients with abnormal Papanicolaou smear for selective treatment

1. Cytology PaP IIID — IVa
2. Histology CIN I — III
3. Cervical canal unaffected
4. Epithelial boundary on the ectocervix is readily visible
5. Lesion on the ectocervix is not too extensive
6. Patient will cooperate with follow-up care

Fig. 4. Preconditions for surface therapy of premalignant lesions of the uterine cervix

topped papules of brownish-black pigmentation (Abell and Gosling 1961; Wade et al. 1979; Ikenberg et al. 1983, Gross et al. 1985 a). Their clinical appearance may resemble that of seborrheic keratosis or pigmented nevi. Although grossly, the lesion appears benign; the histological appearance is quite similar to that of Bowen's disease, giving rise to surgical overtreatment.

In skilful hands surface-treatment procedures have proven to be powerful nonmutilating therapeutic tools and may replace conization and vulvectomy in a large proportion of cases, but they do not combat the HPV infection and are capable only of removing the visible morphological manifestations of the oncogenic virus.

Due to the lack of experimental systems allowing the reproduction of HPV in vitro, systemic treatment of HPV-associated disease has not been developed to date. The nonsurgical therapeutic methods available today fall into two categories, those based on cell destruction and those that take an immunologi-

cal approach. Some encouraging results have been achieved by administering interferons locally or systemically. The effectiveness of interferons is obviously based on different mechanisms:
- Antiviral activity
- Antiproliferative effect
- Immune modulation.

Recombined interferons alpha 2a, 2c and gamma have proven highly effective in condylomatous lesions: 20%–90% of condylomas, show complete remission compared with remission rates of only 30%–60% in intraepithelial neoplasia. There is no correlation between remission rate and the degree of atypia in epithelial dysplasia, but some correlation was found with the type of HPV involved. Lesions affected by HPV 6/11 show significantly better response to low-dose interferon treatment than those associated with HPV 16/18 (Ikic 1983; Vesterinen et al. 1984; Choo et al. 1986; Gall et al. 1986; Kirby et al. 1986; Schonfeld et al. 1984; Schneider 1987).

Using interferon (intra- and sublesionally) following laser vaporization has also been successful in reducing the recurrences after laser treatment alone (Hatch 1989). Gross et al. (1989) have recently reported on the results of local application of recombined interferon α2c hydrogel as an effective adjuvant after surgical treatment of genital warts.

From these preliminary results it can be stated that interferons although effective in genital HPV infection, are no appropriate means for the treatment of high-grade cervical and vulvar dysplasia, and local destructive methods applied singly or in combination with cytotoxic drugs are at present the most reliable and adequate treatment modalities in noninfiltrating precancerous lesions of the lower genital tract.

References

Abell MR, Gosling G jr (1961) Intraepithelial and infiltrative carcinoma of the vulva: Bowen's type. Cancer 14:318–329

Boshart M, Gissmann L, Ikenberg H, Kleinheinz A, Scheurlen W, zur Hausen H (1984) A new type of papillomavirus DNA, its presence in genital cancer biopsies and in cell lines from cervical cancer. EMBO J 3:1151–1157

Burk RD, Kadish AS, Calderin S, Romney SL (1986) Human papillomavirus infection of the cervix detected by cervicovaginal lavage and molecular hybridization: correlation with biopsy results and Papanicolaou smear. Am J Obstet Gynecol 154:982–989

Burke L (1982) The use of carbon dioxide laser in the therapy of cervical intraepithelial neoplasia. Am J Obstet Gynecol 144:337–340

Choo YC, Seto WH, Hsu C, Merigan TC, Tan YH, Ma HK, Ng NH (1986) Cervical intraepithelial neoplasia treated by perilesional injection of interferon. Br J Obstet Gynaecol 93:372–379

Cole ST, Streek RE (1986) Genome organization and nucleotide sequence of human papillomavirus type 33, which is associated with cervical cancer. J Virol 58:991–995

Crum CP, Richart RM (1980) The out-patient approach to the patient with an abnormal Papanicolaou smear. In: Grundman E (ed) Cancer of the uterine cervix. Fischer, Stuttgart (Cancer campaign, vol 8)

Della Torre G, Pilotti S, de Palo G, Rike F (1978) Viral particles in cervical condylomatous lesions. Tumori 64:549–553
De Villiers EM, Schneider A, Gross G, zur Hausen H (1986) Analysis of benign and malignant urogenital tumors for human papillomavirus infection by labelling cellular DNA. Med Microbiol Immunol 174:281–286
Dürst M, Kleinheinz A, Hotz M, Gissmann L (1985) The physical state of human papillomavirus type 16 DNA in benign and malignant genital tumors. J Gen Virol 66:1515–1525
Ferenczy A, Braun L, Shah KV (1981) Human papillomavirus (HPV) in condylomatous lesions of the cervix. Am J Surg Pathol 5:661–670
Ferenczy A, Mitao M, Naga N, Silverstein SJ, Crum CP (1985) Latent papillomavirus and recurring genital warts. N Engl J Med 313:784–788
Fu YS, Braun L, Shah KV, Lawrence W, Robboy S (1983) Histologic, nuclear DNA, and human papillomavirus studies of cervical condylomas. Cancer 52:1705–1711
Fukushima M, Yamakawa Y, Shimano S, Hashimoto M, Sawada Y, Fujinaga K (1987) The physical state of the human papillomavirus type 16 (HPV 16) DNA in carcinoma of the uterine cervix. In: Lancaster WD, Jenson AB (eds) 6th International Papillomavirus Workshop. Georgetown University, Washington
Gall SA, Hughes CE, Mounts P, Segritti A, Weck PK, Whisnant JK (1986) Efficacy of human lymphoblastic interferon in the therapy of resistant condyloma acuminata. Obstet Gynecol 67:643–651
Gross G, Hagedorn M, Ikenberg H (1985 a) Bowenoid papulosis. Presence of human papillomavirus (HPV) structural antigens and of HPV 16-related DNA sequence. Arch Dermatol 121:858–863
Gross G, Schneider A, Hauser-Brauner B, Wagner D, Ikenberg H, Gissmann L (1985 b) Transmission of genital papillomavirus infections: a study of sexual partners. J Cell Biochem [Suppl] 69:71
Gross G, Roussaki A, Pfister H (1989) Recurrent vulvar Buschke-Löwenstein's tumorlike condyloma acuminata and Hodgkin's disease effectively treated with recombinant interferon α2c gel as adjuvant to electrosurgery. In: Fritsch P, Schuler G, Hintner H (eds) Immunodeficiency and skin. Karger, Basel, pp 178–184 (Current problems in dermatology, vol 18)
Guillet G, Braun L, Shah KV, Ferenczy A (1983) Papillomavirus in cervical condylomas with and without associated cervical intraepithelial neoplasia. J Invest Dermatol 81:513–516
Hatch GD (1989) Interferon treatment of human papillomavirus. Colposcopist 21:1–4
Heinzl S (1981) Die Behandlung benigner und prämaligner Zervixveränderungen mit CO_2-Laser. Gynäkologe 14:245–251
Heinzl S (1988) Destructive methods in the treatment of cervical and vulval intraepithelial neoplasia. In: Stegner H-E, Coppleson M (eds) Colposcopy in diagnosis and treatment of preneoplastic lesions. Springer, Berlin Heidelberg New York
Hilgarth HH, Hillemanns G, Göppinger A (1988) Therapy of intraepithelial neoplasia. In: Stegner H-E, Coppleson M (eds) Colposcopy in diagnosis and treatment of preneoplastic lesions. Springer, Berlin Heidelberg New York
Hills E, Laverty CR (1979) Electron-microscope detection of papillomavirus particles in selected koilocytic cells in routine cervical smears. Acta Cytol 23:53–56
Ikenberg H, Gissmann L, Gross G, Grussendorf-Conen E, zur Hausen H (1983) Human papillomavirus type 16-related DNA in genital Bowen's disease and in bowenoid papulosis. Int J Cancer 32:563–565
Ikenberg H, Spitz C, Schworer D, Pfleiderer A (1987) Human papillomavirus DNA in genital carcinomas. Correlation with clinical stage, histology and clinical course of the disease. In: Lancaster WD, Jenson AB (eds) 6th International Papillomavirus Workshop. Georgetown University, Washington
Ikic D (1983) Intralesional therapy. In: Sikora K (ed) Interferon and cancer. Plenum Press, New York

Kirby P, Wells D, Kiviat N, Corey L (1986) A phase-I trial of intramuscular recombinant gamma interferon for refractory genital warts. J Invest Dermatol 86:485
Kurman RJ, Shah KV, Lancaster WD, Jenson AB (1982) Immunoperoxidase localization of papillomavirus antigens in cervical dysplasia and vulvar condylomas. Am J Obstet Gynecol 140:931–935
Kurman RJ, Jenson AB, Lancaster WD (1983) Papillomavirus infection of the cervix. II. Relationship to intraepithelial neoplasia based on the presence of specific viral structural proteins. Am J Surg Pathol 7:39–52
Lorincz AT, Quinn AP, Lancaster WD, Temple GF (1987) A new type of papillomavirus associated with cancer of the uterine cervix. Virology 159:187–190
MacNar JCM, Walkinshaw SA, Cordiner JW, Clements JB (1986) Human papillomavirus in clinically and histologically normal tissue of patients with genital cancer. N Engl J Med 315:1052–1058
McCance DJ, Campion MJ, Clarkson PK, Chesters RM, Jenkins D, Singer A (1985) Prevalence of human papillomavirus type 16 DNA sequences in cervical intraepithelial neoplasia and invasive carcinoma of the cervix. Br J Obstet Gynaecol 92:1101–1105
Meisels A, Morin C, Casas-Cordero M (1982) Human papillomavirus infection of the uterine cervix. Int J Gynecol Pathol 1:75–94
Morin C, Meisels A (1980) Human papillomavirus infection of the uterine cervix. Acta Cytol 24:82–84
Morin C, Braun L, Casas-Cordero M, Shah KV, Roy M, Fortier M, Meisels A (1981) Confirmation of the papillomavirus etiology of condylomatous cervical lesion by the peroxidase-antiperoxidase technique. JNCI 66:831–835
Oriel JD (1981) Genital warts. Sex Transm Dis 8:326–329
Powell L (1978) Condyloma acuminatum. Recent advances in development, carcinogenesis and treatment. Clin Obstet Gynecol 21:1061–1079
Pratili MA, Ledoussal V, Harvey P, Laval C, Bertrand F, Jibard N, Croissant O, Orth G (1986) Human papillomavirus in the epithelial cells of the cervix uteri: frequency of types 16 and 18. Preliminary results of a clinical, cytologic and viral study. J Gynecol Obstet Biol Reprod (Paris) 15:45–50
Reid R, Laverty CR, Coppleson M, Isarangkul W, Hills E (1980) Noncondylomatous cervical wart virus infection. Obstet Gynecol 55:476–483
Sato S, Okagaki T, Clark BA, Twiggs LB, Fukushima M, Ostrow RS, Faras AJ (1987) Sensitivity of koilocytosis, immunocytochemistry, and electron microscopy as compared to DNA hybridization in detecting human papillomavirus in cervical and vaginal condyloma and intraepithelial neoplasia. Int J Gynecol Pathol 5:297–307
Schneider A (1987) Morphologische und klinische Aspekte humaner Papillomavirus-Infektionen im unteren Genitaltrakt. Habilitationsschrift, University of Ulm
Schneider A, Schuhmann R, de Villiers E-M, Gissmann L (1986) Klinische Bedeutung von humanen Papillomavirus-(HPV)-Infektionen im unteren Genitaltrakt. Geburtshilfe Frauenheilkd 46:261–266
Schonfeld A, Nitke S, Schattner A, Wallach D, Crespi M, Hahn T, Levavi H, Yarden O, Shoham J, Doerner T, Revel M (1984) Intramuscular human interferon-beta injections in treatment of condylomata acuminata. Lancet 1:1038–1042
Schwarz E, Freese UK, Gissmann L, Mayer W, Roggenbuck B, Stremlau A, zur Hausen H (1985) Structure and transcription of human papillomavirus sequences in cervical carcinoma cells. Nature 314:111–114
Seedorf K, Krämmer G, Dürst M, Suhai S, Röwekamp WG (1985) Human papillomavirus type 16 DNA sequence. Virology 145:181–185
Singer A, Jordan JA (1978) The management of premalignant cervical disease. Clin Obstet Gynaecol 5:629
Syrjänen KJ (1983) Human papillomavirus lesions in association with cervical dysplasias and neoplasias. Obstet Gynecol 62:617–624

Tase T, Okagaki T, Clark BA, Twiggs LB, Ostrow RS, Faras AJ (1989) Human papillomavirus DNA in adenocarcinoma in situ, microinvasive adenocarcinoma of the uterine cervix, and coexisting squamous intraepithelial neoplasia. Int J Gynecol Pathol 8:8–17

Wade TR, Kopf AW, Ackerman AB (1979) Bowenoid papulosis of the genitalia. Arch Dermatol 115:306–308

Wagner D, Ikenberg H, Boehm N, Gissmann L (1984) Identification of human papillomavirus in cervical swabs by deoxyribonucleic acid in situ hybridization. Obstet Gynecol 64:767–772

Warhol MJ, Pinkus GS, Rice RH, El-Tawil GH, Lancaster WD, Jenson AB, Kurman RJ (1984) Papillomavirus infection of the cervix. III. Relationship of the presence of viral structural proteins to the expression of involucrin. Int J Gynecol Pathol 3:71–81

Wickenden C, Steele A, Malcolm AD, Coleman DV (1985) Screening for wart virus infection in normal and abnormal cervices by DNA hybridization of cervical scrapes. Lancet 1:65–67

Woodruff JD, Braun L, Cavalieri R, Gupta P, Pass F, Shah KV (1980) Immunologic identification of papillomavirus antigen in condyloma tissues from the female genital tract. Obstet Gynecol 56:727–732

Yoshikawa H, Matsukura T, Yamamoto E, Kawana T, Mizuno M, Yoshike K (1985) Occurrence of human papillomavirus type 16 and 18 DNA in cervical carcinomas from Japan: Age of patients and histological type of carcinomas. Gann 76:667–671

Cervical HPV Diagnosis: Colposcopy, Cytology, Histology

D. Wagner

Introduction

The application of HPV-DNA cloning and hybridization techniques in the clinical diagnosis of cervical, vaginal and vulvar cancer and their precursor lesions has provided strong evidence that HPV infections constitute an essential factor for the genesis of these neoplasias.

During the past few years, highly sensitive molecular hybridization techniques have been used to analyse large numbers of cervical squamous epithelial carcinomas for HPV-DNA integrated in the genome of cancer cells. It was shown that roughly 95% of cervical squamous carcinomas contain HPV–DNA.

HPV-DNA typing in a series of 600 cervical carcinomas revealed that 60% were positive for HPV-type 16, 15% were positive for HPV-type 18, 20% were positive for HPV-types 31, 33, 35, 39, 51, 55, and 5% were negative (Table 1). Similar results have been demonstrated in up to 90% of cervical intraepithelial neoplasias, depending on the severity of the lesion (Table 2). Furthermore, it was shown that up to 15% of clinically healthy women who were screened by means of in situ filter hybridization of cervical smears were HPV positive (de Villiers et al. 1987).

These observations demonstrate that HPV infections of the female genital tract are common in gynaecology and that, for reasons unknown thus far, cervical cancer and its precursors are likely to occur in HPV-positive women. Therefore the development of a new screening system for cervical cancer would seem valuable which is based exclusively on the application of highly sensitive hybridization tests on cellular or tissue samples to detect HPV-positive women. Colposcopy, cytology and histology can then be used in a

Table 1. HPV-types in 600 cases of cervical carcinoma

Type	Percent
6/11	0
16	60
18	15
31, 33, 35, 39, 51, 52, 53	20
Negative	5

Table 2. Frequency of HPV-DNA Positivity in CIN

Grade	Percent
CIN I	78
CIN II	87
CIN III	89

Genital Papillomavirus Infections
Eds.: Gross/Jablonska/Pfister/Stegner

second step to localize and to define a precancerous lesion in women at risk in order to determine the appropriate treatment.

Minimal Morphological Changes in HPV-Positive Women

Unfortunately, such a cervical cancer screening method is presently not realistic. There are no highly sensitive hybridization techniques that can be easily employed in the gynaecologist's office.

Therefore, we must still focus on cytology and colposcopy as the most reliable morphological methods for early cancer detection. When the described recently hybridization techniques were *combined* with these conventional methods, however, the very early detection of precancerous lesions of the cervix, vagina and vulva was improved significantly. The gynaecologist is now aware of minimal morphological changes related to HPV infection which were previously regarded as "unspecific inflammatory reactions".

Colposcopy

Colposcopy is the traditional method of cervical cancer detection in Germany. The classical findings of precancerous lesions were first described in 1926 by Hinselmann, who lived and worked in Hamburg.

He defined them as *Grund* (punctation), *Felderung* (mosaic) and *Leukoplakie* (leucoplakia). Most of these lesions strongly correlate with dysplasia or carcinoma in situ, or – as we now call it – cervical intraepithelal neoplanin (CIN) grades I–III.

If we apply hybridization techniques to cells or tissue taken from those lesions, most of them are found to be HPV positive, inplicating that the HPV infection is a causative factor in the development of this specific growth pattern of the cervical epithelium. In the transformation zone of the uterine cervix HPV does *not* form the condyloma acuminatum typical for the skin, but rather a flatter lesion (condyloma planum) identical to the growth pattern of CIN. Since we are *not* able to differentiate colposcopically between an HPV-positive or -negative CIN lesion, we subsume them all under the designation "flat condyloma" (Fig. 1).

Furthermore, by combining colposcopy and HPV hybridization techniques, we focus our attention on small, nonproliferating acetic-white and iodine-negative lesions within the transformation zone. These had previously been defined as metaplasia following unspecific inflammation (Fig. 2a, b).

In a study group of 66 randomly selected cases presenting with such lesions, 28 (42%) were HPV positive when we applied the filter in situ (FIS) hybridization technique. With the more sensitive Southern-blot technique on cervical biopsies of those lesions HPV-positive findings were obtained in up to

The gynaecologist should recognize the transformation zone of the uterine cervix as a site of special affinity for the papillomavirus, from where it is likely to spread to other areas of the female genital tract.

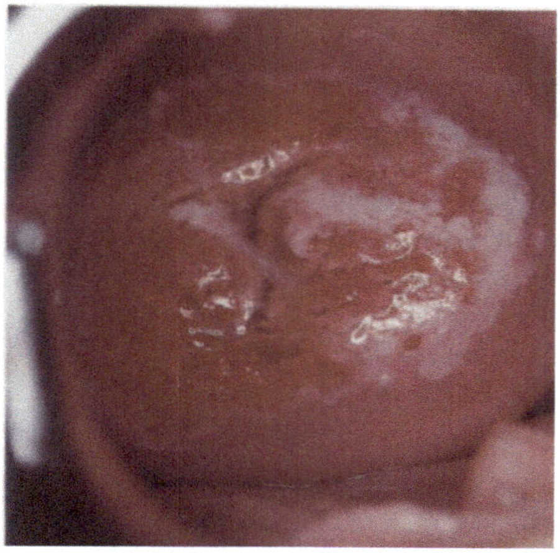

Fig. 1. Acetic white epithelium with irregular mosaic formation in transformation zone of the anterior lip. Histology: CIN III; FIS hybridization: HPV 16/18+. ×0.6

a 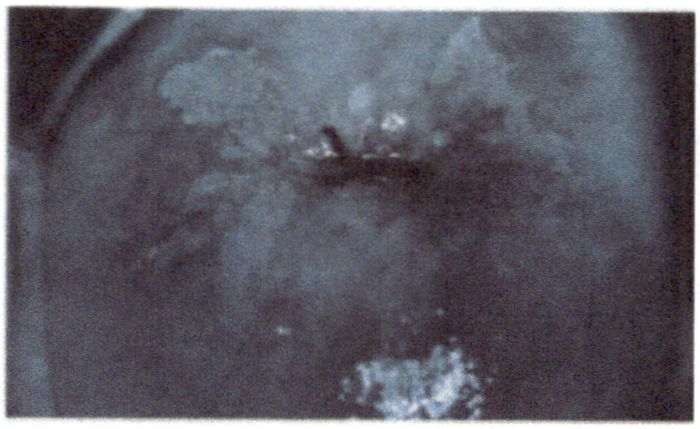 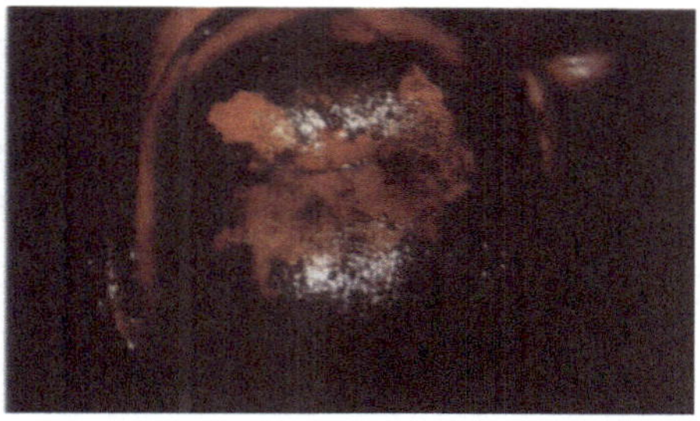 b

Fig. 2. a Nonproliferating acetic white epithelium showing regular mosaic formation on anterior and posterior lip. Histology: CIN I; FIS hybridization: HPV 16/18+. Green filter, ×0.6. **b** Same lesion as in **a**, showing negative iodine reaction of the HPV-infected epithelium. ×0.4

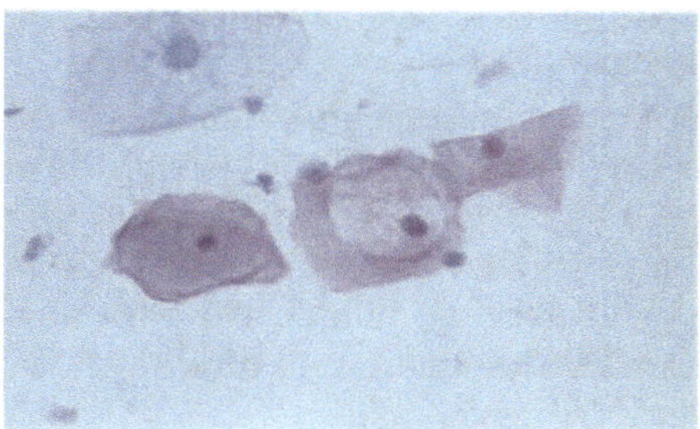

Fig. 3.

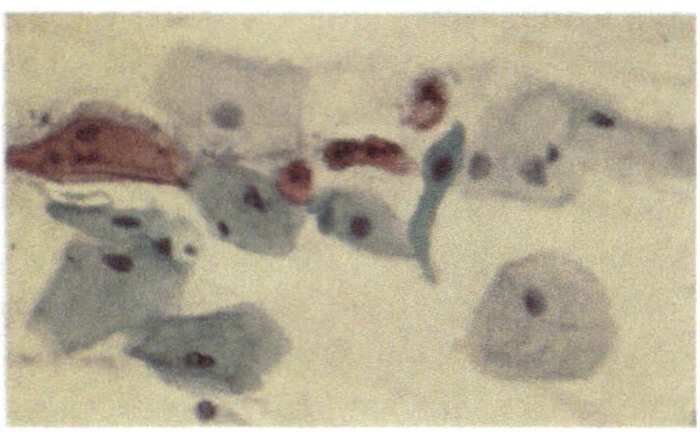

Fig. 4.

Fig. 3. Superficial cells with perinuclear halo, without nuclear atypia. FIS hybridization: HPV 16/18+. Pap staining, ×800

Fig. 4. Group of dyskeratotic cells surrounded by normal superficial cells. FIS hybridization: HPV 16/18+. Pap, ×800

Cytology

Cytologically HPV infections may be easily diagnosed when koilocytes as the classical "marker" cells are present in the cervical smear. This is a rare event, however.

As in colposcopy, correlation studies of cytology and HPV-DNA hybridization revealed a number of minor morphological cell alterations in the HPV-infected cell, which up to now were considered to be caused by an unspecific inflammatory reaction (Wagner et al. 1985).

The following, mostly cytoplasmic changes caused by virus interference in the differentiating squamous epithelial cells have been observed:

1. Perinuclear halo without nuclear atypia (Fig. 3)
2. Dyskeratotic cells (Figs. 4 and 5)
3. Giant superficial cells (Fig. 6)
4. "Cracked" cytoplasm, caused by cytoplasm disintegration (Fig. 7)
5. Double and multinucleation of superficial cells (Fig. 8).

Schneider et al. (1987) showed that the combination of several of these "nonclassical" signs could correctly identify 84% in a group of HPV-positive women, as compared with only 15% when the "classical" signs of koilocytosis and dyskeratosis were applied. The cytopathologist who uses all available criteria might be able to detect HPV-related precursor conditions of cervical cancer more frequently and earlier than has been possible till now.

Histopathology

Like colposcopists and cytologists, histopathologists can also detect early signs of an HPV-infected tissue section which they should incorporate in their diagnosis, even if the histologic feature of condyloma acuminatum is *not* present on the slide. If they accept koilocytosis as a "classical" sign of HPV infection, they may recognize those cell alterations in almost all cases of mild and moderate dysplasia and should correctly label them as "consistent with an HPV infection" (Fig. 9). Most German pathologists have accepted the term "koilocytotic dysplasia" introduced by Stegner in 1981.

But even in severe dysplasia or carcinoma in situ, sometimes small areas of koilocytosis may suggest virus replication in this part of the epithelium, while sensitive hybridization techniques show integrated HPV-DNA in the undifferentiated part of the epithelium (Fig. 10). Furthermore, the observation of very mild cytological alterations like "halo cells" or mild dyskeratosis in a slightly abnormal epithelium warrants careful examination of additional tissue sections for more advanced lesions nearby (Figs. 11, 12).

Conclusions

The introduction of molecular hybridization techniques into clinical pathology has been very helpful in the detection and evaluation of precancerous le-

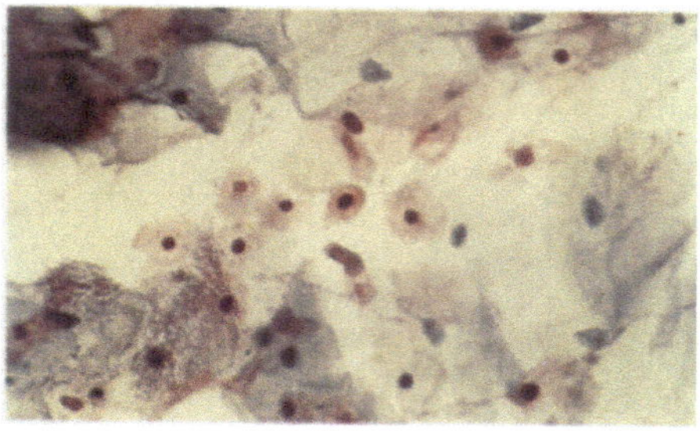

Fig. 5. Small superficial cells (dys-keratocytes) surrounded by normal-sized basophilic superficial cells, partly "clue" cells. FIS hybridization: HPV 16/18+. Pap, ×800

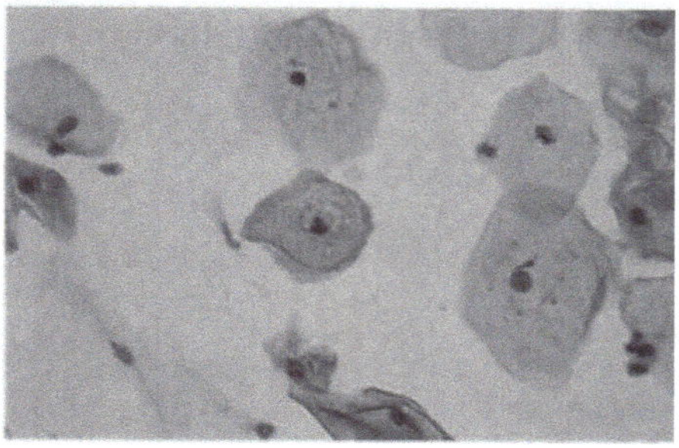

Fig. 6.

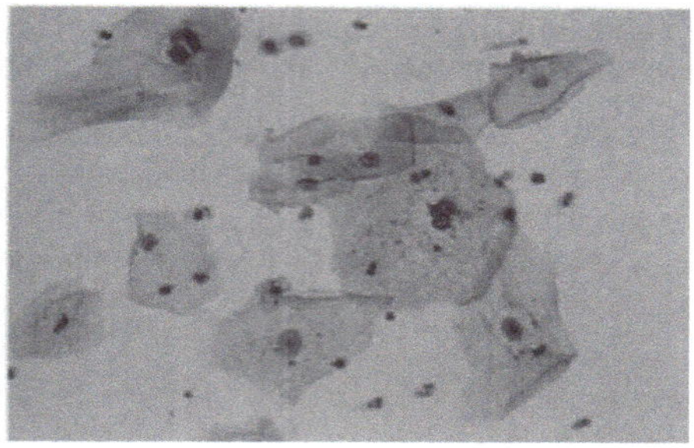

Fig. 7.

Fig. 6. Giant superficial cells, marked anisocytosis. FIS hybridization: HPV 6/11+. Pap, ×500

Fig. 7. "Cracked" cytoplasm in two superficial cells. FIS hybridization: HPV 6/11+. Pap, ×800

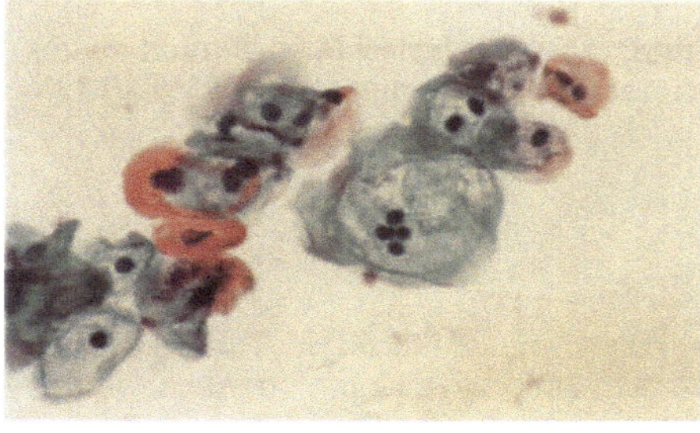

Fig. 8. Double and multinucleated superficial cells, perinuclear halo, anisocytosis. FIS hybridization: HPV 16/18+. Pap, ×500

sions at a very early subclinical stage. The combination of the traditional morphological methods such as colposcopy, cytology and histopathology with in situ hybridization of cytological material and tissue sections has now made possible the more reliable preselection of a high-risk group of women being the most likely candidates for the development of a precursor lesion of genital cancer at any time in their future life.

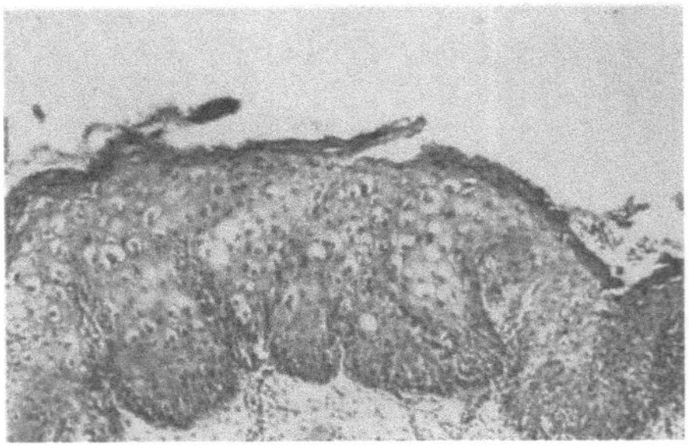

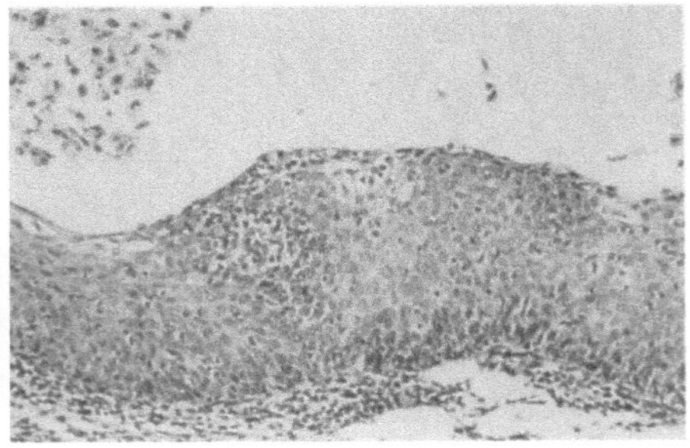

Fig. 9. **Fig. 10.**

Fig. 9. Mild dysplasia with koilocytes in middle and upper layers of the epithelium. FIS hybridization: HPV 16/18+. H&E, ×375

Fig. 10. Severe dysplasia with a small superficial spot of koilocytosis, indicating virus replication in this area. FIS hybridization: HPV 16/18+. H&E, ×375

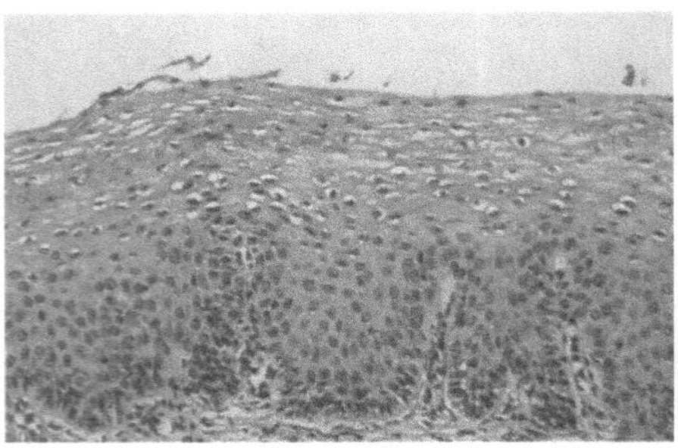

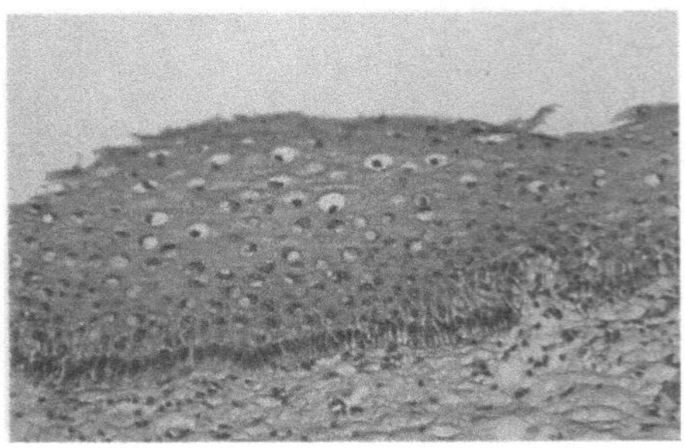

Fig. 11. **Fig. 12.**

Fig. 11. Proliferating epithelium containing halo cells with no nuclear atypia. FIS hybridization: HPV 16/18+. H&E, ×375

Fig. 12. Halo cells in a slightly atypical epithelium. (Severe dysplasia was found in additional tissue-sections of the cervical cone specimen.) FIS hybridization: HPV 16/18+. H&E, ×375

References

De Villiers EM, Wagner D, Schneider A, Wesch H, Miklaw H, Wahrendorf J, Papendick U, zur Hausen H (1987) Human papillomavirus infections in women with and without abnormal cervical cytology. Lancet 2:703–706

Hinselmann H (1926) Zur Kenntnis der praecancerösen Veränderungen der Portio. Zentralbl Gynäkol 1:90

Schneider A, Meinhardt G, de Villiers EM, Gissmann L (1987) Sensitivity of the cytologic diagnosis of cervical condyloma in comparison with HPV-DNA hybridization studies. Diagn Cytopathol 3:250–255

Stegner HE (1981) Zur Klassifikation virusbedingter Dysplasien der Cervix uteri in der zyto-histologischen Routinediagnostik. Gynäkologe 14:252–253

Wagner D, de Villiers EM, Gissmann L (1985) Der Nachweis verschiedener Papillomvirustypen in zytologischen Abstrichen von Präkanzerosen und Karzinomen der Cervix uteri. Geburtshilfe Frauenheilkd 45:226–231

Vulvar Dystrophy and Other Premalignant Lesions of the Vulva – The View of the Gynecologist

H. F. Nauth

Introduction

Two entirely different vulvar entities seem to contribute to carcinogenesis, i.e., intraepithelial HPV infection, on the one hand, and stromal, chiefly subepithelial, dystrophy, mainly occurring in the form of hyperplasia, on the other. Their frequency rate is high and steadily rising [2, 10, 23].

These two conditions are initially benign and have a long latency period before they may progress into a precancerous lesion (atypical condyloma, bowenoid papulosis, carcinoma in situ, dystrophy with atypia, dystrophy with dysplasia, etc.). Finally, they may develop into invasive carcinoma, a process more often than not diagnosed in a late stage [14]. The prognosis of vulvar carcinoma is, in most cases, still unfavorable [24].

It seems paradoxical that invasive carcinoma of the uterine cervix is more easily detected than the more accessible vulvar carcinoma [5]. The relatively long latency period should enable precancerous vulvar lesions to be noticed at least as early as cervical preinvasive lesions, all the more since the frequency pattern is not particularly age-specific (Fig. 1).

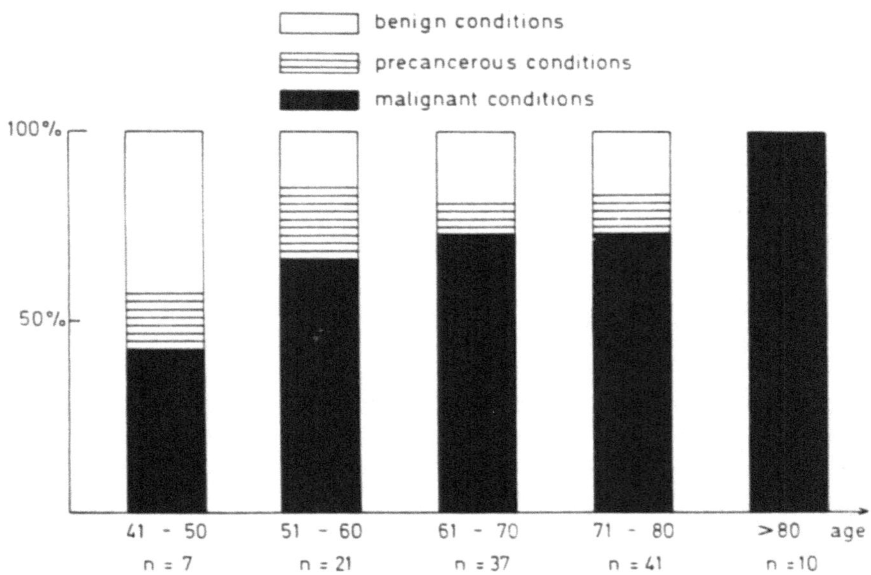

Fig. 1. Age distribution of 120 malignant, precancerous, and benign vulvar conditions

Genital Papillomavirus Infections
Eds.: Gross/Jablonska/Pfister/Stegner

Table 1. Vulvar diagnostic methods

1. Anamnesis
2. Inspection, palpation
3. Photography
4. Magnifying methods (lense, colposcope)
5. Infection tracing
A) Phase contrast cytology
B) Cultures (Mycosis, gonorrhea, HSV)
C) Serology (venereal disease)
D) DNA hybridization (HPV)
6. Collin's test
7. Cytology
8. Biopsy

One of the reasons why vulvar carcinoma is seldom detected in its early stage is that it occurs late in life and older women are more reluctant than younger ones to undergo regular examination. More importantly, however, the clinical picture of precancerous vulvar lesions is so uncharacteristic that physicians often fail to recognize the malignant nature and initially misinterpret them as "kraurosis" until an advanced carcinoma develops. There is a wide range of diagnostic methods for vulvar lesions and each such method, on its own, yields nonspecific results. It is therefore of paramount importance that a strict diagnostic concept be adhered to, correlating all the different methods and results (Table 1). The use of all the relevant diagnostic methods should improve early detection of vulvar cancer [1].

Simple Clinical Examinations

History

With vulvar disease it is often difficult to obtain an accurate history [7]. While historical information usually correlates with the clinical findings in younger patients (injury, malformation, infection, tumor), it often does not in older women (diabetes, chronic incontinence, allergy, psychoneurosis, early menopause, plastic surgery, radiation, special dermatological disorders). Moreover, some older women are averse to repeated and thorough genital examinations and may also have a poor memory for anamnestic details. The major symptoms, such as pruritus, burning paresthesia, and painful intercourse can be but must not be concomitant with visible changes. Conversely, macroscopic lesions do not necessarily give cause for subjective complaints [11, 12]. The patient should be questioned about the approximate duration of her complaint and about the treatment received so far.

Inspection and Palpation

Many vulvar changes can be confirmed by a single macroscopic examination (malformations, injury, cysts, benign tumors, simple atrophy, classic type lichen sclerosis, or condylomata acuminata). The majority of lesions, however, exhibit striking variability and multicentric location [6]. This means that the same basic process may give rise to different macroscopic pictures and that, conversely, the same picture may be caused by various diseases. A correct interpretation is even more difficult, if not impossible, when two or more diseases lead to a mixture of changes (dystrophy, infection, precancer). For this reason, benign lesions, precancerous lesions, and even early invasive cancers are often clinically indistinguishable from one another (Figs. 2–6). Diagnosis at first sight must be avoided in the presence of such "decoy" conditions; instead, a simple anatomic description should be made pending further examinations. The macroscopic description should include: (1) color – white (leukoplakia), red (erythroplakia), other (pigmented, etc.); (2) skin level –

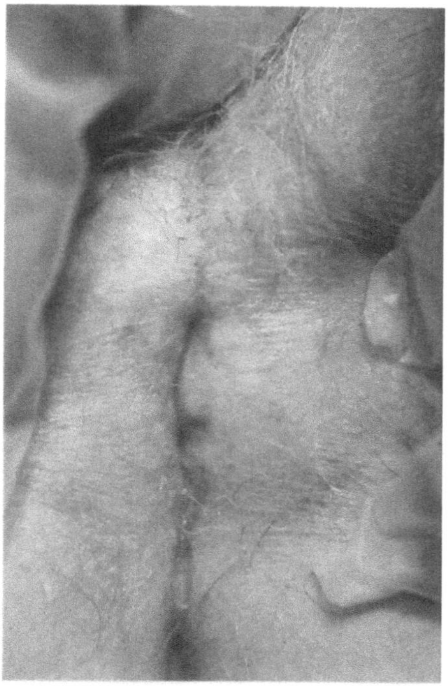

Fig. 2.

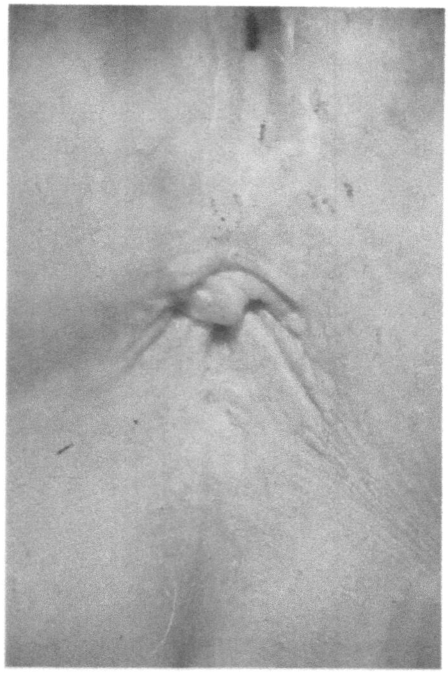

Fig. 3.

Fig. 2. Advanced vulvar lichen sclerosus with atrophy of the labia minora and clitoris, stenosis of the introitus, diffuse leukoplakia, microhematomas, vulnerability and bleeding rhagade at the posterior commissure. This finding could be mistaken for a carcinoma

Fig. 3. Mild dysplasia of the vulva with leukoplakia and rhagades in the perineum region. This finding could be confused with anal excema

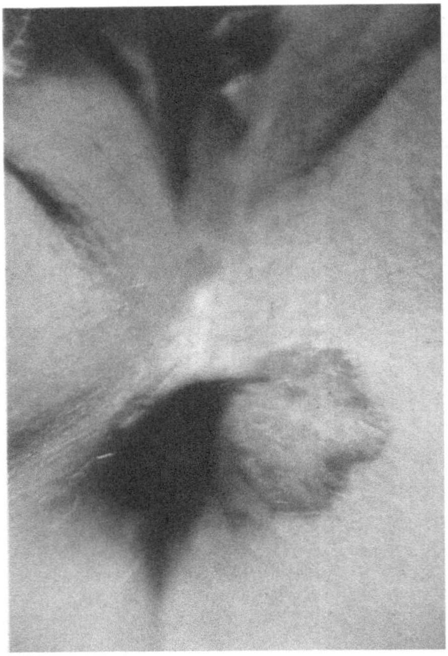

Fig. 4.

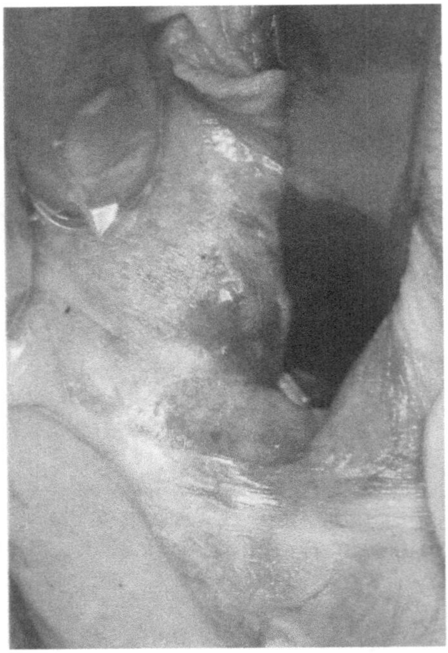

Fig. 5.

Fig. 4. Carcinoma in situ with prominent pigmentation in the perianal region. This could be mistaken for a benign pigmented wart

Fig. 5. Keratinized squamous Ca of the vulva with erythroplakia at skin level and increased vascularization. Could be confused with an atrophic inflammation

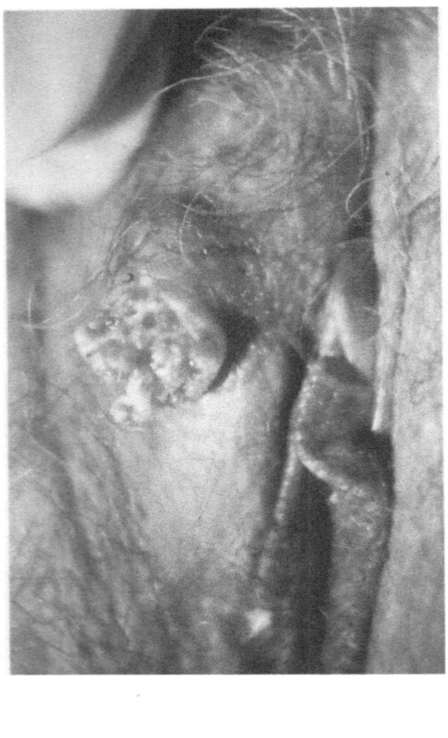

Fig. 6. Benign papilloma of the vulva. Could be mistaken for a carcinoma

raised, tumorous, atrophic, or ulcerated; (3) extension – size (localized or spreading); and (4) location.

Simple Clinical Supplements

Following the Lesion

Vulvar lesions not disappearing within several weeks, either spontaneously or as an effect of therapy, must be closely followed for persistence, progression, or regression. A subjective description of the condition is insufficient. Photographic documentation is the most effective method of following the lesion [1], using close-up lenses, macrotubes, or bellows on a constant graphic scale.

Magnification

Epithelial changes can be detected earlier and recognized with a greater degree of accuracy by using a magnifying lens or colposcope rather than relying on the naked eye [22]. It must be noted that the color of a lesion is not specific to any particular disease. The skin color depends on four parameters (Fig. 7).

The inflamed epidermic stroma (and sometimes precancerous and cancerous stroma) is superficially red. Dystrophic lesions, however, are pale because of diminished vascularization due to sclerosis.

The nonkeratinized portion of the epithelium acts as a first light filter and may swell for various reasons (mechanical irritation, chronic inflammation, hyperplastic dystrophy, condyloma, precancer, cancer), resulting in white coloration. Atrophy is marked by a reduction of this epithelial layer, with ensuing redness.

The keratinized layer acts as a second filter and may be increased or reduced by the same conditions described for the nonkeratinized epithelium. The reaction of the two epithelial layers is, however, not always predictable as it depends on a variety of influencing factors.

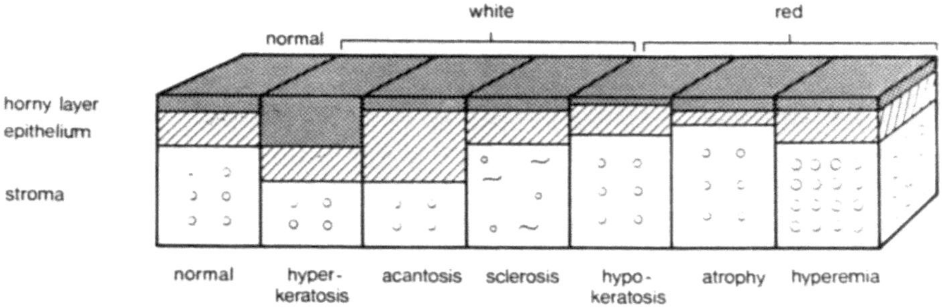

Fig. 7. Seven of 27 possible combinations of the three epidermic layers, entailing variation in surface coloring

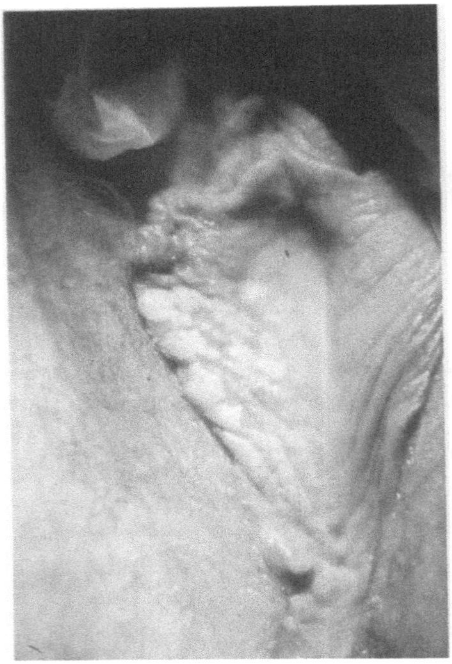

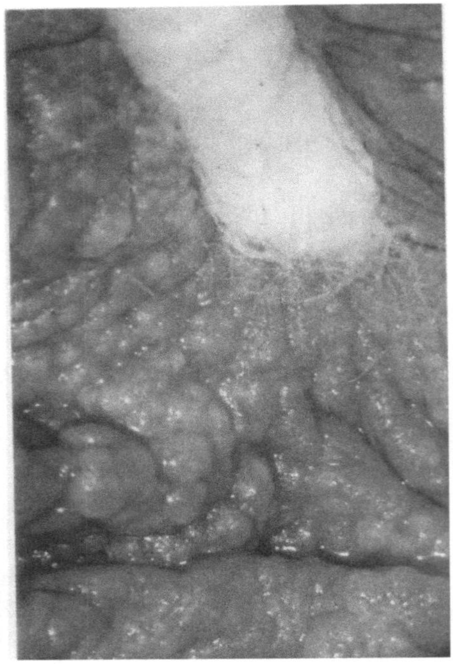

Fig. 8. **Fig. 9.**

Fig. 8. Carcinoma in situ of the vulva with prominent patchy leukoplakia of the right labium minus

Fig. 9. Condylomata acuminata of the vulva. Colposcopic picture after acetic-acid test, showing a so-called aceto-white epithelium

Pigmentation is an additional filter and also influences superficial coloring. Apart from varying pigment distribution, some lesions cause depigmentation (vitiligo) while others produce hyperpigmentation, as is often the case with carcinoma in situ (see Fig. 5). Since one disease can cause various reactions in different skin layers and one part of the skin can react in the same way to a variety of diseases, macroscopic diagnosis on the basis of skin color alone is difficult, if not impossible.

Certain conclusions can, however, be reached with the aid of magnification, provided the optical impression is not limited by the above-mentioned filter effects. Thus, erythroplakic lesions show regular vascularity. An inflammatory reaction can be distinguished from a malignant condition by the vessel branching and caliber reduction. Broad prominent leukoplakic lesions tend to occur more frequently in precancer and cancer than in benign lesions, although this is not a general rule (Fig. 8). The well-known colposcopic features of the cervix (mosaicism, punctation, etc.) are not present in vulvar skin because of the absence of glandular structures. HPV infections which are often associated with intraepithelial neoplasia frequently produce clinically occult focal lesions that are colposcopically detectable only after acetic-acid application (so-called aceto-white epithelium, see Fig. 9).

Simple Tests for Infection

Phase contrast microscopy of material obtained using a premoistened cotton swab is the most important method for detecting bacterial and fungal infections. A mild coccal colonization is physiologic. The amount of bacteria, the detritus, and the quantity of leukocytes characterize acute inflammation. Fungal and bacterial cultures are obligatory supplements to infection studies and sometimes are positive despite negative phase contrast cytology.

Special Supplementary Methods

Collins Test

A 2% toluidine blue dye is applied to the skin and removed with a 3% acetic acid solution after about 3 min. This dye will stain a parakeratotic epidermal portion [1]. Since many skin diseases are accompanied by parakeratosis or dyskeratosis, a positive reaction with this test is not specific; it is, however, of semiquantitative value as the intensity of staining usually correlates with the malignant potential of the lesion (Figs. 10–12). If the Collins test is negative, a

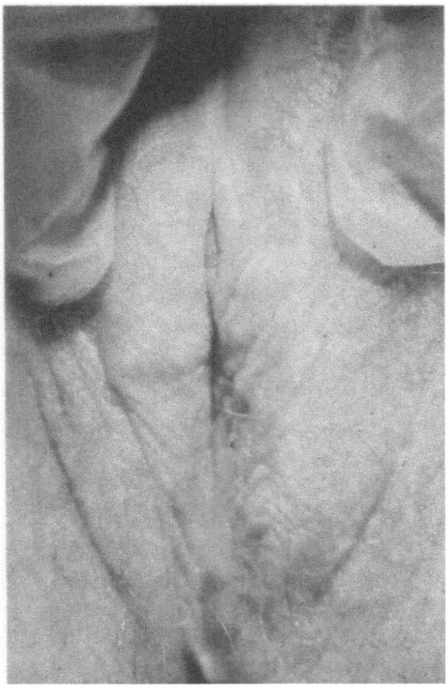

Fig. 10. Squamous cell hyperplasia with an area of mild dysplasia: diffuse atrophy and leukoplakia without focus formation

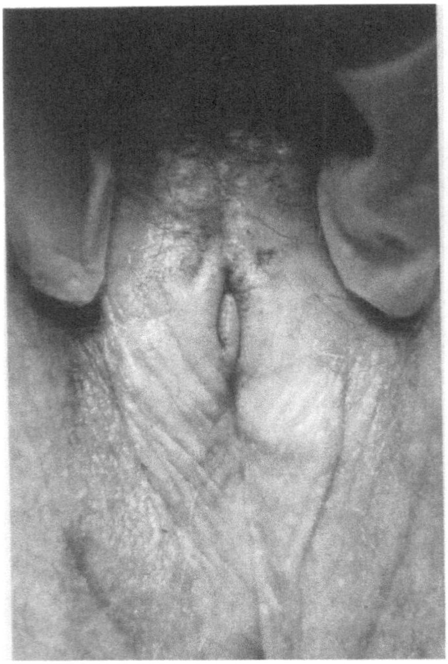

Fig. 11.

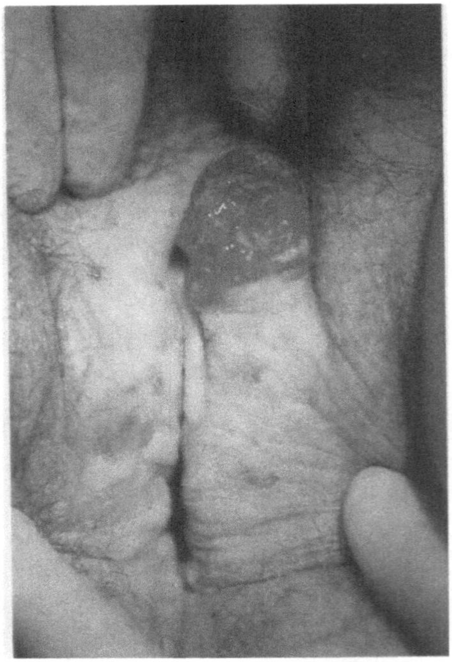

Fig. 12

Fig. 11. Same case as in Fig. 14. Toluidine blue test partially positive. The area of stain accretion marks the biopsy site

Fig. 12. Keratinized squamous carcinoma of the vulva. Diffuse leukoplakia with erythroplakic tumor on the left labium minus

precancerous or cancerous lesion is highly improbable. The method is especially worthwhile for demarcation before local excision, ensuring complete removal of the lesion. The test is also indicated as a follow-up procedure. Another reason for using this test is to localize the most severe parakeratosis with a view to taking smears at a later date. Any epithelial defect (ulceration, wounds, etc.) and the unkeratinized epithelium of the vaginal introitus will stain blue.

Advanced Tests for Infection

Special tests have to be performed if infections by microorganisms not detectable with the customary methods are present. The serologic diagnosis of sexually transmitted diseases relies upon techniques described in the dermatological literature [9]. Identification of human papillomavirus with DNA hybridization and virus typing probably will be important in the future; at present it is only practicable at large medical centers. With the use of this technique, "low-risk" infections may be separated from "high-risk" ones, since virus HPV types 6 and 11 are usually related to benign condylomata acuminata, whereas HPV types 16 and 18 are associated with precancerous and cancerous lesions and seem to initiate the malignant process [2, Schneider, this volume]. Infection with herpes simplex virus is confirmed by cell culture and virus typing with monoclonal antibodies [19].

Cytology

Reservations about vulvar cytology voiced in the past were usually the result of ignorance about the proper sampling technique, uncertainty about the representative cellular material, and the special criteria of malignancy. Today, the smear is rightly considered a valuable technique for vulvar lesions.

Sampling Technique

Smear preparation presents no problem if one follows definite rules. No ointment should be applied locally for at least 2 days before cell sampling. Commercially produced cotton swabs are more suitable than homemade ones because of possible contamination by anucleate squames from the fingers. The cotton swab has to be premoistened with one or two droplets of saline since dry swabs have no, or poor, adhesive properties, whereas overly wet swabs have an undesirable rinsing effect. The vulvar skin should be repeatedly rubbed. Contact with the unkeratinized epithelium of the vaginal introitus must be avoided. Repeated rolling on the slide will improve cellular yield. After fixation, staining according to Papanicolaou is applied for at least 20 min, using orange-G-6.

Representative Cellular Material

Representative cellular material consists mainly of anucleate or "orthokeratotic" cells (Fig. 13a). About 10%–20% of the squames contain a nucleus and therefore are called "parakeratotic" (Fig. 13b). Unkeratinized squamous epithelial cells are found in various quantities; they do not, however, represent native material but have migrated from the vaginal region. These cells should therefore be ignored in the diagnosis of vulvar conditions. This cell type will occur more frequently in premenopausal women, due to greater vaginal discharge, than in older women. Important and easily recognizable distinguishing features exist between these unkeratinized squamous cells and the squames (Fig. 14) [16]. Keratinized vulvar cells are generally smaller than vaginal epithelial cells and are of angular shape. The cytoplasm is densely structured, sometimes containing keratin granules, and occasionally showing small cellular protrusions (pseudopodias). The cytoplasm stains orangeophilic or flavophilic, and sometimes a chromophobic ectoplasm is found. In the case of

a

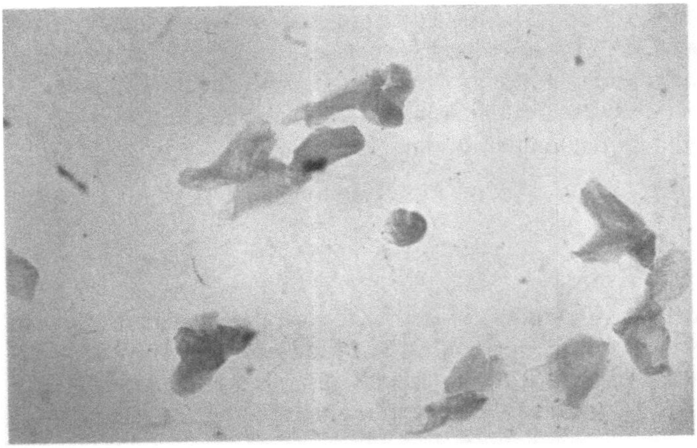

b

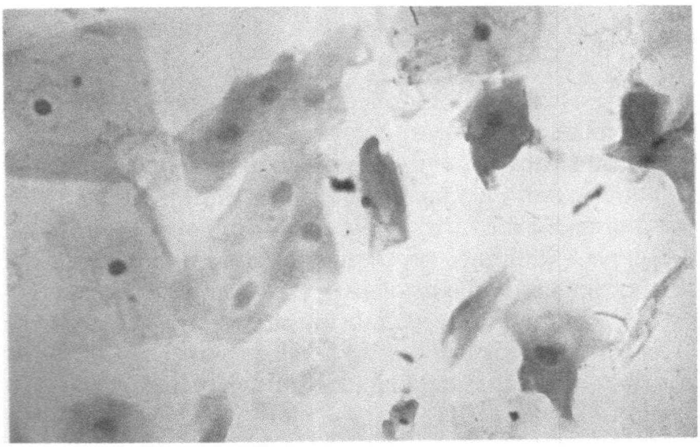

Fig. 13. a Anucleated keratinized squames (orthokeratotic cells; Papanicolaou stain, × 100). **b** Nucleated keratinized cells (parakeratotic cells; Papanicolaou stain, × 100)

nuclear absence, formation of a "nuclear vacuole" is possible. If the nucleus is preserved, signs of severe nuclear degeneration (pyknosis, lysis, or rhexis) are generally present. Vaginal cells from the superficial layer are larger than vulvar cells, they are circular in shape, and have a homogeneous, translucent cytoplasm staining eosinophilic or basophilic cells. Their nucleus can be vital or pyknotic.

Physiologic Reactions

The quantitative relation between orthokeratotic and parakeratotic cells depends on age and menstrual cycle. The percentage of parakeratotic cells is the "parakeratotic index" [15] and is about 7% during the first and last phase and about 17% in the middle phase of the cycle (Fig. 15). A peak occurs at

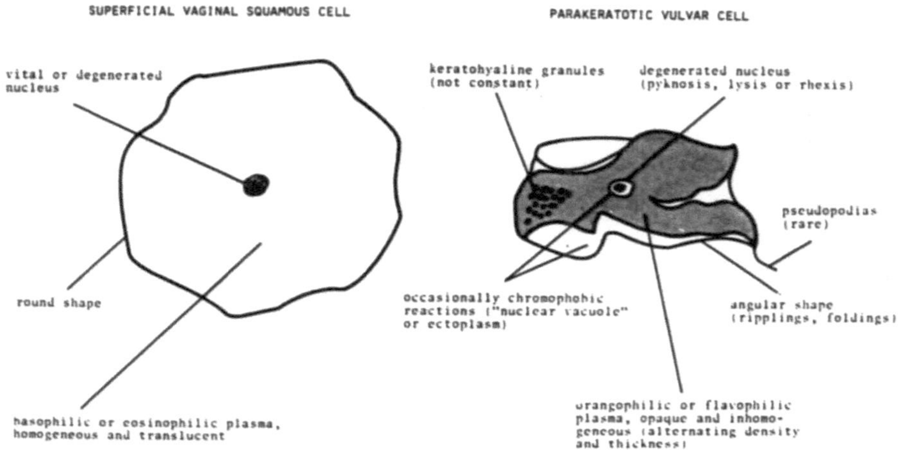

Fig. 14. Comparison of the features of a superficial vaginal squamous cell with those of a parakeratotic vulvar cells

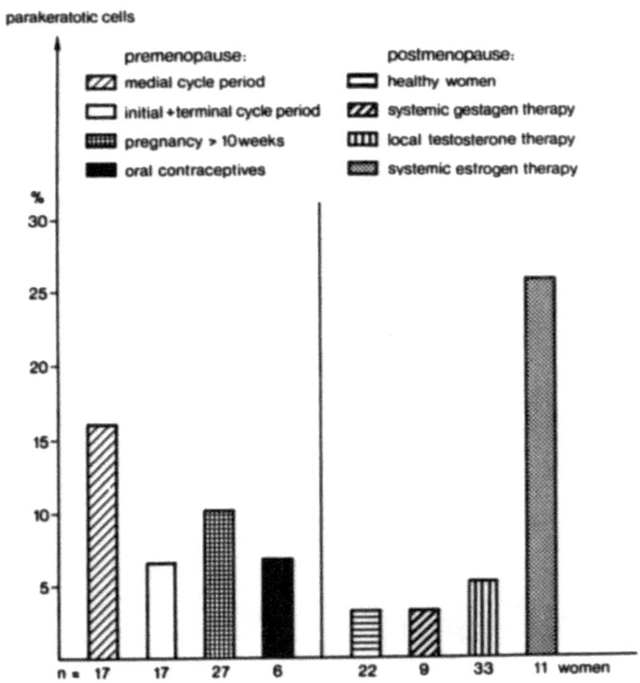

Fig. 15. Parakeratotic index at different ages and in different hormonal situations

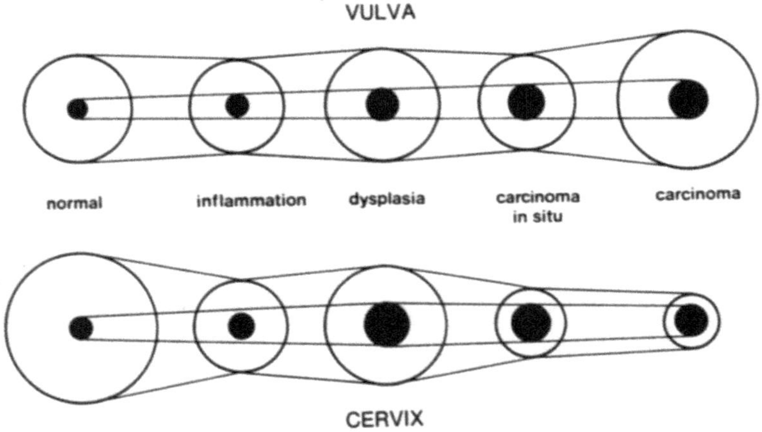

Fig. 16. Standardized schematic presentation of the planimetric data on exfoliated cells from various vulvar lesions in comparison with those from cervical lesions (cervical data from Stanley F. Patten)

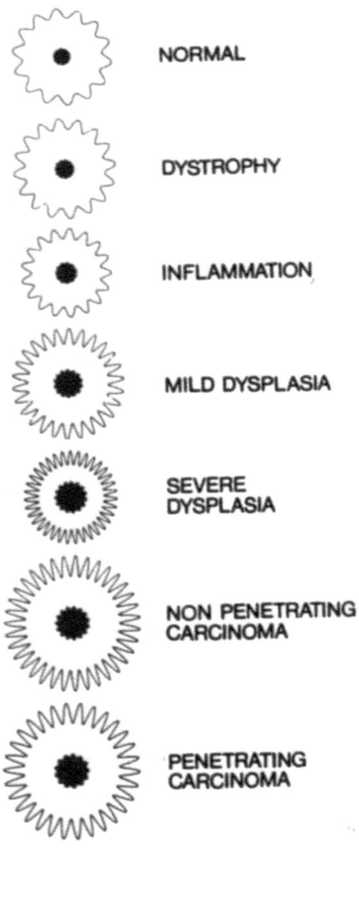

Fig. 17. Standardized schematic representation of the planimetric data (mean values) on nucleated keratinized cells exfoliated from various vulvar lesions

ovulation. Pregnancy and oral contraceptives cause no significant change in the average index. The parakeratotic index is at its lowest in postmenopausal women (about 6%) and highest when estrogen is applied exogenously (25%). Gestagen and androgen do not change the physiologic index but significantly lower the pathologically elevated index.

Criteria for Malignancy

Cytologic examination of vulvar smears has to take into account important morphologic features which completely differ from the well-established criteria for cervical cytology (Fig. 16) [17, 18]. Evaluation of the chromatin pattern is usually impossible because of complete nuclear degeneration. With increasing degree of malignancy, the cell size increases almost in proportion to nuclear size. Thus, the nuclear: cytoplasmic ratio remains relatively constant, with some exception in carcinoma in situ. Increasing cytoplasmic pleomorphism with increasing degree of malignancy is an additional feature of vulvar cytology (Fig. 17). If nuclear enlargement and cytoplasmic pleomorphism are present in a squame, the condition is no longer termed "parakeratotic" but is referred to as "dyskeratotic." Because of these special cytologic criteria for vulvar malignancy, vulvar cytology is more complex than cervical cytology. Not until the anaplastic portion of a vulvar carcinoma penetrates the superficial keratinized layer will the classic malignant features become obvious. Immature tumor cells with clear-cut malignant features will then be observed with increasing frequency.

Cytologic Differentiation of Vulvar Diseases

The parakeratotic index is extremely elevated in all inflammatory lesions and sometimes in dystrophic conditions (above 40%). Inflammation can also lead to some nuclear enlargement; thus, differentiation from dyskeratotic cells may be difficult (Fig. 18a). Uniformity of nuclear enlargement and of cytoplasmic shape suggests a benign process.

Condylomatous lesions also cause a rise in the parakeratotic index as well as occasional nuclear swelling. In contrast to intravaginal condylomas, however, no koilocytes occur. It would appear that the koilocyte, as a virus-specific phenomenon, is confined to the stratum granulosum and "heals" with the keratinizing process. The genetically determined process of keratinization thus seems to be enhanced by the virus, as HPV infection is often concurrent with hyperkeratosis.

Mild precancerous conditions show a proportionate enlargement of both, nucleus and cytoplasm, with some evidence of pleomorphism (Fig. 18b). Severe dysplasia and carcinoma in situ, however, show marked cytoplasmic pleomorphism combined with an increase in the relative nuclear area, similar to that found in mild cervical dysplasia.

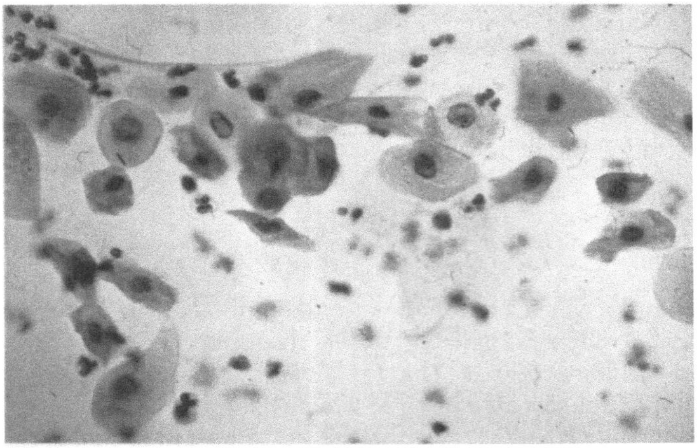

a

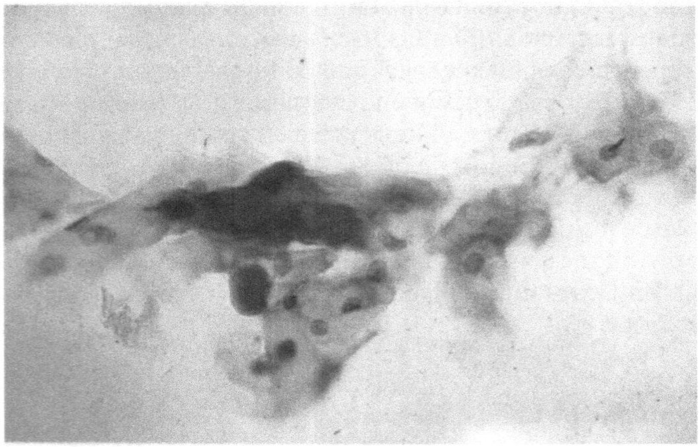

b

Fig. 18. a, b

Fig. 18. a Increased parakeratosis in mycotic vulvar infection, with evidence of some nuclear enlargement (Papanicolaou stain, ×250). **b** Mildly dyskeratotic cells in mild vulvar dysplasia (Papanicolaou stain, ×250). **c** Severely dyskeratotic cells in non-penetrating vulvar cancer (Papanicolaou stain, ×250). **d** Anaplastic tumor cells beside severely dyskeratotic cells in penetrating vulvar cancer (Papanicolaou stain, ×250)

Invasive vulvar cancer frequently has dyskeratotic cells only, without evidence of anaplastic tumor cells. The nuclear:cytoplasmic ratio of these cells usually is not elevated but reduced. The cytoplasmic pleomorphism, however, is extremely high (Fig. 18c). In carcinoma with penetration by the anaplastic tumor portion through the superficial dyskeratotic horny layer, immature tumor cells reach the skin surface and are detectable in the cytologic smear in varying quantities. These immature cells are not easily recognized because of their poor preservation. However, if well preserved, they exhibit all the classic signs of malignancy (circular shape, basophilic and fragile cytoplasm, enlargement of absolute and relative nuclear areas, atypical chromatin pattern, and nucleolus formation; Fig. 18d). Figure 19 shows the different possibilities of

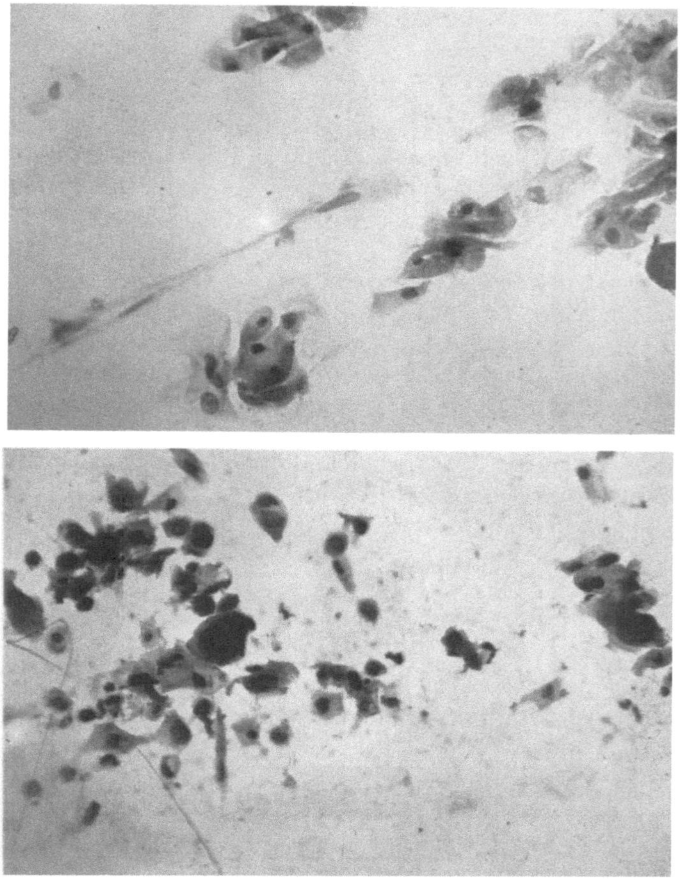

c

d

Fig. 18. c, d

reaction of the stratum corneum both in the histologic section and in the cytologic smear. Vulvar cytology yields comparatively reliable results, considerably more so than endometrial cytology, while it does not quite reach the level of accuracy of cervical cytology.

Biopsy

All vulvar lesions should be biopsied, especially if they tend to take a chronic course [3]. The procedure is very simple, rapidly done, and almost painless; no elaborate equipment, shaving, or wound care are required. If need be, several biopsy specimens may be taken immediately or successively. For skin biopsies the punch biopsy is more suitable than a forceps biopsy which is preferred for the uterine cervix. After desinfection and local anesthesia with the smallest possible needle, a cylinder is punched out with a rotary motion and the tissue

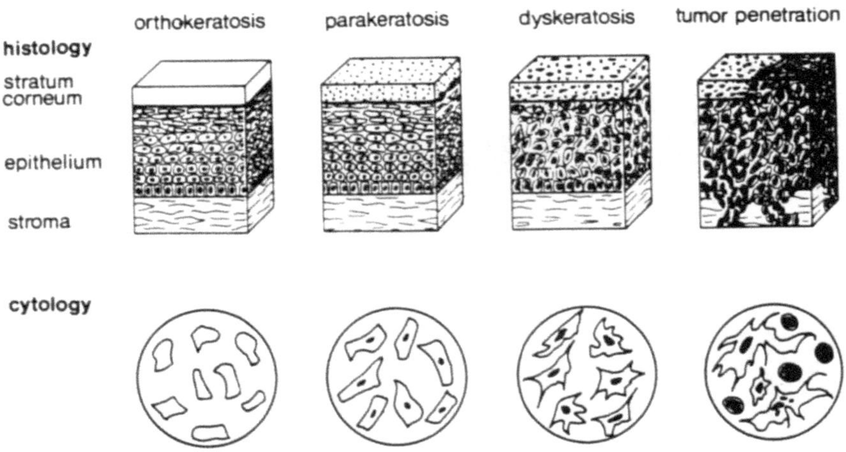

Fig. 19. Schematic representation of the various histologic conditions of the stratum exfoliativum of the vulva and their cytologic counterparts

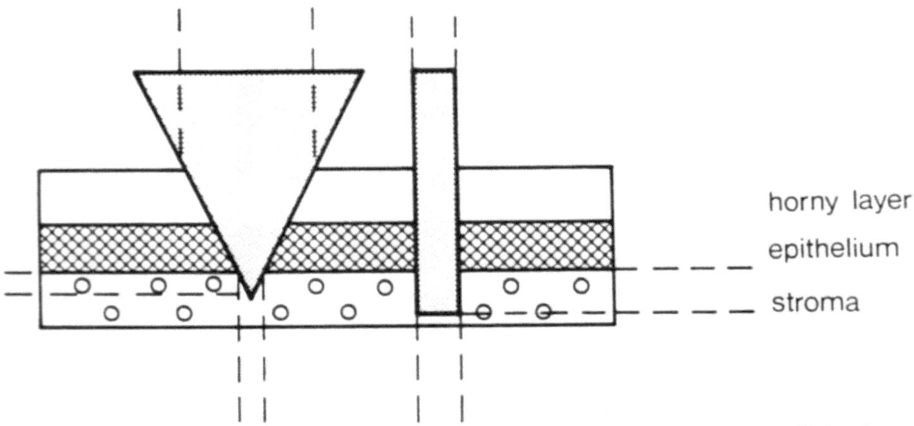

Fig. 20. Forceps and punch biopsy of the vulva: the punch biopsy allows sufficiently deep excision despite a relatively small surface defect

obtained is cut off. The wound is closed with catgut. The tissue cylinder is then placed on a piece of cardboard, stromal side down for orientation, and fixed in formalin. This enables the pathologist to cut at right angles to the epithelial surface rather than in an oblique fashion. With the punch technique the skin defect is always constant despite varying depth; this is not so with the forceps biopsy (Fig. 20). It must be taken into account, however, that in a diffuse lesion, a biopsy may miss the area with the most typical change. Small vulvar lesions are therefore best removed completely by scalpel.

Histopathologic diagnosis should use the standard criteria and nomenclature of the International Society for the Study of Vulvar Disease [8, 25] (Table 2).

Table 2. Terminology of Dystrophies and Preneoplasias of the vulva (as established 1989 by the International Society for the Study of Vulvar Disease, and the International Society of Gynecological Pathologists)

1. Non-neoplastic epithelial disorders A) Lichen sclerosus B) Squamous cell hyperplasia C) Other dermatoses 2. Intraepithelial neoplasia A) Mild dysplasia B) Moderate dysplasia C) Severe dysplasia D) Carcinoma in situ	The following terms were *deleted from the* terminology: 1. Kraurosis vulvae, hyperplastic vulvitis, neurodermatitis, mixed dystrophy, hyperplastic dystrophy 2. Leukoplakia, leukokeratosis, leukoplakic vulvitis 3. Morbus Bowen, erythroplasia of Queyrat, carcinoma simplex

Management

Dystrophy

For dystrophy, topical testosterone is the treatment of choice (2% testosterone propionate in ointment form) [13]. There are but few side effects which only in rare instances necessitate discontinuation of the therapy. Often a lifelong medication, albeit at reduced doses, is required. In premenopausal women and in children testosterone must, however, be used with caution, as its side effects may then be considerable and entail, e.g., clitoris hypertrophy. Because of its teratogenic potential, contraception is essential. For this reason topical treatment with progesterone is nowadays increasingly used in the premenopausal age group, although positive results are so far too few to conclusively prove its value. In the case of squamous cell hyperplasia, topical treatment with hydrocortisone, possibly alternated with testosterone, may yield good results. Local anaesthetics, anti-itching substances, and antihistamines are applied with varying success. Estrogen application is now considered to be of little use, yielding only short-term, if any, results. It has its merit, however, when applied either systemically or topically in postmenopausal women to complement androgen therapy. Recently, good results with etretinate (systemic) and thymus extract (topical) have been reported [4, 20]. Alcohol injection under general anesthesia is chosen as a last-resort treatment when conventional methods fail. It effects local disinnervation. Although it may cause considerable local reactions (edema, necrosis, ulceration), it is still preferable to radical vulvectomy which is considered obsolete and is no longer practiced in dystrophy treatment.

Condyloma

HPV infects the germinal layer of the squamous epithelium. Bacterial or mycotic genital infections enhancing dermal permeability thus favor HPV infection. The first and foremost step in treating a condyloma is therefore to eliminate such genital infections. When dealing with small condylomas, local podophyllin application has proved successful. To avoid local skin reactions this must, however, be done with utmost care. Sizeable condylomas are removed either surgically, electrically, or by laser. Reports about the success of topical or systemic interferon therapy vary [21, Gross, this volume; Saastamoinen, this volume; Tiedemann, this volume; Vance, this volume].

Precancerous Lesion

Today, in marked contrast to past practice, a waiting attitude is adopted in the case of precancerous lesions up to and including carcinoma in situ (CIS). The reason for this is the long latency period (by far exceeding that of cervical intraepithelial neoplasia, CIN) before the eventual manifestation of an invasive carcinoma. Furthermore, there are increasing reports of spontaneous regressions of such lesions. While a circumscribed lesion should be excised completely, multifocal lesions with moderate atypia may be followed for 1–2 years, keeping the patient under strict observation. For high-grade precancerous lesions laser therapy has proved a useful tool. This requires, however, previous exclusion of an invasive process by way of biopsy. Interferon treatment seems to be only moderately successful for severe precancerous lesions. Topical fluorouracil may cause considerable discomfort and the therapeutic effectiveness is equally questionable. A final therapeutic possibility is a limited resection (e.g., skinning vulvectomy, labia resection, partial vulvectomy, possibly combined with plastic surgery).

Summary

Dystrophy and intraepithelial neoplasia of the vulva are difficult to distinguish clinically. It is therefore imperative to use all diagnostic methods available to arrive at a specific pathologic diagnosis. In certain cases it may be necessary to solicit the advice of a dermatologist. On no account is a macroscopic diagnosis alone sufficient, and the misleading term "kraurosis" should be avoided. Therapy depends on the diagnosis and is increasingly individualized, replacing a run-of-the-mill approach. Considerations of cosmeticis and "quality of life" take absolute precedence over immediate and radical removal of a lesion. For this reason a policy of long-term follow-up and conservative treatment should be adopted for dystrophy and precancerous lesions. Topical hormone treatment and laser therapy have now replaced radical surgery and irradiation therapy, while chemotherapy of HPV-induced changes (e.g., interferon treatment) is still in the developmental phase.

References

1. Friedrich EG (1976) Vulvar disease. In: Friedmann EA (ed) Major problems in obstetrics and gynecology, 9th edn. Saunders, Philadelphia
2. Gissmann L (1984) Papilloma viruses and their association with cancer in animals and in man. Cancer Surv 3:161
3. Golstein AJ, Kent DR (1975) All vulvar lesions should be biopsied. Am J Obstet Gynecol 121:173
4. Hagedorn M (1987) Lichen sclerosus et atrophicus: Behandlung mit Thym-Uvocal-Creme. Aktuell Dermatol 13:30–33
5. Hoffman P (1970) Histologie und Prognose des Vulvarkarzinoms. Geburtshilfe Frauenheilkd 30:452
6. Hughes RR (1971) Early diagnosis and management of premalignant lesions and early invasive cancers of vulva. South Med J 64:1490
7. Janovski NA, Douglas CP (1972) Diseases of the vulva. Harper and Row, Hagerstrown
8. Kaufman RH, Woodruff JD (1976) Historical background in developmental stages of the new nomenclature. J Reprod Med 17:133
9. Korting GW (1981) Spezielle Dermatologie, Chaps 44–49. Thieme, Stuttgart (Dermatologie in Praxis und Klinik, vol 4)
10. Meisels A, Morin C (1981) Human papilloma virus and cancer of uterine cervix. Gynecol Oncol 12:111
11. Nauth HF (1982) Vulva-Diagnostik. Fortschr Med 10:396
12. Nauth HF (1982) Vulva-Diagnostik. Fortschr Med 11:478
13. Nauth HF (1982) Zur lokalen Testosterontherapie des Lichen sclerosus. Geburtshilfe Frauenheilkd 42:476–481
14. Nauth HF (1986) Vulvazytologie. Thieme, Stuttgart
15. Nauth HF, Haas M (1985) Cytologic and histologic observations on the sex hormone dependence of the vulva. J Reprod Med 30:667–674
16. Nauth HF, Schilke E (1982) Cytology of the exfoliative layer in normal and diseased vulvar skin: correlation with histology. Acta Cytol 26:269
17. Nauth HF, Neumann GK, Feilen KD (1987) Structural and morphometric analysis of para- and dyskeratotic cells exfoliated in various vulvar lesions: correlation with data of cervical cytology. Anal Quant Cytol 9:243–252
18. Papanicolaou GH, Traut HF (1947) Diagnosis of uterine cancer by vaginal smear. Commonwealth Fund, New York
19. Pereira L, Klassen T, Barringer JR (1980) Type-common and type-specific monoclonal antibody to herpes simplex virus type 1. Infect Immunol 29:724
20. Romppanen U et al. (1986) Orale Behandlung der Dystrophie der Vulva mit einem aromatischen Retinoid, Etretinat. Geburtshilfe Frauenheilkd 46:242–247
21. Schneider A et al. (1987) Interferon treatment of human genital papillomavirus infection: importance of viral type. Int J Cancer 40:610–614
22. Seidl S (1974) Praktische Karzinomfrühdiagnostik in der Gynäkologie. Thieme, Stuttgart
23. Sexually Transmitted Diseases (1980) Extract from the Annual Report of the Chief Medical Office of the Department of Health and Social Security for the Year 1978. Br J Vener Dis 56:178
24. Volk M, Schmidt-Matthiesen H (1983) Therapieergebnisse beim Vulvakarzinom. Arbeitsgem Gynaekol Onkol 4:13
25. Wilkinson EJ (1989) New nomenclature in vulvar disease. The Calposcopist 21:1–3

Vulvar Dystrophies and Lichen Sclerosus – the View of the Dermatologist

M. Hagedorn

Introduction

Vulvar dystrophies comprise a heterogeneous group of diseases characterized by pruritus, chronicity, therapeutic resistance, and unknown etiology (Hagedorn 1988). A further problem is caused by the fact that both the gynecologist and dermatologist are involved in vulvar diseases, each using different nomenclatures. Therefore, a multitude of terms exist, which consequently results in different opinions about the prognosis. It was the merit of the International Society for the Study of Vulvar Diseases (ISSVD) to introduce a classification of vulvar dystrophies in 1976, which was based on histological criteria (Friedrich 1976). It is quite interesting that this nomenclature is mentioned in the newer German textbooks of gynecology and pathology, although only 50% of pathological-anatomical diagnoses are made in accordance with it (Hagedorn 1986). Dermatologists in the German-speaking regions do not use this classification nor do they see any necessity for it. In a recently published review thus was discussed, resulting in wider usage so as to clarify the practicability of such a classification (Hagedorn 1988).

The ISSVD Classification

In the ISSVD classification of 1976 hyperplastic dystrophy (in 1983 the term was changed to hypertrophic dystrophy) without and with cellular atypia is differentiated from classical lichen sclerosus (et atrophicus is omitted because initially there is no atrophy) and from mixed dystrophy with and without cellular atypia. Mild, moderate, and severe atypia is now classified as vulvar intraepithelial neoplasia (VIN) I, II and III.

Hypertrophic dystrophy is macroscopically characterized by the triad epithelial thickening, hyperkeratosis, and inflammation. In this group the well-known lichen simplex is subsumed, which is not acceptable to dermatologists. Lichen simplex (lichen Vidal) is a clear-cut dermatosis caused by traumatic irritation which belongs to a separate group of chronic vulvar diseases. Sanchez and Mihm (1982) add to lichen simplex, psoriasis, lichen planus, and eczematous dermatitis.

Hypertrophic dystrophy used to be termed leukoplakia, leukokeratosis, or kraurosis. Histologically these lesions show ortho- and/or parakeratosis, acanthosis, papillomatosis, and a cellular infiltrate consisting of lymphocytes

Genital Papillomavirus Infections
Eds.: Gross/Jablonska/Pfister/Stegner

and macrophages. Hypertrophic dystrophy with atypia and its grading can only be diagnosed on the basis of histological evaluation of biopsy material.

Lichen sclerosus clinically shows a tight atrophy which can lead to complete destruction of the architecture of the vulva, sometimes combined with superficial hemorrhagias. Macroscopically there is an immense variation in the color and surface of the skin. Histologically lichen sclerosus can be diagnosed easily by the homogeneous collagen tissue in the papillary dermis, atrophy of the epidermis, vacuolar alteration of the basal cell layer and the surrounding lympho-histiocytic infiltrate. Hyperkeratosis of follicular openings are not as pronounced as in extravulvar sites. Atypias are absent.

The third group is mixed dystrophy in which lichen sclerosus appears with foci of epithelial hyperplasia both macroscopically and microscopically. Exactly as in hypertrophic dystrophy there is a subdivision into lesions with and without atypias.

Diagnostic Procedure

The first step in diagnosis requires a smear of the vulva to determine mycotic or bacterial infections. Second a biopsy is necessary, because as mentioned above, the proper diagnosis of vulvar dystrophy and its grading is only possible on the basis of histology. The site of biopsy is of great importance in order to discover epidermal atypias. We use 5% acetic acid for 2–5 min, resulting in a milky discoloration of the suspicious areas. Sometimes several biopsies are necessary to establish the exact diagnosis.

General Considerations of Vulvar Dystrophies

Some 17%–19% of diseases affecting the vulva are dystrophies (Friedrich et al. 1979; Hagedorn 1988). A differentiation of the different types of dystrophy

Table 1. Frequencies of vulvar dystrophies (in %)

	Kaufman et al. (1974) $n=127$		Friedrich et al. (1979) $n=232$		Hagedorn and Thomas (1987) $n=90$		Hagedorn (1989) $n=60$	
I Hypertrophic vulval dystrophy	48	90 n a 10 w a	51	90 n a 10 w a	61	38 n a 62 w a	22	62 n a 38 w a
II Lichen sclerosus	38		38		15		70	
III mixed dystrophies	14	83 n a 17 w a	11	90 n a 10 w a	24	45 n a 55 w a	8	80 n a 20 w a

w a, with atypia; n a, no atypia

Table 2. Age distribution of vulvar dystrophies (n = 276)[a]

	Hypertrophic dystrophy		Lichen sclerosus	Mixed dystrophy	
	na	wa		na	wa
0–10	1	–	8	–	–
11–20	1	–	1	–	–
21–30	14	3	9	1	–
31–40	16	5	10	2	–
41–50	20	3	12	1	–
51–60	15	7	32	7	2
61–70	8	15	19	7	4
71–80	5	10	7	8	6
81–90	4	2	2	1	4
Over 90	–	–	2	–	–
Total	84	45	102	27	16

na, no atypia; wa, with atypia
[a] Kaufmann et al. (1974) $n = 127$; Hagedorn and Thomas (1987) $n = 89$; Hagedorn (1989) $n = 60$

is shown in Table 1. The differences might be due to inhomogeneous groups of patients, classified differently by gynecologists and dermatologists.

Table 2 shows the differences in the age distribution mainly between hypertrophic dystrophy and lichen sclerosus. Hypertrophic dystrophy patients are on average between 20 and 60 years of age, whereas lichen sclerosus patients have two age peaks, one in the first decade and the other between 40 and 70. Dystrophies with atypias in general appear in menopausal or postmenopausal women.

Therapy

One of the main characteristics of all vulvar dystrophies is the therapeutic resistance. For vulvar dystrophies without atypia different possibilities exist (Table 3). The first medications we use are thymus extracts containing oint-

Table 3. Therapy of vulvar dystrophies with no atypia

Hypertrophic dystrophy	Topical corticosteroids; subfocal corticosteroid injections
Lichen sclerosus	Topical testosterone ointment
Mixed dystrophy	Thymus extracts containing ointment; CO_2 laser; cryosurgery

Tabelle 4. Vulvar dystrophies: therapy with Thymuval ($n=40$)

Cases	Response	Lichen sclerosus	Hypertrophic dystrophy	Mixed dystrophies	Length of treatment (in months)
4	Complete	4	–	–	7–18
12	Marked	12	–	–	1–24
10	Partial	6	1	3	2–20
14	No	8	6	–	Up to 6

ments, which have been shown to be successful (Hagedorn 1987). The results are demonstrated in Table 4, showing effectiveness (complete and marked response) in 16 of 40 cases. Afterwards testosterone ointment is applied to menopausal and postmenopausal women, because virilization is probable in the premenopausal years. Corticosteroids administered by subfocal injections provide is some relief. Cryosurgery and CO_2-laser therapy seem also tabe of value.

Vulvar dystrophies with atypias should be excised or treated using cryosurgery or CO_2-laser. Follow-up of these patients is necessary; we observe these patients every 6 months.

References

Friedrich EG Jr (1976) New nomenclature for vulvar disease. Report of the committee on terminology. Obstet Gynecol 47:122–124

Friedrich EG, Burch N, Bahr JP (1979) The vulvar clinic: an eight-year appraisal. Am J Obstet Gynecol 135:1036–1040

Hagedorn M (1986) Die Vulvadystrophien. Med Welt 37:1139–1141

Hagedorn M (1987) Lichen sclerosus et atrophicus: Behandlung mit Thym-Uvocal-Ceme. Aktuel Dermatol 13:30–33

Hagedorn M (1988) Vulvadystrophien. Zentralbl Haut 155:247–255

Hagedorn M (1989) Vulval dystrophies: aspect of clinic and therapy. 10th Int Congress of JSSVD, Washington, October 22–26, 1989

Hagedorn M, Thomas C (1987) Vulvadystrophien. Zentralbl Haut 154:8

Kaufman RH, Gardner HL, Brown D Jr, Beyth Y (1974) Vulvar dystrophies: an evaluation. Am J Obstet Gynecol 120:363–367

Sanchez NP, Mihm MC (1982) Reactive and neoplastic epithelial alterations of the vulva. A classification of the vulvar dystrophies from the dermatologist's view point. I. Am Acad Dermatol 6:378–388

HPV Infection of the External Genitals: Clinical Aspects and Therapy in Dermatovenereology

G. von Krogh

Epidemiology: the "Iceberg Dilemma"

The term "genitoanal papillomavirus infection" (GPVI) refers to lesions caused by HPV types exhibiting a tropism for genitoanal epithelium. Data on the prevalence and incidence among unselected populations are scarce. Clinically evident lesions have traditionally been called condylomata acuminata, venereal warts or genitoanal warts. However, overt wart disease represents merely "the tip of the iceberg," while subclinical lesions predominate and are at least ten times as common (Chuang 1987; de Villiers et al. 1987; Grussendorf-Conen et al. 1987; Syrjänen 1988). This presents a dilemma to the clinician, who faces several intriguing questions, such as when and how to investigate for – or to treat – subclinical disease. In spite of quickly accumulating knowledge on HPV infections during the past few years, very little is yet known about their natural history and, in particular, about the biological significance of subclinical carriage of various HPV types with genitoanal tropism. For the sake of priority, therefore, the clinical approach must be rather pragmatic. The current survey attempts to focus on the role of the dermatovenereologist in the practical management of GPVI lesions.

Overt warts are highly contagious through sexual transmission. About two thirds of patients are 15–30 years of age. The median age is 20–24 years, when the prevalence is at least 300 cases per 100000 population (Oriel 1971a; Chuang 1987).

A steady increase in the prevalence has been reported for genitoanal warts (Table 1). In England the incidence more than doubled during the period 1971–1982 (Department of Health and Social Security 1985). In the USA the incidence tripled during a 10-year period between the late 1960s and the 1970s (Chuang et al. 1984). A significant increase has recently also been reported

Table 1. Yearly incidence of venereal warts per 100000 population (Department of Health and Social Security 1985; Chuang et al. 1984)

	England		USA	
	1971	1972	1965–1969	1975–1978
Men	40	91	30	83
Women	20	53	43	145
Both sexes	30	71	35	106

Genital Papillomavirus Infections
Eds.: Gross/Jablonska/Pfister/Stegner

from Scandinavia (Lassus et al. 1988). The present incidence of clinically overt genital warts in the West is probably about 100 cases per 100000 population.

Nonsexual transmission via objects may possibly occur. HPV-DNA has been identified on medical instruments used for patients with GPVI (McCance et al. 1986), as well as on patients' underwear (Bergeron et al. 1988). The possibility of vertical or nonsexual horizontal transmission must be kept in mind in particular when the question of potential sexual abuse is brought up around children presenting with genitoanal warts.

There are five approaches to the diagnosis of GPVI lesions. They may be evaluated on the basis of their gross appearance, by means of colposcopic magnification, using the acetic acid test, by histology and/or cytology, and virologically by HPV capsid antigen detection or HPV-DNA hybridization assays.

Clinical Significance

Lesions are multifocal, multicentric and multiform; individual cases are never identical. They may be of the classical *"acuminate"* papilliferous, cauliflower-like type (condylomata acuminata), have a more *papular* rounded surface or be rather *flat,* yet visible to the naked eye. Some lesions are completely *sub-clinical* and may coexist with overt warts. Individual patients are frequently afflicted with more than one type simultaneously.

Clinically, overt warts are often of great psychological and cosmetic significance; patients experience them as distasteful and disfiguring, representing a major hindrance to sexual performance. They may bleed when submitted to traumas such as coitus. They may also itch, or be associated with a burning sensation or with dyspareunia.

The most important medical aspect is the potential coexistence in female patients of cervical intraepithelial neoplasia (CIN). Worldwide, cervical cancer is the greatest cancer killer of women under the age of 40 years, and its frequency in young women appears to be increasing (Reid and Campion 1988). Although the incidence of vulvar, penile and anal cancer is much less common, the potential of long-term malignant transformation in these areas must also be kept in mind. The incidences of cervical and outer genital cancer covary, suggesting a field effect by a common causative influence; "high-risk" HPV types such as HPV 16 are considered the most likely pathogenetic candidates (Campion et al. 1985, 1986; Reid and Campion 1988).

Physicians working in the field of sexually transmitted diseases (STDs) must be prepared to focus, above all, on the uterine cervix, which represents a locus minoris with respect to malignant transformation. Accordingly, close educational co-operation needs to be developed with gynaecologists, in particular those possessing knowledge in the use of colposcopy. An optimal program for the examination of women with suspected GPVI entails, in addition to a careful inspection of the vulva, Papanicolaou (Pap) smear samplings and/or a colposcopic evaluation of the cervix subsequent to use of the acetic acid test (see Wagner, this volume).

Overt Wart Disease

Acuminate Warts

The highly vascularized pinkish-red to grayish-white "condylomata acuminata" exhibit a pathognomonic appearance. They afflict predominantly areas submitted to trauma during intercourse (Tables 2, 3). In uncircumcised men the frenulum and the inner aspect of the foreskin are most often afflicted (Fig. 1); in circumcised men the shaft of the penis is often involved (Oriel 1971 a; Chuang et al. 1984). In women warts are most common on the posterior fourchette of the vulva (Fig. 2), in the vestibulum, in the introitus and in the vagina (Fig. 3).

Up to one quarter of men but only a few percent of women have concomitant warts in the urinary meatus (Oriel 1971 a; von Krogh 1981 a; Chuang

Table 2. Site distribution of genitoanal warts in men

Site	Percentage of patients affected	
	Oriel 1971 a ($n=191$) Mostly uncircumcised men	Chuang et al. 1984 ($n=246$) Mostly circumcised men
Glans, frenum, corona	52	10
Prepuce	33	8
Urethra	23	10
Shaft of penis	18	51
Scrotum	2	1
Perineum	0	3
Anus, perianal	8	34

Table 3. Site distribution of genitoanal warts in women

Site	Percentage of patients affected	
	Oriel 1971 a ($n=141$)	Chuang et al. 1984 ($n=500$)
Urethra	8	4
Labium majora	31	66
Labium minora, clitoris	32	
Introitus	73	37
Vagina	15	
Cervix	6	8
Perineum	23	29
Anus	18	23

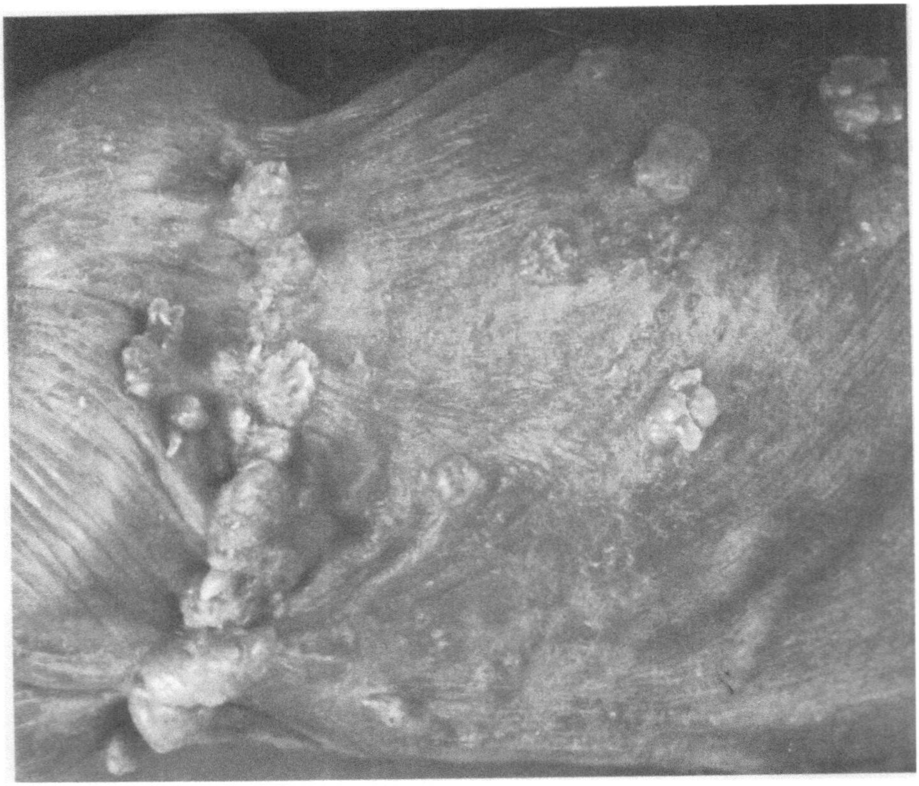

Fig. 1. Typical acuminate warts (condylomata acuminata) in the preputial cavity of an uncircumcised man, with fore-skin retracted

et al. 1984). The proximal border of meatus warts can almost always be delineated by simple digital eversion; the use of a small nose speculum or an otoscopic tube is very helpful. As for other lesions, the magnified illumination of the colposcope improves the diagnostic accuracy. If the proximal border cannot be inspected, investigation by a urologist is indicated. However, this is rarely necessary.

Acuminate warts sometimes afflict intertriginous areas such as the groin. Concurrent warts in the perineum and in the perianal area are common in both sexes (Tables 2, 3) and also occur in strictly heterosexual men (Goorney et al. 1987).

Whenever receptive anal intercourse has been practised, anoscopy should be performed (Oriel 1971 b). About 70% of homosexual men with external anal warts also have internal warts (Carr and William 1977; Sohn and Robilotti 1977; Samenius 1983). Warts are very rare proximal to the "dentate line" (Braun and Raguse 1987) representing the transitional zone between stratified squamous epithelium of the anal canal and stratified cuboidal epithelium of the rectum. This area is encountered about 2 cm above the anus.

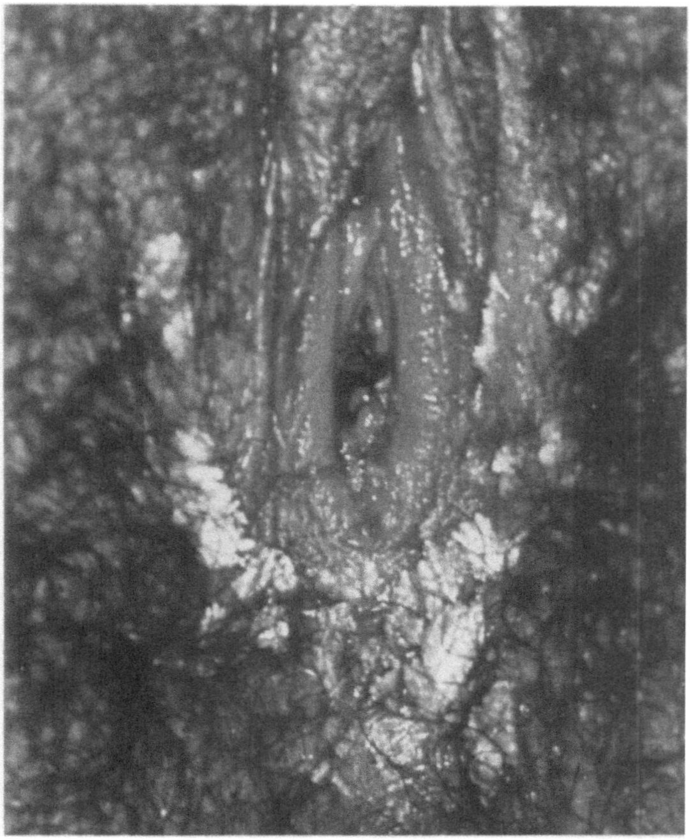

Fig. 2. Disfiguring hyperkeratotic acuminate warts in the posterior fourchette of the vulva and in the perineal area

Papular Warts

On drier genital areas condylomas tend to become more rounded or papular, as on the outer aspect of the foreskin and on the penile shaft (Fig. 4). Papular warts may also be encountered in moist areas. Unless they are covered by a pronounced hyperkeratosis, the capillaries are usually well visible through magnification. The color may be similar to that of acuminate warts but is often more brownish when the warts afflict pigmented sites of the outer genitals such as the penile shaft and the labium majora. However, when conspicuously pigmented warts are encountered, the potential existence of "bowenoid papulosis" must be ruled out (Lloyd 1970; Wade et al. 1978; Ikenberg et al. 1983; Kimura et al. 1987).

Fig. 3. High-power magnification of protuberant acuminate warts on the vaginal wall. Typical capillaries are clearly visible through the hyperplastic epithelium of these highly vascularized tumours (Zeiss Photocolposcopy 12.5 × 20)

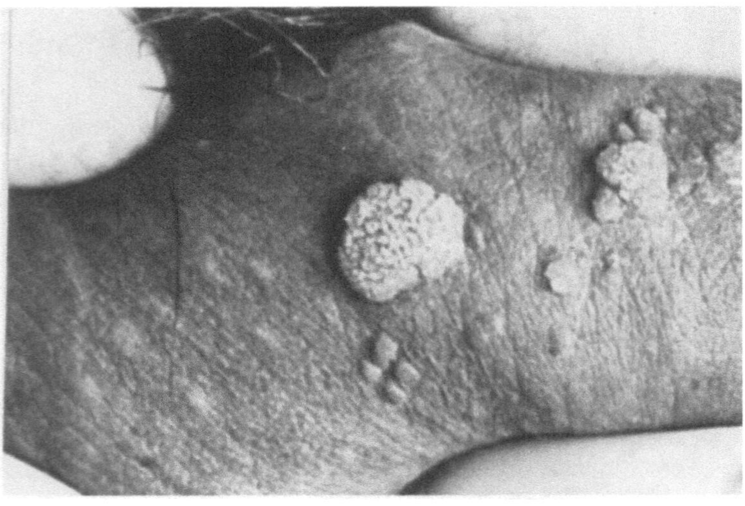

Fig. 4. Coexisting acuminate and papular warts on the outer aspect of the foreskin/penile shaft

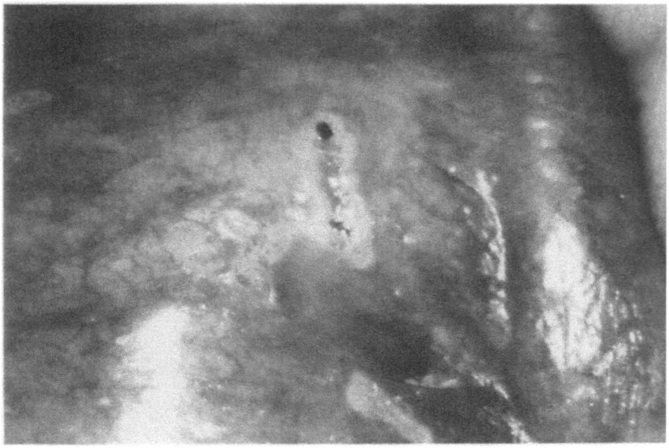

Fig. 5. Subclinical aceto-white lesions afflicting the inner aspect of the foreskin, appearing as well-demarcated epithelial undulations

Flat/Subclinical Lesions

Flat lesions correlate to verruca plana of the skin, while subclinical lesions do not correlate to the original concept of the term "wart." A differentiation between these types is of mostly academic interest, as both are difficult to appreciate by gross examination. Such lesions were first described for the transformation zone (TZ) of the cervix (Meisels et al. 1977; Reid et al. 1980); on magnification they appear as focal areas of epithelial undulation that are better visualized subsequent to the application of 3%–5% acetic acid (Figs. 5, 6). Identical lesions appear to be very common on the outer genitals as well. Most of them are invisible prior to acetic acid application. Others reveal their presence as a colour deviation of the epithelium, varying from pinkish red to brownish red or with a shade of grayish white.

Coexistent Intraepithelial Neoplasia

Benign lesions frequently coexist with epithelial atypia of varying severity (Campion et al. 1985, 1986; Gross et al. 1985; Nash et al. 1986; Syrjänen 1987; Syrjänen 1989). The term "intraepithelial neoplasia" is currently accepted for such changes, and these may be classified according to degree of severity into grades I, II and III. Such dysplastic changes are termed CIN on the cervix, VAIN in the vagina, VIN in the vulva, PEIN in the perineum, PIN on the penis and AIN in the anal area. In the majority of cases, severe dysplasia is associated with the presence of "high-risk" HPV types, most commonly HPV 16. Benign GPVI lesions and varying degrees of intraepithelial neoplasia appear to be part of a morphological continuum (see Syrjänen, this volume).

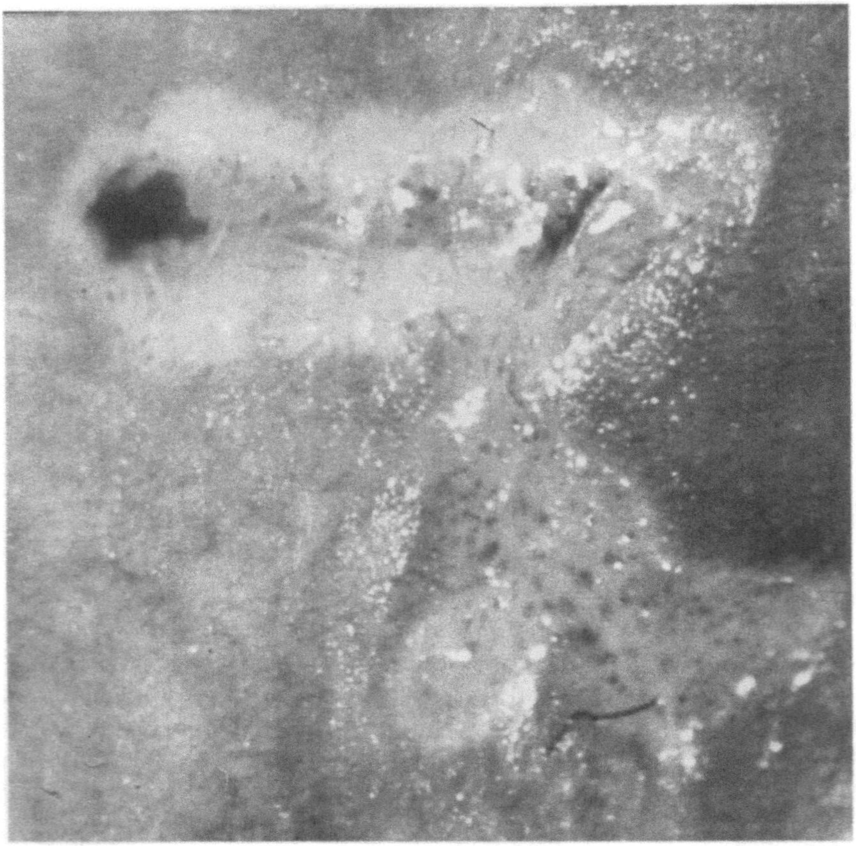

Fig. 6. Same patient as in Fig. 5, seen with high-power colposcopic magnification (penoscopy) subsequent to application of acetic acid. The most useful colposcopic markers are the sharp demarcation, the slightly elevated border and the punctate vascular pattern (*bottom*). The capillary pattern may be temporarily camouflaged immediately following acetic acid application. Sometimes a central "groove" or a fissure is seen (*top*) (Zeiss Photocolposcopy 12.5 × 20)

Dysplastic areas sometimes merge with areas of benign hyperplasia, or they exist directly adjacent to areas of benign hyperplasia. Koilocytosis, a pathognomonic sign of HPV infection, may be present but is frequently absent in areas of severe dysplasia. In the absence of koilocytosis an implicit association with HPV infection cannot be made without the help of virological methods.

This dilemma is illustrated in the case of bowenoid papulosis, the lesions being invariably associated with "high-risk" HPV types although koilocytes are regularly not found. In typical cases bowenoid papulosis is characterized by the occurrence on the outer genitals of 2- to 10-mm maculopapular lesions that are either erythematous, reddish-violaceous, brownish or leukoplakia-like and that reveal a severe intraepithelial neoplasia histopathologically

(Gross et al. 1985 this volume; Obalek et al. 1986; Campion and Singer 1987). A parallel dilemma is the finding that most biopsies from the outer genitals that prove to contain areas of severe epithelial dysplasia tend to remain completely subclinial rather than showing up as clinically conspicuous bowenoid papulosis (G. von Krogh, M.-A. Hedblad, A. Wikström, unpublished work).

The Acetic Acid Test and Colposcopic Magnification

The acetic acid test is performed using a cotton wool swab or gauze moistened with 3%–5% aqueous acetic acid that is applied to the area for 3–5 min. On the cervix 1 min must be allowed to pass prior to evaluation, in order to minimize false-positive results.

When combining the test procedure with magnified illumination of the colposcope, flat/subclinical lesions with highly variable features may be identified. A well-demarcated, slightly elevated border is typical. An endophytic central depression is common and may be associated with epithelial fissuring. A reliable criterion is the presence of a typical punctate, "pin-pointed" vascular pattern that is usually best visualized in the central part of lesions (Fig. 6). However, many lesions exhibit a smooth, velvety surface with a hyperkeratosis that may conceal the vascular pattern. Also, the influence of the acetic acid may temporarily camouflage the capillaries. Some "flat" lesions will show up in the colposcope as having an undulating exterior, sometimes with "microspikes" interspersed.

The potential use of a colposcopic index system for the differentiation between benign and premalignant lesions has been proposed (Reid et al. 1984), aiming at a proper selection for biopsy of representative areas exhibiting intraepithelial neoplasia. However, although certain colposcopic hallmarks of severe dysplasia have been suggested, there are no reproducible and universally accepted colposcopic criteria for distinction between mild, moderate and severe dysplasia (Barasso 1988; Barasso et al. 1987; Singer 1988).

Flat/subclinical lesions may be associated in both sexes with a superfical epithelial fissuring. In women, an epithelial discontinuity occurs not infrequently in the posterior fourchette of the vulva and in the perineal area. This may sometimes give rise to a marked dyspareunia and is nowadays considered one major cause of vulvodynia (Bodén et al. 1988a, b; McKay 1989). Similar fissures are also seen in men (Fig. 5) and are sometimes associated with itching, burning, balanoposthitis and/or dyspareunia (G. von Krogh, M.-A. Hedblad, A. Wikström, unpublished work).

Histopathology

The following histopathological criteria are characteristic: a varying degree of either exophytic or endophytic epidermal hyperplasia with acanthotic elongation of the rete lists, a more or less pronounced hyper- and parakeratosis, the presence of koilocytosis and dilated vessels of corium papillae. Some degree of nonspecific dermal inflammation is common. Furthermore, as previously indicated, a varying degree of epithelial dysplasia may be seen (intraepithelial neoplasia). In the absence of koilocytosis the biopsy can merely be classified as concordant with but not diagnostic for HPV influence.

When and How to Investigate for Subclinical Infection

Specificity Problems of the Acetic Acid Test

It needs to be emphasized that the acetic acid test is *not* "the gold standard" for diagnosing GPVI; the test may turn out false positive in a high number of cases, a fact that has been appreciated only recently. A false-positive rate of up to 25% seems to exist (Schultz and Skelton 1988; G. von Krogh, M.-A. Hedblad, A. Wikström, unpublished work). As a consequence hereof, pretreatment of aceto-white lesions with a topical anti-inflammatory agent is advantageous prior to further therapy such as surgery. A complete or partial disappearance of the aceto-white colour subsequent to topical anti-inflammatory therapy is not uncommon (G. von Krogh, M.-A. Hedblad, A. Wikström, unpublished work).

False-positive results are one reason why the acetic acid test cannot be recommended at present for indiscriminate screening purposes in presumably healthy individuals. However, additional problems also exist.

Indiscriminate Screening for GPVI not Recommended

Subclinical GPVI appears to be a very common condition. Using koilocytosis as the major cytological criterion of active infection, the prevalence appears to be 2%–4% among sexually active women (Meisels et al. 1982; de Villiers et al. 1987; Syrjänen et al. 1988). These figures represent low estimates; using virological techniques, the prevalence of infection is in the range of 8%–10% or higher among individuals 15–35 years of age (de Villiers et al. 1987; Grussendorf-Conen et al. 1987). Thus, the problem is of a magnitude signifying that current clinical management, in the absence of antiviral therapy, must incline to a pragmatic concept of priority. Most importantly, general screening for subclinical GPVI cannot be recommended at present (International Titisee Conference 1987).

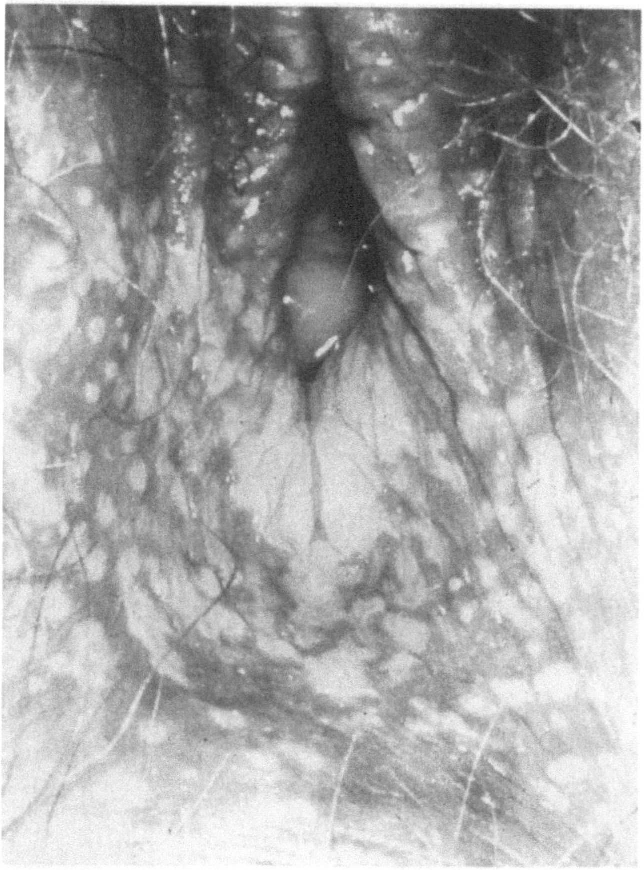

a

b

Fig. 7 a, b. Aspects following acetic acid test in the vulva (**a**) and on the penis (**b**) of two presumably healthy individuals. (Courtesy of Eva Rylander, M.D., Associate Professor, Dept. of Gynaecology, University Hospital of Umeå, Sweden)

When to Intervene for Subclinical Infections

Acetic acid screening will identify subclinical lesions of the outer genitoanal area among a high number of asymptomatic and healthy sexually active individuals (Fig. 7a, b). Once an individual becomes aware that he or she has the condition, anguish emerges in a significant proportion of cases. At present, no reliable routine therapy exists for these lesions. Furthermore, crucial questions such as whether these lesions are contagious or have other biological significance cannot be answered on an individual basis. This type of clinical approach cannot distinguish between individuals who are infected with "low-risk" and those infected with "high-risk" HPV types. Finally, it is not possible to predict whether the lesions are prone to disappear spontaneously, in time, through immunomodulatory rejection (see Jablonska, this volume).

My personal approach to handling this dilemma of priority is as follows: Firstly, the acetic acid test is useful when a protracted course of "recurrences" or "reoccurrences" of overt warts exists, aimed at identifying the full extension of subclinical lesions that may potentially transform into overt warts (Fig. 8). Secondly, the test is valuable for a demarcation of papular and flat warts prior to therapy. Thirdly, the test is extremely useful when GPVI-associated symptoms are suspected. Finally, individuals of both sexes who present actively as partners of patients with diagnosed GPVI, or men presenting because cervical

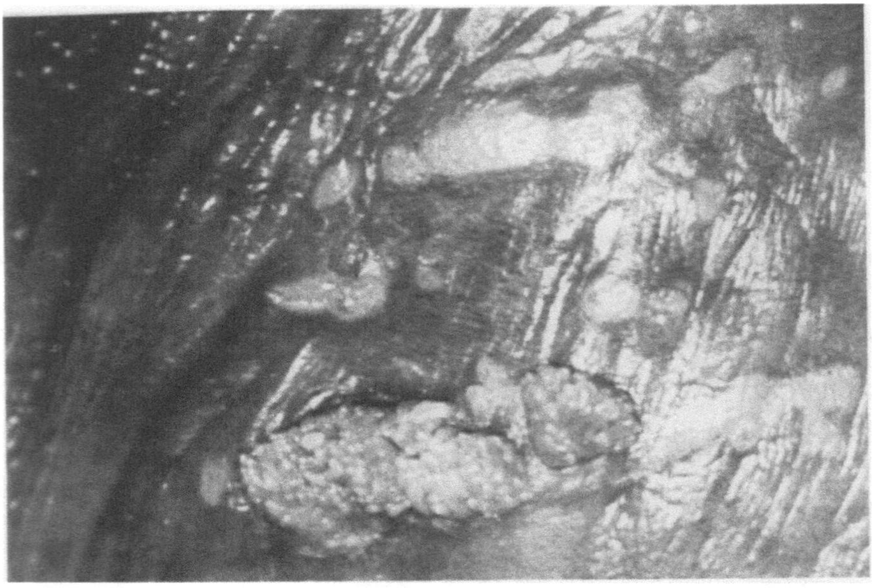

Fig. 8. Acetic acid test performed on a man who had experienced numerous recurrences of penile warts. The aim was to determine whether subclinical lesions might induce the recurrence of overt warts. Prior to the test a solitary acuminate wart was visible (*below*); subsequent to the test adjacent subclinical lesions were detected as well (*above*)

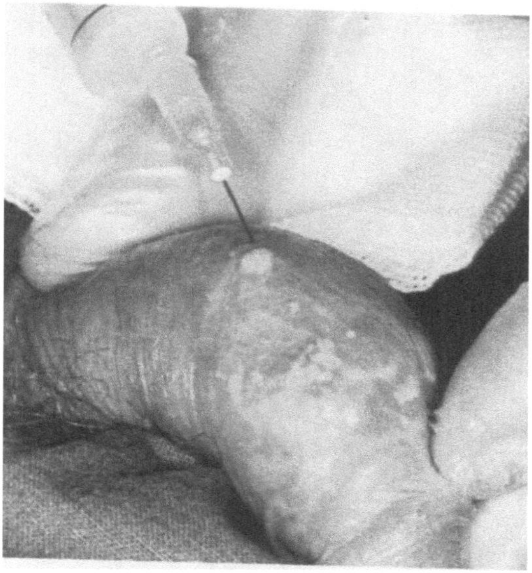

Fig. 9. A 35-year-old man presented without a prior history of overt genital warts. He wanted a thorough examination because his wife had been diagnosed as having CIN III, and he had learned through the media that there is a potential for subclinical penile lesions. He was afraid he might be a carrier of "high-risk" HPV types. He was diagnosed as having aceto-white lesions, and eradication of the infection was attempted with diathermy during local infiltration anaesthesia

intraepithelial neoplasia has been diagnosed in their female partner expect a thorough investigation of the genitals, including the use of acetic acid (Fig. 9).

Virology

Three unequivocal epidermal fingerprints exist for HPV infection: koilocytes, HPV capsid antigen and HPV-DNA. As for koilocytosis, capsid antigen can be detected in only about half of the cases; HPV-DNA, on the other hand, is almost consistently present. The DNA may be identified through various types of hybridization assays. In situ hybridization on paraffin-embedded biopsy material has the advantage that specific HPV types and micromorphological changes can be detected simultaneously (Fig. 10).

With the exception of typical bowenoid papulosis (Fig. 11), the gross appearance of GPVI lesions offers limited value for predicting which individuals are carriers of "high-risk" HPV types. In a series of 91 penile lesions showing positive in the in situ hybridization test (Syrjänen et al. 1987; von Krogh et al. 1988) it appeared that although "low-risk" HPV types 6 and 11 were most common in the acuminate wart type, they were found in 86% of papular and 77% of flat lesions as well. The oncogenic HPV types 16 or 18 existed in 23%

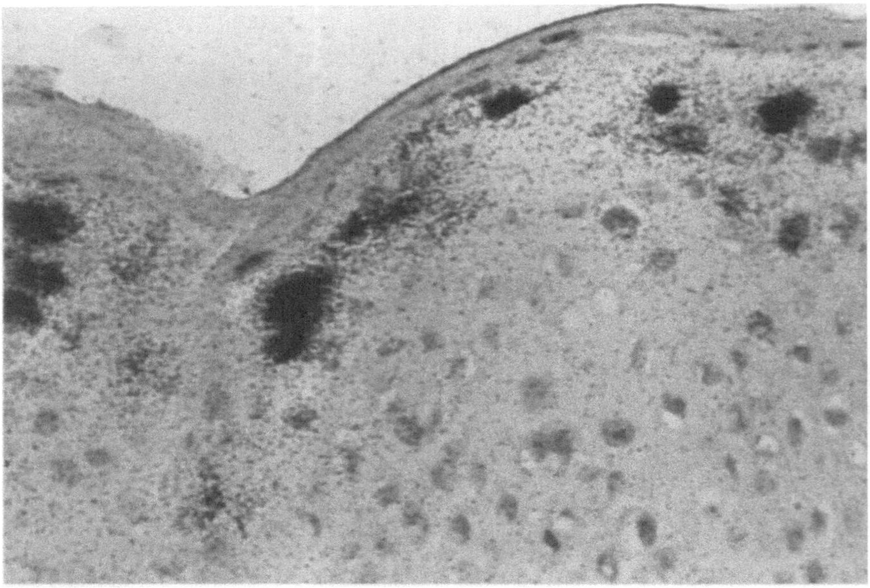

Fig. 10. In situ hybridization assay has detected HPV 11-specific radioactive signals (*black grains*) in the uper part of the epidermis. This method is extremely valuable for proving the viral nature of lesions lacking koilocytes, and it may reveal the specific HPV type(s) present in the biopsied lesion. (Courtesy of Stina Syrjänen, Associate Professor, Dept. of Oral Pathology and Radiology, Institute of Dentistry, University of Kuopio, Finland)

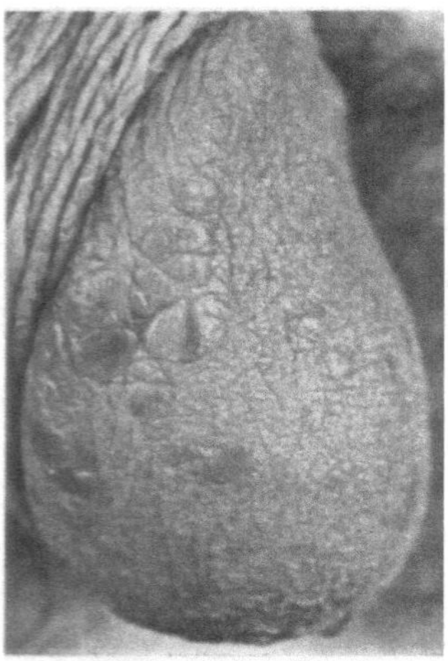

Fig. 11. Bowenoid papulosis in the 28-year-old partner of a woman diagnosed as having CIN III. Bowenoid papulosis typically appears on the outer genitoanal area in young adolescents, in contrast to Bowen's disease and Queyrat's erythroplasia, which occur in patients 40 years and older. These conditions are generally easy to distinguish clinically: In Bowenoid papulosis there are usually multiple papules 5–10 mm in diameter; Bowen's disease usually appears as a solitary bright red plaque more than 10 mm in diameter

of flat lesions and in 14% of papular and 6% of acuminate warts. Accordingly, although most "high-risk" HPV types were detected in flat lesions they were associated with any macromorphological GPVI types. This clinical dilemma is further confounded by the fact that most lesions associated with oncogenic HPV types tend to occur in a completely subclinical form (Reid and Campion 1988).

Similarly, with the exception of lesions exhibiting severe intraepithelial neoplasia, histological appraisal of dysplasia does not provide conclusive information regarding underlying HPV type(s). In our series (Syrjänen et al. 1987) HPV 16 was most common in warts exhibiting some degree of dysplasia ($p<0.001$). HPV 16- and 18-associated lesions lacked any dysplasia in only 2% of cases. Both of two biopsies showing carcinoma in situ contained HPV 16. However, HPV types 6 or 11 were detected in 81% of biopsies exhibiting either a mild or a moderate degree of epithelial dysplasia; the corresponding figure for HPV 16 and 18 was 19%. Although lack of dysplasia correlated with the presence of HPV 6 and/or 11 ($p<0.01$), the absence of dysplasia did not preclude the presence of "high-risk" HPV types. Accordingly, although macro- and micromorphology may be indicative of the underlying HPV type(s), DNA hybridization assays are necessary to distinguish between individuals infected with "low-risk" and those infected with "high-risk" HPV types. These tests are yet not available for routine clinical application.

Therapy

The "iceberg" dilemma raises a number of crucial questions that cannot yet be properly addressed. The following recommendations for the dermatovenereologist are based on common sense in the light of available resources.

When to Treat

The following situations clearly require active therapy:

1. Patients who present with disfiguring overt genital warts feel unclean and psychologically embarassed, and the condition represents a true hindrance to sexual performance (Fig. 1).
2. Some patients have associated symptoms such as bleeding and pain during intercourse due to fissuring and associated inflammatory responses (Fig. 5). This is probably more common in women than in men.
3. The sexual partner of a patient with warts presents for examination (Fig. 9). In this situation the question is whether or not such individuals should be routinely examined for subclinical infection. The potential existence of subclinical diseases is now rather generally known among laymen, due to extensive media coverage during the past few years. Once diagnosed, such patients more or less demand treatment.
4. Any premalignant lesions such as bowenoid papules (Fig. 11) or cervical dysplasia should be removed.

How to Treat

Several approaches have been proposed for the removal of wart growths. It must be kept in mind that the disappearance of visible lesions does not mean that the virus has been eradicated. Available modalities fall into four categories: a) cytodestructive drugs; b) surgery; c) immunomodulation and d) vaccines and antiviral agents. Only the first two constitute standard therapy. It should be emphasized that recurrence rates following initial therapy are high, varying from 10% to 60%. Most recurrences are seen within the first 2–3 months of follow-up.

Podophyllin

The first choice of therapy was previously podophyllin, a crude and non-standardized extract from the roots of either of the two *Podophyllum* plants: *P. peltatum* and *P. emodi* (Kaplan 1942; Culp and Kaplan 1944; von Krogh 1978). This remedy is potentially very toxic, both locally and systemically (von Krogh 1981 a, 1982). Furthermore, its content of quercetin makes it potentially mutagenic (Brown and Dietrich 1979; Pamukcu et al. 1980). The remedy should not be used for very large warts, as severe systemic side effects may result, including coma, vascular crisis and respiratory failure. Use in the pregnant woman is contraindicated. Owing to the risk of severe local side effects the warts should be painted maximally at weekly intervals and the procedure should be performed by a doctor or a nurse. The preparation must be washed off within a few hours after application. This limitation puts heavy demands on the medical and nursing staff and is a considerable inconvenience to the patient, who may experience unpredictable local irritation varying from one podophyllin batch to another. Furthermore, the efficacy of podophyllin is highly unsatisfactory. In a controlled study using 20% podophyllin solutions applied at weekly intervals, we evaluated the long-term efficacy on penile warts (von Krogh 1978). The cumulative effect of one to two applications was only 38%, although freshly prepared podophyllin solutions were used. Podophyllin is a very unstable remedy, and the therapeutically active ingredients are quickly precipitated or degrade into inactive isomers. This may explain why Simmons (1981) reported that a mean of 11 applications were required for a complete cure. Similarly, Lassus found that only one third of patients were cured after three of four treatments (Lassus et al. 1984; Lassus 1987).

Podophyllotoxin

The therapeutic effect of podophyllin emanates from chemically closely related so-called lignans. Of these, podophyllotoxin is the only one occurring in both plant species. Podophyllotoxin appears to be the most potent cytotoxic agent among the four lignans when applied to the epidermis (von Krogh and Maibach 1982, 1983). This drug exerts an antimitotic and an anti-DNA-synthesis effect on the cells.

Table 4. Podophyllin resin versus purified podophyllotoxin

	Podophyllin	Podophyllotoxin
Systemic toxicity risk	High	Very low
Stability	Very poor	Excellent
Other ingredients	Numerous	None
Standardized product	No	Yes
Mutagenic properties	Yes	No

Clinical use of purified podophyllotoxin for genital warts was first investigated by Sullivan et al. (1948). Considerable progress has been made since this early report (von Krogh 1978, 1981 a, b, 1982, 1986). A method has been developed for self-treatment of outer genital warts using 0.5% podophyllotoxin in 70% ethanol, with methylrosaniline added as a color indicator [1]. The solution is to be applied to each wart twice daily for 3 days. Prior to applications the treated area is washed and dried. This treatment course may be repeated at about weekly intervals on any residual warts in order to accomplish a cumulative effect. Originally, cotton wool swabs were used for application, but a specially designed plastic applicator may also be used, [1] which makes the remedy more economical. Men are informed to keep their foreskin retracted while applying the solution and to do so for several minutes untill the ethanol has evaporated. Treatment in women is facilitated when a mirror and a lamp are used during applications. This type of therapy against vulvar warts should be tried only provided that the women is not pregnant and has an adequate contraceptive.

While the risk of systemic toxicity is high for podophyllin, it is very low for 0.5% podophyllotoxin (Table 4). In contrast to podophyllin, the podophyllotoxin preparation is quite stable; it is a standardized product containing no other ingredients. Finally, no mutagenic properties have been demonstrated for podophyllotoxin (Beutner, this volume).

Warts on moist areas, such as in the preputial cavity of uncircumcised males, react most favourably, in particular when they are acuminate. Such warts disappear in 70% of men after a single 3-day course of home treatment. Efficacy is lower in the urinary meatus and on penile skin; yet a satisfactory cumulative result may be accomplished for the latter type of lesions as well when several courses are given (Beutner, this volume).

The long-term efficacy of podophyllotoxin is highly superior to that of podophyllin. In our series, 22% of the men were completely cured after the first podophyllin application, while the corresponding figure using 0.5% podophyllotoxin was 48% ($p<0.001$). After a second course, the cumulative effect of podophyllin was only 38%, but it was as high as 82% for two courses of self-treatment with podophyllotoxin ($p<0.001$).

---

1 Wartec, Warticon

These results have been confirmed by others. Lassus et al. (1984; Lassus 1987) allocated randomly 100 men with preputial cavity warts to either physician treatment with 20% podophyllin or self-treatment with 0.5% podophyllotoxin twice daily for 3 days. All of the 48 podophyllotoxin-treated patients were cured primarily after a maximum of four treatments. During the follow-up period, 11 of 48 (23%) of these men experienced a relapse; nine of them were permanently cured after another podophyllotoxin treatment. Of the 52 podophyllin-treated men, 37 (71%) were cured primarily after four treatments but 14 (38%) experienced a recurrence during the follow-up period. Of these, seven were retreated successfully with podophyllin. The difference between the two treatment groups was highly significant ($p < 0.001$) in favour of podophyllotoxin with regard to permanent cure rates. Thus, 77% (37/48) of patients treated with 0.5% podophyllotoxin remained clear compared with 44% (23/52) of those treated with podophyllin.

Most patients do not experience any discomfort whatsoever, although the treated epithelium transitorily may appear grayish-white immediately after cessation of therapy. When warts are numerous, superficial epithelial erosions may induce some tenderness – and occasionally some pain – for a few days.

In summary, the use of 0.5% purified podophyllotoxin for the primary routine treatment of penile and vulvar warts has several advantages compared with podophyllin, which can no longer be recommended. A long-term cure is most often achieved when a 3-day course of self-treatment is repeated at weekly intervals. However, more than four to six courses are usually of little benefit, as tachyphylaxis seems to occur over a period of time. Although the bulk of the warts will be reduced significantly at this point, it is recommended that alternative methods such as surgery be used to eradicate residual warts.

Fluorouracil

Recalcitrant warts in the urinary meatus may be treated topically with 5-fluorouracil (5-FU) cream applied with a cotton wool swab subsequent to voiding for a period of at least 2, preferably 3 weeks (von Krogh 1976). Excess medication outside the urinary orifice should be wiped off. After about 1 week erosion occurs causing a dysuria that is usually well tolerated. At least a 70% cure rate may be anticipated. Also, 5-FU has been found valuable for treating intravaginal warts (Ferenczy 1984; Krebs 1986); however, this treatment should be given only by the gynaecologist. It has been advocated that topical 5-FU may also be used as an adjuvant therapy given once or twice weekly subsequent to surgical treatment, but this approach has not been properly documented.

Surgery

Simple surgical measures include scissor excision, electrocautery and cryotherapy. Scissor excision and/or diathermy performed with the patient under local anaesthesia may be used as primary therapy whenever only a few lesions are present on the outer genitals (Thomson and Grace 1978; Gollock et al.

1982; Simmons and Thomson 1986). Intra-anal warts may be electrodesiccated or frozen through the anoscope, but it is preferable to excise them through a bivalve proctoscope, an operation that should be performed by a proctologist (Lindkear Jensen 1985; Khawaja 1986; Samenius 1983).

Flat warts are most suitably treated with diathermy (Graber et al. 1967), and the procedure is most accurately performed using the colposcope for high-power magnification. One should try to leave epithelial bridges between individual lesions in order to facilitate a fast re-epithelialization. Sometimes considerable epithelial defects ensue. Nevertheless, these will heal within 2–4 weeks without leaving any significant scarring if the procedure is performed properly, inducing a superficial destruction of the epidermis and the upper dermis.

During the past two decades convenient cryounits have been developed, using nitrous oxide or liquid nitrogen as the refrigerant; it is delivered in "closed systems" as a continuous flow through flexible cables to and from hollow cryoprobe tips of various sizes. As is reported by Dachow (this volume), "open-system" spray tecniques may also be used. One advantage of cryotherapy is that it may often be used without prior anaesthesia. In carefully performed comparative studies of large patient groups the method has proved to be very valuable in the hands of the experienced physician, with a cure rate comparable to that for other surgical approaches (Bashi 1985; Sand et al. 1987). The average number of cryotherapy sessions required for a complete cure is two to three.

Laser therapy has advantages as well as disadvantages. The method certainly requires special training as well as experience and skill. Also, a separate operating room and precautionary measures such as the use of a suction device are necessary. Therefore, the laser seems to be of limited value for routine clinical use. In a controlled randomised study on patients with recalcitrant genitoanal warts, excision and/or electrocautery appeared to be equally as effective as CO_2 laser treatment, and there was no difference in the incidence of postoperative pain, healing time or rate of scar formation between the two methods (Duus et al. 1985). The main indication for laser treatment on the outer genitals seems to be the presence of particularly extensive and/or recalcitrant lesions. On the other hand, laser is the treatment of choice for cervical lesions, whether these are associated with CIN or not (Anderson and Hartley 1980; Anderson et al. 1984; Rylander et al. 1984; Jordan et al. 1985).

Interferons

As reported by the groups of Vance, van den Berg and Gross (this volume), a number of interesting studies have been performed on the potential role of various interferons (IFNs) in the treatment of GPVI. One major advantage of successful therapy is the absence of ulceration. Yet conflicting results with IFNs do exist; in my own view, this type of therapy should be considered only in particularly recalcitrant cases, or when very large "problem warts" occur. At present, many feel that IFNs do have a place in the therapeutic arsenal for

GPVI, but primarily as adjuvant therapy to other modalities. Dosages and the proper form of administration have yet to be optimized. From a practical point of view, further refinements are required. The approach described by Saastamoinen (this volume) is highly interesting, using topical self-application of an ointment-based preparation.

Conclusion

The biological behaviour of genitoanal papillomavirus infection (GPVI) is highly variable, and its presence is associated with considerable diagnostic and therapeutic problems. Classical acuminate and papular warts are usually well visualized. However, the potential coexistence of flat macular and completely subclinical lesions has attracted limited attention until now. In such cases, the use of topical acetic acid, following by biopsy, may improve the diagnostic sensitivity. Whether HPV-DNA hybridization assays on histological and cytological specimens may become important for routine clinical evaluation and therapeutic programs is yet to be evaluated.

The diversity of therapeutic methods aimed at eradicating GPVI lesions reflects the fact that no single treatment is fully satisfactory. A number of patients require several sessions using a combined chemical and surgical approach. The traditional use of podophyllin has disadvantages, indicating that the remedy should be abandoned. The alternative of self-treatment b.i.d. for 3 days with a standardized 0.5% preparation of purified podophyllotoxin is more effective against outer genital warts and has a high degree of safety. Topical 5-FU for 2–3 weeks may eradicate urinary meatus warts. Many patients can also be treated with simple scissor excision and diathermy. CO_2 lasers are expensive, and their use requires precautionary measures as well as a high degree of training. The adjuvant use of IFN is worth consideration in recalcitrant cases.

References

Anderson MC, Hartley RB (1980) Cervical crypt involvement by intraepithelial neoplasia. Obstet Gynecol 55:546–550

Anderson MC, Horwell D, Broby Z (1984) Outcome of pregnancy after laser vaporization conization. Colposcopy Gynecol Laser Surg 1:35–40

Barrasso R (1988) Letter to the editor (Response to Coppleson M). Gynecol Oncol 29:120–123

Barrasso R, Coupez R, Tonesco M, de Brux J (1987) Human papilloma viruses and cervical intraepithelial neoplasia: the role of colposcopy. Gynecol Oncol 27:197–207

Bashi SA (1985) Cryotherapy versus podophyllin in the treatment of genital warts. Int J Dermatol 24:535–536

Bergeron C, Ferenczy A, Richart R (1988) Underwear contamination by HPV. Poster exhibition, 7th International Papillomavirus Workshop, 16–20 May 1988, Antipolis, France

Bodén E, Eriksson A, Rylander E, von Schoultz B (1988) Clinical characteristics of papillomavirus vulvovaginitis – an entity with oncogenic potential. Acta Obstet Gynecol Scand 67–147–151

Bodén E, Rylander E, Evander M, Wadell G, von Schoultz B (1989) Papilloma virus infection of the vulva. Acta Obstet Gynecol Scand 68:179–184

Braun J, Raguse T (1987) The value of surgical excision in the treatment of anal condylomas. Colo-proctol 9:23–26

Brown JP, Dietrich PS (1979) Mutagenicity of plant flavonols in the Salmonella/Mammalian microsome test. Activation of flavonol glycosides by mixed glycosides from rat cecal bacteria and other sources. Mutat Res 66:223–240

Campion MJ, Singer A (1987) Vulval intraepithelial neoplasia; clinical review. Genitourin Med 63:147–152

Campion MJ, Clarkson R, McCance DJ (1985) Squamous neoplasia of the cervix in relation to other genital tract neoplasia. Clin Obstet Gynecol 12:265–280

Campion MJ, McCance DJ, Jenkins J, Atia W, Singer A, Oriel JD (1986) Subclinical penile human papillomavirus infection: the clue to the high-risk male. J Colposc Gynecol Laser Surg 8:100–110

Carr G, William DC (1977) Anal warts in a population of gay men in New York City. Sex Transm Dis 4:56–57

Chuang T-Y (1987) Condylomata acuminata (genital warts). An epidemiological view. J Am Acad Dermatol 16:376–384

Chuang T-Y, Perry HO, Kurland LT, Ilstrup M (1984) Condyloma acuminatum in Rochester, Minn., 1950–1978. I. Epidemiology and clinical features. Arch Dermatol 120:469–475

Culp OS, Kaplan IW (1944) Condylomata acuminata. Two hundred cases treated with podophyllin. Ann Surg 120:251–256

Department of Health and Social Security (1985) Sexually transmitted diseases for the year 1983. Genitourin Med 61–204–207

de Villiers E-M, Wagner D, Schneider A, Wesch H, Miklaw H, zur Hausen H (1987) Human papillomavirus infection in women with and without abnormal cervical cytology. Lancet II:703–705

Duus BR, Philipsen T, Christensen JD, Lundvall F, Sondergaard J (1985) Refractory condylomata acuminata: a controlled clinical trial of carbon dioxide laser versus conventional surgical treatment. Genitourin Med 61:59–61

Ferenczy A (1984) Comparison of 5-fluorouracil and CO_2 laser for treatment of vaginal condylomata. Obstet Gynecol 64:773–778

Gollock JM, Slatford K, Hunter JM (1982) Scissor excision of anogenital warts. Br J Vener Dis 58–400–401

Goorney BP, Waugh MA, Clarke J (1987) Anal warts in heterosexual men. Genitourin Med 63:216

Graber EA, Barber HRK, O'Rourke JJ (1967) Simple surgical treatment for condylomata acuminatum of the vulva. Obstet Gynecol 29:247–250

Gross G, Hagedorn M, Ikenberg H, Rufli T, Dahlet C, Grosshans E, Gissmann L (1985) Bowenoid papulosis. Presence of human papillomavirus (HPV) structural antigens and of HPV 16-related DNA sequences. Arch Dermatol 121:858–863

Grussendorf-Conen E-I, Meinhof W, de Villiers E-M, Gissmann L (1987) Occurrence of HPV genomes in penile smears of healthy men. Arch Dermatol Res 279:S73–S75

Ikenberg H, Gissmann L, Gross G, Grussendorf-Conan E-I, zur Hausen H (1983) Human papillomavirus type 16-related DNA in genital Bowen's disease and in bowenoid papulosis. Int J Cancer 32:563–565

International Titisee Conference (1987) Papillomavirus in human genital cancer: causative role and practical consequences. 1–4 Oct 1987. Boehringer Ingelheim fonds, Titisee

Jordan JA, Woodman CBJ, Mylotte MJ et al. (1985) The treatment of cervical intraepithelial neoplasia by laser vaporization. Br J Obstet Gynaecol 92:394–398

Kaplan IW (1942) Condylomata acuminata. New Orleans. Med Surg J 94:388–396

Khawaja HT (1986) Treatment of condylomata acuminatum. Lancet 1:208–209
Kimura S, Hirai A, Harada R et al. (1987) So-called multicentric pigmented Bowen's disease. Dermatologica 157:229–237
Krebs H-B (1986) Prophylactic topical 5-fluorouracil following treatment of human papillomavirus-associated lesions of the vulva and vagina. Obstet Gynecol 68:837–841
Lassus A (1987) Comparison of podophyllotoxin and podophyllin in treatment of genital warts. Lancet 2:512–513
Lassus A, Haukka K, Forsström S (1984) Podophyllotoxin for treatment of genital warts in males. A comparison with conventinal podophyllin therapy. Eur J Sex Transm Dis 2:31–33
Lassus J, Pönkä A, Haukka K, Lassus A (1988) Increase in new patients with genital warts attending STD clinics in Helsinki, 1980-6. Genitourin Med 64:205–208
Lever WF, Schaumburg-Lever (1983) Histopathology of the skin, 6th edn, chap 26. Lippincott, London, pp 496–498
Lindkear Jensen S (1985) Comparison of podophyllin application with simple surgical excision in clearance and recurrence of perianal condylomata acuminata. Lancet 2:1146–1148
Lloyd K (1970) Multicentric pigmented Bowen's disease of the groin. Arch Dermatol 101:48–51
McCance DJ, Campion MJ, Baram A, Singer A (1986) Risk for transmission of human papillomavirus by vaginal specula. Lancet 2:816–817
McKay M (1989) Vulvodynia. Arch Dermatol 125:256–262
Meisels A, Fortin R, Roy M (1977) Condylomatous lesions of the cervix. II. Cytologic, colposcopic and histopathologic study. Acta Cytol 21:379–390
Meisels A, Morin C, Casas-Cordero M (1982) Human papillomavirus infection of the uterine cervix. Int J Gynecol Pathol 1:75–94
Nash O, Allen W, Nash S (1986) Atypical lesions of the anual mucosa in homosexual men. JAMA 256:873–876
Obalek S, Jablonska S, Beandenon MB et al. (1986) Bowenoid papulosis of the male and female genitalia; risk of cervical neoplasia. J Am Acad Dermatol 14:433–444
Oriel JD (1971 a) Natural history of genital warts. Br J Vener Dis 47:1–13
Oriel JD (1971 b) Anal warts and anal coitus. Br J Vener Dis 47:373–378
Pamukcu AM, Yalciner S, Hatcher JF, Bryan GT (1980) Quercetin, a rat intestinal and bladder carcinogen present in bracken fern (*Pteridium aquilinum*). Cancer Res 40:3468–3472
Reid R, Campion MJ (1988) the biology and significance of human papillomavirus infections in the genital tract. Yale J Biol Med 61:307–325
Reid R, Laverty CR, Coppleson M et al. (1980) Noncondylomatous cervical wart virus infection. Obstet Gynecol 55:476–483
Reid R, Stanhope R, Herschman BR (1984) Genital warts and cervical cancer. IV. A colposcopic index differentiating subclinical papillomaviral infection from cervical intraepithelial neoplasia. Am J Obstet Gynecol 149:815–823
Rylander E, Isberg A, Joelsson I (1984) Laser evaporization of cervical intraepithelial neoplasia. A five-year follow-up. Acta Obstet Gynecol Scand Suppl 125:33–36
Samenius B (1983) Perianal and ano-rectal condyloma acuminata. Schweiz Rundsch Med Prax 72:1009–1014
Sand PK, Shen W, Bowen L, Ostergard DR (1987) Cryotherapy for the treatment of proximal urethral condyloma acuminatum. J Urol 137:874–876
Schultz RE, Skelton HG (1988) Value of acetic acid screening for flat genital condylomata in men. J Urol 139:777–779
Simmons PD (1981) A comparative double-blind study of 10% and 25% podophyllin in the treatment of anogenital warts. Br J Vener Dis 57:208–209
Simmons PD, Thomson JPS (1986) Scissor excision of penile warts: case report. Genitourin Med 62:277–278
Sohn N, Robilotti JG (1977) The gay bowel syndrome. Am J Gastroenterol 67:478–484

Sullivan M, Friedman M, Hearin JT (1948) Treatment of condylomata acuminata with podophyllotoxin. South Med J 41:336–339

Syrjänen KJ (1987) Papillomaviruses and cancer. In: Syrjänen KJ, Gissman L, Koss L (eds) Papillomaviruses and human disease. Springer, Berlin Heidelberg New York, pp 468–503

Syrjänen KJ (1988) Genital papillomavirus (HPV) infections: clinical significance and associations with genital cancer. In: Oriel JD, Waugh M (eds) Anglo-Scandinavian Conference on Sexually Transmitted Diseases. Royal Society of Medicine Services, London, pp 45–53 International congress and symposium series

Syrjänen KJ (1989) Histopathology, cytology, immunohistochemistry and HPV typing techniques. In: von Krogh G, Rylander R (eds) GPVI – Genitoanal papilloma virus infection. A survey for the clinicians. Conpharm AB, Karlstad, Sweden, Chapter III, pp 33–67

Syrjänen KJ, von Krogh G, Syrjänen SM (1989) Anal condylomata in men. Histopathological and virological assessment. Genitourin Med 65:216–224

Syrjänen SM, von Krogh G, Syrjänen KJ (1987) Detection of human papillomavirus DNA in anogenital condylomata in men using in situ DNA hybridisation applied to paraffin sections. Genitourin Med 63:32–34

Thomson JPS, Grace RH (1978) The treatment of perianal and anal condylomata. J R Soc Med 71:180–185

von Krogh (1976) 5-Fluorouracil cream is the successful treatment of therapeutically refractory condylomata acuminata of the urinary meatus. Acta Derm Venereol (Stockh) 56:297–301

von Krogh (1978) Topical treatment of penile condylomata acuminata with podophyllin, podophyllotoxin and colchicine. A comparative study. Acta Derm Venereol (Stockh) 56:163–168

von Krogh (1981 a) Podophyllotoxin for condylomata acuminata eradication. Clinical and experimental comparative studies on Podophyllum lignans, colchicine and 5-fluorouracil. Acta Derm Venereol [Suppl] (Stockh) 98

von Krogh (1981 b) Penile condylomata acuminata: an experimental model for evaluation of topical self-treatment with 0.5%–1.0% ethanolic preparations of podophyllotoxin for three days. Sex Transm Dis 8:179–186

von Krogh (1982) Podophyllotoxin in serum: absorption subsequent to three-day repeated applications of a 0.5% ethanolic preparation on condylomata acuminata. Sex Transm Dis 9:26–33

von Krogh (1986) Topical self-treatment of penile warts with 0.5% podophyllotoxin in ethanol for four or five days. Sex Transm Dis 14:135–140

von Krogh G, Maibach HI (1982) Cutaneous cytodestructive potency of lignans. I. A comparative evaluation of influence on epidermal and dermal DNA synthesis and on dermal microcirculation in the hairless mouse. Arch Dermatol Res 274:9–20

von Krogh G, Maibach HI (1983) Cutaneous cytodestructive potency of lignans. II. A comparative evaluation of macroscopic-toxic influence on rabbit skin subsequent to repeated 10-day applications. Dermatologica 167:70–77

von Krogh G, Syrjänen SM, Syrjänen KJ (1988) Advantage of human papillomavirus typing in the clinical evaluation of genitoanal warts. Experience with the in situ deoxyribonucleic acid hybridization technique applied on paraffin sections. J Am Acad Dermatol 18:495–503

Wade TR, Kopf AW, Ackerman AB (1978) Bowenoid papulosis of the penis. Cancer 42:1890–1903

HPV Infections of the Urethra

J. D. Oriel

Anogenital infection by human papillomaviruses (HPV) is a common disease. It is predominantly sexually transmitted, with an incubation period of between 1 and 8 months, with an average of 3 months (Oriel 1971). The resulting lesions can be broadly classified as exophytic (condylomata acuminata), papular, flat warts and verruca vulgaris (von Krogh 1987). Condylomata acuminata predominate in moist areas such as the preputial cavity and the anal canal, and were first described in the terminal urethra by Goldenberg (1891). Although it was formerly regarded as rare, urethral HPV infection is now known to be quite common, particularly in men.

Epidemiology

The prevalence of urethral HPV infection in the general population is unknown. In studies of men with genital warts associated urethral lesions have been noted in between 5% and 23% of cases (Table 1). The urethra may be the only site affected (Oriel 1971). Condylomas of the urethra are less often seen in women; in one study only 2 (4%) of 49 women with genital warts were affected at this site (Oriel 1971). The age incidence of the urethral disease has not been studied, but the highest risk group for genital warts is aged 20–24 years (Editorial note 1983) and it is probably similar for urethral lesions.

Condylomata acuminata of the urethral meatus have been described in both male and female children; the route of infection is uncertain, but in one study of 13 boys and 9 girls there was no history of sexual abuse (Redman and Meacham 1973).

Table 1. Prevalence of urethral HPV infection in men with genital warts

Authors	No. infected/No. examined	(%)
Culp and Caplan (1944)	10/200	(5)
Gersh (1945)	7/130	(5)
Oriel (1971)	32/137	(23)
Barrasso et al. (1987)	22/105	(21)

Genital Papillomavirus Infections
Eds.: Gross/Jablonska/Pfister/Stegner

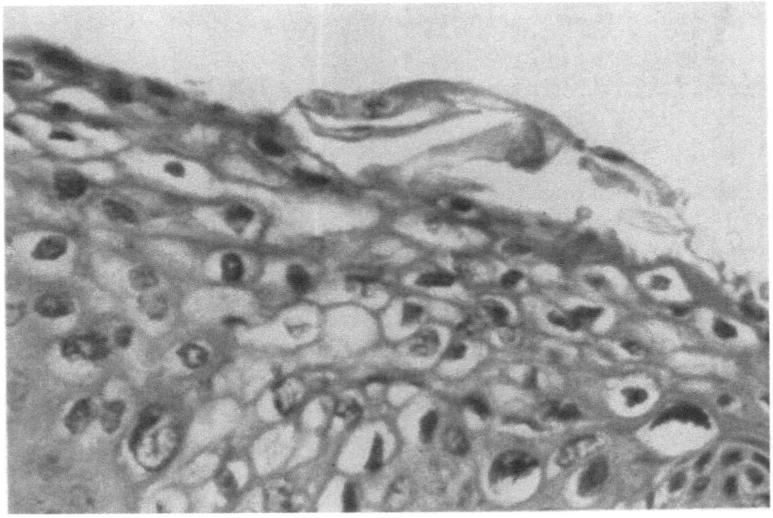

Fig. 1. Condyloma acuminatum of the urethra. Section shows koilocytosis and parakeratosis

Pathology

Urethral condylomas show the histological features of genital warts – papillomatosis, acanthosis, koilocytosis and parakeratosis (Fig. 1). Urethral scrapings from men with condylomas stained using Papanicolaou's method have been reported to show koilocytotic changes (Krebs and Schneider 1987). Dean et al. (1983) examined a series of histologically confirmed urethral condylomas by a peroxidase-antiperoxidase technique; the primary genus-specific antiserum was reactive in 11 (44%) of 25 cases. Because the majority of HPV-induced lesions of the urethra are exophytic condylomas it seems likely that HPV 6 or 11 is present, but data on a large series of such lesions are not yet available (Barrasso et al. 1987).

Clinical Features

Morphology

In the urethra exophytic condylomas predominate, presenting as typical soft fleshy vascular tumours (Fig. 2). Subclinical disease, in which the lesions resemble flat condylomas of the uterine cervix, has been described in men (Levine et al. 1984) and in a woman (Syrjanen and Pyrhoenen 1983).

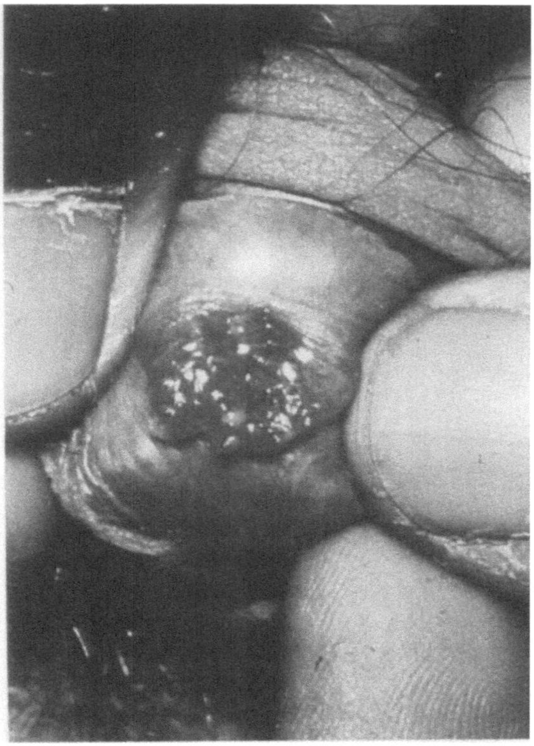

Fig. 2. Condylomata acuminata of the terminal urethra

Distribution

The majority of HPV-associated lesions occur in the terminal 1 cm of the urethra; proximal involvement (Fig. 3) is now regarded as extremely rare (von Krogh 1987), although in the earlier literature such lesions were often reported. Gartman (1956) stated that 30% of cases of urethral condylomas showed extension above the fossa navicularis. In a summary of the literature, Kleiman and Lancaster (1962) reported that among 88 men with urethral HPV infection the meatus was affected in 56, the distal 4 cm of the urethra in 22 and the entire urethra in 5. Other cases of proximal urethral HPV infection have been described by Morrow et al. (1952), Dretler and Klein (1975), Wein and Benson (1977), and Pollack et al. (1978). A case of giant condyloma of the urethra was described by Lindner and Pasquier (1954).

Condylomata acuminata of the bladder (Fig. 4) are very rare. Most of the cases described have followed HPV infection of the external genitalia and urethra in immunosuppressed individuals (Pettersson et al. 1976; Debenedictis et al. 1977). Giant condylomas, the eponymous Buschke-Loewenstein tumours, have also been described in the bladder (Kleiman and Lancaster 1962; Lewis et al. 1962; Hotchkiss and Rouse 1968).

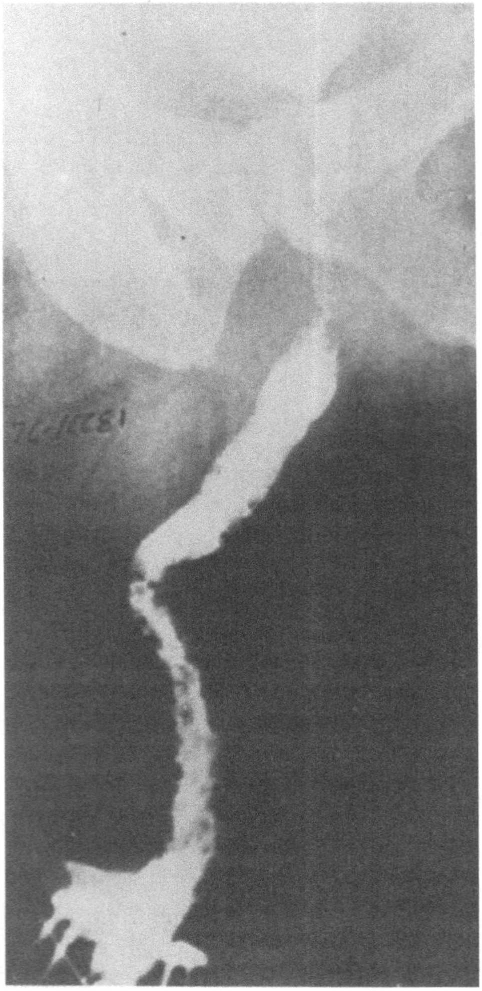

Fig. 3. Retrograde urethrogram reveals entire urethra to be carpeted with innumerable verrucous filling defects. (From Pollack et al. 1978)

Diagnosis

Genital HPV infection is often symptomless, and unless the lesions protrude from the meatus or cause bleeding urethral condylomas are often unnoticed by the patient. Subclinical lesions, or condylomas in the proximal urethra, will certainly not be suspected. Unrecognised urethral HPV infection is important epidemiologically, as has emerged in several recent studies (Levine et al. 1984; Krebs and Schneider 1987; Barrasso et al. 1987; see Table 2).

Examination of the fossa navicularis may be difficult if the meatus is narrow. Urethral dilatation with straight sounds and the use of a small nasal speculum (von Krogh 1987) make the examination easier. The application of 5% acetic acid, which causes whitening of HPV-infected epithelia, together

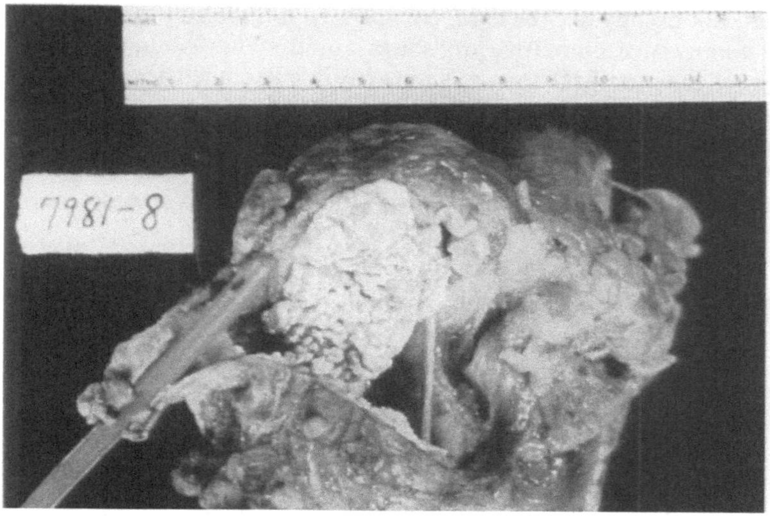

Fig. 4. Condylomata acuminata of the bladder

with the use of a colposcope or binocular loupe, are also useful diagnostic aids. A decision has to be made as to whether to investigate the proximal urethra. Despite the high figures reported in early studies, it is not now thought that HPV infection of the proximal urethra is sufficiently common to justify routine examination. If there is a reduced urine stream, a mass can be felt along the line of the urethra or there are persistent recurrences after treatment, examination of the proximal urethra is desirable. Retrograde urethrography is open to the objection that infected material might be spread from the distal to the proximal urethra, and because of this Pollack et al. (1978) have recommended voiding urethrography following intravenous con-

Table 2. HPV-associated lesions in male partners of women with cervical disease

Authors	Diagnosis of women	Male partners		
		No. examined	No. (%) with HPV lesions	No. (%) with urethral condylomas
Levine et al. (1984)	CIN	34	18 (53)	1 (3)
Krebs and Schneider (1987)	CIN	127	83 (65)	6 (5)
Barrasso et al. (1987)	CIN/flat condylomas	480	309 (64)	22 (5)

CIN, cervical intraepithelial neoplasia

trast medium. Endoscopy may be used, particularly as a prelude to treatment, and a useful noninvasive alternative procedure is urethral ultrasound.

Cytology of endourethral smears stained using Papanicolaou's method has been advocated by Krebs and Schneider (1987), but routine cytological examination of urine specimens has a very low yield (Levine et al. 1984; Krebs and Schneider 1987). Examination of urethral scrapings for viral antigen or DNA is not yet established as a sensitive or specific diagnostic procedure.

Treatment

Genital warts are a sexually transmitted disease, so before the treatment of urethral HPV infection is started it is important to exclude associated infections by appropriate diagnostic tests. Since genital HPV infection is both multifocal and multicentric a careful search for other HPV lesions is necessary. It is also important to examine the sexual partners of these patients. In women not only genital warts but malignant and premalignant disease may be associated with HPV infection in their male partners (Barrasso et al. 1987), and those who have such partners should clearly be fully investigated by examination of the external genitalia supplemented by cervical cytology and colposcopy, and carefully followed up.

The treatment of urethral HPV infection is not well documented, and there have been few well-controlled studies. All the methods in current use are in some respects unsatisfactory. Podophyllin and its refined derivative podophyllotoxin cannot be applied accurately to condylomas in this site and in any case are washed off with the next urination. They are of little value. The fluorinated pyrimidine 5-fluorouracil (5-FU) has been used with variable results. Dretler and Klein (1975) successfully treated 19 of 20 men with intraurethral warts with 5% 5-FU cream instilled into the urethra after each voiding for 3–8 days. Others have had less success and both Debenedictis (1977) and Cetti (1984) have reported the development of dysuria and retention of urine because of mucosal edema following 5-FU therapy.

The success of local destructive methods for the treatment of urethral condylomas depends in part on the site and extent of the lesions. If the whole of the distal condylomas can be seen, cryotherapy or electrocautery under local anaesthesia may be successful. If they cannot be completely seen, it may be better to dilate the distal urethra or perform a meatotomy before using destructive procedures. It may be difficult to destroy condylomas affecting the proximal urethra through a urethroscope, and exposure and direct treatment through a urethrostomy may be necessary (Cetti 1984). Carbon dioxide laser is satisfactory for the ablation of meatal lesions, but not for proximal lesions unless the urethra is opened. Some surgeons have used 5-FU cream as an adjunct to other modes of therapy. The treatment of urethral condylomas is laborious and often followed by recurrences, and there is a real need for systemic therapy for this disease, as for other types of genital HPV infection.

References

Barrasso R, de Brux J, Croissant O, Orth G (1987) High prevalence of papillomavirus-associated penile intraepithelial neoplasia in sexual partners of women with cervical intraepithelial neoplasia. N Engl J Med 317:916–923
Cetti NE (1984) Condyloma acuminatum of the urethra: problems in eradication. Br J Surg 71:57
Culp OS, Caplan IW (1944) Condylomata acuminata: 200 cases treated with podophyllin. Ann Surg 120:251–256
Dean P, Lancaster WD, Chun B, Jenson AB (1983) Human papillomavirus structural antigens in squamous papilloma of the male urethra. J Urol 129:873–875
Debenedictis TJ, Marmar JL, Praiss DE (1977) Intraurethral condylomata acuminata: management and a review of the literature. J Urol 118:767–769
Dretler SP, Klein LA (1975) The eradication of intraurethral condylomata acuminata with 5 per cent 5-fluorouracil cream. J Urol 113:197–198
Editorial note (1983) Condylomata acuminata – United States 1966–1981. J Am Med Ass 250:336
Gartman E (1956) Intraurethral verruca acuminata in men. J Urol 75:717–718
Gersh I (1945) Condylomata acuminata of the male external genitalia: an effective method of surgical treatment. Urol Cutan Rev 49:432–445
Goldenberg H (1891) Polyps of the male urethra. NY Med J 53:533–535
Hotchkiss RS, Rouse AJ (1968) Papillomatosis of the bladder and ureters, preceded by condylomata acuminata of the vulva: a case report. J Urol 100:723–725
Kleiman H, Lancaster Y (1962) Condyloma acuminatum of the bladder. J Urol 88:52–55
Krebs HB, Schneider V (1987) Human papillomavirus-associated lesions of the penis – colposcopy, cytology and histology. Obstet Gynecol 70:299–304
Levine RU, Crum CP, Herman E, Silvers D, Ferenczy A, Richart RM (1984) Cervical papillomavirus infection and intraepithelial neoplasia: a study of male sex partners. Obstet Gynecol 64:16–20
Lewis HY, Wolf PL, Pierce JM (1962) Condyloma acuminatum of the bladder. J Urol 88:248–251
Lindner HJ, Pasquier CM (1954) Condylomata acuminata of the urethra. J Urol 72:875–879
Morrow RP, McDonald JR, Emmett JL (1952) Condylomata acuminata of the urethra. J Urol 68:909–917
Oriel JD (1971) Natural history of genital warts. Br J Vener Dis 47:1–13
Pettersson S, Hansson G, Blohme I (1976) Condyloma acuminatum of the bladder. J Urol 115:535–536
Pollack HM, de Benedictis TJ, Marmar JL, Praiss DE (1978) Urethrographic manifestations of venereal warts (condylomata acuminata). Radiology 126:643–646
Redman JF, Meacham KR (1973) Condylomata acuminata of the urethral meatus in children. J Pediatr Surg 8:939–941
Syrjanen KJ, Pyrhoenen S (1983) Demonstration of human papilloma-virus (HPV) antigens in a case of urethral condyloma. Scand J Urol Nephrol 17:267–270
Von Krogh G (1987) Treatment of human papillomavirus-induced lesions. In: Syrjanen K, Gissmann L, Koss LG (eds) Papillomaviruses and human disease. Springer, Berlin Heidelberg New York, pp 296–333
Wein AJ, Benson GS (1977) Treatment of urethral condyloma acuminatum with 5-fluorouracil cream. Urology 9:413–415

Bowenoid Papulosis

G. Gross

Definition

Bowenoid papulosis is a disorder characterized by benign appearing partially pigmented macular and papular lesions of the anogenital region of both sexes, exhibiting histologic features of a squamous cell carcinoma in situ (CIS) of Bowen's type. The lesions are mostly multiple and of multicentric origin. There is a spectrum of inconspicuous, noncondylomatous lesions with a smooth and only slight papillomatous surface. The disorder affects predominantly young adults of about 30 years of age, thus following the age distribution of sexually transmitted diseases such as genital herpes and condylomata acuminata with which bowenoid papulosis may be associated [25, 56, 76, 77]. In contrast to Bowen's disease, in which after a long duration invasive carcinoma develops, bowenoid papulosis has a more benign course and tends to spontaneously regress. Malignant conversion has been reported in only very few instances. Etiologically, bowenoid papulosis is regarded today as a sexually transmitted human papillomavirus (HPV) disease mostly linked with HPV type 16 [29].

History

In 1970 Lloyd [46] described a peculiar condition which was histologically identical to ordinary Bowen's disease, but which exhibited different clinical characteristics. In contrast to Bowen's disease, bowenoid papulosis is characterized by multiple and multicentric partially pigmented lesions with bilateral involvement of the anogenital region as well as involvement of younger patients. Due to the histologic similarity the lesions were termed multicentric pigmented Bowen's disease in 1970 [46] before they received the name "bowenoid papulosis" in 1978 by Wade and coworkers [76]. In the meantime and even before the primary description was made similar cases were reported with names such as pigmented penile papulosis with carcinoma in situ changes [40], bowenoid dysplasia of the vulva [72], multicentric vulvar carcinoma in situ [78], reversible vulvar atypia [16], and carcinoma in situ of the genitalia [4] (Table 1). As in a large number of other cases reported spontaneous regression was observed [5, 16, 69] and in women this phenomenon was associated with pregnancy or delivery; hormonal factors were suggested to play an etiologic role. Demonstration of viruslike particles in some cases led to the hypothesis that this disorder is virus associated [47]. In view of anamnestic

Genital Papillomavirus Infections
Eds.: Gross/Jablonska/Pfister/Stegner

data genital herpes simplex virus (HSV) and HPV were the most suspected viral agents [41, 42, 76].

Etiology

There were different speculations on the cause of bowenoid papulosis [80]. Hormones, irritation, and chemicals as well as immunological factors were considered. Viruslike inclusions have been observed in some lesions of Bowen's disease by different investigators [47, 53, 79]. Whether viruses were definitely involved in the etiology of bowenoid papulosis remained, however, unresolved until recently [25, 35]. A serie of reports described an association with HSV type II either by demonstration of HSV II antigen in cytoplasm of dysplastic cells [41] or presence of recurrent genital herpes [76, 77]. Nevertheless, in the series of Wade et al. HSV-I and HSV-II antigen could not be demonstrated either by the use of immunoperoxidase method or by cultural, histologic, and sërologic investigation [77]. Using molecular hybridization techniques a consistent association of bowenoid papulosis, including the pigmented form, with HPV-16 has now been established [25, 35]. In rare instances the presence of HPV-6 and HPV-11 DNA as well as coinfections with HPV-16 and HPV-6 [35] have been described. Less frequently further genital HPV types HPV-18, HPV-33, HPV-34, HPV-39 [3], HPV-40, HPV-42, and HPV 55 were identified in bowenoid papulosis lesions [see 74]. Sequences of the same viruses have been shown in a small percentage of cervical intraepithelial neoplasia (CIN) [74]. Biopsies of Bowen's disease both from genital and from extragenital sites were also shown to harbor HPV-16 DNA [35], which points to common etiologic elements in both entities. Further etiologic factors such as irritation do not seem to play a major role in the development of bowenoid papulosis lesions, which occur predominantly in young circumcised men [56]. This is in contrast to the development of erythroplasia of Queyrat, a condition seen almost exclusively in older uncircumsised men [20]. The role of poor hygiene, maceration and trauma, and of other irritative factors is fairly uncertain. Of far more importance seems to be smoking, which is a regularly associated factor in both female and male patients suffering from bowenoid papulosis [33]. In contrast to condylomata acuminata and common skin warts, bowenoid papulosis lesions exhibit papillomavirus common antigens only in about 5% of cases [9, 25, 32, 59]. Nevertheless, these data indicate shedding of mature virus particles present in sufficient concentrations to be sexually transmitted between partners [22, 29, 51, 52]. Thus, in contrast to benign condylomata acuminata which are associated in more than 90% of cases investigated with HPV-6 and HPV-11 [74], bowenoid papulosis represents a reservoir for HPV-16 [29]. The concurrent presence of HPV-16 DNA in genital carcinoma in situ of sexual partners led to the hypothesis that bowenoid papulosis is a high-risk lesion in cervical carcinogenesis [27, 29, 51]. Additional data have been provided by epidemiological studies on venereal transmission of HPV between partners [21, 22, 29, 33, 45, 51, 52].

Prognosis

The course of bowenoid papulosis is uncertain. The lesions may regress spontaneously, especially in young women after giving birth [5, 16, 69]. Furthermore, regular use of contraceptives and remission of bowenoid papulosis lesions after thyroidectomy led to the hypothesis that hormonal alterations may play an additional role in the course of bowenoid papulosis [56]. Although this disease is regarded as rather benign and self-limited, present especially in younger men and women, the possibility exists that bowenoid papulosis harboring HPV-16 is also a risk factor with respect to the etiology of carcinoma of the lower genital tract and of the anal canal [60]. The risk of malignant change, however, is far higher in the uterine cervix than in the vulva, vagina, penis, and anus [81]. Factors which play the most important role for either the benign or malignant course of anogenital infections with potential oncogenic HPV types are deficient intracellular surveillance mechanisms [81] and immunosurveillance mechanisms directed against the virus-infected cells and virus-induced tumors [48, Jablonska and Majewski, this volume]. This is supported by the about 100-fold increased incidence of anogenital malignant tumors in immunosuppressed patients [57, 58, 61, 63, 67, 68]. Cell-mediated cytotoxicity seems to be mainly responsible for intact immunosurveillance in patients suffering from genital HPV disease and bowenoid papulosis [48, Jablonska and Majewski, this volume]. The role of humoral immunity in the regression of HPV-associated lesions is less clear. Other exogenic factors modulating the outcome of genital HPV disease and of bowenoid papulosis are cigarette smoking and drug addiction [29, 31, 33, 51, 52]. Cofactors seem to be important since only a small number of premalignant lesions convert to invasive disease; furthermore, there is an overall striking difference between the incidence of virus-associated genital cancer in females and in males [7, 8].

Clinical Features

Bowenoid papulosis is primarily a disease of young adults. In a serie of 22 patients, the age of men was found to range from 20 to 42 years (average age 28 years) and that of women from 18 to 39 years (average age 32 years) [25]. This is in line with reports from other groups [56, 76, 77]. Nevertheless, this disorder is also seen in children of 1–2 years of age and in old patients [38]. In contrast, the average age of patients suffering from Bowen's disease is 48 years, and it is uncommon for this disease to begin prior to the age of 20 [19]. Vulvar Bowen's disease generally occurs between the 3rd and 6th decades of life [1]. Every age group – reproductive, menopausal and postmenopausal – is represented; however, there is a tendency towards the older age groups. The course of bowenoid papulosis is often chronic. On average, men and women suffer about 2.4 and 3.6 years, respectively. In some cases of bowenoid papulosis lasting for as long as 8–10 years, transformation into carcinoma is

Table 1. Synonyms for bowenoid papulosis

Title	Author(s)	Ref.
Intraepithelial and infiltrative carcinoma of the vulva: Bowen's type	Abell and Gosling	[1]
Multicentric pigmented Bowen's disease (MPBD)	Lloyd	[46]
	Kimura et al.	[42]
Reversible vulvar atypia	Friedrich	[16]
Bowenoid atypia of the vulva	Skinner et al.	[69]
Multicentric Bowen's disease of the genitalia	Berger and Hori	[5]
Pigmented penile papules with carcinoma in situ changes	Katz et al.	[40]
Early vulvar carcinoma	Kunschner et al.	[43]
Bowenoid papulosis of the penis	Wade et al.	[76]
Vulvar neoplasia in the young	Hilliard et al.	[34]
Bowenoid papulosis of the genitalia	Wade et al.	[77]
Carcinoma in situ of the vulva	Buscema et al.	[10]
Intraepithelial carcinoma of the vulva	Kaplan et al.	[39]
Multicentric vulvar carcinoma in situ	Wilkinson et al.	[78]
Bowenoid dysplasia of the vulva	Ulbright et al.	[72]
Vulvar intraepithelial neoplasia (VIN)	Crum et al.	[14]
Penile intraepithelial neoplasia (PIN)	Levine et al.	[45]

seen [60]. This is a rare event in multifocal lesions of the vulva, which are also termed as early vulva carcinoma [43], carcinoma in situ of the vulva [10], or vulvar intraepithelial neoplasia (VIN) [14] and which are all in line with bowenoid papulosis (Table 1).

Incidence According to Sex and Race

There seems to be a slight predominance in men, especially regarding the papular lesions. In almost all reports there is a predominance of Caucasians. Of 72 patients reported by Patterson et al. [56] only 11 where non-Caucasians.

Sites of Predilection

Bowenoid papulosis lesions affect the anogenital skin and adjacent mucosal sites in males and females. In our series the glans penis and the vaginal vestibulum as well as the labia minora and majora have been the most commonly involved sites [25, 29, 56, 76, 77]. Other sites of skin and mucosa are rarely involved. However, similar papular lesions with histologically proven bowenoid dysplasia located at the axilla and at the neck were described in a few instances [42 and E. I. Grussendorf-Conen, personal communication].

Clinical Description

The clinical appearance of bowenoid papulosis includes a spectrum of clinically inconspicuous, noncondylomatous manifestations [1, 25] (Fig. 1): There are lichenoid pink-colored or reddish-brown papules, and sometimes these lesions coalesce to form small plaques. There are also macular erythematous lesions, especially seen on the glans penis, at the inner aspect of the foreskin or at the introitus vaginae. Furthermore, there are leukoplakialike lesions and subclinical lesions which are hardly detectable, sometimes only after use of acetic acid (Fig. 1), also presenting with histological signs of carcinoma in situ. The lesions range from 0.2 cm to about 3.5 cm in diameter (average about 1.0 cm). Mostly the lesions are multiple, sometimes appearing as plaques. But there are also patients with single lesions. Pigmentation is frequently seen. In contrast to condylomata acuminata, the surface of bowenoid papulosis lesions is smooth or only very slightly papillomatous. In contrast to erythroplasia of Queyrat the surface of the lesions is mostly intact. In Bowen's disease and in erythroplasia of Queyrat single lesions predominate in 50% of the patients and eczematous lesions are common, and the plaques appear often fissured, crusted and eroded, or even ulcerated [56]. Scaling is also mostly absent in bowenoid papulosis. Altogether the clinical appearance of bowenoid papulosis is that of a totally benign disease immitating a number of other skin disorders.

Symptoms

The lesions usually are asymptomatic, but occasionally pruritic. During the course of disease, enlargement of lesions and inflammation are seen. Hemorrhage is an extremely rare event.

Diagnosis

In spite of their name, bowenoid papulosis lesions mostly lack morphologic characteristics of papilloma. In view of the fact that this disease is relatively unknown among clinicians and that its course is symptomless, the diagnosis is usually made by microscopic analysis.

In order to diagnose bowenoid papulosis clinical description of the lesion, history, and histologic investigation are all necessary. Demonstration of viral DNA can be done by in situ hybridization. Demonstration of structural antigens of papillomaviruses is not sufficient since only about 5% of lesions have shown to be positive [9, 25, 32, 59]. In most cases the diagnosis can be made clinically (macroscopy and histology). Colposcopy or peniscopy is necessary for the detection of subclinical lesions which also can exhibit histologic features of a carcinoma in situ or of a severe intraepithelial neoplasia. This is done by applying 3%–5% acetic acid over 5–10 min; this

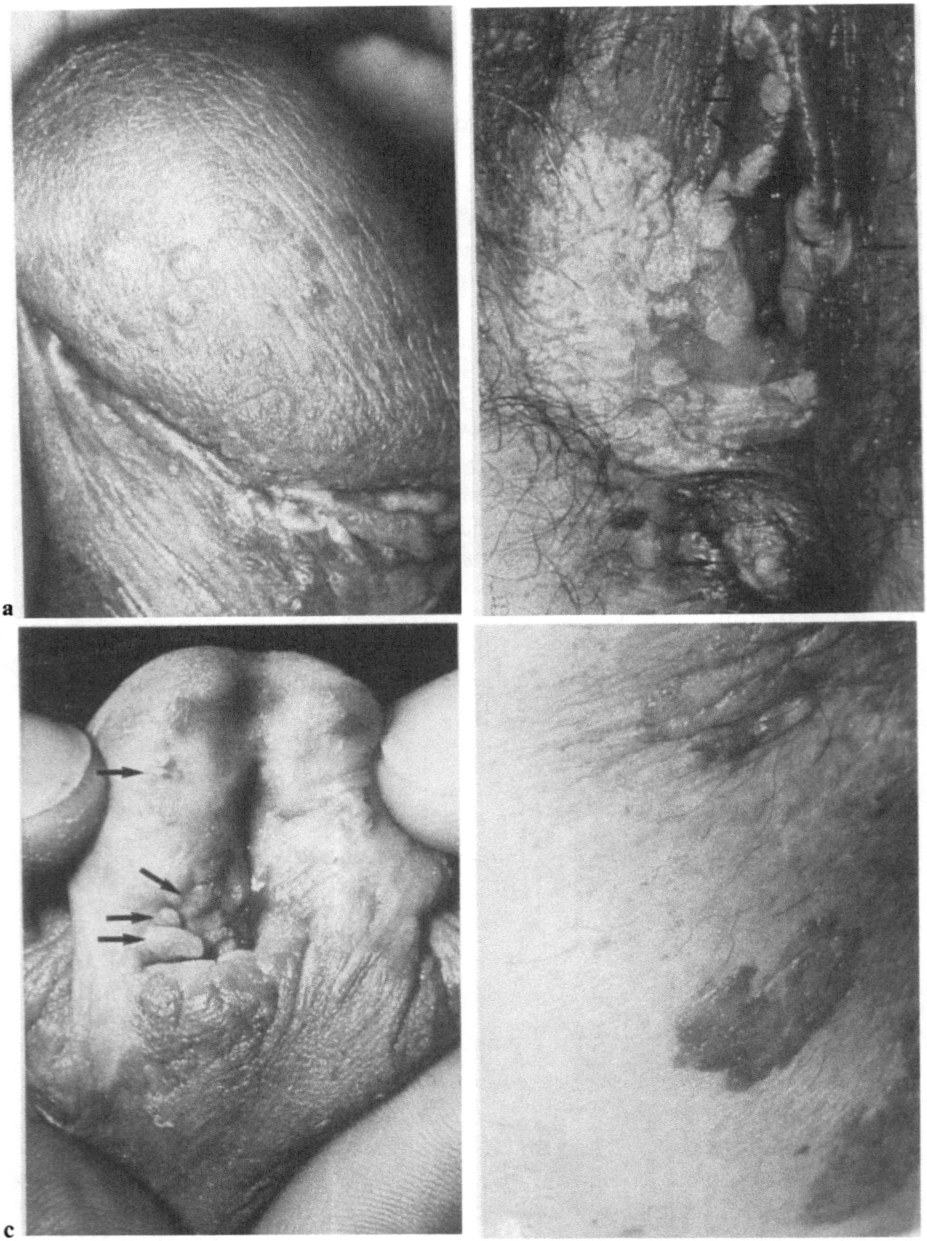

a b c d

Fig. 1 a–d. Bowenoid papulosis: gross appearance. **a** Multiple lichenoid papules on the glans penis (HPV 16 DNA present, Southern blot hybridization). **b** Acetic acid – white confluent plaques of maculopapular lesions at the vulva. Multiple lesions (*arrows*) also present at the perianal skin. **c** Multiple slightly pigmented papules on the meatus urethrae externus in a young heroin addict after various surgical therapies for meatal condylomata acuminata (HPV 16 DNA and HPV 11 DNA present, Southern blot hybridization). **d** Multicentric slightly pigmented papules and plaques at the perianal skin (*arrows*) leads to a white epithelium and to the visualization of pathologic vessels characteristic of HPV infection. Furthermore, application of 2% Toluidin blue (Collin's assay) which is rinsed with acetic acid helps to confine the area from which a biopsy should be taken. Positive results are indicative of cancerous or inflammatory diseases in which nucleated cells are on the surface of the epithelium. Cytological techniques such as Papanicolaou staining of cells from the vulva or penis are inadequate for the diagnosis of vulvar or penile HPV infections as not enough intact cells can be recovered.

Differential Diagnosis

Bowenoid papulosis may appear as pigmented papular lesions or flat condylomalike lesions which must be differentiated from Bowen's disease (Table 2) and from a large number of other dermatological disorders, e.g.,

Table 2. Bowenoid papulosis and Bowen's disease

Characteristics	Bowenoid papulosis	Bowen's disease
Age of onset (years)	About 30	Over 45
Number of lesions	Multiple	Solitary
Distribution of lesions	Skin and mucous membranes	Skin
Spontaneous regression	+	–
Symptoms	None (slight pruritus)	Pruritus in about 50% of cases
Clinical appearance of disease	Lichenoid papules	Slightly raised erythematous crusted, scaly plaque
	Pigmented papules	(MPBD)
	Erythematous macules	
	Leukoplakialike lesions	
Color of lesions	Pink/reddish	Red
	Grayish/white	White
	Brown/black	
Microscopic features	Hyperkeratosis	Variable hyperkeratosis
	Parakeratosis	Parakeratosis
	Irregular acanthosis	Acanthosis
	Loss of polarity in rare case scattered dysplastic cells („salt and pepper“ features) together with orderly epithelial maturation	Loss of polarity
	Abnormal mitoses rare	Dyskeratotic cells and abnormal mitoses involving the whole thickness of epithelium
	Spared acrotrichium	Involvement of acrotrichium
	Acrosyringium may be involved	Spared acrosyringium

MPBD, multicentric pigmented Bowen's disease

condyloma acuminatum, psoriasis, lichen ruber, seborrheic keratosis balanitis, vulvitis, lichen sclerosus et atrophicus, basal cell carcinoma, and even malignant melanoma. Furthermore penile lentigo, penile melanosis, and melanosis must be differentiated [44]. The same is true for melanoacanthoma of the penile shaft [75]. In such cases in situ hybridization is useful for the detection of HPV DNA. In some cases Southern blot hybridization should be done in order to clarify the etiology of such atypical pigmented penile macules and papules.

Relationship to Malignancy

Female patients with bowenoid papulosis or vulvar intraepithelial neoplasia are at high risk for cervical carcinoma. In the majority of sexual partners of patients with bowenoid papulosis cervical human papillomavirus infection is found, and severe dysplasia or carcinoma in situ may be present [22, 27, 51, 52, 62]. Patients with multifocal or early vulvar carcinoma have a high incidence of cervical neoplasia [23, 25, 51, 52]. The same is true for bowenoid intraepithelial neoplasia in women treated previously for cervical carcinoma [51, 52]. Long duration of bowenoid papulosis with tendency to dissemination of disease should always prompt a thorough investigation of the patient in order to prevent transformation to vulvar carcinoma [6]. There are instances of other gynecologic disorders such as pelvic inflammatory disease, uterine bleeding, cervicitis, and cervical dysplasia which may precede or accompany the development of bowenoid papulosis [56]. In a number of patients suffering from bowenoid papulosis coincidentally to carcinoma in situ of the cervix, various other neoplasias were present: Hodgkin's disease, carcinoma of the breast which had metastasized to the ovaries, and neoplasias due to chemotherapy with cyclophosphamide and corticosteroids, systemic lupus erythematous [56], and malignant melanoma together with carcinoma of the breast which had metastasized to axillary lymph nodes [22].

Histopathology

Due to the symptomless course and the inconspicuous clinical appearance biopsies are rarely done. The most common diagnoses submitted with the biopsy specimens are condyloma acuminatum, common wart, psoriasis, lichen ruber, melanosis, nevus, seborrheic keratosis, and penile lentigo. Less frequently fungal infections, balanitis, vulvitis, lichen simplex chronicus, and basal cell carcinoma and malignant melanoma are suspected [21, 29, 56, 76, 77].

Hematoxylin eosin-stained tissue sections show hyperkeratosis, parakeratosis, granulosis, vacuolated keratinocytes, irregular acanthosis, papillomatosis, and inflammation in varying intensity. A very prominent feature is the epidermal maturation with koilocytes present in the outmost

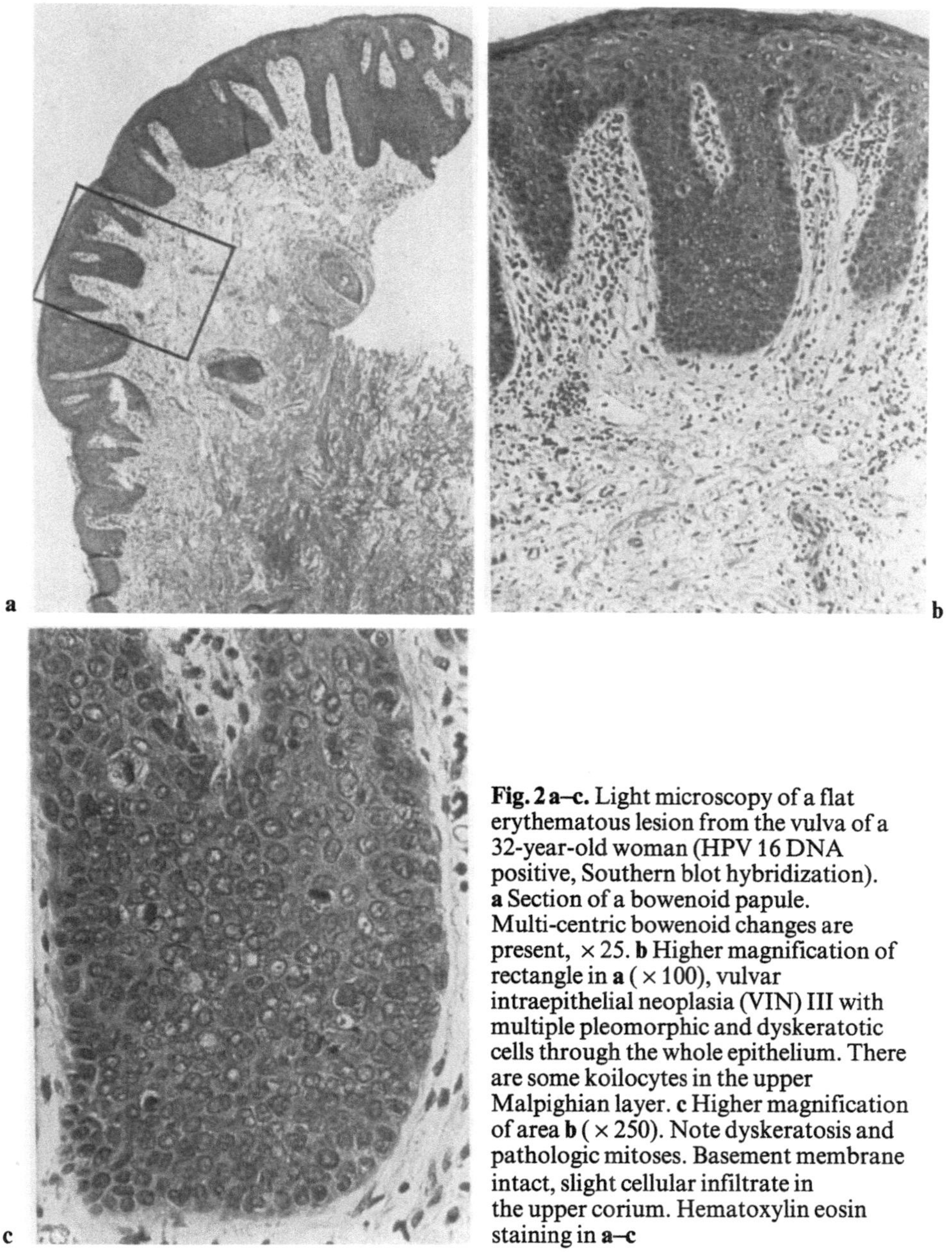

Fig. 2 a–c. Light microscopy of a flat erythematous lesion from the vulva of a 32-year-old woman (HPV 16 DNA positive, Southern blot hybridization). **a** Section of a bowenoid papule. Multi-centric bowenoid changes are present, × 25. **b** Higher magnification of rectangle in **a** (× 100), vulvar intraepithelial neoplasia (VIN) III with multiple pleomorphic and dyskeratotic cells through the whole epithelium. There are some koilocytes in the upper Malpighian layer. **c** Higher magnification of area **b** (× 250). Note dyskeratosis and pathologic mitoses. Basement membrane intact, slight cellular infiltrate in the upper corium. Hematoxylin eosin staining in **a–c**

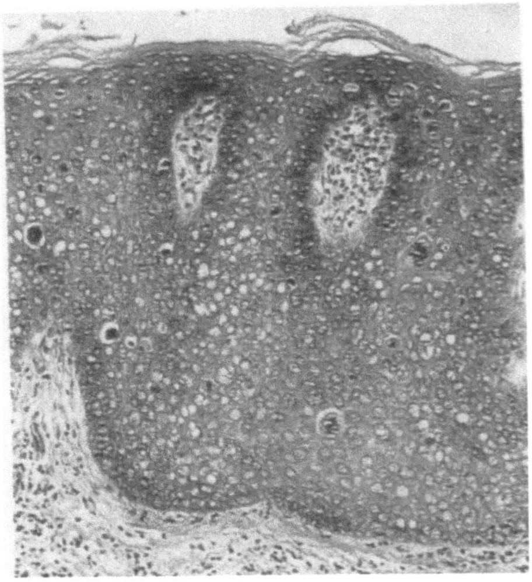

Fig. 3. Bowenoid papulosis with dysplastic in situ cell features of "salt and pepper" and marked perinuclear vacuolization of squamous cells (HE ×10)

epithelial layers together with the papillomavirus common antigen within the nuclei of these cells [9, 25, 32, 59]. In close connection benign acanthosis and features of Bowen's atypia are seen, frequently abnormal mitotic figures in the whole epidermis and scattering of dysplastic keratinocytes with hyperchromatic nuclei at all levels of the epidermis or mucosa (Figs. 2, 3). A most characteristic feature is the salt and pepper appearance of in situ dysplastic cells together with orderly keratinocyte maturation of epidermis and mucosa together with pigment being free and within melanophages in the papillary dermis (Figs. 2, 3) [56]. Plasma cells are rarely seen. In the stratum corneum there are parakeratotic cells with foci of follicular hyperkeratosis and marked basophilic rounded nuclei surrounded by a halo, resembling koilocytes [26] or vacuolated cells of HPV infection seen in common warts, flat warts, or plantar warts [23]. By definition basement membranes at dermoepidermal and mucosal-submucosal junctions are intact. Due to investigations of Patterson et al. [56] in bowenoid papulosis acrotrichium is spared and acrosyringium may be involved by dysplastic cells; this is in contrast to Bowen's disease where involvement of acrotrichium is regularly seen but the acrosyringium is usually spared. Further histologic features which may help to differentiate between bowenoid papulosis and Bowen's disease are the presence of vesicular chromatin within dysplastic cells of bowenoid papulosis and the lesser degree of cytologic atypia in bowenoid papulosis (Table 2).

Electron Microscopy

Using transmission electron microscopy Faber and Hagedorn [15] described dysplastic cells with enlarged nucleoli, free of keratohyalin with widened inter-

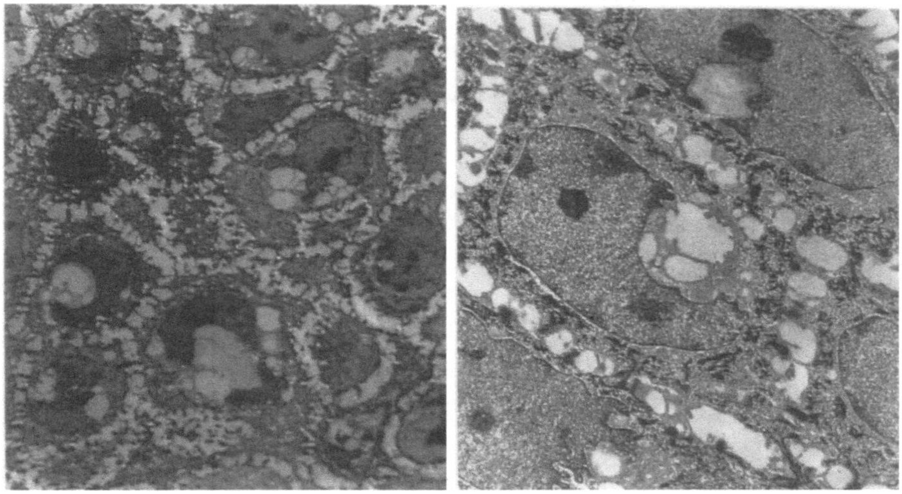

a b

Fig. 4 a, b. Electron microscopy. **a** Ultrathin section of dyskeratotic keratinocyte with prominent perinuclear vacuolization and widened intercellular spaces, short tonofilaments, and a decrease of desmosomes. **b** Perinuclear vacuolization with invagination and deformation of the nucleus to a kidneylike shape

cellular spaces, short and partially aggregated tonofilaments, and decreased numbers of desmosomes (Fig. 4). A further remarkable finding is a peculiar type of perinuclear vacuolization of keratinocytes with invagination of the nucleus or with deformation of the nucleus to a kidneylike shape (Fig. 4 B). Chromatin appears to be dispersed with irregular condensation at the periphery of the nucleus and of the nucleoli of dysplastic cells.

Treatment

There are a large number of therapeutic modalities in HPV-associated precancerous lesions of the genital tract [21, 52, 66]. It is generally accepted that radical surgical methods should be avoided in bowenoid papulosis [71], because the course of this disease is often benign. Nevertheless, due to ignorance of this peculiar entity there are still young women with scars and psychological disorders as a consequence of such therapies. Management of bowenoid papulosis in young patients should be conservative (i.e., superficial surgery-electrocautery, therapy with carbondioxid laser, or cryotherapy). In addition topical 5-fluorouracil or preparations containing vitamin A acid have been proposed. In circumscribed disease surgical excision is generally indicated. It is important to know that like condylomata acuminata, bowenoid papulosis lesions tend to recur after ablative therapy or after initially successful topical treatment. Circumcision is not indicated when the penis is affected because an association has been shown between previous circumcision and development

Table 3. Interferon treatment of CIN, VIN, VAIN, and PIN

Diagnosis	Type of interferon	Route of administration	Number of evaluable patients	Success of therapy CR	PR	NC	Authors	Ref.
CIN II–III	Alpha	Gel	15	3	–	12	Ikic et al.	[36]
CIN II–III	Alpha	Gel	6	3	3	–	Möller et al.	[49]
CIN II–III	Beta	Intralesional + gel	11	6	2	5	de Palo et al.	[54]
VIN III	Beta	Intralesional + gel	2	1	–	1	de Palo et al.	[54]
VAIN I–II	Alpha	Gel	8	3	2	3	Vesterinen et al.	[73]
PIN I[a]	Beta	Intralesional + gel	1	1	–	–	Gross et al.	[24]
CIN II–III	Beta	Perilesional	16	8	2	6	de Palo et al.	[55]
CIN II–III	Beta	Gel	7	2	4	1	Choo et al.	[12]
CIN II–III	Alpha + Beta	Intralesional	12	9	–	3	Choo et al.	[13]
CIN II–III	Alpha	Gel	13	3	–	10	Byrne et al.	[11]
PIN III[b]	Alpha	Subcutaneous	3	1	2	–	Gross et al.	[28]
CIN I–III	Alpha	Subcutaneous + gel	6	3	2	1	Schneider et al.	[64]
VIN III	Alpha	Subcutaneous + gel	3	–	3	–	Schneider et al.	[64]
VIN II[b]	Alpha	Subcutaneous + gel	1	1	–	–	Slotman et al.	[70]
CIN II–III	Gamma	Gel	24	10	9	5	Schneider et al.	[65]
CIN II–III	Beta	Perilesional + gel	24	14	3	7	Neis et al.	[50]

CIN, cervical intraepithelial neoplasia; VIN, vulvar intraepithelial neoplasia; VAIN, PIN, penile intraepithelial neoplasia; CR, complete remission; PR, partial remission; NC, no change.
[a] Initially CR, later recurrence of flat penile lesions (histology: PIN III, virustype HPV 16 (Southern blot).
[b] Low-dose interval therapy.

of bowenoid papules [5, 40, 56, 77]. Most important is the question of what will happen to the lesions if they are left untreated. As bowenoid papulosis must be considered a sexually transmitted disease, lesions should be removed or at least transmission must be prohibited by use of condoms. Furthermore, the partner has to be followed-up thoroughly as well.

Widespread bowenoid papulosis lesions and subclinical papillomavirus infections with histologically proven features of a carcinoma in situ or severe intraepithelial neoplasia must be treated, especially in immunocompromised patients.

There is an increasing body of data from clinical studies which supports the hypothesis that interferon therapy is useful in HPV diseases such as viral warts, laryngeal papillomatosis, and condylomata acuminata [2; Gross, this

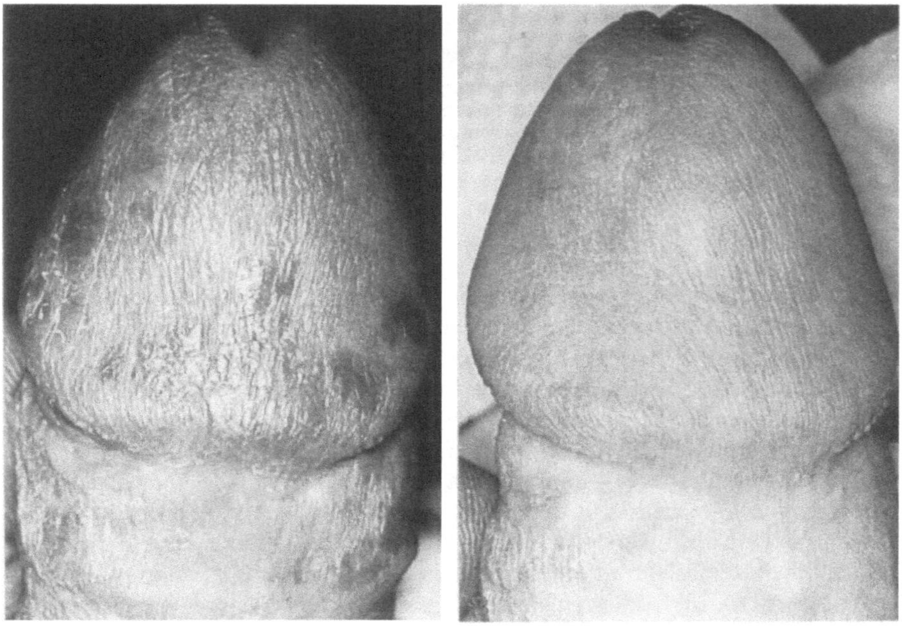

a b

Fig. 5 a, b. Interferon gamma-induced remission of bowenoid papulosis lesions. **a** Flat, erythematous lesions on the glans and on the prepuce. **b** After successful interferon gamma therapy (patient belonged to group A – continuous therapy). Flat bowenoid lesions have regressed completely

volume]. Clinical improvement was also reported in flat condylomata of the uterine cervix, CIN, VIN, VAIN (vaginal intraepithelial neoplasia) and also in bowenoid papulosis which responded to parenteral administration of interferon (Table 3). The disadvantage of this approach is the long duration of treatment. Complete remissions were seen in one out of three patients and three out of six patients treated with interferon alpha-2 (daily dose 5 MU) [28] and interferon gamma (daily dose 4 MU, respectively; Gross, unpublished data). Recently published preliminary results of an open randomized trial of recombinant interferon gamma show that in contrast to condylomata acuminata bowenoid papulosis lesions respond better to continuous than to intermittent interferon gamma injections [30]. Recombinant interferon gamma was given subcutaneously to 12 patients at a daily dose of 4 MU by injection. Four patients each were assigned to one of three treatment groups consisting of continuous therapy with three subcutaneous injections per week for 13 weeks, intermittent block therapy with four 6-week cycles consisting of five injections on days 1, 3, 5, 7, and 9 of each cycle, and intermittent single dose therapy with six 4-week cycles consisting of only one subcutaneous injection on day one of each cycle. At the 26th week after onset of therapy complete responses were seen in three out of four patients of treatment group A (Fig. 5), whereas in the treatment group B and C only one patient each responded partially. Alternatively interferon can be given as adjuvant both systemically and topically as gel to treat bowenoid papulosis after superficial surgery, especially colposcopy guided CO_2-laser. Topical interferon treatment is thought to prevent recurrence of disease by inducing an antiviral state in basal epithelial cells, and it seems likely that this approach is independent of the general immune system.

Follow-up

As bowenoid papulosis presents a risk for the partner infection with HPV-16 as well as a risk with respect to the development of carcinoma of the vulva, of the cervix, and of the penis, a consequent follow-up is necessary. With persistent bowenoid papulosis lesions a thorough clinical investigation is necessary in order to exclude presence or development of malignant disease such as breast carcinoma, malignant melanoma, or other neoplasias.

Follow-up studies of patients with Bowen's disease and erythroplasia of Queyrat provide marked differences from patients suffering from bowenoid papulosis [56]. Both Bowen's disease and erythroplasia of Queyrat reveal an increased potential for the development of invasive carcinoma and distant metastases [18–20]. Malignant transformation is seen in both genital and extragenital Bowen's disease. A large study showed that 5% of patients suffering from Bowen's disease presented with invasion of primary lesion [18]. This is in contrast to data from the same authors, demonstrating 10% with invasive squamous cell carcinoma in primary penile erythroplasia of Queyrat lesions [20].

A further factor as related to development of bowenoid papulosis is the effect of podophyllum resin, which is rather frequently used to treat genital warts of outpatients. The application of podophyllum resin leads to epithelial atypia very similar to bowenoid papulosis [7, 17, 37, 56] with clumping, pyknosis, and distortion of the epithelial nuclei and formation of epithelial giant cells resulting from incomplete mitoses. In contrast to true bowenoid papulosis such lesions, together with the epithelial changes, disappear within 1–2 weeks after application. So far no concise data exist on the outcome of such lesions in the long term.

Summary and Conclusions

It has been the merit of molecular biological methods introduced into clinical investigations to reveal the etiology of bowenoid papulosis. The origin of this disease was not clear until the early 1980s.

Today it seems likely that bowenoid papulosis is a reservoir especially of HPV 16 since in about 80% of cases investigated DNA of this HPV type could be identified using Southern blot hybridization [25, 35]. This is of importance with regard to the male, who possibly plays a key role in the genesis of cervical carcinoma [29, 51, 52, 80]. Recent investigations have yielded increasing evidence that bowenoid papulosis is only one out of a number of noncondylomatous lesions at the external genitalia which harbor HPV 16 DNA and further HPV DNA of types different from HPV 6 and HPV 11, the latter being present in more than 90% of classical cases of condylomata acuminata. Such lesions may be hardly visible since they are often flat and macular. Application of low concentrations of acetic acid and observation with a colposcope or a peniscope allow them to be visualized. Histologically both bowenoid features and features similar to condylomata acuminata may be present. So far it is not clear whether these subclinical papillomavirus infections (SPI) are precursors of bowenoid papulosis or of condylomata acuminata or whether they are a special entity associated with distinct virus types. Furthermore, the natural course of bowenoid papulosis and of SPI is not yet fully understood. Although in general bowenoid papulosis has a fairly benign course with spontaneous regression after several months, malignant transformation has been observed in a number of cases. Thus, bowenoid papulosis bears a risk with respect to the development of squamous cell carcinoma of the lower genital tract of both sexes.

Early diagnosis is important since bowenoid papulosis lesions usually harbor HPV 16 or other "higher-risk virus types" in the form of possibly infectious virus particles, as demonstrated by the presence of virus capsid antigens in about 5% of the lesions investigated.

Management of bowenoid papulosis is unsatisfactory, because most of the therapeutic methods are ineffective and radical treatment is still being used for a disease which regresses spontaneously in a large number of cases.

Recent in vitro data of Jablonska and Majewski (this volume) imply that immunomodulation might be helpful. So far, at least interferon given parenterally or topically as adjuvant to CO_2-laser or electrocautery seems to prevent recurrence of bowenoid papulosis (unpublished data). Preliminary results of a controlled study with recombinant interferon gamma given subcutaneously point to a beneficial effect. In contrast to condylomata acuminata, however, bowenoid papulosis lesions respond better to the continuous than to the interval regimen. Thus, apparently the proliferation of bowenoid papulosis lesions and possibly of other intraepithelial neoplasias in the anogenital region is inhibited more by antiproliferative than by immunomodulating effects. The latter effect is quite certain to be responsible for the high response rate seen in condylomata acuminata treated with low doses of interferon given at intervals (Brzoska, this volume; Gross, this volume).

References

1. Abell MR, Gosling JR (1961) Intraepithelial and infiltrative carcinoma of the vulva. Bowen's type. Cancer 14:318–329
2. Androphy EJ (1986) Papillomaviruses and interferon. In: Ciba Foundation Symposium 120: papillomaviruses. Wiley, New York, pp 221–228
3. Beaudenon S, Kremsdorf D, Obalek S et al. (1987) Plurality of genital human papillomaviruses: characterization of two new types with distinct biological properties. Virology 161:374–384

4. Bender ME, Katz I, Posalaky Z (1980) Carcinoma in situ of the genitalia. JAMA 243:145–147
5. Berger BW, Hori Y (1978) Multicentric Bowen's disease of the genitalia. Spontaneous regression of lesions. Arch Dermatol 114:1698–1699
6. Bergeron C, Naghashfar Z, Canoan C et al. (1987) Human papillomavirus type 16 in intraepithelial neoplasia (bowenoid papulosis) and coexistent invasive carcinoma of the vulva. J Gynecol Pathol 6:1–11
7. Blaustein A (1977) Pathology of the female genital tract. Springer, Berlin Heidelberg New York, pp 138–139
8. Bracken RB (1981) Genitourinary cancer. Cancer Chemother 2:199–242
9. Braun L, Farmer ER, Shah KV (1983) Immunoperoxidase localization of papillomavirus antigen in cutaneous warts and bowenoid papulosis. J Med Virol 12:187–193
10. Buscema J, Woodruff JD, Petros P (1980) Carcinoma in situ of the vulva. Obstet Gynecol 64:16–20
11. Byrne MA, Möller BR, Taylor-Robinson D (1986) The effect of interferon on human papillomavirus associated with cervical intraepithelial neoplasias. Br J Obstet Gynecol 93:1156–1174
12. Choo YC, Hsu C, Seto WH et al. (1985) Intravaginal application of leukocyte interferon gel in the treatment of cervical intraepithelial neoplasia (CIN). Arch Gynecol 237:51–54
13. Choo YC, Seto WH, Hsu C et al. (1986) Cervical intraepithelial neoplasia treated by perilesional injection of interferon. Br J Obstet Gynecol 93:372–379
14. Crum CP, Liskow A, Petros P (1984) Vulvar intraepithelial neoplasia (severe atypic and carcinoma in situ): a clinicopathologic analysis of 41 cases. Cancer 54:1429–1434
15. Faber M, Hagedorn M (1981) A light and electron microscopic study of bowenoid papulosis. Acta Derm Venereol (Stockh) 61:397–403
16. Friedrich EG (1972) Reversible vulvar atypia. A case report. Obstet Gynecol 39:137–141
17. Friedrich EG (1983) Vulvar disease, 2nd edn. Saunders, Philadelphia, pp 59–60 (Major problems in obstetrics and gynecology, vol 9)
18. Graham JH, Helwig EB (1961) Bowen's disease and its relationship to systemic cancer. Arch Dermatol 83:738–758
19. Graham JH, Helwig EB (1972) Premalignant cutaneous and mucocutaneous diseases. In: Graham JH, Johnson WC, Helwig EB (eds) Dermat Pathology. Harper and Row, Hagerstown, pp 581–606
20. Graham JH, Helwig EB (1973) Erythroplasia of Queyrat: a clinicopathologic and histochemical study. Cancer 32:1396–1414
21. Gross G (1987) Lesions of the male and female external genitalia associated with human papillomaviruses. In: Syrjänen K, Gissmann L, Koss LG (eds) Papillomaviruses and human disease. Springer, Berlin Heidelberg New York, pp 197–234
22. Gross G (1988) Partnerdiagnostik der sexuell übertragbaren Papillomviruskrankheit – Fallbericht. Hautarzt [Suppl VIII]:73–75
23. Gross G, Pfister H, Hagedorn M et al. (1982) Correlation between human papillomavirus (HPV) type and histology of warts. J Invest Dermatol 78:160–164
24. Gross G, Ikenberg H, Gissmann L (1984) Bowenoid dysplasia in human papillomavirus 16 DNA positive flat condyloma during interferon beta treatment. Lancet I:1467–1468
25. Gross G, Hagedorn M, Ikenberg H et al. (1985) Bowenoid papulosis. Presence of human papillomavirus (HPV) structural antigens and of HPV 16-related DNA sequences. Arch Dermatol 121:858–863
26. Gross G, Ikenberg H, Gissmann L et al. (1985) Papillomavirus infection of the anogenital region: correlation between histology, clinical picture and virus type. Proposal of a new nomenclature. J Invest Dermatol 85:147–152

27. Gross G, Wagner D, Hauser-Brauner B et al. (1985) Bowenoide Papulose und Carcinoma in situ der Cervix uteri bei Sexualpartnern. Hautarzt 36:465–469
28. Gross G, Villiers EM, Roussaki A et al. (1986) Successful treatment of condylomata acuminata and bowenoid papulosis with subcutaneous injections of low-dose recombinant interferon alpha. Arch Dermatol 122:749–750
29. Gross G, Ikenberg H, de Villiers EM et al. (1986) Bowenoid papulosis: a venerally transmissible disease as reservoir for HPV 16. In: Zur Hausen H, Peto P (eds) Origin of female genital cancer: virological and epidemiological aspects (Banbury report). Cold Spring Harbor Laboratories, Cold Spring Harbor, pp 149–165
30. Gross G, Roussaki A, Papendick U (1990) Efficacy of interferons on Bowenoid papulosis and other precancerous lesions. J Invest Dermatol (in press)
31. Gross G, Roussaki A, Ikenberg H, Drees N (1990) Genital warts do not respond to systemic recombinant interferon alpha 2a treatment during cannabis consumption. Dermatologica (submitted for publication)
32. Guillet GY, Braun L, Masse R et al. (1984) Bowenoid papulosis: demonstration of human papillomavirus (HPV) with anti-HPV immune serum. Arch Dermatol 120:514–516
33. Hauser B (1987) Untersuchungen zur sexuellen Übertragbarkeit genitaler und analer Papillomvirusinfektionen. Inangural-Dissertation, Universität Freiburg
34. Hilliard GD, Massey FM, O'Toole RV (1979) Vulvar neoplasia in the young. Am J Obstet Gynecol 135:185–188
35. Ikenberg H, Gissmann L, Gross G et al. (1983) Human papillomavirus type-16-related DNA in genital Bowen's disease and in bowenoid papulosis. Int J Cancer 32:563–565
36. Ikic D, Kirchmajer V, Maricic Z et al. (1972) Application of human leukocyte interferon in patients with carcinoma of the uterine cervix. Lancet I:1027–1030
37. Kaminetzky HA (1960) Human cervical epithelial changes produced by podophyllin. Am J Obstet Gynecol 80:1055–1060
38. Kao GF, Graham JH (1982) Bowenoid papulosis. Int J Dermatol 21:445–446
39. Kaplan AL, Kaufmann RH, Birken RA (1981) Intraepithelial carcinoma of the vulva with extension to the anal canal. Obstet Gynecol 58:368–371
40. Katz H, Posalaky Z, McGinley D (1978) Pigmented penile papules with carcinoma in situ changes. Br J Dermatol 99:155–162
41. Kaufmann RH, Dreesmann GR, Burek J et al. (1981) Herpesvirus induced antigens in squamous cell carcinoma in situ of the vulva. N Engl J Med 303:483–488
42. Kimura S, Hirai R, Harada R et al. (1978) So-called multicentric Bowen's disease. Report of a case and a possible etiologic role of human papillomavirus. Dermatologica 157:229–237
43. Kunschner A, Kenbour A, David B (1978) Early vulvar carcinoma. Am J Obstet Gynecol 132:599–602
44. Leicht S, Youngberg G, Dias-Miranda C (1988) Atypical pigmented penile macules. Arch Dermatol 124:1267–1270
45. Levine RM, Crum CP, Herma F (1984) Cervical papillomavirus infection and intraepithelial neoplasia: a study of male sexual partners. Obstet Gynecol 64:16–20
46. Lloyd KM (1970) Multicentric pigmented Bowen's disease of the groins. Arch Dermatol 101:48–51
47. Lupulescu A, Mehregan AH, Rahbari H et al. (1977) Venereal warts vs. Bowen's disease: a histologic and ultrastructural study of five cases. JAMA 237:2520–2522
48. Malejczyk J, Majewski S, Jablonska S (1987) Natural cell-mediated cytotoxicity in patients with anogenital lesions induced by potentially oncogenic human papillomaviruses. In: Steinberg BM, Brandsma JC, Taichman LB (eds) Papillomaviruses. Cold Spring Harbor Laboratories, Cold Spring Harbor, pp 381–385 (Cancer cells, vol 5)
49. Möller BR, Johannesen E, Osther K et al. (1983) Treatment of dysplasia of the cervical epithelium with an interferon gel. Obstet Gynecol 62:625–629

50. Neis KJ, Tesseraux M, Claußen C et al. (1989) Lokale Therapie cervikaler intraepithelialer Neoplasien mit natürlichem beta-Interferon. Arch Gynecol 245:550
51. Obalek S, Jablonska S, Orth G (1985) HPV associated intraepithelial neoplasia of external genitalia. Clin Dermatol 3:104–113
52. Obalek S, Jablonska S, Beaudenon MB et al. (1986) Bowenoid papulosis of the male and female genitalia: risk of cervical neoplasia. J Am Acad Dermatol 14:433–444
53. Olson RL, Norquist RE, Everett MA (1968) An electron microscopic study of Bowen's disease. Cancer Res 28:2078–2085
54. de Palo G, Stefanon B, Rilke F et al. (1984) Human fibroblast interferon in cervical and vulvar intraepithelial neoplasia associated with papilloma virus infection. Int J Tissue React VI (6):523–525
55. de Palo G, Stefanon B, Rilke F et al. (1985) Human fibroblast interferon in cervical and vulvar intraepithelial neoplasia associated with viral cytopathic effects. J Reprod Med 30:404–408
56. Patterson JW, Kao GF, Graham JH (1986) Bowenoid papulosis. A clinicopathologic study with ultrastructural observations. Cancer 57:823–836
57. Pelissier CI, Yaneva M, Joubert E (1981) Cancers genitaux et lesions precancereuses chez les femmes transplantees renales. A propos de l'etude de 90 femmes greffees. Gynecologie 32:359–365
58. Penn I (1986) Cancers of the anogenital region in renal transplant recipients. Analysis of 65 cases. Cancer 58:611–616
59. Penneys NS, Mogollon RJ, Nadji M et al. (1984) Papillomavirus common antigens. Papillomavirus antigen in verruca, benign papillomatous lesions, trichilemmoma, and bowenoid papulosis: an immunoperoxidase study. Arch Dermatol 120:859–861
60. Planner RS, Andersen HE, Hobbs JB (1987) Multifocal invasive carcinoma of the vulva in a 25-year-old women with bowenoid papulosis. Aust NZ J Obstet Gynecol 27:291–295
61. Porecco R, Penn I, Droegmueller W et al. (1975) Gynecologic malignancies in immunosuppressed organ homograft recipients. Obstet Gynecol 45:359–364
62. Rüdlinger R (1987) Bowenoid papulosis of the male and female genital tracts: risk of cervical neoplasia. J Am Acad Dermatol 16:625–627
63. Schneider V, Kay S, Lee HM (1983) Immunosuppression as a high-risk factor in the development of condyloma acuminatum and squamous neoplasia of the cervix. Acta Cytol 27:220–224
64. Schneider A, Papendick U, Gissmann L et al. (1987) Interferon treatment of human genital papillomavirus infection: importance of viral type. Int J Cancer 40:610–614
65. Schneider A, Kirchmayr R, Wagner D et al. (1989) Efficacy trial of topically applied gamma interferon in cervical intraepithelial neoplasia. In: Gross G (ed) Genital papillomavirus infections. Advances in modern diagnostic and therapy, number 57, Hamburg
66. Schwarz RA (1988) Bowen's disease. In: Schwartz RA (ed) Skin cancer: recognition and management. Springer, Berlin Heidelberg New York, pp 26–35
67. Sillmann FH, Stanek A, Sedlis A et al. (1984) The relationship between human papillomavirus and lower genital intraepithelial neoplasia in immunosuppressed women. Am J Obstet Gynecol 150:300–308
68. Sillmann FH, Sedlis A (1987) Anogenital papillomavirus infections and neoplasia in immunodeficient women. Obstet Gynecol Clin North Am 14:537–558
69. Skinner MS, Sternberg WH, Ichinose H (1973) Spontaneous regression of bowenoid atypia of the vulva. Obstet Gynecol 39:173–181
70. Slotman BJ, Helmerhorst TJM, Wijermans PW et al. (1988) Interferon-alpha in treatment of intraepithelial neoplasia of the lower genital tract: a case report. Eur J Obstet Gynecol Reprod Biol 27:327–333
71. Taylor DR Jr, South DA (1981) Bowenoid papulosis: a review. Cutis 27:92–98

72. Ulbright TM, Stehmann FB, Roth IM (1982) Bowenoid dysplasia of the vulva. Cancer 50:2910–2919
73. Vesterinen E, Meyer B, Cantell K et al. (1984) Topical treatment of flat vaginal condyloma with human leukocyte interferon. Obstet Gynecol 64:535–538
74. de Villiers EM (1989) Heterogeneity of the human papillomavirus group. J Virol 63:4898–5903
75. Vion B, Merot Y (1989) Melanoacanthoma of the penis shaft. Report of a case. Dermatologica 179:87–89
76. Wade TR, Kopf AW, Ackermann AB (1978) Bowenoid papulosis of the penis. Cancer 42:1890–1903
77. Wade TR, Kopf AW, Ackermann AB (1979) Bowenoid papulosis of the genitalia. Arch Dermatol 115:306–308
78. Wilkinson EJ, Friedrich EG, Fu YS (1981) Multicentric vulvar carcinoma in situ. Obstet Gynecol 58:69–74
79. Zelickson AS, Prawer SE (1980) Bowenoid papulosis of the penis. Demonstration of intranuclear viral-like particles. Am J Dermatopathol 2:305–308
80. Zur Hausen H (1977) Human papillomaviruses and their role in squamous cell carcinomas. Curr Top Microbiol Immunol 78:1–30
81. Zur Hausen H (1986) Intracellular surveillance of persisting viral infections. Human genital cancer results from deficient cellular control of papillomavirus gene expression. Lancet II:1370–1372

Oral Manifestations of HPV Infections *

S. M. Syrjänen and J. Kellokoski

Introduction

Sixty types of human papilloma virus (HPV) are currently recognized. Recent ultrastructural, morphological, and immunocytochemical data suggest that squamous cell papillomas of the oral cavity, focal epithelial hyperplasia, condyloma acuminatum, verruca vulgaris, and epithelial hyperplasia are associated with or caused by HPV. Much of the recent interest in HPV as a disease-inducing agent in man is due to its suggested associations with epithelial atypias and squamous cell carcinomas (zur Hausen 1977; Ludwig et al. 1981; Meisels et al. 1982). This concept also seems to be pertinent to lesions of the oral cavity, where some dysplastic changes and squamous cell carcinomas have been ascribed to HPV infections, especially those by the high risk type HPV 16. In addition, recent data clearly suggest that more than one HPV type is involved in the development of the squamous cell lesions in the oral cavity. So far, HPV types 1, 2, 6, 7, 11, 13, 16, 18, 32, and 57 have been found in different oral lesions (Petzoldt and Pfister 1980; Lutzner et al. 1982; Pfister et al. 1983; Syrjänen et al. 1984 a, 1987, 1988; Löning et al. 1985; de Villiers et al. 1986).

The above observations, although still incomplete, clearly suggest that HPV infections are found on the oral mucosa. Interestingly, in the majority of cases the HPV types in oral lesions are those seen in the genital and laryngeal areas or in the skin. Thus, there is no reason to suspect that the behavior of oral mucosa would be different from that of the other mucous membranes, e.g., those of the larynx and genital tract. At the moment, no reliable data exist on the prevalence and incidence of oral HPV infections. This is mainly because of the diagnostic difficulties encountered by clinicians; the appearance of HPV lesions in the oral cavity is rarely distinct enough to be readily diagnosed by gross morphology only. In fact, remarkable morphological similarities (at the light microscopic level) exist between the different oral epithelial lesions such as papillomas, condylomas, and verrucas, which have been conventially described in textbooks for oral medicine under different subheadings like oral neoplasias, tumors, proliferative changes, or viral diseases.

* The original studies included in this review were supported in part by a research grant from the Finnish Cancer Society, by PHS grant number 5R01 CA 42010-02 awarded by the National Cancer Institute, DHHS, and a research grant from the Social Insurance Institution of Finland

Genital Papillomavirus Infections
Eds.: Gross/Jablonska/Pfister/Stegner

The present communication summarizes the current evidence available on HPV infections associated with oral squamous cell lesions in the human. The diagnostic techniques available as well as the problems encountered in distinguishing these lesions are also discussed in brief.

Squamous Cell Papilloma

Squamous cell papilloma (Fig. 1) is a common benign epithelial tumor originating from the oral epithelium. Most oral pathology texts list this entity in the category of benign epithelial neoplasia (Abbey et al. 1980; Shafer et al. 1983; Lucas 1984) and some authors raise the question as to whether the lesion is a reaction to injury rather than a true neoplasia (Batsakis 1979). Most authors agree, however, that papilloma is a fairly common lesion in the oral cavity (Knapp 1971), while albeit some authors have regarded intraoral papilloma as a relatively rare tumor (Spouge 1973; Axell 1976).

Papillomas occur at any age, but they are most frequently seen in patients in the third and fifth decades (Greer and Goldman 1974; Abbey et al. 1980). The majority of papillomas are located on the palatal complex, dorsum and lateral borders of the tongue, and lower lip, respectively (Abbey et al. 1980). In a review of 464 oral papillomas, Abbey and coworkers found a slight male predominance (53.8%), and white people comprised the majority of the patients (87.5%; Abbey et al. 1980).

Papilloma is an exophytic growth, presenting with a broadly based ovoid swelling with a size of 2 mm or larger. The surface may show small finger-like projections, resulting in a rough or cauliflower-like verrucous surface. The color of the lesions varies from whitish to pink depending on the degree of

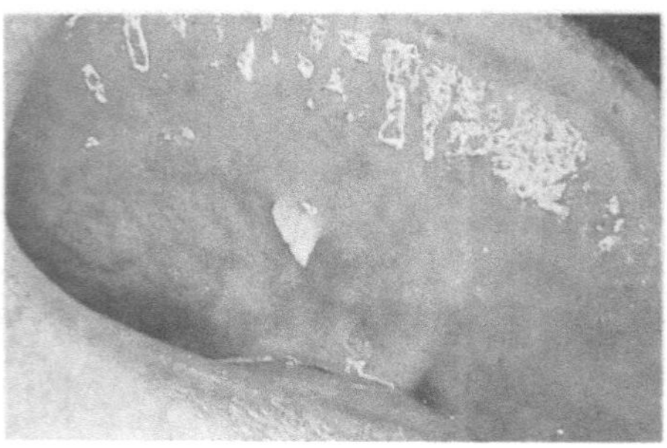

Fig. 1.

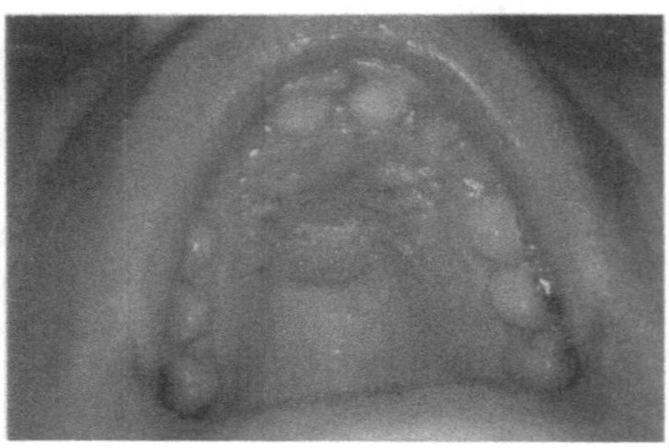

Fig. 2.

Fig. 1. A palatal papilloma in a 25-year-old, healthy, caucasian women. The lesion was excised and there has been no recurrence. (Courtesy of Dr. M. A. Lamberg)

Fig. 2. Tiny hyperkeratinized papillary growth in the border of the tongue from a 25-year-old healthy man. Histologically, hyperparakeratotic squamous epithelium was arranged in deep papillary folds and projections also show acanthosis and koilocytosis. HPV 11 DNA was found in the lesion by in situ hybridization keratinization and vascularization (Shafer et al. 1983; Lucas 1984). The lesion is almost invariably solitary; multiple papillomas have been reported to be extreme rarities (Abbey et al. 1980). Reccurence of oral papillomas after radical excision is regarded as an exceptional event.

On light microscopy, oral papillomas are indistinguishable from the squamous cell papillomas at other locations. They consist of thin, long finger-like projections of the squamous epithelium supported by a connective tissue core (Syrjänen et al. 1983b). The epithelium is usually hyperkeratotic and shows acanthosis (Shafer et al. 1983; Lucas 1984). Histological studies by Abbey and associates (1980) have revealed a tendency for hyperkeratotic lesions to arise from the nonkeratinized oral sites. Most authors emphasize the benign character of oral papillomas, although some, notably Shklar (1965), have considered this lesion as a potentially premalignant one. Greer and Goldman (1974) examined 110 lesions and could not detect dysplasia in any of them. More recently, Syrjänen and colleagues (1984b) found, however, changes consistent with mild dysplasia in 20% (14/70) of oral squamous cell papillomas. This is in agreement with the figures reported by others (MacDonald and Rennie 1975; Abbey et al. 1980).

Evidence of HPV Involvement

The papillomavirus etiology of oral papillary lesions in animals has been well established (Shope 1962; Cheville and Olson 1964). Until recently, this tempting hypothesis has gained little attention for human oral papillomas. Although Frithiof and Wersäll demonstrated viral particles closely resembling HPV in oral papillomas as early as in 1967, doubt about the viral etiology of these lesions has been repeatedly expressed (Greer and Goldman 1974; Shafer et al. 1983).

During the 1980s, however, immunocytochemical techniques first enabled researchers to identify HPV group-specific capsid antigens in oral papillomas. Subsequently, DNA hybridization methods have unequivocally shown the presence of HPV 6 and 11 in these lesions (for review, see S. Syrjänen 1987). Even with the most sensitive DNA hybridization techniques, some of the oral papillomas still remain entirely HPV DNA-negative. It can be argued that these lesions might represent regressing HPV lesions, where the viral DNA has disappeared but the clinical lesion is still present. An additional explanation is that there are oral squamous cell papillomas with different etiology; one which is a truly proliferative lesion caused by nonspecific irritation, and the other one caused by HPV infection.

Condyloma Acuminatum

Condyloma acuminatum (venereal wart) is generally regarded as a sexually transmitted disease affecting the skin and mucous membranes of the anogenital tract. Recent data suggest that genital condyloma is epidemic in young sexually active women with a prevalence of as high as 5%–10% (Ludwig et al. 1981; Meisels et al. 1982; K. Syrjänen 1987; see the chapter by K. Syrjänen in this volume). Although oral-genital sex is known to be a fairly common practice currently, genital warts transmitted into the oral cavity have been infrequently reported (Knapp and Uohara 1967; Doyle et al. 1968; Summers and Booth 1974; Shaffer et al. 1980; Choukas and Toto 1982). As early as in 1901, Heidingsfield described a case of a "puella publica" believed to have acquired condylomata in her tongue as a result of "coitus illegitum" (1901). Thus, two possible routes of transmission for oral condylomas are to be considered; sexual transmission and autoinoculation. The role of a simultaneous (or preceding) mucosal trauma in oral epithelium has been emphasized as a possible trigger of HPV infection.

At onset, oral condylomas usually present with multiple small, white or pink nodules, which proliferate and coalesce to form soft sessile or pedunculated papillary growths (Knapp and Uohara 1967; Doyle et al. 1968; Shaffer et al. 1980). The surface contour in most cases is more cauliflower-like than papillomatous. The lesions are scattered or diffusely involving the tongue, buccal mucosa, palate, lips, or alveolar ridge. Based on the cases reported in the literature, no conclusions about their sex and age predilection can be drawn. The lesions are usually asymptomatic, and not infrequently, regression in between 2 weeks to several months has been detected.

When viewed on the light microscope, the stratified squamous epithelium is disposed in deep papillary folds and projections making up to verrucoid lesion (Fig. 2). The surface of the epithelium generally shows parakeratosis but may be nonkeratinized as well. There is marked acanthosis with thickening and elongation of the rete pegs. In some rare instances, the epithelial organization is disturbed enough to be misinterpreted as a verrucous carcinoma. Vacuolization of the epithelial cells with deeply hyperchromatic round or ovoid nuclei is a characteristic finding (Praetorius-Clausen 1972).

Evidence of HPV Involvement

Most authors agree that condyloma acuminatum in the oral cavity is an infectious disease (Pindborg 1985; Shafer et al. 1983). Evidence for HPV etiology has been provided by detection of viral particles on electron microscopy (Gysland et al. 1976; Shaffer et al. 1980) as well as by demonstration of HPV structural antigens by immunocytochemical means (Jenson et al. 1982a, b; Syrjänen et al. 1983b, 1984b). Viral DNA extraction was performed once by Lutzner and coworkers (1982), but HPV typing failed – most probably due to the small quantity of viral DNA available. In subsequent series, HPV types found in genital condylomas (e.g., HPV 6 and 11) have been identified in oral condylomas using DNA hybridization procedures (see S. Syrjänen 1987).

Verruca Vulgaris

Verruca vulgaris, also frequently referred to as the common wart, is the most common HPV lesion of the skin. The current literature offers conflicting views regarding the existence of verruca vulgaris on oral mucosa (Hertz 1972; Shafer et al. 1983; Lucas 1984). The histological and clinical similarities between verruca vulgaris, squamous cell papilloma, and condyloma acuminatum are well recognized, to the extent that the validity of differential diagnosis between these lesions has been seriously questioned (Waldron 1970; Hertz 1972). The majority of oral warts are seen in children, who also have warts in their fingers (Fig. 3; Pindborg 1985; Shafer et al. 1983).

Clinically, oral warts usually appear as firm, whitish, sessile, papillomatous, rough-surfaced lesions. They may be located at any site, although the lips and the tongue are the preferred sites. On light microscopy, oral warts are characterized by a papillomatous surface with conspicuous hyperkeratinization. The granular layer is frequently most pronounced in the grooves between the papillomatous elevation, which are formed around the thin, elongated connective tissue papillae. The rete ridges are elongated and very characteristic in their inward bending at the margin of the verruca, pointing radially toward the center.

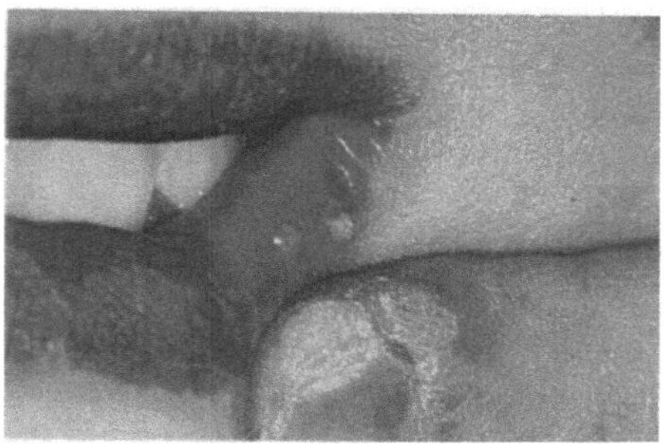

Fig. 3.

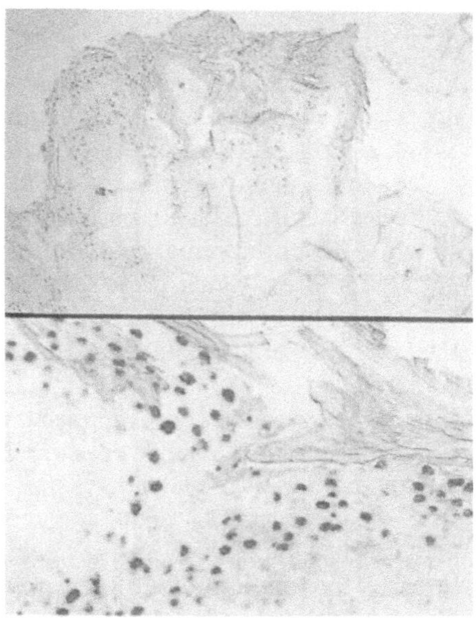

Fig. 4.

Fig. 3. Labial warts (verruca vulgaris) in a 10-year-old healthy girl also having warts in her fingers

Fig. 4. In situ hybridization of the labial lesion shown in Fig. 3. Intensive positive signals for biotinylated HPV 2 DNA are seen in the upper layer of the epithelium. No counterstain is used

Evidence of HPV Involvement

Both the genital tract HPV types, i.e., HPV 6 and 11 and the skin types, e.g., HPV 1, 2, 4, and 7 have been reported in oral warts (Fig. 4; Lutzer et al. 1982; Eversole et al. 1987; for review see Syrjänen 1987). In a strict sense, however, only those lesions containing the skin HPV types should be called verrucas (S. Syrjänen 1987). These lesions are mostly located on the lips or in the keratinized oral mucosa. Interestingly, HPV 7 seems to be the type most frequently found in oral lesions of HIV-infected patients.

Focal Epithelial Hyperplasia

The term focal epithelial hyperplasia (FEH) was first introduced by Archard et al. in 1965 to signify certain multiple nodular elevations of the oral mucosa observed among American Indians in the United States and Brazil as well as in an Eskimo boy from Alaska. Since then, numerous additional cases have been reported. It also seems apparent by now that a number of authors had previously described a similar, if not identical, clinical entity (Helms 1894; Stern 1922; Estrada 1956, 1960; Soneira and Fonseca 1964).

Most of the reported FEH cases have derived from various parts of South and Central America, Greenland, and Alaska, but individual cases have also been described from other countries. The highest prevalence (33.8%) has been found in Indian children in Venezuela (Soneira and Fonseca 1964). Other reports on American Indians living in Colombia (Estrada 1956; Gomez et al. 1969), Brazil, El Salvador, and Guatemala (Witkop and Nismander 1965), as well as in Paraguay (Fischman 1969) have established the prevalence of FEH to be less than 3.5%. Similar studies among the Eskimo populations have shown FEH to be remarkably frequent, the prevalence varying from 7% to 36% (Praetorius-Clausen 1972, 1973; Praetorius-Clausen et al. 1970). In these works, the highest figures are found among the Greenlandic Eskimos on the East coast, where the population is only insignificantly mixed with caucasians (Praetorius-Clausen 1973). In contrast, very low prevalence figures or only isolated cases have been reported in caucasian populations. Praetorius-Clausen found just a single case of FEH among 322 caucasian Danes living in Greenland, i.e., a prevalence of only 0.3%. Notably, he did not find a single case of FEH among 3000 conscript Danish soldiers living in Denmark (Praetorius-Clausen 1973). However, in a study on oral lesions in an adult Swedish population of 20333 subjects, Axell (1976) found a remarkably higher prevalence of FEH, 0.11%. During the past few years, this lesion has been reported more frequently in other countries as well (Petzoldt and Pfister 1980; Petzoldt et al. 1982; Pfister et al. 1983, Syrjänen et al. 1984a). A strong familial history has been suggested for FEH patients by several authors (Archard et al. 1965; Perriman and Uthman 1971; Gomez et al. 1969; Buchner and Mass 1973).

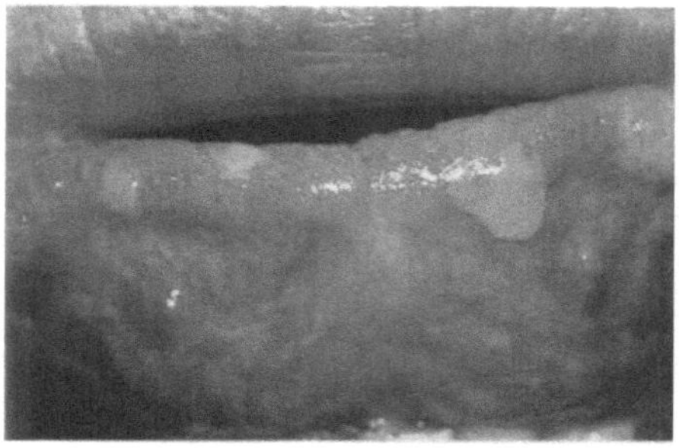

Fig. 5.

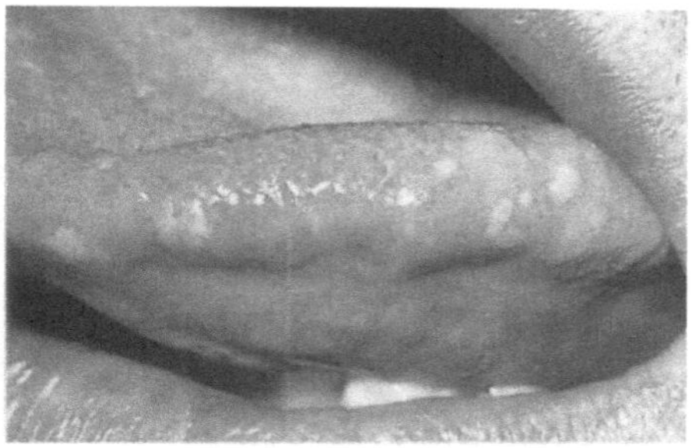

Fig. 6.

Fig. 5. Focal epithelial hyperplasia (FEH) from apex and the border of the tongue in a 43-year-old Eskimo male in southwest Greenland. (Courtesy of Dr. F. Praetorius)

Fig. 6. A 43-year-old Eskimo male showing typical FEH lesion on the edge of the tongue. He was living in the same village as the man shown in Fig. 5. (Courtesy of Dr. F. Praetorius)

Clinically, FEH appears as multiple, soft, flat or rounded slightly elevated nodules (Figs. 5, 6). The lesions are asymptomatic, their color ranging from pale to that of the normal adjacent mucosa. When the involved mucosa is stretched, the lesions tend to disappear. The surface of the papules is smooth, with or without weblike markings. Their size is usually between 1 and 5 mm (Archard et al. 1965; Gomez et al. 1969; Praetorius-Clausen 1972). Although most commonly seen in the lower lip, FEH lesions sometimes extend to the vermilion border, the next most frequent sites being the buccal mucosa, commissure, and the upper lip. The lesions may persist for several years without changing to malignant. Finally, they tend to undergo a spontaneous regression. The lesions recur only infrequently. FEH appears to occur predominantly in children less than 18 years of age, although an increasing number of adults with this lesion have been described lately (Praetorius-Clausen 1973; Petzoldt and Pfister 1980; Pfister et al. 1983; Syrjänen et al. 1984 a).

The most typical FEH lesions are localized nodular elevations in the oral epithelium. Varying degree of acanthosis or thickening of the spinous cell layer and mild parakeratosis are frequently seen. The rete pegs show thickening, elongation and fusion by horizontal outgrowth. Nuclear degeneration with swelling of the cells as well as presence of intranuclear inclusion bodies have been demonstrated. One of the prominent features of FEH, the so called FEH cells have been first described by Praetorius-Clausen (1969). Dyskeratosis or epithelial atypia have never been found in FEH lesions (Archard et al. 1965; Waldman and Shelton 1968; S. Syrjänen 1987).

Evidence of HPV Involvement

Papilloma virus-like particles have been repeatedly detected in FEH lesions, thus early on suggesting its viral etiology (Praetorius-Clausen 1969;

Praetorius-Clausen and Willis 1971; Hanks et al. 1972; Petzoldt and Pfister 1980). Subsequently, immunohistochemical techniques were successfully used to demonstrate HPV structural antigens in FEH (Lutzner et al. 1982; Syrjänen et al. 1983 b, 1984 b; Praetorius-Clausen et al. 1985). These HPV antigens have been found in the nuclei of the superficial cells (with morphological changes) in a high percentage (80%) of cases.

With the DNA-hybridization techniques, HPV type 1 DNA could be first identified in a FEH lesion (Petzoldt and Pfister 1980), soon followed by discovery of HPV 13 and one of its subtypes (Pfister et al. 1983; Syrjänen et al. 1984 a). So far, HPV 13 DNA has been found exclusively in oral FEH lesions, suggesting that this is the HPV type most closely related to FEH. Quite lately, however, a new HPV type 32 was reported in FEH (Beaudenon et al. 1987). Thus, these data strongly suggest that FEH is an HPV infection in the oral cavity, with a possible hereditary background as also established for epidermodysplasia verruciformis (EV) lesions in the skin.

Oral Precancer Lesions and Cancer

The etiology of oral precancer lesions and oral cancer is poorly understood. A variety of factors have been considered, including smoking, alcohol, dental restoratives, mechanical irritation, systemic diseases such as scleroderma, sideropenic dysphagia, and Candida infection (Pindborg 1980). Smoking, however, is by far the most frequently mentioned single etiological factor.

Leukoplakias as the most frequent precancerous lesions of the oral cavity have been extensively studied and reviewed by numerous investigators (Shafer and Waldron 1961; Pindborg 1980; Burckhard and Maerker 1981; Banoczy 1982 and Mackenzie et al. 1980; and many others). The term leukoplakia originally proposed by a Hungarian dermatologist, Ernst Schwimmer in 1877, has remained a subject of continuing dispute. This is because of the dual usage of the term; (a) as a clinical concept to designate a white patch, and (b) as a histopathological entity equivalent to an oral precancer lesion. According to the WHO Cancer Units definition (1978), however, the term leukoplakia should be exclusively used in reference to a purely clinical diagnosis. In alignment with this definition, leukoplakia is a whitish patch or plaque that cannot be classified clinically or histologically as any other lesion and is not associated with any physical or chemical causative agent except the use of tobacco (Axell et al. 1984).

The prevalence of leukoplakia in the world literature varies considerably, from 0.4% to 11.7% (Pindborg 1980; Banoczy 1982). The concept of oral leukoplakia as a premalignant lesion is based on the fact that a substantial percentage of oral squamous cell carcinomas appear to have arisen within the areas of leukoplakia. The frequency of malignant transformation in oral leukoplakias has been estimated to fall between 0.13% and 6% (Pindborg 1980).

On biopsy, leukoplakia is known to present with a variety of epithelial and stromal changes ranging from harmless epithelial hyperplasia with hyperkeratosis to various degrees of epithelial dysplasia, including the in situ and early invasive carcinoma. It should be emphasized that the clinical appearance of leukoplakias is not a reliable index for predicting the seriousness of the microscopic changes. Thus, the need for a diagnostic biopsy in all suspicious lesions should be underscored (Waldron and Shafer 1975; Burkhardt and Merker 1981; Fischman et al. 1982; Lind et al. 1986). This is exemplified by the study of Waldron and Shafer (1975), who reported that of the lesions clinically diagnosed as cancer, apparently equal proportions were histologically benign (33%), premalignant (36%), and malignant (31%). Similarly, of the lesions evaluated as leukoplakia or hyperkeratosis by the clinicians, 13% proved to be frankly invasive carcinomas on biopsy (Waldron and Shafer 1975).

Evidence of HPV Involvement

The possible role of HPV infection in the etiology of oral leukoplakias and oral cancer was not appreciated until 1983, when Syrjänen and coworkers described cytopathic changes of HPV (koilocytosis) in oral cancers identical to those previously found in precancer lesions and carcinomas of the uterine cervix (for review see S. Syrjänen 1987). These morphological findings were further confirmed by immunohistochemical demonstration of HPV capsid antigens in these lesions (Syrjänen et al. 1983c).

Quite lately, more direct evidence on HPV involvement in oral cancer has been provided by using the DNA-hybridization techniques (Scully et al. 1985; Löning et al. 1985; Milde and Löning 1986; de Villier et al. 1986; Lookingbill et al. 1987; Adler-Strothz et al. 1986; Ostrow et al. 1987; Dekmezian et al. 1987; Syrjänen et al.1988). Löning and coauthors (1985) reported HPV 11 and 16 DNA in three of six oral cancers analysed. De Villiers et al. (1985) detected HPV 2 in one and HPV 16 in two of their seven tongue carcinomas. Similarly, using the in situ hybridization technique, Syrjänen and collaborators found HPV 16 DNA in a buccal carcinoma (1986). HPV 11 DNA was recently demonstrated in the lymph node metastasis of an intraoral squamous cell carcinoma (Dekmezian et al. 1987). Tentative evidence on HPV etiology of oral leukoplakias has been provided as well, when HPV 11 and 16 sequences were identified in such lesions (Löning et al. 1985; Syrjänen et al. 1986; Gassenmaier and Hornstein 1988). In our recent series, HPV DNA sequences were found in 12 of the 73 oral precancer lesions and carcinomas analysed by in situ DNA hybridization (Syrjänen et al. 1988).

The established models of the precancer to cancer sequence at genital sites are easily applied to the oral cavity as well. The squamous epithelium of the lower female genital tract and oral cavity, although quite distinct anatomically, share a number of features in common with regard to early neoplastic changes. The oral cavity is continuously exposed to various microtraumas, micro-organisms, and chemical factors (e.g., smoking and alcohol) (Pindborg 1980), which might interact with HPV leading to malignant transformation. Indeed, the role of synergistic factors (either chemical, physical, or other infectious agents) has been seriously considered by the current concepts of HPV-induced malignant transformation (zur Hausen 1982).

Oral Manifestations of HPV in Women with Genital HPV Infections

As pointed out before, there are no reliable data on the prevalence of oral HPV infections. Similarly, the frequency of invisible or latent HPV infections in the oral cavity is totally unknown. With these facts in mind, an extensive study was started in 1987 to assess the oral manifestations of HPV in women prospectively followed up for genital HPV infections at Kuopio University, Finland. So far, a total of 339 women have been studied. The epidemiological data are collected by questionnaire, specially focused on sexual behavior, smoking habits as well as alcohol intake. On clinical examination, photographs on oral mucosa are taken before and after 3% acetic acid application. A surgical biopsy is performed whenever clinical changes are found on oral mucosa, to be processed for routine histology and DNA hybridization. In addition, an extensive mucosal scraping is taken for routine cytology and for HPV DNA detection by dot blot hybridization. Saliva samples are taken for analysis of HPV antibodies.

Our preliminary (unpublished) data show that slight oral mucosal changes are common in these patients (approx. 60%) (see Fig. 7). However, only less than 1% of the lesions are clinically consistent with oral condylomas, and 3% fulfill the criteria of leukoplakias. After acetic acid application, 45% of the patients had aceto-white lesions on their oral mucosal membrane. Using the correlation coefficient analysis, these aceto-white lesions showed a statistically significant correlation with degenerative, vacuolized cells (not koilocytes!) and dyskeratotic changes found in mucosal biopsy material. Thus, it is important to realize that aceto-white lesions in the oral cavity by no means are specific for HPV infections, but rather reflect changes caused by nonspecific irritation. Recent observations suggest that the same is also true with the aceto-white lesions in the male genitalia as well as in the female external genital tract. Based on the dot blot hybridization analysis of the oral scrapings with HPV 2, 6, 7, 11, 13, 16, 18, 31, and 33, only 10 HPV-positive samples have been detected so far. Interestingly, the oral infection in most of these cases is not caused by the same HPV type as found in the genital area. Based on the dot blot analysis, it can be concluded, that oral HPV infection is rare in patients with genital HPV infections. In addition, degenerative changes found on oral mucosa frequently simulate koilocytotic atypia. Thus, caution should be exercised to avoid overdiagnosis of HPV infection in the oral cavity. Studies on possibly latent HPV infections in the oral cavity are now under way, with samples being analysed using the Southern blot and polymerase chain reaction (PCR) methods.

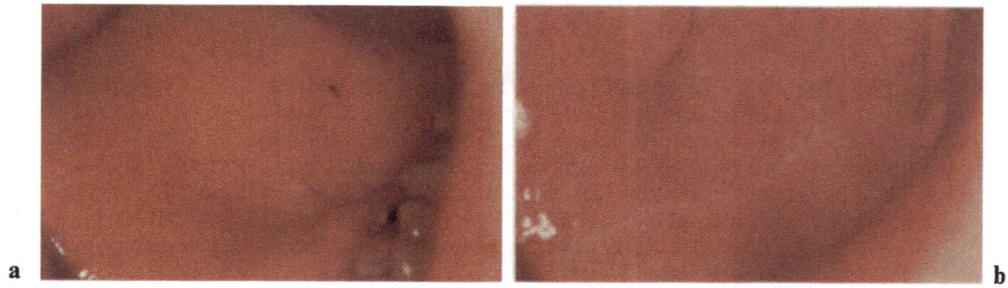

Fig. 7. a Buccal mucosa in a woman with genital HPV infection. A few biting traumas are seen. **b** The same mucosa after acetic acid application. Slight acetowhiteness is seen in entire mucosa. In addition, linea alba is prominently stained. No HPV DNA was found in the mucosal biopsy specimen

Conclusions

It is apparent that keratotic, papillary lesions in the oral cavity are usually small and easily overlooked. The gross appearance of these viral lesions is not distinct enough to be readily diagnosed by clinicians. As pointed out, remarkable morphological similarities exist between oral squamous cell papilloma, condyloma, and verruca. Lesions of a solitary FEH share morphological features in common with fibromas. Many of the previous reports most probably describe identical clinical entities with different names and classifications precluding the proper evaluation of HPV involvement in these series. It should be emphasized that light microscopy using conventional staining procedures is not an adequate means for scrutinizing oral lesions for HPV etiology.

Due to the rapid advances made in recombinant-DNA technology in the past few years substantial progress has been made in the diagnosis of viral diseases in particular. Consequently, in future research oral squamous cell lesions should be analysed for the presence of HPV DNA, and one should not merely apply a confusing nomenclature based on morphology alone. By identifying the HPV type, also the site of viral transmission can be traced. The mucosa of the genital tract and of the oral cavity are similar in many respects, and it is to be expected that HPVs found in the former are also present in the latter. However, the overall frequency of transmission of a genital HPV infection to oral mucosa seems to be extremly low. On the other hand, dermatotrophic HPV types such as 1, 2, 4, and 7 can be expected to affect (by autoinoculation) the oral mucosa as well, albeit only faint evidence for this is available at the moment. Whether additional HPV types exclusively confined to the oral cavity exist (e.g., HPV 13 and its subtypes), remains to be seen.

Acknowledgements. The author wants to extend special thanks to Prof. Dr. Lutz Gissmann and Prof. Dr. Harald zur Hausen, Deutsches Krebsforschungszentrum, Heidelberg, F.R.G., for providing the HPV DNA probes that we used.

References

Abbey LM, Page DG, Sawyer DR (1980) The clinical and histopathologic features of a series of 464 oral squamous cell papillomas. Oral Surg 49:419–428
Adler-Storthz K, Newland JR, Tessin BA, Yendall WA, Shillitoe EJ (1986) Human papillomavirus type 2 DNA in oral verrucous carcinoma. J Oral Pathol 15:472–475
Archard HO, Heck JW, Stanley HR, Gallup NM (1965) Focal epithelial hyperplasia: an unusual oral mucosal lesion found in Indian children. Oral Surg 20:201–212
Axell T (1976) A prevalence study of oral mucosal lesions in an adult Swedish population. Thesis. Odontol Revy 27 [suppl 36]
Axell T, Holmstrup P, Kramer IRH, Pindborg JJ, Shearn M (1984) Intraoral seminar on oral leukoplakia and associated lesions related to tobacco habits. Community Dent Oral Epidemiol 12:146–154
Banoczy J (1982) Oral leukoplakia. Nijhoff, Hague
Batsakis JG (1979) Tumor of the head and neck. Clinical and pathological considerations, 2nd edn. Williams and Wilkins, Baltimore
Beaudenon S, Praetorius F, Kremsdorf D (1987) A new type of human papillomavirus associated with oral focal epithelial hyperplasia. J Invest Dermatol 88:130–135
Buchner A, Mass E (1973) Focal epithelial hyperplasia in an Israeli family. Oral Surg 36:507–511
Burkhardt A, Maerker R (1981) A colour atlas of oral cancers. Wolf, London
Cheville NF, Olson C (1964) Cytology of the canine oral papilloma. Am J Pathol 45:849
Choukas NC, Toto PD (1982) Condylomata acuminatum of the oral cavity. Oral Surg 54:480–485
de Villiers E-M, Weidauer H, Otto H, zur Hausen H (1985) Papillomavirus DNA in human tongue carcinomas. Int J Cancer 36:575–578
de Villiers E-M, Neumann C, Le J-Y, Weidauer H, zur Hausen H (1986) Infection of the oral mucosa with defined types of papillomavirus. Med Microbiol Immunol 174:287–294
Dekmezian RH, Batsakis JG, Goepfert H (1987) In situ hybridization of papillomavirus DNA in head and neck squamous cell carcinomas. Arch Otolaryngol Head Neck Surg 113:819–821
Doyle JL, Grodjesk JE, Manhold JH Jr (1968) Condyloma acuminatum occurring in the oral cavity. Oral Surg 6:434–440
Estrada L (1956) Aporte al estudio odontologico de los Indios Katios. Heraldo Dental 2:5–11
Estrada L (1960) Estudio medico y odontologico de los Indies Katios del Choco. Temas Odontologicas 7:198–210
Eversole LR, Laipis PJ, Merrell P, Choi E (1987) Demonstration of human papillomavirus DNA in condyloma acuminatum. J Oral Pathol 16:266–270
Fischman SL (1969) Focal epithelial hyperplasia. Oral Surg 28:389–393
Fischman SL, Ulmansky M, Sela J, Bab I, Gazit D (1982) Correlative clinicopathological evaluation of oral premelignancy. J Oral Pathol 11:283–289
Frithiof L, Wersäll J (1967) Virus-like particles in human oral papilloma. Acta Otolaryngol 64:263–266
Gassenmaier A, Hornstein OP (1988) Presence of human papillomavirus DNA in benign and precancerous oral leukoplakias and squamous cell carcinomas. Dermatologica 176:224–233
Gomez A, Calle C, Arcilla G, Pindborg JJ (1969) Focal epithelial hyperplasia in a half-breed family of Colombians. J Am Dent Assoc 79:663–667
Greer RO, Goldman HM (1974) Oral papillomas. Clinico-pathologic evaluation and retrospective examination for dyskeratosis in 110 lesions. Oral Surg 38:435–440
Gysland WB, Reimann BEF, Shaffer EL Jr (1976) The virus in oral condyloma acuminatum. Ann Proc Soc 246–247

Hanks CT, Arbor A, Stuart M, Fischman L, de Guzman MN (1972) Focal epithelial hyperplasia. Oral Surg 33:934–941
Heidingsfield ML (1901) Condylomata accuminata linguata. J Cutan Genitourin Dis 19:226–234
Helms O (1894) Syfilis i Grönland. Ugeskr Laeger 5, 1:265–276
Hertz RS (1972) The occurrence of a verruca vulgaris on an intraoral skin graft. Oral Surg 34:934–942
Jenson AB, Lancaster WD, Hartman DP, Shaffer EL Jr (1982a) Frequency and distribution of papillomavirus structural antigens in verrucae, multiple papillomas, and condylomata of the oral cavity. Am J Pathol 107:212–218
Jenson AB, Link CC, Lancaster WD (1982b) Papillomavirus etiology of oral cavity papillomas. In: Hooks J, Jordan G (eds) Viral infections in oral medicine, 1st edn. Elsevier, Amsterdam, pp 133–146
Knapp MJ (1971) Oral disease in 181,338 consecutive oral examinations. J Am Dent Assoc 83:1288–1293
Knapp MJ, Uohara GI (1967) Oral condyloma acuminatum. Oral Surg 23:538–545
Lind P, Syrjänen S, Syrjänen K, Koppang HS, Aas E (1986) Immunoreactivity and human papillomavirus (HPV) on oral precancer and cancer lesions. Scand J Dent Res 94:419–426
Löning T, Ikenberg H, Becker J, Gissmann L, Hoepfer I, zur Hausen H (1985) Analysis of oral papillomas, leukoplakias, and invasive carcinomas for human papillomavirus type related DNA. J Invest Dermatol 84:417–420
Lookingbill DP, Kreider JW, Howett MK, Olmstead PM, Conner GH (1987) Human papillomavirus type 16 in Bowenoid papulosis, intraoral papillomas, and squamous cell carcinoma of the tongue. Arch Dermatol 123:363–368
Lucas RB (1984) Pathology of tumors of oral tissues, 4th edn. Churchill Livingstone, New York
Ludwig ME, Lowell DM, Livolsi VA (1981) Cervical condylomatous atypia and its relationship to cervical neoplasia. Am J Clin Pathol 76:255–264
Lutzner M, Kuffer R, Blanchet-Bardon C, Groissant O (1982) Different papillomaviruses as the causes of oral warts. Arch Dermatol 118:393–399
Mackenzie IC, Dabelsteen E, Squier C (1980) Oral premalignancy. University of Iowa Press, Iowa City
MacDonald DG, Rennie JS (1975) Oral epithelial atypia in denture induced hyperplasia, lichen planus and squamous cell papilloma. Int J Oral Surg 4:40–45
Meisels A, Morin C, Casa-Cordero M (1982) Human papillomavirus infection of the uterine cervix. Int J Gynecol Pathol 1:75–94
Milde K, Löning T (1986) Detection of papillomavirus DNA in oral papillomas and carcinomas: application of in situ hybridization with biotinylated HPV 16 probes. J Oral Pathol 15:292–296
Ostrow RS, Manias DA, Fong WJ, Zachow KR, Faras AJ (1987) A survey of human cancers for human papillomavirus DNA by filter hybridization. Cancer 59:429–434
Perriman A, Uthman A (1971) Focal epithelial hyperplasia. Oral Surg 31:221–225
Petzoldt D, Pfister H (1980) HPV 1 DNA in lesion of focal epithelial hyperplasia Heck. Short communications. Arch Dermatol Res 268:313–314
Petzoldt D, Dennin R, Pfister H, Hoffmann C (1982) Fokale epitheliale Hyperplasie Heck. Hautarzt 33:201–205
Pfister H, Hettich I, Runne U, Gissmann L, Chilf G (1983) Characterization of human papillomavirus type 13 from focal epithelial hyperplasia Heck lesions. J Virol 47:363–366
Pindborg JJ (1980) Oral cancer and precancer. Henry Ling, Dorset Press, Dorchester
Pindborg JJ (1985) Atlas of diseases of the oral mucosa, 4th edn. Munksgaard, Copenhagen
Praetorius-Clausen F (1969) Histopathology of focal epithelial hyperplasia. Evidence of viral infection. Tandlaegebladet 73:1013–1022

Praetorius-Clausen F (1972) Rare oral viral disorders (molluscum contagiosum, localized keratoacanthoma, verrucae, condyloma acuminatum, and focal epithelial hyperplasia). Oral Surg 34:604–618
Praetorius-Clausen F (1973) Geographical aspects of oral focal epithelial hyperplasia. Pathol Microbiol 39:204–213
Praetorius-Clausen F, Willis JM (1971) Papova virus-like particles in focal epithelial hyperplasia. Scand J Dent Res 79:362–365
Praetorius-Clausen F, Mogeltoft M, Roed-Petersen B, Pindborg JJ (1970) Focal epithelial hyperplasia of the oral mucosa in a south-west Greenlandic population. Scand J Dent Res 78:287–294
Praetorius F, Praetorius Clausen P, Mögeltoft M (1985) Immunohistochemical evidence of papilloma virus antigen in focal epithelial hyperplasia. Tandlaegebladet 89:589–625
Schwimmer E (1877) Die idiopathischen Schleimhautplaques der Mundhöhle (leukoplakia buccalis). Arch Dermatol Syph 9:511–570
Scully C, Prime S, Maitland N (1985) Papillomaviruses: their possible role in oral disease. Oral Surg 60:166–174
Shafer WG, Waldron CA (1961) A clinical and histopathologic study of oral leukoplakia. Surg Gynecology 411–420
Shafer WG, Hine MK, Levy BM (1983) A textbook of oral pathology, 4th edn. Saunders, Philadelphia
Shaffer EL Jr, Reimann BE, Gysland WB (1980) Oral condyloma acuminatum. J Oral Pathol 9:163–173
Shklar G (1965) The precancerous oral lesions. Oral Surg 20:58–70
Shope RE (1962) Are animal tumor viruses always virus like. J Gen Physiol [Suppl] 45:143–151
Soneira A, Fonseca N (1964) Sobre una lesion de la mucosa oral en los ninos Indios de la Mision Los Angeles de Tokuko. Venezuela Odontol 29:109–119
Spouge JD (1973) Oral pathology. Mosby, St Louis, p 388
Stern E (1922) Multiple weiche Warzen der Mundschleimhaut. Dermatol Wochenschr 74:274–276
Summers L, Booth DR (1974) Intraoral condyloma acuminatum. Oral Surg 38:273–278
Syrjänen S, Syrjänen K, Lamberg MA (1986) Detection of human papillomavirus DNA in oral mucosal lesions using in situ DNA hybridization applied on paraffin sections. Oral Surg 62:660–667
Syrjänen K (1987) Papillomavirus infections and cancer. In: Syrjänen K, Gissman L, Koss L (eds) Papillomaviruses and human disease. Springer, Berlin Heidelberg New York, pp 467–503
Syrjänen S (1987) HPV infections in oral cavity. In: Syrjänen K, Gissman L, Koss L (eds) Papillomaviruses and human disease. Springer, Berlin Heidelberg New York, pp 104–137
Syrjänen K, Syrjänen SM, Lamberg MA, Pyrhönen S (1983a) Human papillomavirus (HPV) involvement in squamous cell lesions of the oral cavity. Proc Finn Dent Soc 79:1–8
Syrjänen KJ, Pyrhönen S, Syrjänen SM, Lamberg MA (1983b) Immunohistochemical demonstration of human papilloma virus (HPV) antigens in oral squamous cell lesions. Br J Oral Surg 21:147–153
Syrjänen K, Syrjänen S, Lamberg M, Pyrhönen S, Nuutinen J (1983c) Morphological and immunohistochemical evidence suggesting human papillomavirus (HPV) involvement in oral squamous cell carcinogenesis. Int J Oral Surg 12:418–424
Syrjänen S, Syrjänen K, Ikenberg H, Gissmann L, Lamberg M (1984a) A human papillomavirus closely related to HPV 13 found in a focal epithelial hyperplasia lesion (Heck disease). Arch Dermatol Res 276:199–200
Syrjänen K, Happonen RP, Syrjänen S, Calonius B (1984b) Human papilloma virus (HPV) antigens and local immunologic reactivity in oral squamous cell tumors and hyperplasias. Scand J Dent Res 92:358–370

Syrjänen SM, Syrjänen K, Happonen R-P, Lamberg MA (1987) In situ DNA hybridization analysis of human papillomavirus (HPV) sequences in benign oral mucosal lesions. Arch Dermatol Res 279:543–549

Syrjänen SM, Syrjänen KJ, Happonen R-P (1988) Human papillomavirus (HPV) DNA sequences in oral precancerous lesions and squamous cell carcinoma demonstrated by in situ hybridization. J Oral Pathol 17:273–278

Waldman GH, Shelton DW (1968) Focal epithelial hyperplasia (Heck's disease) in an adult Caucasian. Oral Surg 26:124–127

Waldron CA (1970) Oral epithelial tumors. In: Gorlin RJ, Goldman HM (eds) Oral pathology, 6th edn. Mosby, St Louis

Waldron CA, Shafer WG (1975) Leukoplakia revisited. A clinicopathologic study of 3,256 oral leukoplakias. Cancer 36:1386–1392

WHO Collaborating Center for Oral Precancerous Lesions (1978) Definition of leukoplakia and related lesions: an aid to studies on oral precancers. Oral Surg 46:518–539

Witkop CJ, Niswander JD (1965) Focal epithelial hyperplasia in Central and South American Indians and Ladinos. Oral Surg 20:213–217

zur Hausen H (1977) Human papillomaviruses and their possible role in squamous cell carcinomas. In: Current topics in microbiology and immunology, vol 78. Springer, Berlin Heidelberg New York, pp 1–30

zur Hausen H (1982) Human genital cancer. Synergism between two virus infections or synergism between a virus infection and initiating events? Lancet II:1370–1372

Trends and Pitfalls of In Situ Hybridization of Oral Lesions *

K. Milde-Langosch, R.-P. Henke, and T. Löning

Introduction

In 1967, Frithiof and Wersäll first described papillomavirus-like particles in papillomatous lesions of the human oral cavity. Today it is recognized that papillomaviruses are clearly associated with benign oral papillomas and focal epithelial hyperplasias (FEH), and they are increasingly implicated in the etiology of premalignant leukoplakias and invasive squamous cell carcinomas of the oral cavity, larynx, and pharynx. But as in genital tumors, the role of virus infection in the development to the malignant state is far from being understood. Additional factors, i.e., smoking habits, environmental influences, chewing of tobacco, alcohol consumption, or genetic factors may play a role in the etiology of oral cancers (Pindborg 1980; Lipkin et al. 1985; Kabat and Wynder 1989).

The similarity between many oral lesions (papillomas, leukoplakias, carcinomas) and those of the lower genital tract suggests that the same HPV types and the same pathogenetic mechanisms might lead to the development of these diseases. Kreider et al. (1987) found a similar susceptibility of skin grafts from the human cervix and vocal cord to transformation with HPV 11 DNA, whereas abdominal skin was resistant to transformation to the malignant state. In fact, "genital" HPV types (HPV 6, 11, 16, and 18) have been found in lesions of the larynx and oral cavity by different authors (Naghashfar et al. 1985; Löning et al. 1985; de Villiers et al. 1985, 1986; Stremlau et al. 1987; Syrjänen et al. 1988). For the juvenile laryngeal papillomas, a vertical transmission from the HPV-infected mother to the child during birth is discussed (Mounts et al. 1982). On the other hand, specific HPV types which cannot be found in genital lesions were isolated from oral and laryngeal HPV-associated diseases (Pfister et al. 1983; Beaudenon et al. 1987; Kahn et al. 1986).

We used in situ hybridization with nonradioactive, biotinylated probes for the detection of HPV DNA in tissue samples of the oral cavity, pharynx, and larynx. In addition, dot blot hybridizations were performed in several cases. The experiments should answer the following questions:

1. How many papillomas, leukoplakias, focal epithelial hyperplasias, hairy leukoplakias, and carcinomas bear detectable amounts of HPV DNA?

---

* This study was supported by grants from the Deutsche Forschungsgemeinschaft (Lo 285/2-3) and the Hamburger Stiftung zur Förderung der Krebsbekämpfung (I 208)

Genital Papillomavirus Infections
Eds.: Gross/Jablonska/Pfister/Stegner

2. Which HPV types are present, do they represent the well-known "genital" types or special "oral" HPV types?
3. Which associations can be found between viral infection and special histological features?

Apart from these questions, we will also discuss the experimental pitfalls leading to false results in the diagnosis of viral infection by in situ hybridization.

Materials and Methods

Tissue Samples

Biopsy specimens from oral papillomas, leukoplakias, or squamous cell carcinomas were fresh frozen in liquid nitrogen and stored at −80° C before sectioning. In the FEH cases and the cases of HIV infected patients, paraffin-embedded material was used. The FEH specimens were collected by Dr. Hanna Stromme Koppang, Department of Oral Pathology and Section for Forensic Odontology, Oslo.

Viral Probes

Viral DNA of HPV 6, 11, 13, 16, and 18 cloned into the plasmid pBR 322 was kindly provided by Drs. H. zur Hausen und L. Gissmann, German Cancer Research Center, Heidelberg. HPV 32 DNA, cloned into the plasmid pSP62, was a gift of Dr. O. Croissant, Institut Pasteur, Paris. The plasmids were propagated in *Escherichia coli* and labelled by nick translation with biotinylated deoxyuridine triphosphate (Bio-11-dUTP) and a nick translation reagent kit (Gibco/BRL, Eggenstein). Biotinylated HPV 1, EBV, and CMV DNA were commercially available from Enzo (New York).

In Situ Hybridization

Paraffin and cryostat sections were induced to adhere to aminoalkylsilane-treated glass slides (Rentrop et al. 1986). Frozen sections were air-dried, fixed in methanol/acetic acid (3:1) for 5 min, immersed in boiling phosphate-buffered saline (PBS, pH 7.4) for 15 s, cooled in ice water, and postfixed in methanol for 3 min before air-drying.

Paraffin sections were dewaxed in xylene, hydrated in decreasing concentrations of alcohol, and incubated with 1 mg/ml Pronase from *Streptomyces griseus* (Calbiochem, Frankfurt) in 50 m*M* TRIS-HCl, pH 7.6, 5 m*M* EDTA for 6 min. The proteolytic reaction was stopped by incubation of the slides in 2 g/l glycin in PBS for 5 min, followed by postfixation in 4% paraformaldehyde in PBS for 5 min, washing in PBS, dehydration, and air-drying.

For hybridization, each section was covered with 20 μl of the following hybridization mixture: 2 × SSC (1 × SSC = 0.15 *M* NaCl, 0.015 *M* trisodium citrate, pH 7.2), 10% dextran sulfate, 0.1 mg/ml herring sperm DNA, 1–2 μg/ml biotinylated viral DNA, and 20% and 45% deionized formamide for nonstringent (T_m-34° C) and stringent (T_m-17° C) hybridization conditions, respectively. After application of the coverslips and sealing with rubber cement, nucleic acids were denatured by heating in a steel box floating in a 90° C water bath for 10 min, followed by hybridization at 37° C overnight.

After hybridization and careful removal of the coverslips, the sections were washed for 2 × 5 min in 1 × SSC, 20% formamide, 37° C (nonstringent conditions) or in 1 × SSC, 45% formamide, 37° C (stringent conditions) followed by several washes in 2 × SSC at room temperature.

The detection of the hybridized biotinylated probe was performed either using the SAAP (*s*trept*a*vidin *a*lkaline *p*hosphatase), method or the SGSS (*s*treptavidin *g*old *s*ilver *s*taining) procedure, both of which are described in detail by Rivière et al. (this volume).

Dot Blot Hybridizations

The DNA extraction and subsequent dot blot hybridizations were performed according to Löning et al. (1987).

Results

The results of the HPV detection by in situ hybridization under nonstringent conditions in papillomas, leukoplakias, focal epithelial hyperplasias, and squamous cell carcinomas of the oral cavity, pharynx, and larynx are summarized in Fig. 1. HPV DNA was detected in 10/10 (100%) papillomas, 7/9 (78%) leukoplakias, 15/17 (88%) focal epithelial hyperplasias, and 12/20 (60%) squamous cell carcinomas.

Oral and Laryngeal Papillomas

These lesions showed the typical cytopathic effects of papillomavirus infection: koilocytosis, multinuclear cells, and dyskeratosis. Strong nuclear staining was observed in the ballooned keratinocytes of the upper and intermediate epithelial layers (Fig. 2) with the typical aspects of koilocytosis. HPV typing was performed in only three of these cases and resulted in two cases of HPV 6/11 and one case of HPV 16/18 infection.

Leukoplakias

The results for these lesions were similar to those of the papillomas. All cases were of oral origin and did not show epithelial dysplasia. Of the seven cases

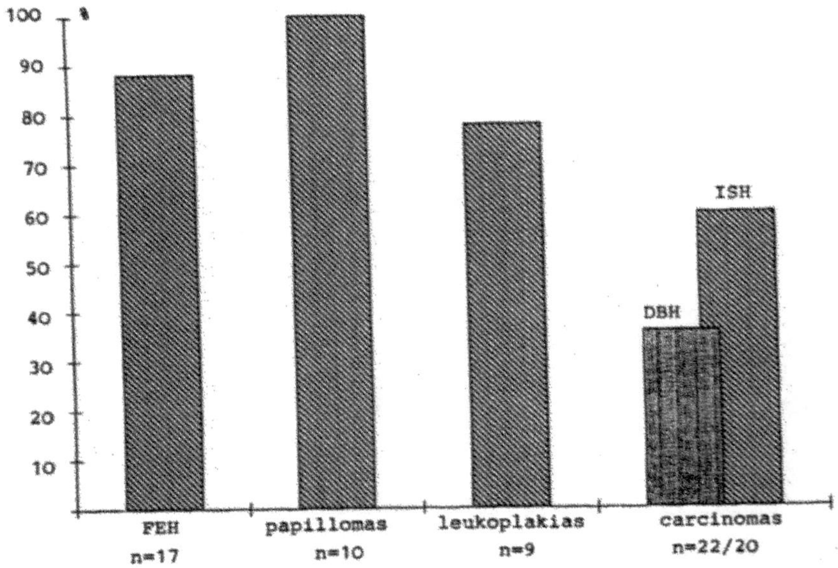

Fig. 1. Prevalence of HPV DNA in lesions of the oral cavity, pharynx, and larynx as determined by in situ hybridization. ISH, In situ hybridization; DBH, dot blot hybridization

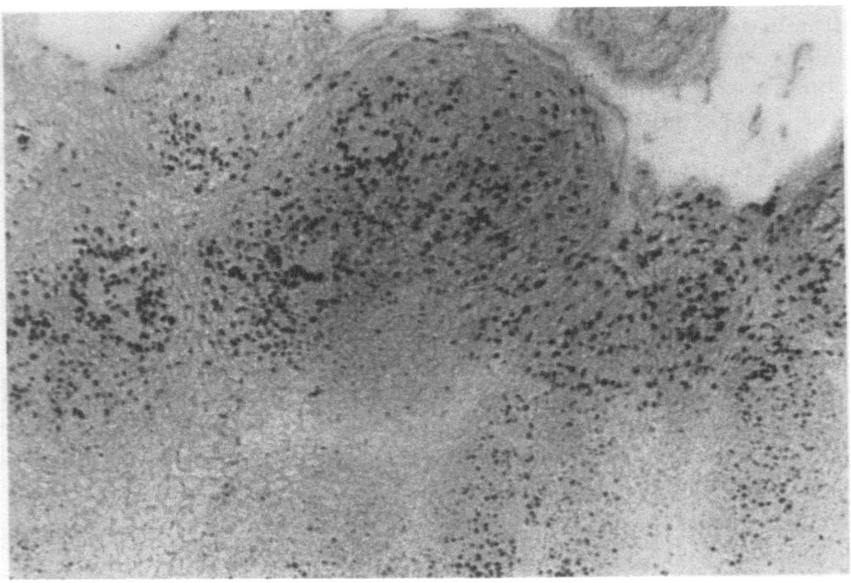

Fig. 2. Papilloma of the tonsillary mucosa. In situ hybridization with biotinylated HPV 16 DNA under nonstringent conditions. Strong nuclear staining in keratinized areas of the epithelium. SAAP detection system, counterstaining with eosin (× 63)

which were positive under nonstringent conditions, one hybridized with HPV 6/11 and one with HPV 16/18 at higher stringency. The nuclear hybridization signals were mainly found in koilocytic cells. Leukoplakias without koilocytosis were negative for HPV DNA.

Squamous Cell Carcinomas

Of the 16 oropharyngeal and 4 laryngeal carcinomas which were examined by in situ hybridization under nonstringent conditions, 12 (60%) were positive for HPV DNA. Nuclear staining was observed in differentiated, keratinized regions of the tumor, partly showing koilocytic cells (Fig. 3). Twenty-two carcinomas (13 cases of the oral cavity including the tongue, two of the nasopharynx, three of the hypopharynx, and four of the larynx) were studied by dot blot hybridization. In eight tumors, HPV DNA was found under conditions of reduced stringency (T_m-34° C). At more stringent hybridization conditions (T_m-17° C), HPV 16/18 DNA was identified in four cases and HPV 6/11 DNA in one case. The HPV-positive carcinomas included 2/4 G1, 5/11 G2, and 1/7 G3 tumors. In four cases, hybridizations were performed not only with DNA extracted from the tumor, but also with DNA extracted from tissue of the resection margin. In three cases, concordant results were obtained, two being positive for HPV DNA. In one case, viral DNA was found only in the tumor itself.

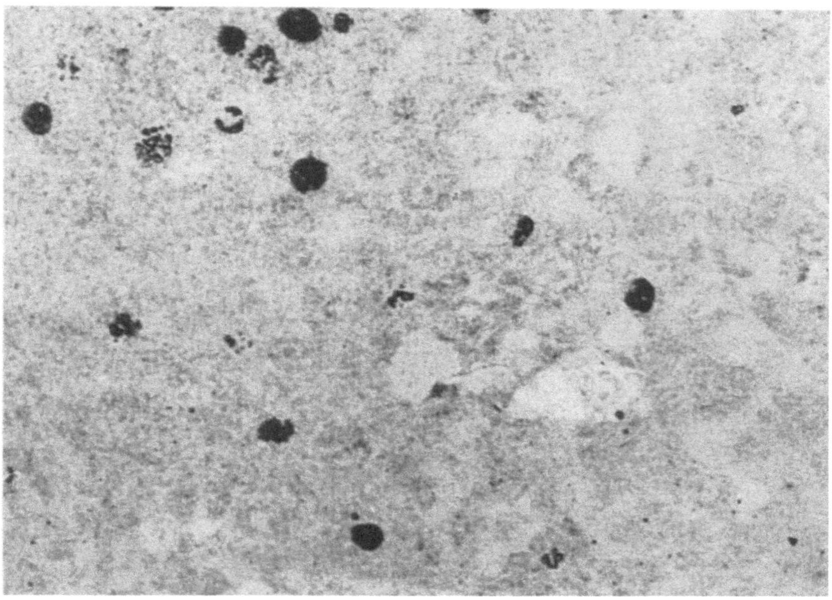

Fig. 3. Pharyngeal carcinoma. In situ hybridization with a biotinylated HPV 16 probe under nonstringent conditions. Staining of few nuclei in the tumor. SGSS detection system (× 320)

Focal Epithelial Hyperplasia

In another study, 17 cases of the oral focal epithelial hyperplasia were examined (FEH; Heck's disease), a benign disease frequently found among Eskimos and American Indians (Praetorius-Clausen 1972, 1973). Clinically, these lesions appear as painless, soft, circumscribed papules of the lip, tongue, or buccal mucosa (Praetorius-Clausen 1972). Except for one patient of Finnish-Laplandish origin, all patients of our study were caucasians from Norway. The lesions were located at the lip (nine cases), the tongue (five cases), or the buccal mucosa (three cases). Histologically, acanthosis and parakeratosis of the epithelium as well as koilocytosis were observed in all cases. In situ hybridizations were performed with HPV 6/11, 16/18, 13, and 32 under stringent conditions and with a mixture of HPV 1/6/11/13/16/18/32 under nonstringent conditions. HPV 13 DNA was found in nine cases and HPV 32 DNA in five cases, whereas one case was positive for both viral types. In two cases, no HPV DNA could be detected even under nonstringent conditions. The nuclear staining was most prominent in the superficial layers of the hyperplastic epithelium, but was also observed in deeper layers of the stratum spinosum in some specimens (Fig. 4).

Hairy Leukoplakia and Oral Mucosa from HIV-Infected Patients

Immunosuppressed patients (organ transplant recipients or persons infected with the human immunodeficiency virus, HIV) were shown to have an increased incidence of papillomavirus-induced warts and squamous cell carcinomas (Gassenmeier et al. 1986; Hoxtell et al. 1977) as well as other viral infections (Armstrong 1984). On the other hand, Reichart et al. (1987) who examined 110 HIV-infected patients, did not find any HPV-related oral lesions like warts, papillomas, or squamous cell carcinomas among these persons. We examined punch biopsy specimens from the oral cavity of 34 HIV-infected persons for the presence of HPV 6/11/13/16/18, Epstein-Barr virus (EBV), and cytomegalovirus (CMV). In 17 of these cases, hairy leukoplakia had been diagnosed. On the basis of immunofluorescence and electron microscopy, Greenspan et al. (1985) demonstrated the presence of papillomaviruses and Epstein-Barr viruses in these lesions, and on the basis of Southern blot hybridizations, they found high amounts of EBV DNA in all of 13 cases of hairy leukoplakia.

In our in situ hybridization study, none of the tested HPV types could be demonstrated under stringent conditions, whereas EBV DNA could be detected in 17 specimens. Thirteen of these EBV-positive cases were classified as hairy leukoplakia according to clinical and histological criteria. Additionally, in four specimens of apparently normal oral mucosa EBV DNA was demonstrated.

Histologically, all EBV positive cases were characterized by cytopathic changes which are different from, but resemble those of HPV infection. In

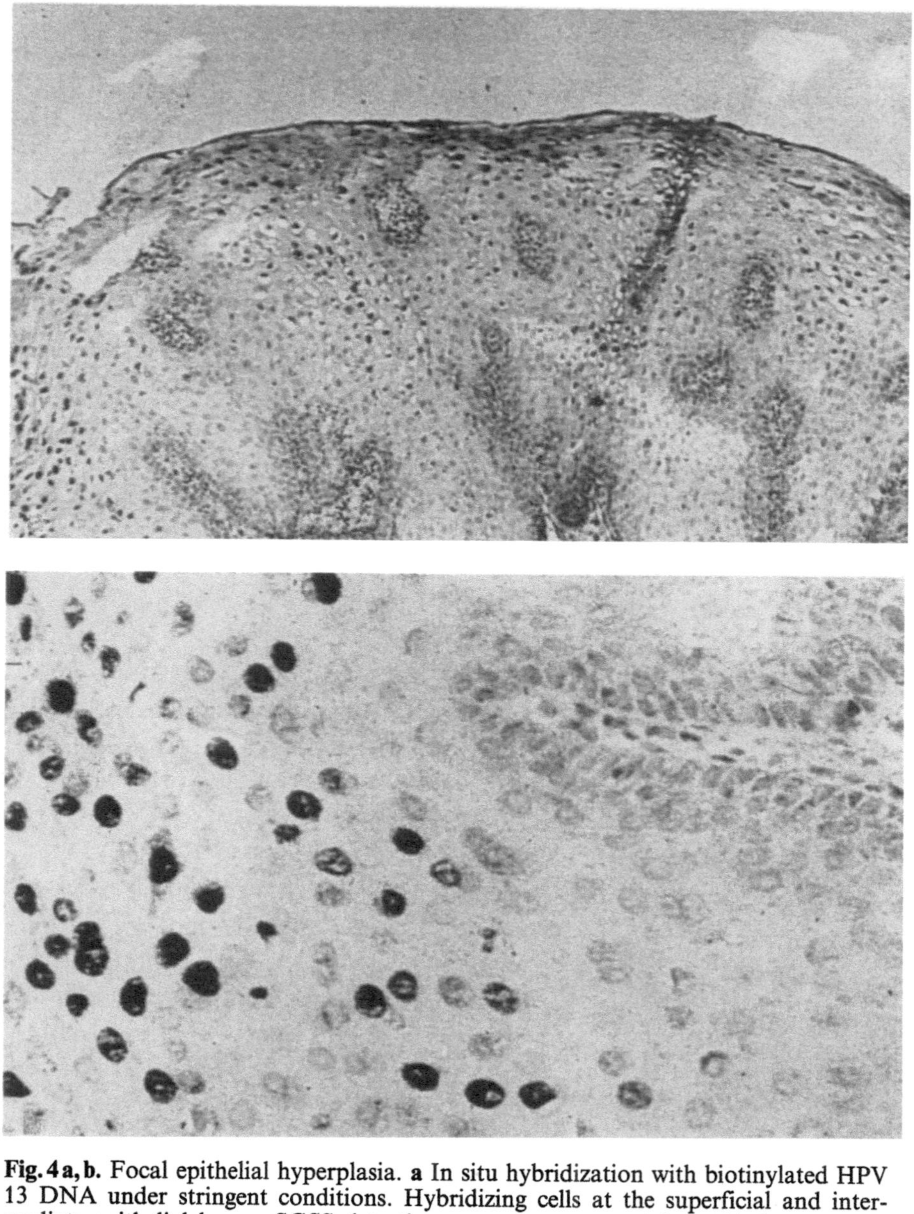

a

b

Fig. 4 a, b. Focal epithelial hyperplasia. **a** In situ hybridization with biotinylated HPV 13 DNA under stringent conditions. Hybridizing cells at the superficial and intermediate epithelial layers. SGSS detection system, counterstaining with hematoxylin (× 80). **b** Higher magnification (× 320)

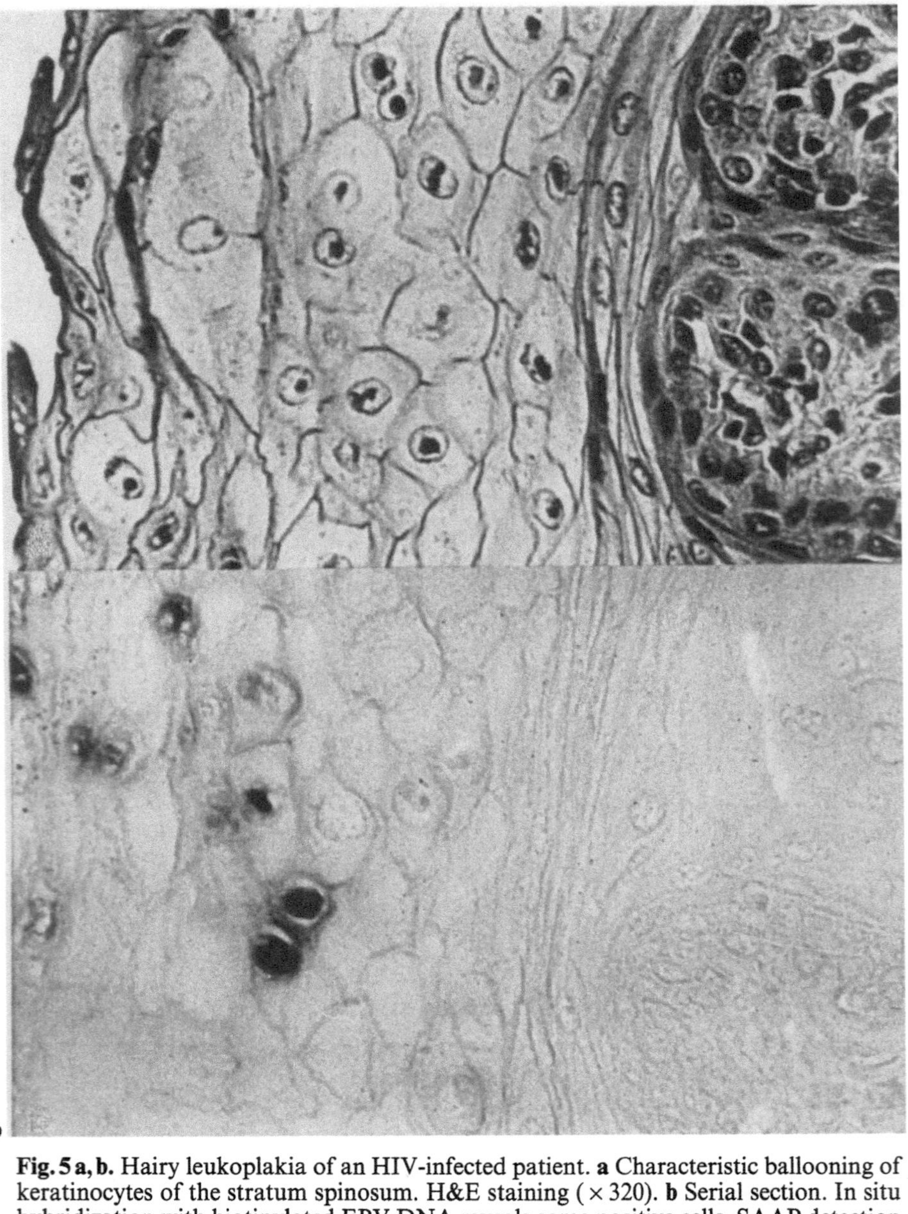

a

b

Fig. 5 a, b. Hairy leukoplakia of an HIV-infected patient. **a** Characteristic ballooning of keratinocytes of the stratum spinosum. H&E staining (× 320). **b** Serial section. In situ hybridization with biotinylated EBV DNA reveals some positive cells. SAAP detection system (× 320) particular, strong "ballooning" (intracellular edema) of the cells of the stratum spinosum was observed. These spongy-looking cells were often sharply delineated from the three to four layers of flat parakeratotic keratinocytes at the surface of the epithelium. In situ hybridization with biotinylated EBV DNA resulted in strong nuclear staining preferentially in the upper stratum spinosum. Neither basal nor stromal cells were positive for EBV DNA (Fig. 5).

Hybridizations with CMV or HPV 6/11/13/16/18 DNA under stringent conditions did not show any positive results. Under nonstringent conditions (T_m-40° C) with a mixture of the five HPV types, weak nuclear hybridization signals were observed in some cases, indicating that additional HPV types which were not used in our experiments might be present.

Discussion

Viral Etiology of Oral Lesions

Based on our in situ hybridization results, the oral lesions we investigated can be divided into three groups:

1. Lesions bearing HPV types which are only found in the oral region (HPV 13/32)
2. Lesions bearing HPV types otherwise found in anogenital lesions (HPV 6/11/16/18)
3. Lesions not bearing any of the tested HPV types but instead Epstein-Barr virus DNA

Group 1 concerns FEH only. Our results of 86% HPV 13 and 32 infections in 17 cases of FEH are in accordance with those of other investigators (Beaudenon et al. 1987) who found HPV 32 in six and HPV 13 in four out of ten cases of FEH. Previous studies reported the presence of HPV 13 in oral lesions other than FEH or the existence of other HPV types in FEH lesions in sporadic cases (Syrjänen et al. 1987; Petzoldt and Pfister 1980; de Villiers et al. 1986). In our study, none of the "genital" HPV types 6, 11, 16, and 18 was found in Heck's disease. The strong association between FEH and the HPV types 13 and 32 indicates that these types may play a substantial role in the pathogenesis of this disease.

Group 2 comprises the oropharyngeal and laryngeal papillomas, leukoplakias, and squamous cell carcinomas. Our results are in accordance with previous studies which found HPV 6, 11, 16, and 18 in these lesions (de Villiers et al. 1985; de Villiers et al. 1986; Stremlau et al. 1987; Nagashfar et al. 1985; Syrjänen et al. 1988). Yet, in contrast to the results of zur Hausen (1985) for genital lesions, clear correlations between HPV type and the degree of malignancy were not seen in oral lesions. We found HPV 16/18 DNA in an oral papilloma as well as HPV 6/11 DNA in an invasive carcinoma. The high number of oral lesions for which no exact determination of the HPV type could be done, point to additional HPV types not examined in this study. In fact, HPV 7, 13, and 32 in oral papillomatosis (de Villiers et al. 1986), HPV 2 in oral carcinomas (de Villiers et al. 1985, 1986; Adler-Storthz et al. 1986), and HPV 30 in a laryngeal carcinoma (Kahn et al. 1986) were observed.

Group 3 comprises the oral hairy leukoplakias of HIV-infected patients. Our failure to detect HPV 1/6/11/13/16/18 DNA in addition to EBV DNA does not exclude the presence of other HPV types in these lesions. Greenspan et al. (1985) could demonstrate HPV antigens in 43/67 hairy leukoplakias, whereas Eversol et al. (1988) found HPV 16 DNA by in situ hybridization in 1/20 cases. Our in situ hybridizations under nonstringent conditions resulted in weak nuclear signals in some cases.

In Situ Hybridization for the Study of HPV Infections

For all manifestations of viral infection described in groups 2 and 3 further studies including a broad spectrum of HPV types are necessary in order to elucidate the role of different papillomaviruses in the pathogenesis of oral disease. In situ hybridization with nonradioactive probes can be a valuable instrument in these studies. The procedure can be carried out on very small biopsy or cytologic preparations, and results are obtained within 1 to 2 days. Additionally, positive staining can be correlated with certain cytopathic or histological changes. As in situ hybridization allows the detection of very few infected cells in relatively large tissue specimens, the sensitivity of HPV detec-

Table 1. Trouble-shooting for in situ hybridization

Observation	Cause	Strategy
Floating of sections from the slides	Mechanical and chemical stress during ISH	Aminoalkylsilane-pretreatment of the slides
No signal in positive control	Loss of DNA by nucleases	Rapid fixation of the tissue specimen after biopsy; hybridization under sterile conditions
	DNA is inaccessible for the probe	Pretreatment with proteases, detergents, etc.
Nonspecific staining in negative control	Nonspecific hybridization of the viral or plasmid DNA	Use of isolated viral insert; higher stringency in hybridization or washing steps
	Endogeneous alkaline phosphatase	Levamisole in substrate mixture; use of other detection systems
	Endogeneous biotin	Preincubation with biotin/avidin
	Nonspecific antibody binding	Use of higher antibody dilutions; blocking with human serum, BSA, etc.

tion can be very high compared with Southern blot or dot blot techniques. Yet, there are a lot of methodical pitfalls, and some experience in in situ hybridization is necessary to avoid false results. Most of the problems which can arise are summarized in Table 1. Positive and negative controls should always be included in the experiments, i.e.:

1. Tissue sections from a case which was clearly positive in former experiments
2. Hybridization with biotinylated plasmid DNA or a probe of a virus usually not found in this tissue
3. Exclusion of any biotinylated DNA from the hybridization mixture

By in situ hybridization under stringent conditions, an identification of the papillomavirus type is possible in most cases. However, it might be difficult to distinguish between some HPV types showing a high degree of homology, i.e., HPV 6 and 11, or HPV 16 and 31 (Pfister 1986). In these cases where cross-hybridizations between two HPV types can simulate a double infection, an exact typing needs Southern blot hybridization experiments.

Acknowledgements. The authors wish to thank Ms. I. Orlt and Ms. H. Hinze for invaluable technical assistance.

References

Adler-Storthz K, Newland JR, Tessin BA, Yendall WA, Shillitoe EJ (1986) Human papillomavirus type 2 DNA in oral verrucous carcinoma. J Oral Pathol 15:472–475

Armstrong D (1984) The acquired immune deficiency syndrome: viral infections and etiology. Prog Med Virol 30:1–13

Beaudenon S, Praetorius F, Kremsdorf D, Lutzner M, Worsaae N, Pehau-Arnaudet G, Orth G (1987) A new type of human papillomavirus associated with oral focal epithelial hyperplasia. J Invest Dermatol 88:130–135

de Villiers EM, Weidauer H, Otto H, zur Hausen H (1985) Papillomavirus DNA in human tongue carcinomas. Int J Cancer 36:575–578

de Villiers EM, Weidauer H, Le JY, Neumann C, zur Hausen H (1986) Papillomviren in benignen und malignen Tumoren des Mundes und des oberen Respirationstrakts. Laryngol Rhinol Otol 65:177–179

Eversole LR, Stone CE, Beckman AM (1988) Detection of EBV and HPV DNA sequences in oral "hairy" leukoplakia by in situ hybridization. J Med Virol 26:271–277

Frithiof L, Wersäll J (1967) Virus-like particles in papillomas of the human oral cavity. Arch Gesamte Virusforsch 21:31–44

Gassenmeier A, Fuchs P, Schell H, Pfister H (1986) Papillomavirus DNA in warts of immunosuppressed renal allograft recipients. Arch Dermatol Res 278:219–223

Greenspan JS, Greenspan D, Lennette ET, Abrams DI, Conant MA, Petersen V, Freese UK (1985) Replication of Epstein-Barr virus within epithelial cells of oral "hairy" leukoplakia, an AIDS-associated lesion. N Engl J Med 313:1564–1571

Hoxtell EO, Mandel JS, Murray SS, Schumann LM, Goltz RW (1977) Incidence of skin carcinoma after renal transplantation. Arch Dermatol 113:436–438

Kabat GC, Wynder EL (1989) Type of alcohol beverage and oral cancer. Int J Cancer 43:190–194

Kahn T, Schwarz E, zur Hausen H (1986) Molecular cloning and characterization of the DNA of a new human papillomavirus (HPV 30) from a laryngeal carcinoma. Int J Cancer 37:61–65

Kreider JW, Howett MK, Stoler MH, Zaino RJ, Welsh P (1987) Susceptibility of various human tissues to transformation in vivo with human papillomavirustype 11. Int J Cancer 39:459–465

Lipkin A, Miller RH, Woodson GE (1985) Squamous cell carcinoma of the oral cavity, pharynx, and larynx in young adults. Laryngoscope 95:790–793

Löning T, Ikenberg H, Becker J, Gissmann L, Hoepfner I, zur Hausen H (1985) Analysis of oral papillomas, leukoplakias, and invasive carcinomas for human papillomavirus type related DNA. J Invest Dermatol 84:417–420

Löning T, Meichsner M, Milde-Langosch K, Hinze H, Orlt I, Hörmann K, Sesterhenn K, Becker J, Reichart P (1987) HPV DNA detection in tumours of the head and neck. A comparative light microscopy and DNA hybridization study. Oto Rhino Laryngology 49:259–269

Mounts P, Shah KV, Kashima H (1982) Viral etiology of juvenile- and adult-onset squamous papilloma of the larynx. Proc Natl Acad Sci USA 79:5425–5429

Naghashfar Z, Sawada E, Kutcher MJ, Swancar J, Gupta J, Daniel R, Kashima H, Woodruff JD, Shah K (1985) Identificatioin of genital tract papillomaviruses HPV-6 and HPV-16 in warts of the oral cavity. J Med Virol 17:313–324

Petzoldt D, Pfister H (1980) HPV 1 DNA in lesions of focal epithelial hyperplasia Heck. Arch Dermatol Res 268:313–316

Pfister H (1986) Papillomaviren und Tumorkrankheiten des Menschen. Ber Pathol 103:177–186

Pfister H, Hettich I, Runne U, Gissmann L, Chilf GN (1983) Characterization of human papillomavirus type 13 from lesions of focal epithelial hyperplasia Heck. J Virol 47:363–366

Pindborg JJ (1980) Oral cancer and precancer. Wright, Bristol

Praetorius-Clausen F (1972) Rare oral viral disorders (molluscum contagiosum, localized keratoacanthoma, verrucae, condyloma acuminatum, and focal epithelial hyperplasia). Oral Surg 34:604–618

Praetorius-Clausen F (1973) Geographical aspects of oral focal epithelial hyperplasia. Pathol Microbiol 39:204–210

Reichart PA, Gelderblom HR, Becker J, Kuntz A (1987) Int J Oral Maxillofac Surg 16:129–153

Rentrop M, Knapp B, Winter H, Schweizer J (1986) Aminoalkylsilane-treated glass slides as support for in situ hybridization of keratin cDNAs to frozen tissue sections under varying fixation and pretreatment conditions. Histochem J 18:271–276

Stremlau A, Zenner HP, Gissmann L, zur Hausen H (1987) Nachweis und Organisationsstruktur der DNS menschlicher Papillomviren beim Kehlkopf- und Hypopharynxkarzinom. Laryng Rhinol Otol 66:311–315

Syrjänen SM, Syrjänen K, Happonen RP, Lamberg MA (1987) In situ hybridization analysis of human papillomavirus (HPV) sequences in benign oral mucosal lesions. Arch Dermatol Res 279:543

Syrjänen SM, Syrjänen KJ, Happonen RP (1988) Human papillomavirus (HPV) DNA sequences in oral precancerous lesions and squamous cell carcinoma demonstrated by in situ hybridization. J Oral Pathol 17:273–278

zur Hausen H (1985) Genital papillomavirus infections. Prog Med Virol 32:15–21

Human Papillomaviruses in Anogenital Condylomas and Squamous Cell Cancer: In Situ Hybridization Study with Biotinylated Probes and Comparative Investigation of Different Detection Protocols

A. Rivière, R.-P. Henke, and T. Löning

Introduction

Tumors are classified on histological and cytological grounds in order to distinguish between benign and malignant lesions and to recognize different grades of malignancy. The borderline between these categories is often not sharp and is not likely to be better defined on the basis of the morphological phenotype only.

A promising instrument for increasing our knowledge of the actual state and the putative behavior of a given tumor has become available with the construction of DNA and RNA probes for extrinsic (Burck et al. 1988) and intrinsic (Höfler 1988) pathogens and the development of protocols for the detection of the labeled nucleic acid hybrids. In the case of tumors of squamous epithelia, especially of the anogenital sites, genotyping has been mostly directed to the investigation of exogenous viral DNA. Up to now, human papillomaviruses (HPV), their heterogeneity, and their pathogenic role represent one of the most challenging fields in gynecology and dermatology (Pfister 1984; Broker and Botchan 1986; Zur Hausen 1985). The clinical need for a rapid and reliable method for detecting and differentiating between HPV genotypes has been met with the development of radioactive dot blot (e.g., filter in situ) hybridization techniques (Schneider et al. 1985; De Villiers et al. 1986). The major disadvantage of this method, however, is the loss of information at the cellular level, which still represents the gold standard of diagnosis and the basis of clinical strategies.

In order to relate genomic alterations to the morphological phenotype, attention has been paid to the generation of nonisotopic in situ hybridization protocols, which obviate the experience and equipment which are necessary for the labeling and detection of radioactive nucleotides (Brigati et al. 1983; Beckmann et al. 1985). Several nonisotopic labeling and detection systems are now available, even commercially, and we recently, compared different DNA labeling protocols (nick-translated biotinylated versus sulfonated DNA probes) and found similar results in terms of sensitivity and specificity. In the present study, all hybridizations were carried out with biotinylated DNA probes for different HPV types (HPV 6, 11, 16, 18). Comparative experiments were conducted using different detection protocols (immunoenzyme/immunogold protocols) in order to ascertain the most convenient and at the

Genital Papillomavirus Infections
Eds.: Gross/Jablonska/Pfister/Stegner

same time efficient and reliable hybridization approach in routine pathology. We applied the in situ hybridization technology to the study of HPV infections of routinely processed acuminate and giant condylomas, verrucous carcinomas, and highly differentiated squamous cell carcinomas, since we further hoped to obtain information on the delineation of these tumor categories by means of HPV type-specific hybridizations.

Materials and Methods

Tissues

The report is based on a study of 17 patients with penile acuminate condylomas (mean age 20 years), six with anal acuminate condylomas (five males, one female, mean age 37 years), three with anal giant condylomas, two with verrucous carcinomas, and eight with squamous cell carcinomas (see Table 1). All cases were selected after a histological review of the files from the Department of Pathology at the University of Hamburg.

Viral Probes

For DNA hybridization, cloned DNA of HPV types 6, 11, 16, and 18 was used. These probes, cloned into the *Bam*HI (or *Eco* RI, in the case of HPV 18) restriction cleavage site of the plasmid pBR322, were a kind gift from Prof. H.

Table 1. Clinical data of the hybridized giant condylomas and carcinomas

Case no.	age	sex	Diagnosis	HPV			
				6	11	16	18
1	57	m	Giant condyloma of penis	+	–	–	–
2	34	m	Giant condyloma of penis	+	+	–	–
3	51	m	Giant condyloma of penis	–	–	–	–
4	41	m	Verrucous carcinoma	+	+	–	–
5	66	m	Verrucous carcinoma	+	–	–	–
6	28	m	SCC	–	–	+	+
7	71	f	SCC	–	–	+	–
8	58	m	SCC	–	–	–	–
9	63	f	SCC	–	–	–	–
10	53	m	SCC	–	–	–	–
11	78	f	SCC	–	–	–	–
12	68	f	SCC	–	–	–	–
13	77	f	SCC	–	–	–	–

SCC, squamous cell carcinoma; +, positive; –, negative

zur Hausen and Prof. L. Gissmann (German Cancer Research Center, Heidelberg, FRG). After propagation in *E. coli,* plasmids were harvested from cesium chloride gradients. They were labeled employing a nick-translation procedure with biotinylated deoxyuridine triphosphate (Bio-11-dUTP) and a nick-translation reagent kit (Gibco/BRL, Eggenstein, FRG).

In Situ Hybridization

In situ hybridization was performed as described in Löning and Milde (1987). Adjacent 4–6 μm sections of each paraffin block were adhered to aminoalkylsilane-treated glass slides, dewaxed in xylene, incubated in ethanol, and air-dried; this was followed by a 15 min digestion with 0.3 mg/ml pronase (Calbiochem, Frankfurt, FRG) in 50 m*M* Tris HCl and 5 m*M* EDTA, at pH 7.4 and room temperature. In order to digest contaminating nucleases, the pronase solution was preincubated for 4 h at 42° C. Pronase treatment was stopped by soaking the slides twice for 5 min in 0.01 *M* Tris HCl (pH 7.5), containing 0.1 *M* NaCl and 2 mg/ml glycine. When peroxidase was used in the detection system, endogenous peroxidase was inhibited by incubating the sections in methanol–1% H_2O_2 for 30 min. The slides were then dehydrated through a graded series of alcohol and air-dried.

For hybridization, each section was covered with 20 μl of the following freshly prepared hybridization solution: 2 × SSC (1 × SSC = 0.15 *M* NaCl, 0.015 *M* trisodium citrate, pH 7.2) 20% (v/v) deionized formamide, 10% (w/v) dextran sulfate, 0.1 mg/ml herring sperm DNA, and 1.0 μg/ml biotinylated HPV DNA.

The sections were covered with 22 × 22 mm coverslips and, after being sealed with rubber cement, were denatured by heating in a 90° C water bath for 10 min followed by hybridization at 37° C overnight in a humidified chamber. After hybridization, the coverslips were carefully removed by immersion in 2 × SSC and sections were washed twice for 10 min in 1 × SSC, 45% formamide, 37° C (stringent conditions, temperature 17° C under melting temperature of HPV hybrids), followed by three washes for 5 min each in 2 × SSC.

Detection of Hybrids

Four different protocols were used for the detection of hybridized HPV probes (Fig. 1). Antibodies and streptavidin conjugates were all incubated for 1 h at 37° C. All washing steps were done for 5 min at room temperature. All the products needed were supplied by Merck (Darmstadt, FRG) unless indicated otherwise.

Streptavidin–Alkaline Phosphatase (SAAP)

The SAAP protocol has been previously described by Henke et al. (1987). Briefly, a polyclonal rabbit antibiotin antibody (Enzo; 1/500 in

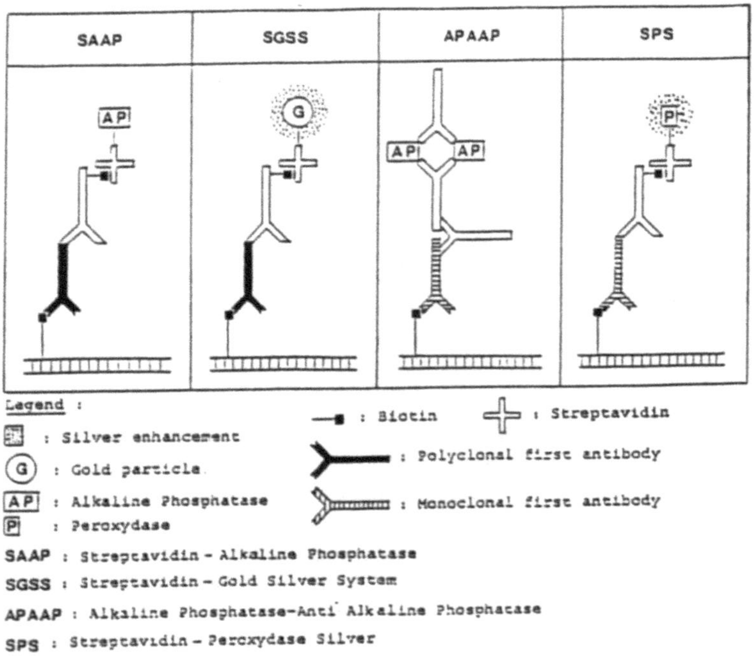

Fig. 1. Representation of the systems for detecting hybrids compared in this study. (▣, silver enhancement; ⓖ, gold particle; AP, alkaline phosphatase; P, peroxidase; —▪, biotin; ✛, streptavidin; ⤚, polyclonal first antibody; ⤚, monoclonal first antibody; *SAAP,* streptavidin-alkaline phosphatase; *SGSS,* streptavidin-gold silver system; *APAAP,* alkaline phosphatase-antialkaline phosphatase; *SPS,* streptavidin-peroxidase silver)

2 × SSC–human serum 2/1) was applied to the sections. After three washes in 2 × SSC, the sections were incubated with a biotinylated antirabbit IgG (Vector Lab.; 1/500 in 2 × SSC–human serum 2/1). After three washes in buffer 1 (0.1 *M* Tris HCl, pH 7.5; 0.15 *M* NaCl), a SAAP conjugate (BRL; 1/1000 in buffer 1) was added to the sections. Following two washes in buffer 1 and three others in buffer 2 (0.1 *M* Tris HCl, pH 9.5; 0.1 *M* NaCl; 50 m*M* $MgCl_2$), the enzymatic reaction with 0.033% nitroblue tetrazolium (NBT; BRL) and 0.016% 5-bromo-4-chloro-3-indolylphosphate (BCIP; BRL) in buffer 2 resulted in blue precipitates. The reaction was stopped with buffer 3 (20 m*M* Tris HCl, pH 7.5; 0.5 m*M* EDTA) after 15–30 min. The slides were then mounted in gelatin.

Streptavidin–Gold-Silver System (SGSS)

The SGSS protocol includes three incubation steps followed by an intensifying step with silver salts (Henke et al. 1989). After the posthybridization washes, the slides were incubated for 5 min each in Lugol's iodine solution and distilled water, and 10 min in a 2.5% sodium thiosulfate solution in order to enhance the signal at the intensifying step (Holgate et al. 1983). After having been washed for 10 min in distilled water, the slides were incubated with the two antibodies described in the previous protocol, rinsed twice in buffer 1, twice in buffer 4 (20 m*M* Tris HCl, pH 8.2; 150 m*M* NaCl), and incubated with a complex formed between streptavidin and gold particles 5 nm in diameter (Janssen; 1/100 in buffer 4). Three washes with buffer 4 and two others with phosphate-buffered saline (PBS) were done prior to postfixation with 2% glutaraldehyde in PBS for 15 min. After washing in PBS and in distilled water had been performed, silver galvanization around the gold particles was allowed to take place (Intense II intensifying kit; Janssen). The slides were counterstained with hematoxylin, then dehydrated and mounted in Eukitt.

Alkaline Phosphatase–Antialkaline Phosphatase (APAAP)

After the posthybridization washes, the sections were covered with a monoclonal mouse antibiotin antibody (DAKO) diluted 1/100 in buffer 5 [RPMI 1640 (Seromed) together with 0.1% bovine serum albumin (Hoechst Behring) and 0.1% sodium azide], rabbit antimouse IgG (DAKO; 1/50 in buffer 5) and the alkaline phosphatase antialkaline phosphatase complex (DAKO; 1/50 in buffer 5). After each step, three washes with buffer 6 (50 m*M* Tris HCl, pH 7.5; 150 m*M* NaCl) were done. As substrates, new fuchsin and naphthol AS biphosphate were used [solution 1, 175 ml 0.285 *M* Tris, pH 8.75, 0.85 *M* NaCl, 62.5 ml 2.1% propanediol solution, and 100 mg levamisole (Sigma); solution 2, 125 mg naphthol AS biphosphate (Sigma) solubilized in 1.5 ml dimethylformamide]. Solutions 1 and 2 were added to solution 3 [1.25 ml 4% sodium nitrite solution and 0.5 ml 5% new fuchsin (Serva) in 2 *N* HCl]. After filtration, this mixture was added to the slides. The alkaline phosphatase reaction, resulting in red precipitates, was stopped with buffer 3 after 15–30 min.

Streptavidin–Peroxidase Silver (SPS)

After the sections had been incubated with mouse monoclonal antibiotin antibodies (DAKO; 1/100 in buffer 5), biotinylated sheep antimouse antibodies (Amersham; 1/200 in buffer 5), and the streptavidin-peroxidase complex (Dianova; 1/1000 in buffer 5) including washes with buffer 6 in between, the peroxidase was revealed using diaminobenzidine (0.05%), H_2O_2 (0.05%) in Tris HCl, pH 8.3). After the reaction had been stopped with distilled water, the signal was enhanced by a silver salt reaction (as recommended by the supplier, Amersham).

Microscopic Observation

All sections were evaluated using conventional light and/or interference reflection microscopy (Verschueren 1985), the latter being used for the silver enhancement protocols (SGSS and SPS).

Controls

The specificity of in situ hybridization was controlled using hybridization with biotinylated plasmid DNA, whereby the probe was omitted in the hybridization cocktail, or the slide were passed directly to 37° C without preceding denaturation.

Results

General Remarks

The results of in situ hybridization are shown in Table 2. Two out of 17 cases of penile condylomas reacted with the HPV 6 probe, three were positive for HPV 11, and seven hybridized to both HPV 6 and 11. Neither HPV 16 nor 18 were demonstrated in these lesions. With the exception of one case, the degree of koilocytosis was notably lower in the HPV negative specimens than in the 12 positive cases (Fig. 2).

All six anal condylomas were positive for HPV : four cases were positive for both HPV types 6 and 11, one each for HPV 6 or HPV 11. In one case of anal papillomatosis, including acuminate condylomas and bowenoid papulosis, the hybridization signal was confined to the biopsies of classical condylomas (Fig. 3).

Three giant condylomas and two verrucous carcinomas were studied (Figs. 4, 5). Among these, four harbored detectable HPV DNA sequences; two cases (one giant condyloma and one verrucous carcinoma) were positive for HPV 6. One other giant condyloma and one verrucous carcinoma hybridized with HPV 6 and 11, although the signal achieved with HPV 11 was considerably poorer than the strong response to HPV 6. Out of the eight intra-anal squamous cell carcinomas studied, two were positive; one for HPV 16, the other for both HPV 16 and 18. A focal accentuation of the hybridization

Table 2. Frequency and distribution of HPV types in all lesions examined

Diagnosis	Number	HPV						
		6/11	6	11	16/18	16	18	Negative
Penile condyloma	17	7	2	3	–	–	–	5
Anal condyloma	6	4	1	1	–	–	–	–
Giant condyloma	3	1	1	–	–	–	–	1
Verrucous carcinoma	2	1	1	–	–	–	–	–
SCC	8	–	–	–	1	1	–	6

SCC, Squamous cell carcinoma; HPV 6/11 and HPV 16/18, positive cases for two virus types (hybridizations conducted separately for the respective HPV DNAs)

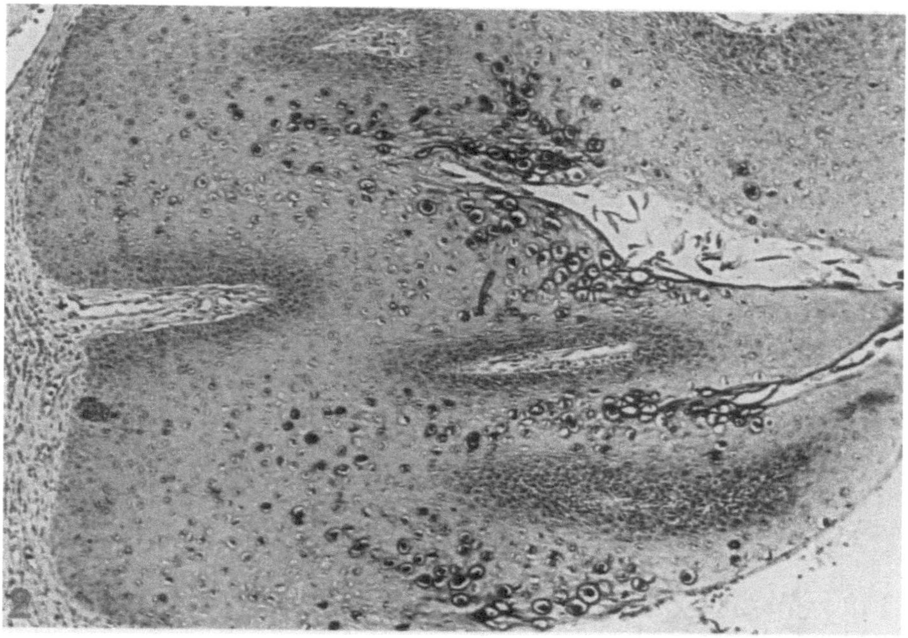

Fig. 2. Penile condyloma with stained koilocytic fields. SAAP technique, × 80

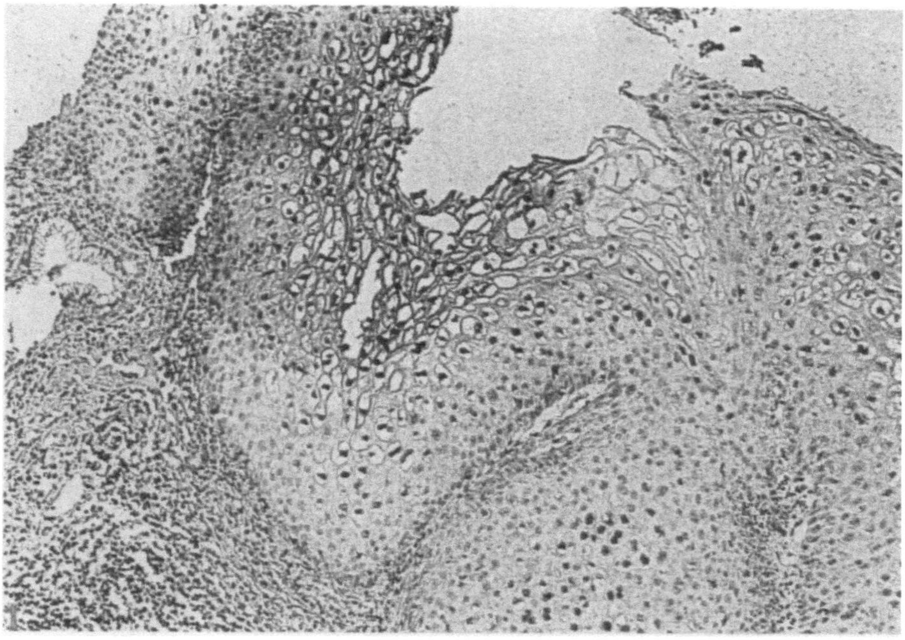

Fig. 3. Anal condyloma immediately at the junctional zone of the rectal mucosa and the squamous epithelium of the anal canal. Note stained nuclei at intermediate and superficial cell layers, especially in koilocytic cells. SGSS technique, × 80

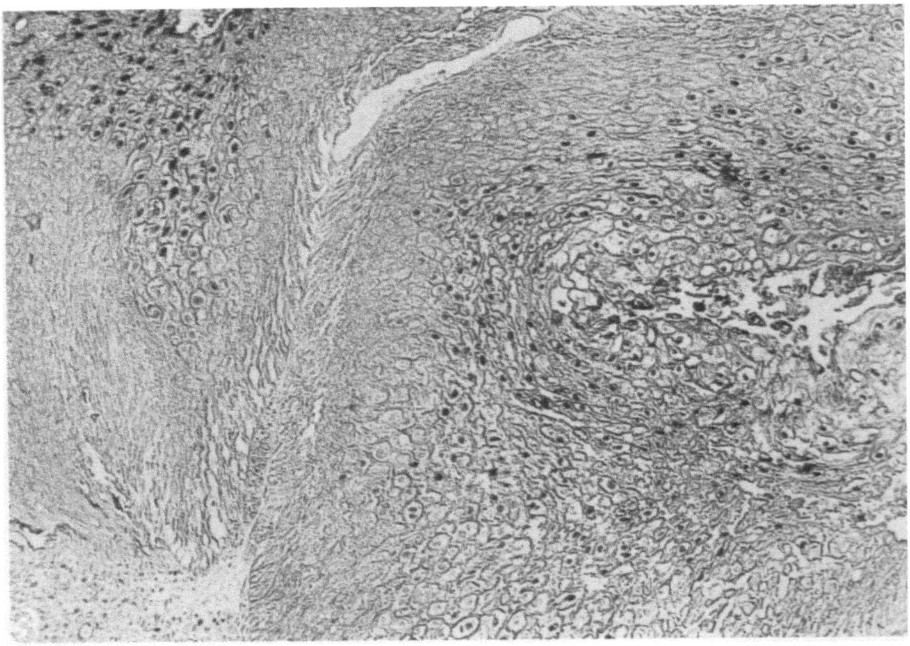

Fig. 4. Giant condyloma of the anal canal. Note stained koilocytes especially at the superficial cell layers. SGSS technique, × 80

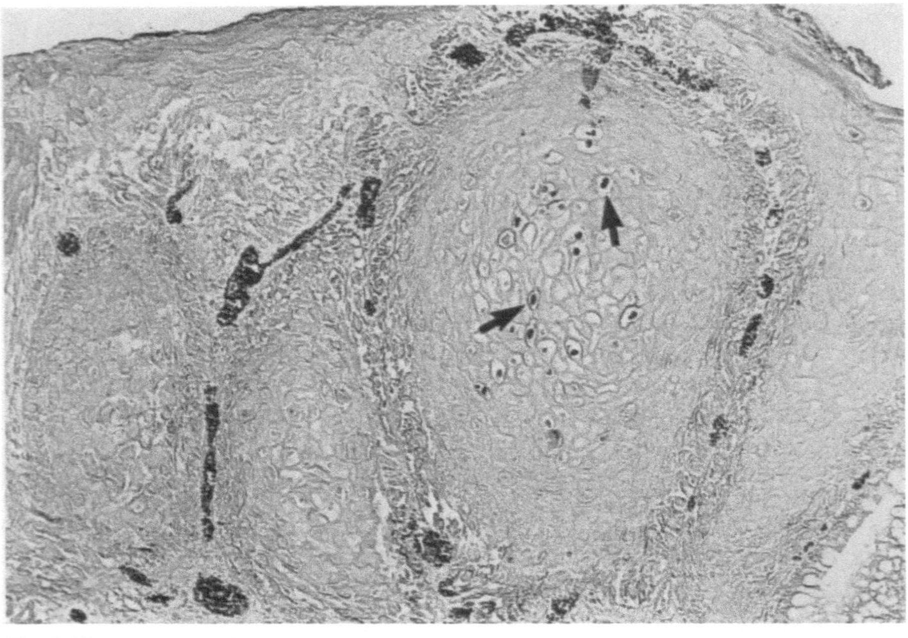

Fig. 5. Verrucous carcinoma. Note few scattered positive koilocytes within keratinized tumor areas (*arrows*), surrounded by capillaries with erythrocytes (*black deposits*). SGSS technique, × 80

signal was a constant finding in all HPV positive tissue samples, with heavily stained nuclei sometimes lying close to unlabeled ones. Staining was most intense in nuclei of superficial cell layers, in fields of koilocytosis, or in dyskeratotic cells. In strongly positive cases, staining was also seen in suprabasal cell layers of the epithelium.

Hybridization signals in stromal fibroblasts were noted only rarely. When these experiments were redone under increasing formamide concentrations (up to 50% formamide), no stromal reaction was seen. Controls were negative in all cases.

Technical Remarks

A comparison of the different systems for detecting hybrids is summarized in Table 3. The immunoenzyme detection methods of choice were those employing bovine alkaline phosphatase. An extremely low background was observed in comparison with results obtained with horseradish peroxidase and subsequent silver enhancement (SPS).

Although the number of positive cases did not change, hybrid detection protocols involving streptavidin-conjugated gold particles (SGSS) made recognition of a higher number of infected keratinocytes in the particular case possible when additionally evaluated using interference reflection microscopy (Fig. 6).

A three-step detection system appeared to be necessary and sufficient to obtain maximum sensitivity; the omission of the first antibody steps in methods involving streptavidin conjugates resulted in a decrease of stained cases or stained cells in a particular case.

With respect to the APAAP technique, experiments were done to enhance the signal by repeating incubations with bridging antibodies and the APAAP complex. However, only the intensity of the signal on a given cell was increased, but not the number of stained cells on a section. Comparison of the two commonly used substrates of alkaline phosphatase resulted in a much more intensive signal with NBT and BCIP than with new fuchsin and naphthol AS biphosphate.

Table 3. Comparison of detection systems of the hybrids

	SPS	SGSS	SAAP	APAAP
Intensity of the signal	High	Very high	High	High
Specificity of the signal	Low	Very high	High	High
Facility of use	Low	High	Very high	Very high

SPS, streptavidin-peroxidase silver; SGSS, streptavidin-gold silver system; SAAP, streptavidin-alkaline phosphatase; APAAP, alkaline phosphatase-antialkaline phosphatase

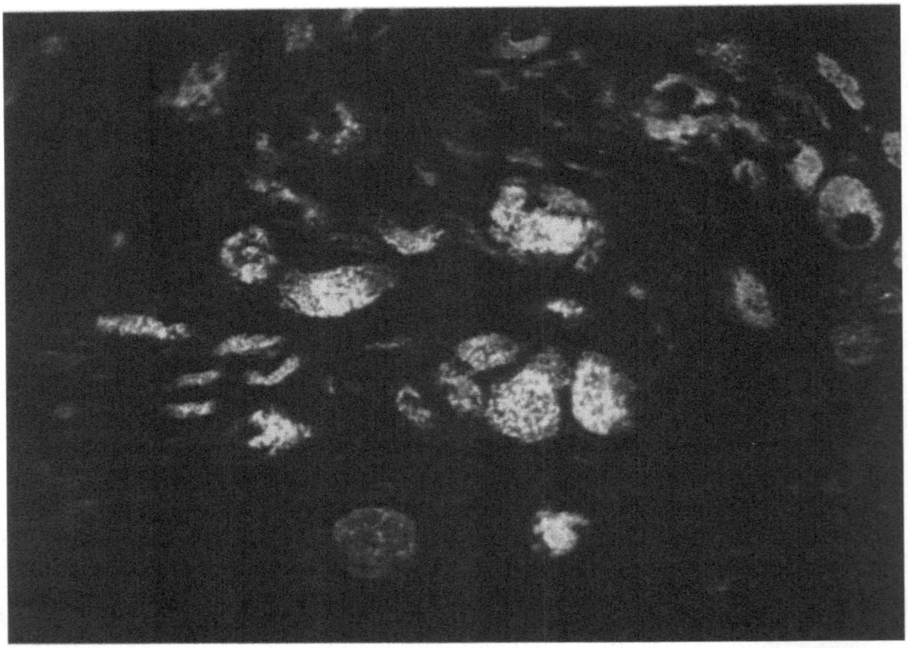

Fig. 6. Interference reflection microscopy demonstrating different degrees of nuclear staining, ×800

Discussion

In situ hybridization with biotinylated HPV probes followed by immunoenzyme or immunogold detection yielded comparable results on paraffin-embedded material in terms of the number of positive cases. Detection rates declined from approximately 80% (18/23) for penile/anal condylomas to 40% (4/10) in the case of invasive cancer.

Specificity was high in the case of hybridizations with distantly related HPV probes (e.g., HPV 6, 16, 18). In contrast, infections with HPV 6 and 11, which have 85% homology (Pfister 1984), cannot always be clearly distinguished by in situ hybridization even under the most stringent conditions. Stromal cell reactions have been reported in the literature (Ostrow et al. 1985; Del Mistro et al. 1987). When we performed hybridizations of increasing stringency on the tumors showing such a signal, the signal disappeared. Hybridizing cells were clearly recognized immediately above the basal cell layer, infections of which are also beyond the detection limit of radioactive probes. The major advantages of nonradioactive hybridizations include the considerably high signal:noise ratio, and the excellent subcellular resolution with detection systems employing alkaline phosphatase or silver galvanized gold particles.

Since the SAAP method including NBT and BCIP as substrates reconciled the demands for acceptable sensitivity and most convenient handling, we preferred this technique for routine in situ hybridizations.

The SGSS protocols permitted recognition of a higher amount of infected keratinocytes in the particular case, especially when supported by interference reflection microscopy (Verschueren 1985).

Apart from the technical perspectives of this study, some interesting information emerged from our pathoanatomical results. The close association of HPV 6 and/or 11 with benign lesions and of HPV 16 and/or 18 with malignant tumors is widely accepted (Pfister 1984; Broker and Botchan 1986), but exceptions to the rule can be increasingly observed. Buschke-Löwenstein's tumors and verrucous carcinoma have been reported to contain episomal, but also integrated or rearranged HPV 6 DNA (Gissmann et al. 1982; Zachow et al. 1982; Okagaki et al. 1984; Rando et al. 1986; De Villiers et al. 1986). Although in situ hybridization does not provide information about the physical state of HPV in the cell, it was noteworthy that two condylomas with cellular atypia, two giant condylomas, and two invasive verrucous carcinomas hybridized with HPV 6. Moreover, progression of one of the two giant condylomas into a SCC was later observed. These data support the concept of Bogolometz et al. (1985) that giant condylomas, Buschke-Löwenstein tumors, and verrucous carcinomas represent a biological continuum. Diagnosis is still a matter of classical, surgical pathology and is not aided by HPV DNA hybridizations employing full-length DNA probes.

References

Beckmann AM, Myerson D, Daling JR, Kiviat NB, Fenoglio CM, Mc Dougall JK (1985) Detection and localization of human papillomavirus DNA in human genital condylomas by in situ hybridization with biotinylated probes. J Med Virol 16:256–273

Bogomoletz WV, Potet F, Molas G (1985) Condylomata acuminata, giant condyloma acuminatum (Buschke-Löwenstein tumour) and verrucous carcinoma of the perianal and anorectal region: a continous precancerous spectrum? Histopathol 9:1155–1169

Brigati DJ, Myerson D, Leary JJ, Spalholz B, Travis SZ, Fong CKY, Hsiung GD, Ward DC (1983) Detection of viral genomes in cultured cells and paraffin embedded tissue sections using biotin-labelled hybridization probes. Virology 126:32–50

Broker TR, Botchan M (1986) Papillomaviruses: retrospectives and prospectives. In: Botchan M, Grodzicker T, Sharp PE (eds) DNA tumor viruses. Cold Spring Harbor Laboratory, Cold Spring Harbor, NY, pp 17–36 (Cancer Cells, vol 4)

Burck KB, Liu ET, Larrick JW (1988) Oncogenes. An introduction to the concept of cancer genes. Springer, Berlin Heidelberg New York Tokyo, pp 38–66

Del Mistro A, Braunstein JD, Halwer M, Koss LG (1987) Identification of human papillomavirus types in male urethral condylomata acuminata by in situ hybridization. Hum Pathol 18:936–940

De Villiers EM, Schneider A, Gross G, zur Hausen H (1986) Analysis of benign and malignant urogenital tumors for human papillomavirus infection by labelling cellular DNA. Med Microbiol Immunol 174:281–286

Gissmann L, De Villiers EM, zur Hausen H (1982) Analysis of human genital warts (condylomata acuminata) and other genital tumors for human papillomavirus type 6 DNA. Br J Cancer 29:143–147
Henke RP, Milde K, Löning T, Strömme-Koppang H (1987) HPV 13 and focal epithelial hyperplasia: DNA hybridization on paraffin embedded specimens. Virchows Arch A 411:193–198
Henke RP, Guérin-Reverchen I, Milde-Langosch K, Strömme-Koppang H, Löning T (1989) In situ detection of human papillomavirus types 13 and 32 in focal epithelial hyperplasia of the oral mucosa. J Oral Pathol Med 18:419–421
Höfler H (1988) Prognosis related oncogene expression in malignant tumors. In: Hübner K (ed) Pathologie der Zelldifferenzierung. Fischer, Stuttgart, pp 174–187 (Verhandlung der Deutschen Gesellschaft für Pathologie, vol 72)
Holgate CS, Jackson P, Lowen PN, Bird CC (1983) Immunogold silver staining: new method of immunostaining with enhanced sensitivity. J Histochem Cytochem 31:938–944
Löning TH, Milde K (1987) Viral tumor markers. In: Seifert G (ed) Morphological tumor markers. Curr Top Pathol 77:339–365
Okagaki T, Clark BA, Zachow KR, Twiggs LB, Ostrow RS, Pass F, Faras AJ (1984) Presence of human papillomavirus in verrucous carcinoma (Ackerman) of the vagina. Arch Pathol Lab Med 108:567–570
Ostrow RS, Zachow K, Weber D, Okagaki T, Fukushima M, Clark BA, Twiggs LB, Faras AJ (1985) Presence and possible involvement of HPV DNA in premalignant and malignant tumors. In: Howley PM, Broker TR (eds) Papillomaviruses: molecular and clinical aspects. UCLA Symposium on Molecular Cell Biology. Alan R Liss, NY, vol 32, pp 101–124
Pfister H (1984) Biology and biochemistry of papillomaviruses. Rev Physiol Biochem Pharmacol 99:112–181
Rando RF, Sedlacek TV, Hunt J, Jenson AB, Kurman RJ, Lancaster WD (1986) Verrucous carcinoma of the vulva associated with an unusual type 6 human papillomavirus. Obstet Gynecol 67:70S–75S
Schneider A, Kraus H, Schuhmann R, Gissmann L (1985) Papillomavirus infection of the lower genital tract: detection of viral DNA in gynecological swabs. Int J Cancer 35:443–448
Verschueren H (1985) Interference reflection microscopy in cell biology: methodology and applications. J Cell Science 75:279–301
Zachow KR, Ostrow RS, Bender M, Watts S, Okagaki T, Pass F, Faras AJ (1982) Detection of human papillomavirus in anogenital neoplasias. Nature 300:771–773
Zur Hausen H (1985) Genital papillomavirus infections. Prog Med Virol 32:15–21

Genitoanal HPV Infections in Immunodeficient Individuals *

R. Rüdlinger, P. Buchmann, R. Grob, F. Colla, R. Steiner, and M. Meandzija

Introduction

Until two decades ago the number of patients who were immunosuppressed was rather low. Since the advent of organ mainly kidney transplantation the situation has quite changed. In Zürich renal transplantation began in 1964 and by 1986 a total of 1039 kidneys had been transplanted [1]. In Germany there are far more than 15000 patients with renal allografts. The situation in other countries is similar. The other increasing group of patients who are immunodeficient are HIV-infected individuals. Hence the number of immunosuppressed patients is rising steadily. Patients from either group are prone to various skin diseases of both infectious and noninfectious origin [2–7].

Renal Transplant Recipients

The author of an editorial in the *Lancet* stated that "For patients on renal dialysis awaiting a properly matched donor kidney, the prospect of dermatological problems years posttransplant will no doubt seem a very minor inconvenience at most. However, for long-term survivors, the reality may be somewhat different" [8]. In one study from our hospital, 75% of 205 patients were registered as suffering from some kind of dermatological problem [9]. There are three main types of cutaneous diseases in these patients: persistent viral warts, premalignant actinic keratoses, and frank cutaneous malignancy. Reports on the high incidence of warts and skin cancer in renal transplant recipients (RTRs) were noted in several studies from various countries [2, 3, 10–20]. Warts in RTRs are not a banality. They can grow to grotesque forms and cover large areas.

In a large study, 30% of RTRs surviving 10 years had developed skin cancer [19]. In general skin cancers in RTRs fit a common picture: the tumors occur primarily on sun-exposed skin, but some unfortunate patients develop almost generalized squamous cell carcinoma (SCC) of the skin. Most authors report more SCC than basal cell carcinoma (BCC), which is remarkable since BCCs outnumber SCCs in the general population by a ratio of 4:1 [21]. SCCs

* This work was supported by grant No. 3.995-0.86 of the Swiss National Science Foundation, Stiftung zur Förderung der AIDS Forschung in der Schweiz und Krebsliga des Kantons Zürich

Genital Papillomavirus Infections
Eds.: Gross/Jablonska/Pfister/Stegner

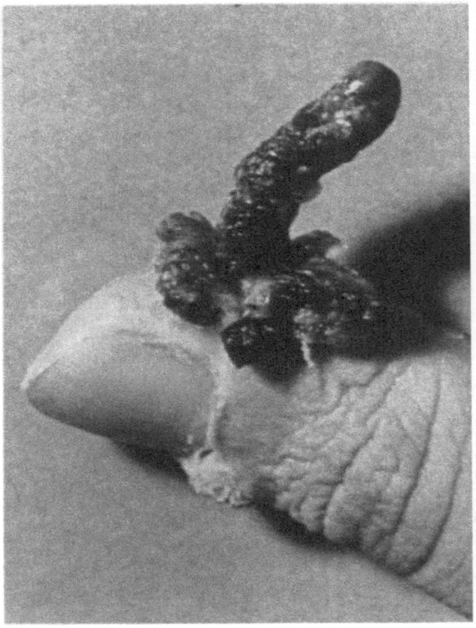

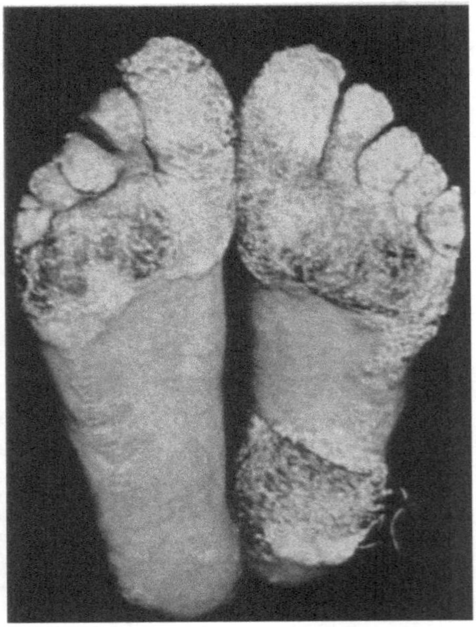

Fig. 1. **Fig. 2.**

Fig. 1. A 16-year-old girl who was a renal transplant recipient (RTR) with HPV 2 induced wart on thumb, presenting as cutaneous horn

Fig. 2. A 25-year-old female RTR with massive HPV 2 induced plantar warts

tend to occur in young transplant recipients, they tend to be multiple, aggressive, and prone to recurrence and metastasis, and even cause death [3, 11–13, 15–18, 20].

HPVs are among the suspected causes for the high incidence of skin cancers in RTRs [22]. For skin cancer at extraanogenital sites this association is not yet clear [23; and R. Rüdlinger et al. submitted]. For anogenital cancer the evidence is in general well recognized [24]. However, for RTR anogenital warts and cancer and HPV types found in these tumors the available data are scarce and most of them anecdotal [25, 26]. Van der Leest report that of 36 patients 3 suffered from condylomata acuminata [27] and in 120 Edinburgh patients who we studied randomly, 5 of 50 females suffered from anogenital warts, including a patient who developed an HPV 16 positive vulval SCC [28].

An increased incidence of anogenital cancer in RTRs has been reported, although to what extent remains uncertain. Sheil et al. reported a 1.8 × increased risk for RTRs to develop cervical cancer and stated that it was the second most frequent cancer in female RTRs after skin cancer [29]. The same authors, however, only 1 year later thought that the risk was indeed much higher: 37.5 × for cervical carcinoma in situ and 16.6 × for invasive cervical carcinoma [16].

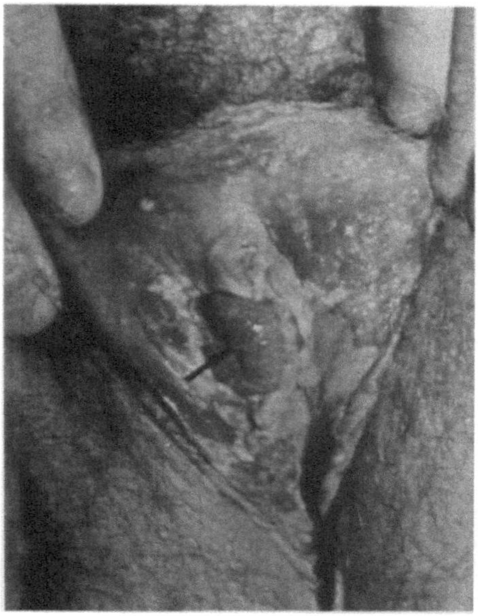
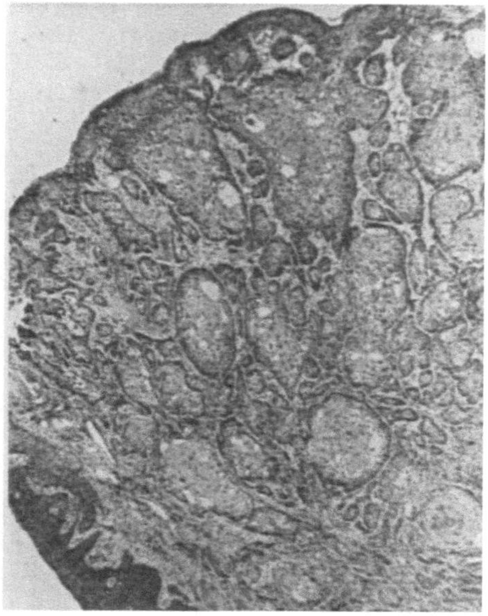

Fig. 3. **Fig. 4.**

Fig. 3. A 35-year-old RTR with widespread genital warts of the flat condylomatous type. *Arrow* indicates a squamous cell carcinoma (SCC) which developed rapidly. It was shown by Southern blot hybridization to harbor HPV 16 DNA

Fig. 4. Histology (HE, ×16) of vulval carcinoma (shown in Fig. 3) of a RTR. Numerous tumor cell nests composed of atypical squamous cells are visible throughout the whole dermis

The results of our Edinburgh RTR study on HPV infections prompted a study of cervical HPV infection in female RTRs [30]. Biopsy specimens from 49 women with renal allografts and 69 nonimmunosuppressed controls were analyzed. They were assessed for colposcopic appearance, cytological and histological diagnosis, and the presence of human papillomavirus types 6/11 and 16/18 DNA sequences. Women with renal allografts who were receiving routine follow up at the transplantation unit were recruited to attend the colposcopy clinic in Edinburgh. Controls were women who had been admitted to a gynecological ward for elective operations. They had no history of cervical intraepithelial neoplasia (CIN), vulval warts, or abnormal results of cervical smear tests. Controls were matched for parity, age of first intercourse, number of sexual partners, smoking habits, and current contraceptive practice. Colposcopy was performed, and cytological smears and biopsy specimens were taken from each patient. If no colposcopic abnormality was detected, biopsy specimens were taken at random from the transformation zone. Figure 5 shows the results obtained.

Another interesting feature in this study is the fact that 16 out of 24 patients with CIN would have been missed had only smears been taken and no

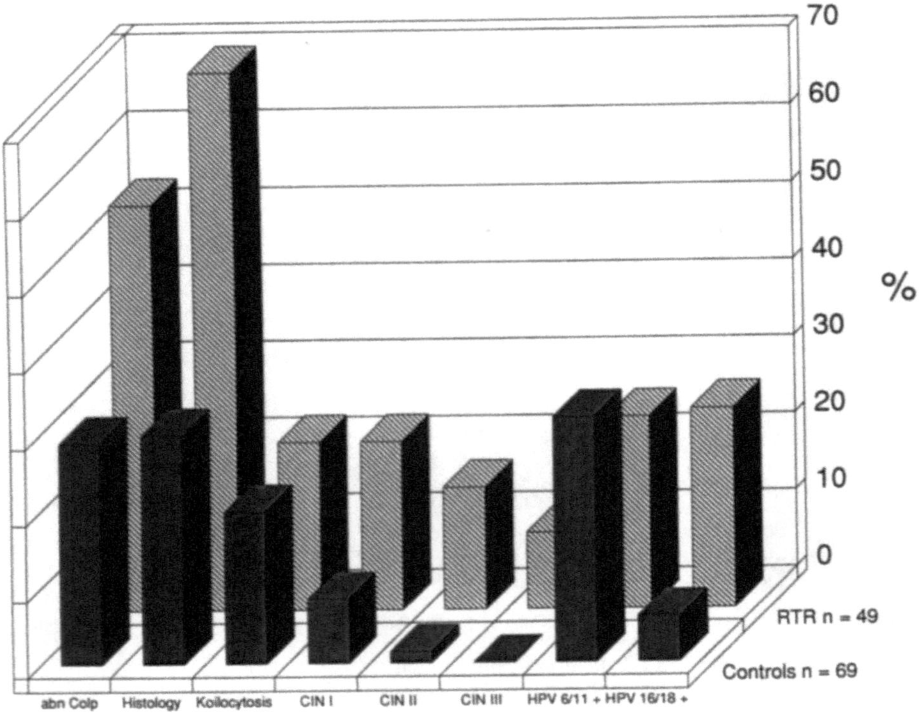

Fig. 5. HPV infection and cervical intraepithelial neoplasia (CIN) in RTRs from Edinburgh (kind permission of Alloub et al. [30]. More than 50% of the RTR group had abnormal colposcopy findings compared with 29% of the controls ($p<0.01$). A significant number of RTRs had abnormal histology ($p<0.0001$). The same is reflected in the number of patients who were diagnosed as suffering from CIN. No significant difference between the groups in the rate of total HPV detection was detected. The presence of type HPV 16/18, however, was significantly higher in the women with allografts than the controls ($p<0.005$). The authors conclude that this difference is directly related to the high grades of CIN observed in these immunosuppressed women

biopsies performed. Thus, cervical smear tests did not provide a reliable screen in these women, a finding that has clear implications for the doctors attending these patients.

From the present information one must conclude that immunosuppressed RTRs should be examined on a regular basis by dermatologists for premalignant and malignant skin tumors and by gynecologists for cervical neoplasia.

HIV-Positive Individuals

Anogenital Warts

Another important group of immunosuppressed individuals are HIV-infected persons. In our experience skin warts can be a problem in HIV + patients but

Table 1. General clinical features of condylomata acuminata in HIV+ patients

Often widespread, multiple
Often circular (anal, urethral)
Often discomfort
Often bleeding (anal)
Urine retention (urethral)

Table 3. Risk, history, and age of HIV+ patients with anogenital warts

26 Males	13 Homosexuals 10 I. v. drug addicts 3 Both risks
8 Females	All i. v. drug addicts
Age (years)	
Males, 28.3 ± 5.7	Females, 26.9 ± 4.3

Table 2. HIV stage at time of diagnosis of anogenital warts

CDC II	22	(64.7%)
CDC III	5	(14.7%)
CDC IV	7	(20.6%)

No significant difference between sexes or risk groups

not, or not yet, to the extent seen in RTRs. Also, skin cancer or keratoses at extraanogenital sites have only rarely been noticed [31, 33]. Most HIV-positive patients still belong to groups which are at risk for STD. Therefore, it is not too surprising that these individuals often suffer from anogenital HPV infection. Prevalences of anogenital warts as high as 18% were found in HIV+ individuals. We performed a study of anogenital warts in HIV-infected individuals seen by one of us when complaining of anogenital HPV infection. We studied 34 patients with anogenital warts and their general clinical features are listed in Tables 1–3.

Two-thirds of our patients were in group II of the CDC classification (MMWR 35/20, 1986), indicating that according to this classification system their HIV infection was clinically asymptomatic; 33 patients displayed anogenital warts of the condyloma acuminatum type and one patient had flat condylomatous lesions. The general clinical characteristics of acuminate warts were multiplicity, widespreadness (Fig. 6), they were often circular when located in the anal canal (Fig. 7) or in the urethra, bearing the risk of stenosis when surgically removed; they often gave rise to discomfort, even bleeding, and in the case of urethral involvement, urine retention in one patient. It seemed noteworthy that in some patients with warts affecting only the anus no history of anal intercourse was given, although the majority of the homosexual patients with anal warts only had practised receptive anal intercourse. One patient who was in stage IV of the HIV disease and who had been severely ill for months gave no history of sexual contacts for more than 1 year and yet genital warts of the flat type had developed, pointing to a possible latency of HPVs [33].

In a number of patients – most of them with asymptomatic HIV disease – widespread anogenital warts were the only hint of impaired immunity. In

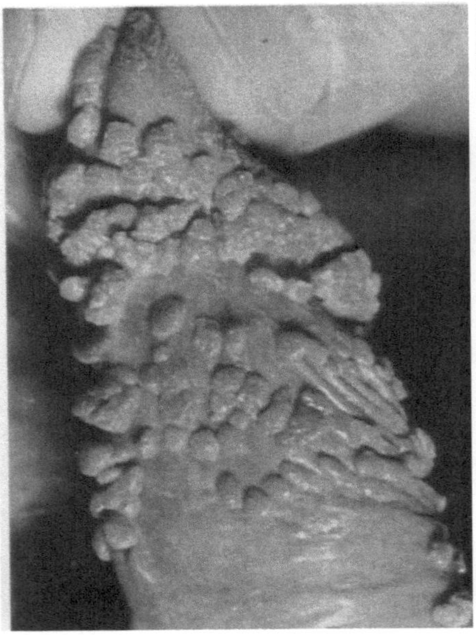

Fig. 6.

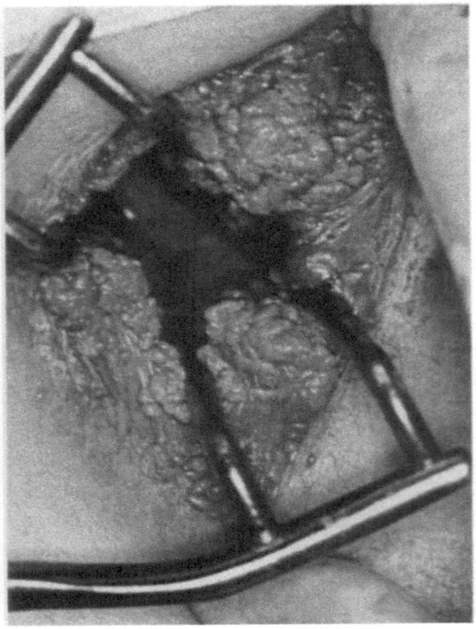

Fig. 7.

Fig. 6. Multiple, widespread condylomata acuminata in HIV+ patient

Fig. 7. Circular anal condylomata acuminata in HIV+ patient

Table 4. Other cutaneous diseases in HIV+ patients with anogenital warts

Other cutaneous problems	
Males, $n=26$ (16)	Females, $n=8$ (4)
Seborrheic dermatitis	Anal fistula
Cold sore	
Oral warts	Syphilis
Verrucae (hands, feet)	Trichomoniasis
Mollusca contagiosa	Candida (oral, vaginal)
Hairy leukoplakia	Verrucae (hands, feet)
Anal fissures	
Kaposi's sarcoma	Candida (oral)
Herpes zoster	
Candida (oral)	Psoriasis
Scrotal abscess	
Anal abscess	

Table 5. Type, localization, histology, and virology of anogenital warts in 34 HIV+ patients. „Anogenital" localization indicates that warts were present anally as well as genitally. Anogenital warts in females grew both anally and in most cases including the vulva, vagina, and cervix. Homosexual patients often suffered from anal warts only and most but not all of them gave a history of receptive anal intercourse. Anal involvement often meant warts perianally, in the anal canal, and the rectal mucosa above the dentate line. *Asteriks* indicate that one patient in either group was affected with urethral condylomas. On the histological level most condylomas were benign viral acanthopapillomas. Slight dysplasia was reported in three cases in each group of males and females. Identification of HPV DNA was performed by Southern blot hybridization in most cases and exceptionally by in situ hybridization only, according to standard procedures, as outlined in detail elsewhere [28, 34]. As reported previously for a smaller group of HIV-positive patients all individuals but one with benign histology were HPV 6/11 positive [35]. In one we detected an HPV type that cross-hybridized with HPV 6/11 and 33 but not 16. As was reported by others, slight atypia did not necessarily correlate with the presence of HPV 16 [36, 37, 38]. The amount of viral DNA that was present in some of the warts was strikingly high. Sometimes autoradiographs could be developed after a 4-h exposure which for us is normally far a far too short time for detecting genital type HPVs

	Males ($n=26$)			Females ($n=8$)
	Homosexuals 13	Drug addicts 10	Both 3	Drug addicts 8
Clinical type	25 Acuminate + 1 flat condyloma			All condylomas acuminate type
Localization:				
Anogenital	2*	6	2	6
Anal	10	2	1	2
Genital	1	2*	–	–
Histology + Virology				
No dysplasia	18 HPV 6/11 (17)/unidentified/HPV type (1)			4 All HPV 6/11
Slight dysplasia	3 HPV 6/11 + 16*/nd/HPV 11a			3 HPV 6a+16/ HPV 11a+16/ HPV 6c
No histology	5	–	–	1 HPV 6/11

Table 6. Various treatments given to 34 HIV+ patients with anogenital warts. The number of male patients who had to be operated on under general anesthesia is 21 and altogether these 21 patients underwent general anesthesia 35 times. All female patients had undergone surgery as well

Males ($n=26$)	Patients	Females ($n=8$)	Patients
Podophyllin 20%	All	Podophyllin + CO_2 laser	2
Mitomycin (urethral)	1	Electrodesiccation	3
Alpha interferon systemically	2	CO_2 laser	2
No general anesthesia	5	CO_2 laser + alpha	1
General anesthesia (electrocauterization, scissor excision)	21	General anesthesia	8

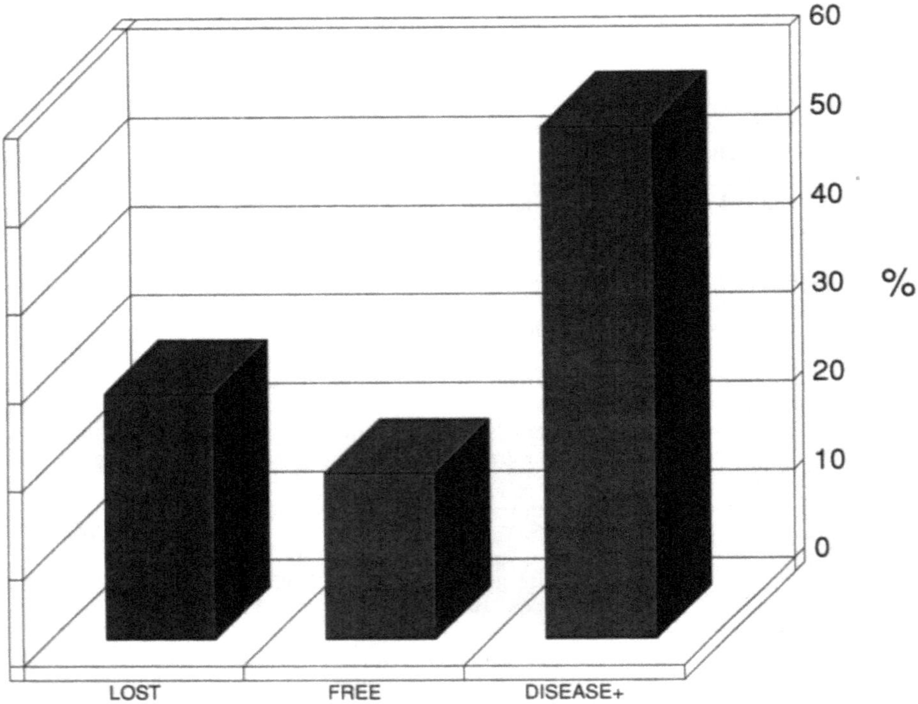

Fig. 8. Follow-up of HIV+ patients with anogenital warts. Patients were grouped into three categories: *lost* for follow-up, *free* of disease for at least 6 months, and *disease+* patients who continued to suffer from their warts

others, additional cutaneous problems were recorded among which were other HPV infections like cutaneous and oral warts.

The patients were treated in various departments. A compilation of the different treatment modalities is given in Tables 7 and 8.

All these treatment efforts were unfortunately not reflected in the outcome of the anogenital warts in our patients (Fig. 8).

Even if one speculated that the patients who were lost for follow-up were probably free of disease the cure rate is less than 50%, a figure that is far beyond cure rates as we see them normally [39]. It must be mentioned at this point that partners were not available for examination, which is of course an unsatisfactory situation. There was no obvious correlation between cure rate and stage of HIV disease.

Bowenoid Papulosis

Three males in stage III (one patient) or IV (two patients) respectively suffered from bowenoid papulosis (BP). One with lesions on the prepuce was circumcised and has been free of disease for more than 4 years. One patient with BP on the penis shaft was lost for follow-up and the third patient with anal BP developed an anal carcinoma (both HPV 16+), which was suspected already on clinical grounds. This patient is described in detail elsewhere [57]. The number of patients with BP studied here is too small to draw definite conclusions. Nonetheless, we were surprised by the transition from BP to invasive anal cancer, an event that is rarely reported in the literature and only in females [40–43].

Anal Cancer

Anal cancer was registered three times. Although the presence of HPV was not documented in all three it was thought appropriate to include these patients here, because anal cancer has been linked with anal intercourse, and HPV and HIV infection [44–55]. One patient with anal carcinoma is mentioned above, he suffered from concurrent BP. A further patient (in stage II) was reported to show condylomatous changes next to the carcinoma. The third patient (stage II also) suffered from a metastatic anal cancer of which he finally died.

Conclusions

On the whole immunosuppressed patients are on the increase and HPV infections in these individuals are among the skin diseases which will continue to be a matter of considerable importance to both patients and medical staff. At present there are marked differences between RTR and HIV+ patients: skin cancer, at extraanogenital sites, whether or not associated with HPV, is a common feature in RTRs. This was not or perhaps not yet observed in HIV+ patients. This may be due to different defects of the immune system in the two groups, but is more likely attributed to the fact that HIV+ patients have not the same survival chances as RTRs, whose skin cancers most commonly arise after prolonged immunosuppression. Anogenital HPV infections also show differences between the two groups: as far as can be concluded from the available data, prevalences of HPV infection at the outer genital and anal areas are higher in HIV+ individuals. HPV infections are sexually transmittable, which offers an acceptable explanation for this fact. Histology and HPV types found in acuminate warts of the outer anogenital areas do not seem to differ between the two groups or the general population. This contrasts with the findings of cervical HPV infection in RTRs, where the presence of HPV 16 was found to be increased. Also, cervical involvement in RTRs is probably not adequately diagnosed by cervical smears. Diagnosis of HPV infection in HIV+ patients was not found to be a problem in this study. However, lesions are probably detected only when widespread and difficult to treat. Hence, earlier diagnosis might help to eliminate the infection more successfully. Partners of HIV+ patients do not seem easily available. Whether or not this was the cause of treatment failure due to constant reinfection in our patients is difficult to say. Partner examination and patient assessment early in the course of HIV disease and regularly thereafter so that lesions can be treated as long as they are limited seems mandatory. Proneness to anogenital cancer in HIV+ patients can not be evaluated conclusively at present and further longitudinal studies are needed. HPV types which are associated with genital cancer, but also those HPV types which are normally not, may acquire an enhanced ability to potentiate malignant growth in the patients [56], facts which should put physicians who care for immunosuppressed patients on the alert.

Acknowledgements. We thank Drs. zur Hausen, Orth, and Lorincz for the generous gift of HPV clones. The secretarial work of Mrs. V. Steinemann is greatly appreciated. Our thanks also go to Mrs. M. Johnson, who was responsible for the photographs.

References

1. Largiadèr F, Buchmann P, Decurtins M, Schneider K, Turina M (1987) The Zürich experience in organ transplantation in 1986. Praxis 76:1129–1133
2. Walder BK, Jeremy D, Charlesworth JA, MacDonald GJ, Pussell BA, Robertson MR (1976) The skin and immunosuppression. Aust J Dermatol 17:94–97
3. Penn I (1981) Depressed immunity and the development of cancer. Clin Exp Immunol 46:459–474
4. Alessi E, Cusini M, Zerboni R (1988) Mucocutaneous manifestations in patients infected with human immunodeficiency virus. J Am Acad Dermatol 19:290–297
5. Civatte J, Janier M (1988) Hauterscheinungen bei HIV Infektionen. Wien Med Wochenschr 19/20:508–516
6. Sindrup JH, Weismann K, Petersen CS, Rindum J, Pedersen C, Mathiesen L, Worm A-M, Kroon S, Sondergaard J, Lange Wantzin G (1988) Skin and oral mucosal changes in patients infected with human immunodeficiency virus. Acta Derm Venereol 68:440–443
7. Schnyder UW (1988) Haut- und Schleimhautveraenderungen bei HIV-Infektionen. Swiss Med 10:35–36
8. Anonymous (1987) Renal transplantation and the skin. Lancet II:1312
9. L'Eplattenier JL, Binswanger U, Ott F, Largiadèr F (1980) Dermatological complications in immunosuppressed patients after kidney transplantation. Schweiz Med Wochenschr 110:1307–1313
10. Koranda FC, Dehmel EM, Kahn G, Penn I (1974) Cutaneous complications in immunosuppressed renal homograft recipients. J Am Med Assoc 229:419–424
11. Hoxtell EO, Mandel JS, Murray SS, Schuman LM, Goltz RW (1977) Incidence of skin carcinoma after renal transplantation. Arch Dermatol 113:436–438
12. Maize JC (1977) Skin cancer in immunosuppressed patients. J Am Med Assoc 237:1857–1858
13. Kinlen LJ, Sheil AG, Peto J, Doll R (1979) Collaborative United Kingdom-Australasian study of cancer in patients treated with immunosuppressive drugs. Br Med J 6203:1461–1466
14. Spencer ES, Andersen HK (1979) Viral infections in renal allograft recipients treated with long-term immunosuppression. Br Med J 6194:829–830
15. Hardie IR, Strong RW, Hartley LC, Woodruff PW, Clunie GJ (1980) Skin cancer in caucasian renal allograft recipients living in a subtropical climate. Surgery 87:177–183
16. Sheil AG, May J, Mahoney JF, Horvath JS, Johnson JR, Tiller DJ, Stewart JH (1980) Incidence of cancer in renal transplant recipients. Proc Eur Dial Transplant Assoc 17:502–506

17. Boyle J, Briggs JD, Mackie RM, Junor RM, Aitchison TC (1984) Cancer, warts and sunshine in renal transplant patients. Lancet I:702–706
18. Blohme I, Larko O (1984) Premalignant and malignant skin lesions in renal transplant patients. Transplantation 37:165–167
19. Sheil AG (1984) Cancer in organ transplant recipients: part of an induced immune deficiency syndrome. Br Med J 6418:659–661
20. Gupta AK, Cardella CJ, Haberman HF (1986) Cutaneous malignant neoplasms in patients with renal transplants. Arch Dermatol 122:1288–1293
21. Freeman RG, Knox JM (1970) Recent experience with skin cancer. Arch Dermatol 101:403–408
22. Lutzner M, Croissant O, Ducasse MF, Kreis H, Crosnier J, Orth G (1980) A potentially oncogenic human papillomavirus (HPV-5) found in two renal allograft recipients. J Invest Dermatol 75:353–356
23. Barr BBB, Benton EC, McLaren K, Bunney MH, Smith IW, Blessing K, Hunter JAA (1989) Human papilloma virus infection and skin cancer in renal allograft recipients. Lancet I:124–129
24. Zur Hausen H (1987) Papillomaviruses in human cancer. Cancer 59:1692–1696
25. Baltzer J, Kuerzel R, Eigler W, Samtleben W, Castro LA, Land W, Gurland HJ, Segerer W, Kuhlmann H, Zander J (1981) Gynaekologische Probleme bei Dialysepatientinnen und Frauen nach Nierentransplantation. Geburtshilfe Frauenheilkd 41:759–764
26. Shokri-Tabibzadeh S, Koss LG, Molnar J, Romney S (1981) Association of human papillomavirus with neoplastic process in the genital tract of four women with impaired immunity. Gynecol Oncol 12:129–140
27. Van der Leest RJ, Zachow KR, Ostrow RS, Bender M, Pass F, Faras AJ (1987) Human papillomavirus heterogeneity in 36 renal transplant recipients. Arch Dermatol 123:354–357
28. Rüdlinger R, Smith IW, Bunney MH, Hunter JAA (1986) Human papilloma virus infections in a group of renal transplant recipients. Br J Dermatol 115:681–692
29. Sheil AG, Mahoney JF, Horvath JS, Johnson JR, Tiller DJ, Kelly GE, Stewart HJ (1979) Cancer following renal transplantation. Aust NZ J Surg 49:617–620
30. Alloub MI, Barr BBB, McLaren KM, Smith IW, Bunney MH, Smart GE (1989) Human papillomavirus infection and cervical intraepithelial neoplasia in women with renal allografts. Br Med J 298:153–156
31. Longo D, Steis R, Lane H (1984) Malignancies in the AIDS patient: native history, treatment strategies and preliminary results. Ann NY Acad Sci 437:421–430
32. Sitz KV, Keppen M, Johnson DF (1987) Metastatic basal cell carcinoma in acquired immunodeficiency syndrome-related complex. J Am Med Assoc 257:340–343
33. Ferenczy A, Mitao M, Nagai N, Silverstein SJ, Crum CP (1985) Latent papillomavirus and recurring genital warts. N Engl J Med 313:784–788
34. Syrjaenen S, Syrjaenen K (1986) An improved in situ DNA hybridization protocol for detection of human papillomavirus (HPV) DNA sequences in paraffin-embedded biopsies. J Virol Methods 14:293–304
35. Rüdlinger R, Grob R, Buchmann P, Christen D, Steiner R (1988) Anogenital warts of the condyloma acuminatum type in HIV-positive patients. Dermatologica 176:277–281
36. Gross G, Wagner D, Schneider A, Ikenberg H, Gissmann L (1985) Sexual transmissibility of papillomaviruses. Z Hautkr 60:1737–1738
37. Gross G (1987) Lesions of the male and female external genitalia associated with human papillomaviruses. In: Syrjaenen K, Gissmann L, Koss LG (eds) Papillomaviruses and human disease. Springer, Berlin Heidelberg New York
38. von Krogh G, Syrjaenen SM, Syrjaenen KJ (1988) Advantage of human papillomavirus typing in the clinical evaluation of genitoanal warts. J Am Acad Dermatol 18:495–503

39. Fuerpasz R (1988) Therapie der Condylomata acuminata unter besonderer Beruecksichtigung der Laserbehandlung an der Gynaekologischen Universitätsklinik Zürich (1983–1986). Thesis, University of Zürich
40. Bergeron C, Naghashfar Z, Canaan C, Shah K, Fu Y, Ferenczy A (1987 a) Human papillomavirus type 16 in intraepithelial neoplasia (bowenoid papulosis) and coexistent invasive carcinoma of the vulva. Int J Gynecol Pathol 6:1–11
41. Bonnekoh B, Mahrle G, Steigleder GK (1987) Transition of bowenoid papulosis (HPV-16) into cutaneous squamous cell carcinoma in two patients. Z Hautkr 62:773–784
42. Kato T, Saijyo S, Hatchome N, Tagami H (1988) Detection of HPV 16 in bowenoid papulosis and invasive carcinoma occurring in the same patient with a history of cervical carcinoma. Arch Dermatol 124:851–852
43. Rüdlinger R, Grob R, Yu XY, Schnyder UW (1989) Human papillomavirus 35 positive bowenoid papuolosis of the anogenital area and concurrent human papillomavirus 35 positive verruca with bowenoid dysplasia of the periungual area. Arch Dermatol 125:655–659
44. Oriel JD, Whimster IW (1971) Carcinoma in situ associated with virus-containing anal warts. Br J Dermatol 84:71–73
45. Oriel JD (1971) Anal warts and anal coitus. Br J Vener Dis 47:373–376
46. Cooper HS, Patchefsky AS, Marks G (1979) Cloacogenic carcinoma of the anorectum in homosexual men. An observation of four cases. Dis Colon Rectum 22:557–558
47. Li FP, Osborn D, Cronin CM (1982) Anorectal squamous carcinoma in two homosexual men. Lancet II:391
48. Austin DF (1982) Etiological clues from descriptive epidemiology: squamous carcinoma of the rectum or anus. NCI Monogr 62:89–90
49. Daling JR, Weiss NS, Klopfenstein LL, Cochran LE, Chow WH, Daifuku R (1982) Correlates of homosexual behaviour and the incidence of anal cancer. J Am Med Assoc 247:1988–1990
50. Peters RK, Mack TM (1983) Patterns of anal carcinoma by gender and marital status in Los Angeles county. Br J Cancer 49:629–636
51. Zachow KR, Ostrow RS, Bender M, Watts S, Okagaki T, Pass F, Faras AJ (1982) Detection of human papillomavirus DNA in anogenital neoplasias. Nature 300:771–773
52. Nash G, Warren A, Nash S (1986) Atypical lesions of the anal mucosa in homosexual men. J Am Med Assoc 256:873–876
53. Frazer IH, Medley G, Crapper RM, Brown TC, Mackay IR (1986) Association between anorectal dysplasia, human papillomavirus and human immunodeficiency virus infection in homosexual men. Lancet II:657–660
54. Gal A, Meyer PR, Taylor CR (1987) Papillomavirus antigens in anorectal condyloma and carcinoma in homosexual men. J Am Med Assoc 257:337–340
55. Birgkigt H-G, Neumann H-J, Willegeroth C (1988) Gleichzeitiges Vorkommen von perianalen Condylomata acuminata und einem Analkarzinom bei einem männlichen Homosexuellen – Zufallsbefund oder erhöhtes Erkrankungsrisiko? Dermatol Monatsschr 174:480–484
56. Milburn PB, Brandsma JL, Goldsman CI, Teplitz ED, Heilman EI (1988) Disseminated warts and evolving squamous cell carcinoma in a patient with acquired immunodeficiency syndrome. J Am Acad Dermatol 19:401–405
57. Rüdlinger R, Buchmann (P (1989) HPV 16-positive bowenoid papulosis and squamous-cell carcinoma of the anus in an HIV-positive man. Dis Colon + Rectum 32:1042–1045

Immunological Aspects

Immunology of Genital Papillomavirus Infections

S. Jablonska and S. Majewski

Introduction

Anogenital carcinomas and benign anogenital warts are associated with specific types of human papillomaviruses (HPVs). HPV 16, 18, 31 and 33 are the most common HPVs associated with anogenital carcinomas and/or with cervical dysplasia (Durst et al. 1983; Ikenberg et al. 1983; Boshart et al. 1984; Beaudenon et al. 1986; zur Hausen 1987; Howley 1987; zur Hausen and Schneider 1987). Also the newly characterized HPVs 35 (Lorincz et al. 1987), 45 (Nagashfar et al. 1987), 51 (Nuovo et al. 1988) and 52 (Shimoda et al. 1988) have been found to be associated with intraepithelial neoplasia and invasive carcinoma of the cervix. In contrast, HPV 6 and 11 are associated, in general, with benign anogenital lesions, i.e. with condylomata acuminata and low-grade cervical intraepithelial neoplasia (de Villiers et al. 1981; Gissmann et al. 1982, 1983). Some other newly characterized HPVs, 42 (Beaudenon et al. 1987) and 44 (Lorincz et al. 1989), can also be regarded as a low risk for the development of genital cancer. However, HPV 6 subtypes have been detected repeatedly in locally destructive anogenital verrucous carcinomas (Okagaki et al. 1984; Lehn et al. 1984; Boshart and zur Hausen 1986; Rando et al. 1986; Guillet et al. 1988; Kasher and Roman 1988). HPV 16 is also associated with bowenoid papulosis (Durst et al. 1983; Gross et al. 1985, 1986; Obalek et al. 1985, 1986), with histological features of Bowen's atypia, usually regressing after several months or years of duration. However, in some cases of bowenoid papulosis malignant transformation may occur (Lloyd 1970; Bergeron et al. 1987).

The question arises what factors are of basic importance for the development of either benign or malignant anogenital lesions after infection with potentially oncogenic HPVs, and why, in some individuals, the HPV infection may persist for several years and/or throughout life.

One possibility, suggested by zur Hausen (1986), is that deficient intracellular surveillance mechanisms may be responsible for the development of anogenital malignancies. In addition, other factors decisive for the development of benign or malignant HPV-induced anogenital lesions may have an immunological basis. The concept of an anti-tumour immunosurveillance mechanism is supported by a) experimental studies in animals, b) mode of regression of benign HPV-induced tumours (warts), and c) the high incidence of HPV-induced tumours in patients with impaired immune responses.

It appears that humoral immunity is of lesser importance for HPV-induced anogenital lesions (Jablonska et al. 1979; Jablonska and Orth 1983).

Genital Papillomavirus Infections
Eds.: Gross/Jablonska/Pfister/Stegner

Recent studies have shown that natural cell-mediated cytotoxicity may be the main mechanism responsible for the maintenance of immune surveillance in patients with HPV infections. We present information on the immunology of anogenital tumours associated with various HPVs, and we focus on the natural killer cell activity against HPV 16 DNA-bearing target cells in patients with HPV-induced anogenital lesions.

Immunology of Papillomavirus Infection in Animals

Two main animal models provide the evidence for the role of an immunosurveillance mechanism in the control of papillomavirus infections: a) the Shope rabbit papilloma-carcinoma complex and b) bovine papillomavirus infections. Both models seem to be of relevance for HPV infections in human disease.

In cottontail rabbits the spontaneous regression of papillomas occurs in about 30% of cases and is accompanied by a dense lymphocytic infiltration (Kreider 1980). The regression seems to be related to the cellular immune reactions, and is diminished by treatment with immunosuppressive agents that decrease leucocytic infiltrations (Kreider and Bartlett 1985). Similar to the human disease, papillomas in this animal model may result in spontaneous regression, persist for several months or years and/or undergo malignant progression. Humoral immunity seems not to play an important role in the control of this viral infection, since similar titres of neutralizing antibodies were found in both animal regressors and progressors (Seto et al. 1977). The immunotherapy of cottontail rabbit papillomas with autologous and homologous papilloma vaccine proved to be successful, whereas injection of virus particles did not augment rejection of the warts. This is suggestive of the role of wart virus-associated antigens in the immune response (Evans and Thomson 1969; Kreider and Bartlett 1981).

The host immunity appears to play a basic role also in the malignant conversion of bovine papilloma virus (BPV)-induced tumours and in their regression. It is known that latent infection with BPV may be activated by immunosuppression (Jarret 1985), and only some papillomas induced by potentially oncogenic BPVs transform into the malignant lesions (Jarret et al. 1980).

Regression of HPV-induced Anogenital Tumours

The regression of HPV-induced anogenital lesions differs considerably with various tumours and depends on the type of HPV involved. Regression of genital warts not infrequently occurs spontaneously after eradication of concomitant infection of the anogenital tract or in women after parturition.

The histological characteristics of regressing genital warts include mononuclear infiltrations around blood vessels and abundant infiltrates in the

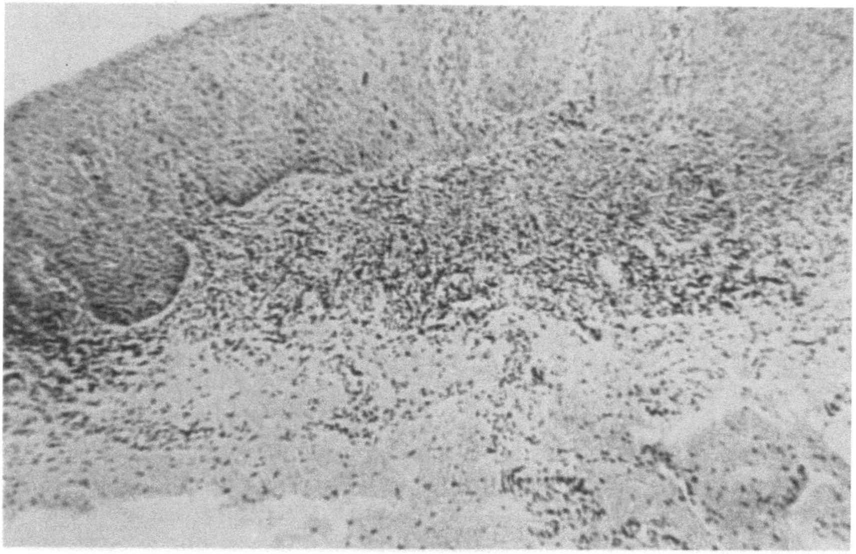

Fig. 1. Spontaneous regression of penile lesion of bowenoid papulosis. Inflammatory infiltrates consisting of lymphocytes and macrophages penetrating into the epidermis, which does not show any more proliferation and atypia

corium, penetrating into the epidermis and destroying proliferating rete ridges. Immunocytochemical studies with the use of monoclonal antibody MoAb 3.9 which reacts with the majority of macrophages showed macrophage infiltration in the cervical intraepithelial neoplasia (Tay et al. 1987). These macrophages are believed to be the first line of defence against the HPV infection. Stationary macrophage-like cells and cell-associated soluble mediators are most likely responsible for the regression of mucosal HPV infections (Jenson et al. 1987), either through a direct anti-virus effect or via nonspecific mechanisms.

HPV 16-induced bowenoid papulosis (BP), a multifocal disease of the external genitalia with histological features of Bowen's atypia, usually regresses after a duration of several months or years (Ikenberg et al. 1983; Durst et al. 1983; Gross et al. 1985; Obalek et al. 1985, 1986). Our histologic examination of spontaneously regressing BP lesions showed lichenoid mononuclear infiltrations invading the epidermis (Fig. 1). A high regression rate was found after conservative surgery in women with BP or after excision of some of the papules in men (Obalek et al. 1985). In contrast, anogenital carcinomas in situ of Bowen's type also induced by HPV 16 and displaying, in general, similar pathologic features have no tendency to spontaneous regression.

Immunosuppression and HPV-induced Anogenital Tumours

Anogenital warts occur in the general population in individuals with no evident defect of the immunosurveillance mechanism. However, in some patients without any signs of immune deficiency, there is a decreased cell-mediated immunity, as assessed by in vitro tests (Seski et al. 1978; Obalek et al.1980). Immunosuppression was found to be a high risk factor in the development of genital warts and cervical neoplasia (Berg and Lampe 1981; Sillman et al. 1984), and this points to the role of immunosurveillance mechanisms in the control of HPV infections.

A 100-fold increase in the incidence of carcinomas of the vulva and anus was reported in renal transplant patients (Blohme and Bryuger 1985; Penn 1986). The neoplasms, either in situ or invasive cancers, occurred late after transplantation, an average of 88 months, and in several women they were multifocal (Penn 1986). The incidence of genital malignancies in the transplant patients followed up for 3 months to 17 years after renal transplantation (mean 4.8 years) was only about 3% (three carcinomas in 90 patients), whereas dysplastic premalignant lesions appeared in a significantly higher number of the patients (Pelissier et al. 1981). The prevalence of cervical condylomas was found to be five times greater than expected (Morin et al. 1981). Schneider et al. (1983) reported that 8.5% of 132 female transplant recipients developed cervical condylomata within about 22 months, and 4.5% developed cervical neoplasia within 3 years. The risk of developing cervical cancer was regarded by Porecco et al. (1975) as 14 times greater in women treated with immunosuppression than in the general population.

The incidence of malignancies depends on the duration of the immunosuppression and the decrase in cell-mediated immunity. Neoplasms in immunodeficient patients occur when the patients are relatively young, tend to persist, progress rapidly and recur (Sillman and Sedlis 1987). In acquired immunodeficiency related to Hodgkin's disease, about 46% of 85 female patients showed evidence of condylomata, dysplasia or carcinoma of the cervix or anogenital region as compared with about 6% of the general female population (Katz et al. 1987). The incidence of HPV-induced anogenital tumours is also significantly higher in HIV patients than in the general population (Rudlinger et al. 1988), which further substantiates the role of immune responses in their development.

The high incidence of condylomata in pregnancy, some with features of carcinoma in situ, followed by spontaneous disappearance after parturition (Skinner et al. 1973; Janovsky and Barchet 1966) might be related to the natural immunosuppression in pregnancy (Hsu 1974) and postpartum immunostimulation. The immunostimulation during parturition may be either systemic or induced by the local trauma which facilitates the contact between viral antigens and immunocompetent cells. It is conceivable that the success of therapy of condylomata with autogenous vaccine (Abcarian and Sharon 1977) was also due to a nonspecific inflammatory reaction induced by injections, since in controlled studies there were comparable rejection frequencies with the use of vaccines from condylomata and patient's normal skin (Malison et al. 1982).

Immunological Findings in Patients with HPV-induced Anogenital Tumours

Humoral Immunity

Antibodies against HPV capsid proteins are induced during the infections and were demonstrated by immunodiffusion, complement fixation, immunofluorescence, and radioimmunoassay (Jablonska et al. 1979, 1982). However, humoral responses in patients with HPV-associated anogenital tumours are usually slight and transitory, since viral particles are present in small quantities and are protected within keratinized cells against the immunocompetent cells. Using various purified virions as antigens it was shown that the humoral responses are HPV-type specific, with no cross-reactivity (Jablonska et al. 1982; Jablonska and Orth 1983). Humoral immunity in patients with anogenital lesions has been explored much less than in patients with cutaneous warts. Using the immunodiffusion and ELISA assays, IgG antibodies to pooled wart antigen were found in a higher percentage in sera of patients with condylomata acuminata than in sera of patients with cutaneous warts (Pyrhonen et al. 1980). It was reported that, using serum against pooled tissue antigens of condyloma, it was possible to detect by indirect immunofluorescence the specific wart-associated antigen (Dunn et al. 1981). With the use of the indirect immunofluorescence method and HPV 6-induced condylomata acuminata as a substrate, a significant increase in IgG antibodies was found in sera from patients with condyloma, cervical intraepithelial neoplasia and invasive cervical cancer (Gross et al. 1986). Using the group-specific HPV antigen, Baird (1983) has disclosed IgG antibodies in sera of a high percentage patients with condylomata, carcinoma in situ, and in invasive cervical carcinomas, with the highest titres in the latter. With the use of group-specific antigen, antibodies were shown in 48.8% of women who had never had warts and in 50% of women with cervical dyslasia (Cubie and Norval 1988). In another study, IgG antibodies to group-specific antigen were detected in 70% women with condylomata and IgM antibodies in 40.7% of the patients. However, IgG antibodies were present in 54.5% of noninfected women and IgM antibodies in 24.2% (Portolani et al. 1987). Thus, detection of these antibodies is of no diagnostic significance. Also, the study of Galloway et al. (1988) showed that 50% of the population possess antibodies against HPV 6 and/or HPV 16 fusion proteins and that no significant differences could be detected between cancer-bearing and control populations.

Antibodies to the CaSki cervical carcinoma cell line (bearing HPV 16 DNA) were found in patients with macroinvasive cervical cancers, whereas cellular immune reactivity in a lymphocyte assay using the same antigen occurred more often in patients with preinvasive cancers (van de Linde et al. 1983). However, 21% of the controls showed positive reactions.

In general, the results of the studies on humoral immunity do not correlate with the clinical findings and give no evidence for a significant role of antibodies in the immunosurveillance against HPV tumors.

More promising are studies on antiviral antibodies directed towards the viral HPV 6 and HPV 16 capsid protein encoded by open reading frame (ORF) L1, detected in sera of patients with genital warts (Li et al. 1987; Jenison'et al. 1989; Galloway et al. 1988). Their practical application is still not clear.

The most important recent finding was that antibodies against the HPV 16 E7 *Escherichia coli* expressed fusion protein in 20% of the patients with cervical cancer and in about 16% of the patients with CIN III. The difference from the control sera (3.4% of positive results) was statistically highly significant. It is conceivable that the anti-HPV16 E7 antibodies might be markers for risk of cervical cancer development (Jochmus-Kudielka et al. 1989).

Cell-mediated Immunity

The histological features of regressing HPV-induced anogenital tumours are characteristic of cell-mediated immune response, suggesting that some HPV-infected or transformed cells are the targets for infiltrating lymphocytes. Various degrees of derangement of the response to nonspecific mitogens of peripheral blood T cells were reported in a proportion of patients with both genital warts and anogenital malignant lesions (Seski et al. 1978; Obalek et al. 1980; Okagaki et al. 1986). Lymphoproliferative responses to some synthetic peptides corresponding to sequences of HPV 16 L1 or E6, found to be specific for HPV 16, were also shown in asymptomatic individuals, which might suggest a widespread HPV 16 infection in the general population (Strang et al. 1990). Also, the T-helper/T-suppressor ratio was found to be decreased in patients with genital warts (Tyring et al. 1986) and in those with HPV-induced anogenital neoplasia (Carson et al. 1984).

In previous studies, we showed that derangement of cell-mediated immunity (CMI) in patients with HPV infections differs depending on the HPV type (Obalek et al. 1980; Jablonska et al. 1982). In our recent studies (Majewski et al., in preparation) we examined T-cell subsets in peripheral blood of these patients by high or low affinity of their receptors for sheep red blood cells (SRBC). We have also studied the effect of theophylline preincubation on the capability of T cells to form rosettes with SRBC. It is known that treatment with theophylline, a phosphodiesterase inhibitor, inhibits the ability of T-suppressor/inducer cells to form E rosettes, whereas this capability of T-helper and T-cytotoxic cells is not affected (Limatibul et al. 1978; Shore et al. 1978; Majewski et al.1986).

We found a most marked decrease of T-cell subsets in patients with HPV 16-induced anogenital carcinomas of Bowen's type, as reflected by a lower percentage of T cells with high and low SRBC receptor affinity (ARFC, active rosette forming cells; TRFC, total rosette forming cells), and by a lower percentage of theophylline-sensitive subpopulation (TRFC theoph.-sensitive). In patients with BP, the significant decrease of T cells was found in all TRFC fractions but not in ARFC subsets. In patients with genital warts, the deficiency of T-cell subsets was less pronounced and was manifested only by a decrease of the TRFC fraction. In contrast to BP, in the patients with condylomata there was a normal percentage of TRFC theoph.-sensitive (T-suppressor/inducer) cells. The theophylline-resistant subset was shown to consist mainly of T-cytotoxic (Leu 2+, Leu 15−), T-inducer (Leu 3+, Leu 8+) and T-helper cells, whereas the theophylline-sensitive subset consisted mainly of T-suppressor/amplifier cells (Skopinska-Rozewska et al. 1988).

Another parameter for the assessment of CMI in vivo is the local skin sensitization to DNCB. This test, reflecting mainly the local immune response, did not disclose any significant defect of CMI in patients with HPV-induced anogenital lesions as compared with healthy controls (Obalek et al. 1980). Also, von Krogh (1983) found no correlation of the DNCB-sensitization test with the persistence of genital warts and the response to the treatment. It is likely that a cutaneous sensitization test does not reflect the immune reactivity of epithelial cells of the genital mucosa, and of the skin of the anogenital area.

In summary, there is no sufficient evidence for a significant role of nonspecific cellular immune mechanisms in the control of anogenital HPV infections.

Natural Cell-mediated Cytotoxicity

There is increasing evidence that the main mechanism responsible for immunosurveillance against virus-infected and -transformed cells is "natural cell-mediated cytotoxicity" (NCMC). Particularly the natural cell-mediated cytotoxicity of natural killer (NK) cells seems to play a crucial role. NK cells are responsible for a spontaneous, major histocompatibility-complex nonrestricted elimination of neoplastic and virus-infected cells (Herberman 1980, 1982); therefore, the activity of NK cells may be of importance for the persistence of HPV-induced anogenital lesions. NK cells belong to the subpopulation of the so-called large granular lymphocytes (LGL), do not express typical surface markers characteristic for T and B cells, and can be specifically detected with the use of antibodies against surface markers, i.e. CD-16 and HNK-1 (Lanier et al. 1983).

Most studies on NK cell activity in human beings have been performed using the human erythroleukaemic K-562 cell line as a target (Herbermann 1980, 1982). It is known that the NK cell population is heterogeneous in respect to recognition of different target cells (Ortaldo and Herberman 1984). Therefore, anti-K-562 cell cytotoxicity studies may not provide any information on NK cell activity against HPV-infected cells in patients with HPV-induced anogenital lesions. Anti-K-562 cytotoxicity of peripheral blood mononuclear cells (PBMC) was found to be nonsignificantly decreased in patients with genital warts, BP, and HPV 16-associated anogenital carcinomas (Malejczyk et al. 1987; Rogozinski et al. 1986). Therefore, to assess NK cell activity in patients with HPV-induced anogenital tumours we used a specific target SK-v keratinocyte cell line. This cell line was established from vulvar BP lesions by Dr. G. Orth (Pasteur Institute, Paris). When transplanted into the nude mouse, the line induces tumours with features of bowenoid carcinomas. Sk-v cells contain multiple integrated copies of HPV 16 DNA (Schneider-Manoury et al. 1987), whereas the original bowenoid lesions contained mainly episomal HPV 16 DNA and only a small portion of integrated viral DNA. This suggests that under in vitro conditions there occurred selection of the cells containing integrated copies of HPV 16 DNA. In vitro, the pluristratified cultures of Sk-v cells display all morphological and ultrastructural features of Bowen's atypia, such as individual cell keratinization, dyskeratosis and multinucleated pleomorphic cells. Thus, the Sk-v cell line retains some properties of in situ transformed keratinocytes; it therefore appeared to be a suitable model for studies on specific immune response against HPV 16-bearing target cells in patients with HPV 16-induced anogenital tumours.

In our studies we found that while normal human keratinocytes were not susceptible to NK cell lysis (Majewski et al. 1990), Sk-v cells could be effectively killed by peripheral blood mononuclear cells (PBMC) of normal donors (Malejczyk et al. 1987, 1989). In contrast, we found that in patients with HPV 16-induced BP and anogenital carcinomas of Bowen's type the anti-Sk-v cytotoxicity was significantly decreased (Fig. 2). Moreover, in healthy controls there was a correlation between anti-K-562 and anti-Sk-v cytotoxicity, whereas in patients with HPV-induced anogenital tumours such a correlation was not detected. The highest degree of decrease of anti-Sk-v cytotoxicity was found in patients with anogenital carcinomas; it was less pronounced in patients with BP. In patients with anogenital carcinomas the degree of Sk-v lysis was related to the degree of malignancy. In patients with HPV 6-induced genital warts and in patients with HPV 6-associated verrucous carcinomas (Jablonska et al. 1988) anti-Sk-v cytotoxicity was found to be within the normal range. We also showed that there was no specific inhibition of Sk-v lysis by PBMC of patients with epidermodysplasia verruciformis, including cases with concomitant carcinomas of Bowen's type (Majewski et al. 1990). The results of our studies in patients with BP and anogenital carcinomas suggest that the persistence of HPV 16-induced anogenital lesions may be associated with a decrease of NK cell activity against HPV 16-bearing target cells. In the studies on the mechanism of this phenomenon we have confirmed that the effector cell in anti-Sk-v cytotoxicity is a typical nonadherent. CD16+ (Leu11b+) NK cell (Malejczyk et al. 1989). Moreover, a "cold target" competitive assay showed that addition of unlabelled cells, either Sk-v or K-562, caused strong, dose-dependent inhibition of 51Cr-labelled Sk-v cell lysis. These results suggest that the same effector cell is responsible for the killing of Sk-v and K-562 cells, and that the different response is dependent on the various receptors on the same effector cell. It should be stressed that the

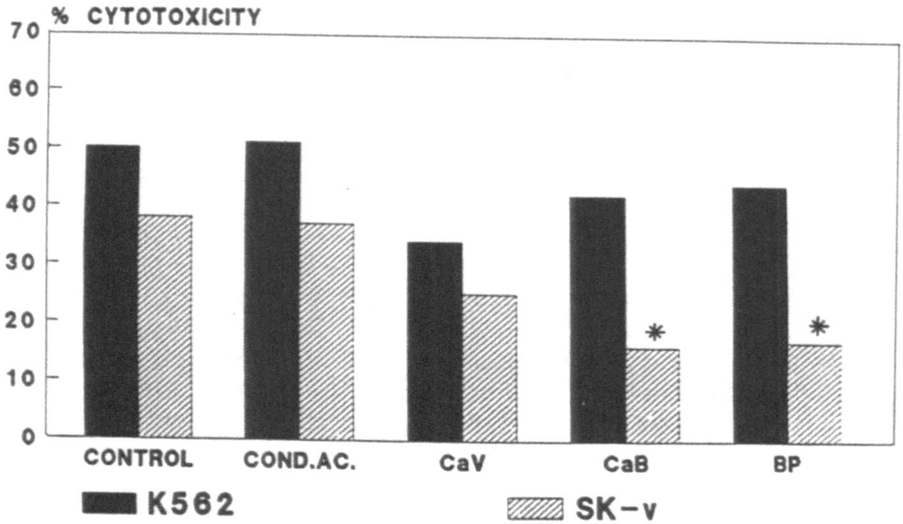

Fig. 2. Comparison of natural anti-Sk-v and anti-K-562 cytotoxicity in patients with benign or malignant anogenital lesions induced by different types of HPVs. *Cond.ac.*, condylomata acuminata, $n=12$ (HPV 6); *CaV*, verrucous carcinoma, $n=3$ (HPV 6, 11); *CaB*, carcinoma of Bowen's type, $n=5$ (HPV 16, 33); *BP*, bowenoid papulosis, $n=9$ (HPV 16, 42); *, statistically significant difference as compared with controls ($n=38$) at $p<0.05$

number of CD16+ cells in patients with HPV 16-induced tumours did not differ significantly from that of healthy controls; i.e. the differences in cytotoxicity are due to the changed activity of the effector cell.

Of special interest are the results of repeated studies in patients with HPV 16-induced anogenital lesions. We found that in cases of spontaneously regressing BP lesions there was a normalization of previously decreased anti-Sk-v cytotoxicity (Majewski et al., in preparation). Similarly, anti-Sk-v cytotoxicity became normal in patients with anogenital carcinomas after effective surgery, whereas in patients treated unsuccessfully the decrease of NK cytotoxicity against Sk-v cells persisted unchanged. Thus this assay may provide useful information on factors affecting the development of HPV-associated neoplasia; it may also be of value for prognostication and for monitoring the disease course.

In further studies, we have found that sera of all studied patients with HPV 16-induced BP and with HPV 16-associated anogenital carcinomas of Bowen's type are capable of inhibiting Sk-v cytotoxicity of NK cells from healthy individuals (Malejczyk et al. 1989). In contrast, the activity against K-562 cells of PBMC from healthy controls was not affected by the patients' sera. Moreover, in patients with anogenital carcinoma the increased suppressive effect of the sera was correlated with decreased levels of anti-Sk-v cytotoxicity and was usually proportional to the degree of malignancy. The nature of the serum factor that inhibits anti-Sk-v cytotoxicity is unknown. The role of some inhibitory factors such as prostaglandins (Bankhurst 1982), NK-specific antibodies (Goto et al. 1980), and immune complexes (Silverman and Cathcart 1980) could be excluded, since all these factors lack "target specificity" and would probably be capable of blocking both anti-Sk-v and K-562 cytotoxicity. Some serum factors that inhibit NK cell activity were reported to be present in cancer patients (Holmes et al. 1986), and their activity was found to correlate with the extent of tumour dissemination.

It is possible that factors specifically inhibiting anti-tumour immune reactivity are derived from tumour cells (Roth 1983). We therefore studied the effects of Sk-v cell-conditioned media on the anti-Sk-v cytotoxicity of normal human mononuclear cells. We found that conditioned media from in vitro-grown Sk-v cells, but not from normal keratinocytes, significantly decreased NK cell activity against Sk-v cells (Malejczyk et al., in preparation). The supernatants from Sk-v cells did not affect the NK cell activity against K-562 cells. The effect of the Sk-v inhibitory factor/s depended on the concentration of conditioned medium. Partial characterization by high-performance liquid chromatography of this inhibitory factor revealed that its activity was associated with two fractions of 2 and 30 kD. Simultaneously, we found in the supernatants a nonspecific stimulatory factor of about 20 kD which augmented NK cell activity against both K-562 and Sk-v cells. Further analysis with the use of ion exchange and reversed-phase chromatography confirmed the existence of two specific inhibitory factors and one nonspecific stimulatory factor in the supernatant from Sk-v cells. Single cell assay revealed that the inhibitory factors affected the binding of effector to target (NK to Sk-v) cells, leading to the block of Sk-v cell lysis. The binding of NK cell to K-562 cell was not affected by the inhibitory factors of the supernatant. It is yet to be established whether in the supernatants of Sk-v cells there are two different inhibitory factors or whether the factor with low molecular weight represents the degradation product of the former. The molecular weight of the stimulatory factor and its nonspecific action on both anti-K-562 and Sk-v cell cytotoxicity may suggest its similarity to the known epidermal cytokine – interleukin-6 (Luger et al. 1985; Kupper 1988).

The mechanism of action of the inhibitory factor/s via blocking of effector to target binding is further supported by our studies on the effect of various immunostimulatory cytokines on anti-Sk-v cytotoxicity in patients with HPV 16-induced anogenital tumours (Majewski et al., in preparation). In the study, PBMC from healthy donors and from patients with HPV-induced premalignant and malignant lesions were preincubated in vitro for 24 h with interleukin-2 (10 U/ml), IL-6 (1 U/ml) and interferon-α (1000 U/ml) and then tested for NK cell activity against K-562 and Sk-v cells. We found normal NK cell activity against K-562 cells of patients' PBMC, which could be significantly stimulated by all three cytokines (Fig. 3). The most potent cytokine was found to be IL-2. In contrast, in patients with HPV 16-induced anogenital lesions the activity against the specific HPV 16-bearing target cell (Sk-v) was dramatically lowered and was not stimulated by any cytokine (Fig. 4).

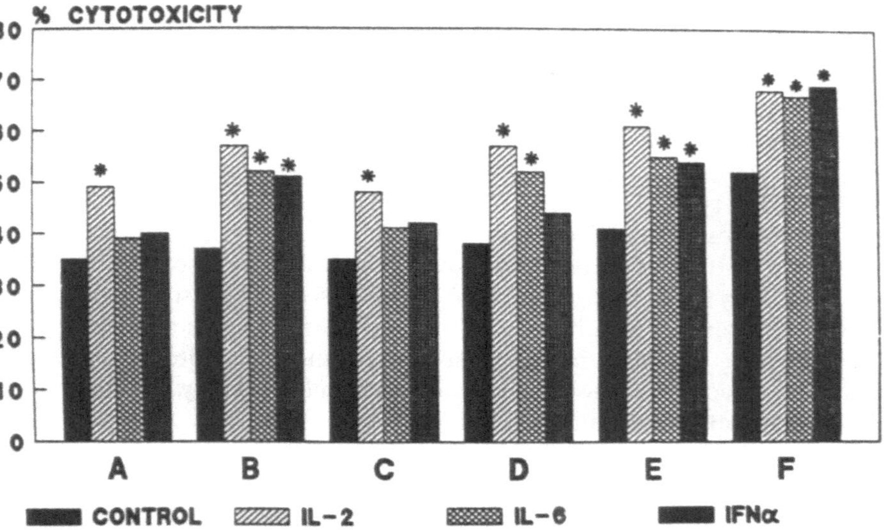

Fig. 3. Effects of immunostimulatory cytokines on natural cytotoxic activity against K-562 cells of peripheral blood mononuclear cells from patients with HPV-induced anogenital lesions. *A*, BP with active lesions; *B*, BP-CaB with active lesions; *C*, CaB with active lesions; *D*, BP after spontaneous regression of the lesions; *E*, BP-CaB after effective surgery; *F*, control; *, statistically significant difference as compared with the healthy controls ($n = 4$) at $p < 0.05$

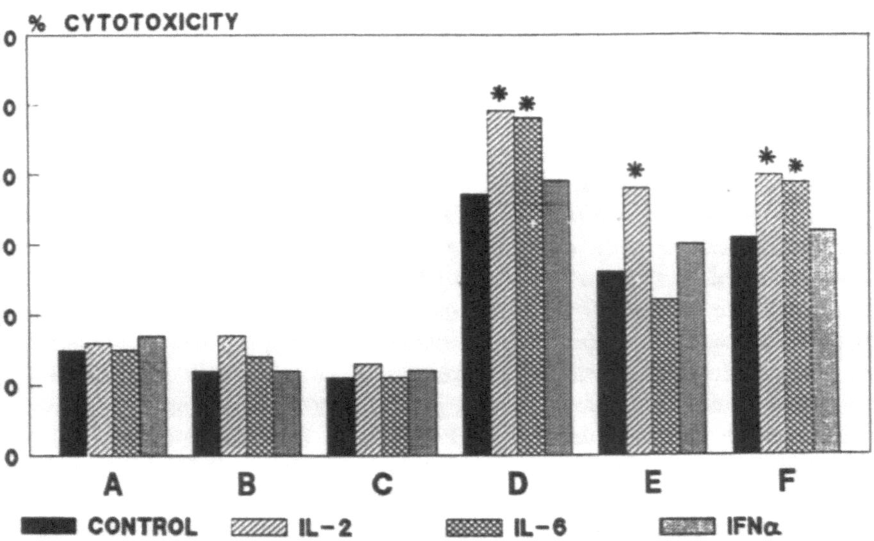

Fig. 4. Effects of immunostimulatory cytokines on natural cytotoxic activity against HPV 16-bearing Sk-v cells of peripheral blood mononuclear cells from patients with HPV-induced anogenital lesions. For abbreviations see Fig. 3

However, NK cell activity against Sk-v cells could be significantly stimulated by these cytokines in healthy individuals and in one patient with HPV 16-induced BP after the spontaneous regression of the lesions. These results suggest that in patients with HPV 16-induced anogenital tumours there exists a specific block of NK cell activity against disease-specific target cells, which probably depends on the inhibition of effector to target cell binding, and which cannot be reversed by stimulation with immunoregulatory cytokines.

Further studies on characterization of the inhibitory factor present in the supernatants from in vitro-grown Sk-v cells would allow its comparison with the inhibitory factors detected in the patients' sera. The tests for inhibitory factors in the sera could be helpful for assessment of some new therapeutic modalities in patients with HPV 16-induced anogenital premalignant and malignant lesions, and the inhibitory factors should probably be a target for a future vaccination programme.

The Role of Local Immune Mechanisms

There is increasing evidence that the skin and mucosa are the sites of induction and elicitation of local immune reactions protecting against persistent viral infections and malignancies. The immune functions of the skin are related to the appearance of skin- and mucosa-associated lymphoid tissue (SALT) (Streilein 1983; Katz 1985; Luger et al. 1985), consisting of Langerhans cells, keratinocytes, melanocytes and immunocompetent lymphocytes. Langerhans cells express HLA-DR antigens and are capable of presenting antigens to immunocompetent T cells (Wolff and Stingl 1983; Choi and Sauder 1986). The antigen presentation is a first step in T-cell activation and generation of T-cytotoxic cells and other effector cells. Activated keratinocytes are the major source of epidermis-derived cytokines capable of both stimulating and inhibiting immune and inflammatory reactions (Luger et al. 1985). Both the Langerhans cells and epidermal or epithelial cytokines may be of basic importance for the local immunosurveillance against HPV-induced tumours.

The studies on infiltrating cells in the uterine cervix showed a low ratio of T-helper to T-suppressor (OKT4 to OKT8) cells (Syrjänen 1984), and this was associated with existing or pending malignancies. Vayrynen et al. (1985) found lowered numbers of T cells and Langerhans cells in the infiltrates, and predominance of Leu10-positive cells (B cells and monocytes), whereas progression was associated with an increase of T-suppressor cells. Depletion of Langerhans cells from the epithelium of patients with anogenital condylomata could result from their migration into the regional lymph nodes. However, an increased number of Langerhans HLA-DR-positive cells was reported in cervical intraepithelial neoplasia by others (Morris et al. 1983). Thus, the results of studies on local immune response differ considerably, probably due to constantly changing in situ conditions and to the interactions between infiltrating lymphoid cells and HPV-infected keratinocytes.

The decisive factor for the development of HPV-induced lesions might be the derangement of local surveillance mechanisms due to generation of various keratinocyte-derived immunoregulatory cytokines. Stimulated epidermal cells are capable of producing various immunostimulatory (interleukin-1, interleukin-3, interleukin-6, tumour necrosis factor-(TNF)-TNFα, interferons α and β; and immunosuppressive factors (transforming growth factor TGFβ, epidermal cell-derived inhibitor of IL-1, and others) (Luger et al. 1985; Kupper 1988). These cytokines play an important role in the modulation of T and B cell activity and NK cell cytotoxicity. The ability of normal keratinocytes and keratinocytes infected or transformed by HPVs to generate immunostimulating cytokines may be an important mechanism in the surveillance against HPV-induced tumours. The significant role in this process may be played by HPV-transformed keratinocytes themselves.

In recent studies (J. Malejczyk et al., in preparation) we found that HPV 16-bearing Sk-v cells are capable of producing spontaneously a high amount of IL-6. This was confirmed not only with the use of a bioassay on an IL-6-dependent B-9 cell line but also by means of Western-blot analysis. IL-6 exerts some antiviral activity, stimulates T and B cells (Kupper 1988) and, most importantly, stimulates NK cells (Luger et al. 1985). This cytokine also enhances the natural cytotoxicity against Sk-v cells of PBMC from healthy individuals.

Another cytokine which is spontaneously generated by Sk-v cells is TNFα. This was confirmed by testing supernatants from Sk-v cultures using the ELISA assay, SDS-PAGE gel electrophoresis and specific antibodies. Moreover, using molecular hybridization (Northern-blot) we found expression of mRNA for this cytokine (J. Malejczyk et al., in preparation). This cytokine seems to be of importance for antiviral and antineoplasia surveillance (Haranaka et al. 1987), and exerts both cytotoxic and cytostatic effects upon transformed cells.

Our preliminary studies have disclosed that IFNα, TNFβ (lymphotoxin) and IFNτ also inhibited proliferation of Sk-v cells. IFNτ, however, exerted the inhibitory effect only at the highest concentrations studied. IL-1 and IL-6 did not affect the cell proliferation. Interferons are generated in the course of viral infection and may present another important factor limiting the growth of HPV-transformed cells. However, clinical studies on the therapeutic value of IFNs in HPV-induced anogenital lesions are, as yet, inconclusive. Since the cytokines act in concordance, it appears that their effectiveness in clinical practice would be much higher if various immunomodulatory cytokines were used simultaneously for the treatment of anogenital HPV-induced tumours.

Summary and Conclusions

The occurrence, persistence and regression of HPV-induced anogenital tumours were convincingly shown to be related to host immunosurveillance mechanisms.

Serological studies are of limited value, since the biological role of antibodies against viral capsid antigens encoded by ORF L1 or L2 of HPV 6 and HPV 16 is unclear, and no correlation could be established between the circulating antibodies and regression of papillomavirus-induced lesions, both in animals and in humans.

In contrast, cell-mediated immune responses appear to be of basic importance for the regression of HPV-associated lesions and prevention of their recurrence. Decreased immune responses correlate with the persistence and progression of warts, and the appearance of various benign and malignant HPV-induced genital tumours in immunosuppressed persons suggests the role of nonspecific cellular immunity in tumour resistance. Certain potentially oncogenic HPVs appear to benefit particularly from impaired cellular immunity, and are responsible for an up to 100-fold increase in the incidence of anogenital carcinomas among the immunosuppressed population.

The most important finding in patients with HPV 16-associated malignancies appears to be abrogation of natural cytotoxicity against the specific HPV 16 DNA-bearing cell line. This abrogation was found to be due to the blocking of specific receptors on the effector cells by the factors derived from the tumours and present in the circulation. The lowered cytotoxic activity against specific target cells could not be stimulated by various cytokines. However, after regression of HPV 16-associated BP, there was a full normalization of the cytotoxic responses. The detected blocking factors appear to be the most important target for future specific immunotherapy of HPV-induced genital tumours.

Since the role of immune factors in genital HPV infections is well established, therapy and prevention should focus on immunomodulating and immunostimulating procedures.

References

Abcarian H, Sharon N (1977) The effectiveness of immunotherapy in the treatment of anal condyloma acuminatum. J Surg Res 22:231–236

Baird PJ (1983) Serological evidence for the association of papillomavirus and cervical neoplasia. Lancet 2:17–18

Bankhurst AD (1982) The modulation of human natural killer cell activity by prostaglandins. J Clin Lab Immunol 7:85–91

Beaudenon S, Kremsdorf D, Croissant O, Jablonska S, Wain-Hobson S, Orth G (1986) A novel type of human papillomavirus associated with genital neoplasias. Nature 321:246–249

Beaudenon S, Kremsdorf D, Obalek S, Jablonska S, Pehau-Arnaudet M, Croissant O, Orth G (1987) Plurality of genital human papillomaviruses: characterization of two new types with distinct biological properties. Virology 161:374–384

Berg JW, Lampe JG (1981) High-risk factors in gynecologic cancer. Cancer 48:429–441

Bergeron C, Naghashfar Z, Canaan C, Shah K, Fu YS, Ferenczy A (1987) Human papillomavirus type 16 in intraepithelial neoplasia (bowenoid papulosis) and coexistent invasive carcinoma of the vulva. J Gynecol Pathol 6:1–11

Blohme I, Bryuger H (1985) Malignant disease in renal transplant patients. Transplantation 39:23–25

Boshart M, zur Hausen H (1986) Human papillomaviruses in Buschke-Loewenstein tumors: physical state of the DNA and identification of a tandem duplication in the noncoding region of a human papillomavirus 6 subtype. J Virol 56:963–966

Boshart M, Gissmann L, Ikenberg H, Kleinheintz A, Scheurlen W, zur Hausen H (1984) A new type of papillomavirus DNA, its presence in genital cancer biopsies and cell lines from cervical cancer. EMBO J 3:1151–1157

Choi KS, Sauder DN (1986) The role of Langerhans cells and keratinocytes in epidermal immunity. J Leukoc Biol 39:349–358

Cubie HA, Norval M (1988) Humoral and cellular immunity to papillomavirus in patients with cervical dysplasia. J Med Virol 24:85–95

De Villiers EM, Gissmann L, zur Hausen H (1981) Molecular cloning of viral DNA from human genital warts. J Virol 40:932–935

Dunn J, Wenstein L, Orogemuller W, Mainke W (1981) Immunological detection of condylomata acuminata-specific antigens. Obstet Gynecol 57:351–356

Durst M, Gissmann L, Ikenberg M, zur Hausen H (1983) A papillomavirus DNA from cervical carcinoma and its prevalence in cancer biopsy samples from different geographic regions. Proc Natl Acad Sci USA 80:3812–3815

Evans CA, Thomsen JJ (1969) Antitumor immunity in the Shope papilloma-carcinoma complex. IV. Search for a transmissible factor increasing the frequency of tumor regression. JNCI 42:477–484

Galloway DA, Jenison SA, Valentine J, Koutsky L, Holmes K, Langenberg A, Corey L, Sherman K, Daling J (1988) Prevalence of antibody to HPV6 and HPV16 in various populations. In: 7th International Papillomavirus Workshop, Antipolis, p 230

Gissmann L, de Villiers E, zur Hausen H (1982) Analysis of human genital warts (condylomata acuminata) and other genital tumors for human papillomavirus type DNA. Int J Cancer 29:143–146

Gissmann L, Wolnik L, Ikenberg H, Koldovsky U, Schnurch HG, zur Hausen H (1983) Human papillomavirus type 6 and 11 DNA sequences in genital and laryngeal papillomas and in some cervical cancers. Proc Natl Acad Sci USA 80:560–563

Goto M, Tanimoto K, Horiuchi Y (1980) Natural cell-mediated cytotoxicity (NCMC) in systemic lupus erythematosus: suppression of NCMC by anti-lymphocyte antibody. Arthritis Rheum 23:1274–1281

Gross G, Hagedorn M, Ikenberg H, Rufli T, Grosshans E, Gissmann L (1985) Bowenoid papulosis. Presence of human papillomavirus (HPV) structural antigens and of HPV16-related DNA sequences. Arch Dermatol 121:858–863

Gross G, Ikenberg H, de Villiers EM, Schneider A, Wagner D, Gissmann L (1986) Bowenoid papulosis: a venereally transmissible disease as reservoir for HPV16. In: Peto R, zur Hausen H (eds) Viral etiology of cervical cancer. Cold Spring Harbor Laboratories, New York, pp 149–165 (Bandbury report, vol 21)

Guillet G, Foucher JM, Le Roy JP (1988) Degenerescence spino-cellulaire d' une tumeur de Buschke-Loewenstein a HPV6. Ann Dermatol Venereol 115:339–343

Haranaka H, Satomi N, Sakurai A, Haranaka R (1987) Antitumor effects of tumor necrosis factor: cytotoxic or necrotizing activity and its mechanism. Ciba Found Symp 131:140–153

Herberman RB (ed) (1980) Natural cell-mediated immunity against tumors. Academic Press, New York

Herberman RB (ed) (1982) NK cells and other effector cells. Academic Press, New York

Holmes E, Sibbitt WL Jr, Bankhurst AD (1986) Serum factors which suppress natural cytotoxicity in cancer patients. Int Arch Allergy Appl Immunol 80:39–43

Howley PM (1987) The role of papillomaviruses in human cancer. In: De Vita VT, Hellman S, Rosenberg SA (eds) Important advances in oncology. Lippincott, Philadelphia, pp 55–73

Hsu CCS (1974) Peripheral blood lymphocyte responses to phytohemagglutin and pokeweed mitogen during pregnancy. Proc Soc Exp Biol Med 146:771–775

Ikenberg H, Gissmann L, Gross G, Grussendorf-Cohen EJ, zur Hausen H (1983) Human papillomavirus type-16-related DNA in genital Bowen's disease and in bowenoid papulosis. Int J Cancer 15:563–565

Jablonska S, Orth G (1983) Human papillomaviruses. In: Rook A, Maibach H (eds) Recent advances in dermatology, vol 6. Churchill Livingstone, London, pp 1–36

Jablonska S, Orth G, Jarzabek-Chorzelska M, Rzesa G, Obalek S, Glinski W, Favre M, Croissant O (1979) Epidermodysplasia verruciformis versus disseminated verrucae. J Invest Dermatol 72:114–119

Jablonska S, Orth G, Lutzner MA (1982) Immunopathology of papillomavirus-induced tumors in different tissues. Springer Semin Immunopathol 5:33–62

Jablonska S, Obalek S, Orth G, Breitburd F, Croissant O (1988) Tumeur geante de Buschke-Lowenstein a HPV6 avec metastases dans la paroi abdominale. Ann Dermatol Venereol 115:1201–1202

Janovsky NA, Barchet S (1966) Multicentric Bowen's disease of the vulva. Obstet Gynecol 28:170–174

Jarret WFH (1985) Bovine papillomaviruses. In: Warts – human papillomaviruses. Clin Dermatol 3:8–19

Jarret WFH, Campo MS, Moar MH, McNell PE, Laird HM, O'Neil BW, Murphy J (1980) Papillomaviruses in benign and malignant tumors of cattle. In: Viruses in naturally occurring cancers. 7th Conference on Cell Proliferation. Cold Spring Harbor, Laboratories, New York

Jenison SA, Yu XP, Valentine J, Galloway DA (1989) Human antibodies react with an epitope of human papillomavirus type 6B L1 open reading frame which is distinct from the type-common epitope. J Virol 62:809–818

Jenson AB, Kurman RJ, Lancaster WD (1987) Tissue effects and host response to human papillomavirus infection. Obstet Gynecol Clin North Am 14:397–406

Jochmus-Kudielka I, Schneider A, Braun R, Kimmig R, Koldovsky U, Schneweis KE, Seedorf K, Gissmann L (1989) Antibodies against the human papillomavirus type 16 early proteins in human sera: correlation of anti-E7 reactivity with cervical cancer. J Natl Cancer Inst 81:1698–1704

Kasher MS, Roman A (1988) Characterization of human papillomavirus type 6B DNA isolated from an invasive squamous carcinoma of the vulva. Virology 165:225–233

Katz RL, Veanattukalathil S, Weiss (1987) Human papillomavirus infection and neoplasia of the cervix and anogenital region in women with Hodgkin's disease. Acta Cytol 31:845–854

Katz SJ (1985) The skin as an immunologic organ. J Am Acad Dermatol 13:530–536

Kreider JW (1980) Neoplastic progression of the Shope rabbit papilloma. In: Viruses in naturally occurring cancers. 7th Conference on Cell Proliferation. Cold Spring Harbor Laboratories, New York, pp 283–300

Kreider JW, Bartlett GL (1981) The Shope papilloma-carcinoma complex of rabbits. A model system of neoplastic progression and spontaneous regression. In: Klein G, Weinhouse S (eds) Advances in cancer research, vol 35. Academic, New York, pp 81–110

Kreider JW, Bartlett GL (1985) Shope rabbit papilloma-carcinoma complex: a model system of human papillomavirus infections. Clin Dermatol 3:20–26

Kupper TS (1988) Interleukin-1 and other human keratinocyte cytokines: molecular and functional characterization. Adv Dermatol 3:293–308

Lanier LL, Le AH, Phillips JH, Warner NL, Babcock GF (1983) Subpopulations of human natural killer cells defined by expression of the Leu-7 (HNK-1) and Leu-11 (NKP-15) antigens. J Immunol 131:1789–1795

Lehn H, Ernst TM, Sauer G (1984) Transcription of episomal papillomavirus DNA in human condylomata acuminata and Buschke-Loewenstein tumors. J Gen Virol 65:2003–2010

Li CC, Shah KV, Seth A, Gilden RV (1987) Identification of the human papillomavirus type 6B L1 open reading frame protein in condylomas and corresponding antibodies in human sera. J Virol 61:2684–2690

Limatibul S, Shore A, Dosch HM, Geleand EW (1978) Theophylline modulation of E-rosette formation: an indicator of T-cell maturation. Clin Exp Immunol 33:503–513

Lloyd KM (1970) Multicentric pigmented Bowen's disease of the groin. Arch Dermatol 101:48–51

Lorincz AT, Quinn AP, Lancaster WD, Temple GF (1987) A new type of papillomavirus associated with cancer of the uterine cervix. Virology 159:187–190

Lorincz AT, Quinn AP, Goldsborough MD, Schmidt BJ, Temple GF (1989) Cloning and partial DNA sequencing of two new papillomavirus types associated with condylomas and low-grade cervical neoplasia. J Virol 63:2829–2834

Luger TA, Kock A, Danner M (1985) Production of distinct cytokines by epidermal cells. Br J Dermatol 113 [Suppl 28]:145–156

Luger TA, Uchida A, Kock A, Colot M, Micksche M (1985) Human epidermal cell and squamous carcinoma cells synthetize a cytokine that augments natural killer cell activity. J Immunol 134:2477–2483

Majewski S, Skopinska-Rozewska E, Jablonska S, Wasik M, Misiewicz J, Orth G (1986) Partial defects of cell-mediated immunity in patients with epidermodysplasia verruciformis. J Am Acad Dermatol 15:966–973

Majewski S, Malejczyk J, Jablonska S, Misiewicz J, Rudnicka L, Obalek S, Orth G (1990) Natural cell-mediated cytotoxicity against various target cells in patients with epidermodysplasia verruciformis. J Am Acad Dermatol 22:423–427

Malejczyk J, Majewski S, Jablonska S, Orth G (1987) Natural cell-mediated cytotoxicity in patients with anogenital lesions induced by potentially oncogenic human papillomaviruses. In: Steinberg BM, Brandsma JL, Taichman LB (eds) Cancer cells, vol 5, papillomaviruses. Cold Spring Harbor Laboratories, New York, pp 381–385

Malejczyk J, Majewski S, Jablonska S, Rogozinski TT, Orth G (1989) Abrogated NK-cell lysis of human papillomavirus (HPV)-16-bearing keratinocytes in patients with precancerous and cancerous HPV-induced anogenital lesions. Int J Cancer 43:209–214

Malison MD, Morris R, Jones LW (1982) Autogenous vaccine therapy for condyloma acuminatum. A double-blind controlled study. Br J Vener Dis 58–62

Morin C, Braun L, Casas-Cordero M, Shah KV, Roy M, Fortier M, Meisels A (1981) Confirmation of the papillomavirus etiology of condylomatous cervix lesions by peroxidase-antiperoxidase technique. JCNI 66:831–835

Morris HHB, Gatter KC, Sykes G, Casemore V, Mason DY (1983) Langerhans cells in human cervical epithelium: effects of wart virus infection and intraepithelial neoplasia. Br J Obstet Gynaecol 90:412–420

Naghashfar ZS, Rosenshein NB, Lorincz AT, Buscema J, Shah KV (1987) Characterization of human papillomavirus type 45, a new type 18-related virus of the genital tract. J Gen Virol 68:3073–3079

Nouvo GJ, Crum CP, de Villiers EM, Levine RU, Silverstein SJ (1988) Isolation of a novel human papillomavirus (type 51) from a cervical condyloma. J Virol 62:1452–1455

Obalek S, Glinski W, Haftek M, Orth G, Jablonska S (1980) Comparative studies on cell-mediated immunity in patients with different warts. Dermatologica 161:73–83

Obalek S, Jablonska S, Orth G (1985) HPV-associated intraepithelial neoplasia of the external genitalia. In: Warts – human papillomaviruses. Clin Dermatol 3:104–113

Obalek S, Jablonska S, Beaudenon MB, Walczak L, Orth G (1986) Bowenoid papulosis of the male and female genitalia: risk of cervical neoplasia. J Am Acad Dermatol 14:433–444

Okagaki T (1986) Human genital papilloma infections: an evaluation of immunologic competence in the genital neoplasia-papilloma syndrome. Am J Obstet Gynecol 155:784–789

Okagaki T, Clark BA, Zachow KR, Twiggs LB, Ostrow RS, Pass F, Faras AJ (1984) Presence of human papillomavirus in verrucous carcinoma (Ackerman) of the vagina. Arch Pathol Lab Med 108:567–570

Ortaldo JR, Herberman RB (1984) Heterogeneity of natural killer cells. Ann Rev Immunol 2:359–394

Pelissier CI, Yaneva M, Joubert E (1981) Cancers genitaux et lesions precancereuses chez les femmes transplantées rénales. A propos de l' étude de 90 femmes greffées.Gynecologie 32:359–365

Penn I (1986) Cancers of the anogenital region in renal transplant recipients. Analysis of 65 cases. Cancer 58:611–616
Porecco R, Penn I, Droegmneller W, Greer B, Makowski E (1975) Gynecologic malignancies in immunosuppressed organ homograft recipients. Obstet Gynecol 45:359–364
Portolani M, Mantovani G, Pietrosemoli P, Cermelli C, Bosetli F (1987) Antibodies to papillomavirus genus antigens in women with genital warts. Microbiologica 10:271–279
Pyrhonen S, Jablonska S, Obalek S, Kuismanen E (1980) Immune reactions in epidermodysplasia verruciformis. Br J Dermatol 102:247–254
Rando RF, Groff DE, Chirikjian JG, Lancaster WD (1986) Isolation and characterization of a novel human papillomavirus type 6 DNA from an invasive vulvar carcinoma. J Virol 57:353–356
Rogozinski T, Klein MK, Bailey D, Schachter RK (1986) Cell-mediated immunity in bowenoid papulosis (bp), multicentric pigmented Bowen's disease (MPBD). Presented at the International Congress of Clinical Immunology, Toronto, 1986
Roth JA (1983) Tumor-induced immunosuppression. Surg Gynecol Obstet 156:233–240
Rudlinger R, Grob R, Buchmann P, Christen D, Steiner R (1988) Anogenital warts of the condyloma acuminatum type in HIV-positive patients. Dermatologica 176:277–281
Schneider V, Kay S, Lee HM (1983) Immunosuppression as a high-risk factor in the development of condyloma acuminatum and squamous neoplasia of the cervix. Acta Cytol 27:220–224
Schneider-Maunoury S, Croissant O, Orth G (1987) Integration of human papillomavirus type 16 DNA sequences: a possible early event in the progression of genital tumors. J Virol 61:3295–3298
Seski JC, Reinhalter ER, Silva J (1978) Abnormalities of lymphocyte transformations in women with condylomata acuminata. Obstet Gynecol 51:188–192
Seto A, Notake K, Kawanishi M, Ito Y (1977) Development and regression of Shope papilloma induced in newborn domestic rabbits. Proc Soc Exp Biol Med 156: 64–67
Shimoda K, Lorincz AT, Temple GF, Lancaster WD (1988) Human papillomavirus type 52: a new virus associated with cervical neoplasia. J Gen Virol 69:2925–2928
Shore A, Dosch MM, Gelfand EW (1978) Induction and separation of antigen-dependent T-helper and T-suppressor cells in man. Nature 247:586–587
Sillman FH, Sedlis A (1987) Anogenital papillomavirus infection and neoplasia in immunodeficient women. Obstet Gynecol Clin North Am 14:537–558
Sillman FH, Stanek A, Sedlis A, Rosenthal J, Lanks KW, Buchhagen D, Nicastri A, Boyce J (1984) The relationship between human papillomavirus and lower genital intraepithelial neoplasia in immunosuppressed women. Am J Obstet Gynecol 150:300–308
Silverman SL, Cathcart ES (1980) Natural killing in systemic lupus erythematosus: inhibitory effects of serum. Clin Immunol Immunopathol 17:219–226
Skinner MS, Sternberg WH, Ichinose H, Collons J (1973) Spontaneous regression of bowenoid atypia of the vulva. Obstet Gynecol 42:40–46
Skopinska-Rozewska E, Majewski S, Blaszczyk M, Wlodarska B, Jablonska S (1988) Theophylline-resistant and theophylline-sensitive "active" and "total" E-rosette-forming lymphocytes in patients with systemic scleroderma. J Invest Dermatol 90:851–856
Strang G, Hickling JK, Mc Indoe GAJ, Howland K, Wilkinson D, Ikeda H, Rothbard JB (1990) Human T cell responses to human papillomavirus type 16 L_I and E_6 synthetic peptides. Identification of T cell determinants, HLA-DR restriction and virus type specificity. J Gen Virol 71:423–431
Streilein JW (1983) Skin-associated lymphoid tissue (SALT): origins and functions. J Invest Dermatol 80 [Suppl]:12–16

Syrjänen KJ (1984) Current concepts on human papillomavirus (HPV) infections in the genital tract and their relationship to intraepithelial neoplasia and squamous cell carcinoma. Obstet Gynecol Surv 39:252–262

Tay SK, Jenkins D, Maddox P, Hogg N, Singer A (1987) Tissue macrophage response in human papillomavirus infection and cervical intraepithelial neoplasia. Br J Obstet Gynaecol 94:1094–1097

Tyring SK, Cauda R, Grossi CE, Tilden AB, Haych KD, Sams WM, Baron S, Whitley RJ (1986) Peripheral blood mononuclear cells from patients with condylomata acuminata exhibit decreased interleukin-2 and interferon-gamma production and depressed natural killer activity. J Invest Dermatol 87:172 (abstr)

Van de Linde AW, Streefkerk M, Schuurman HJ, Te Velde ER, Kater L (1983) Divergence between the occurrence of antibody and cellular immune reactivity to cervical carcinoma cell lines in preinvasive and macroinvasive stages of cervical carcinoma. Br J Cancer 47:147–153

Vayrynen M, Syrjänen K, Mantyjarvi R, Casten O, Saarikoski S (1985) Immunophenotypes of lymphocytes in prospectively followed up human papillomavirus lesions of the cervix. Genitourin Med 61:190–196

von Krogh G (1983) Condylomata acuminata 1983: an up-dated review. Semin Dermatopathol 2:109–129

Wolff K, Stingl G (1983) The Langerhans cell. J Invest Dermatol 80 [Suppl]:17–21

zur Hausen H (1986) Intrăcellular surveillance of persisting viral infections. Human genital cancer results from deficient cellular control of papillomavirus gene expression. Lancet 2:489–491

zur Hausen H (1987) Papillomaviruses in human cancer. Appl Pathol 5:19–24

zur Hausen H, Schneider A (1987) The role of papillomavirus in human anogenital cancer. In: Salzman NP, Howley PM (eds) The papillomaviruses. Plenum, New York, pp 245–263 (The papovaviridae, vol 2)

Immunomodulation of HPV 16 Immortalized Exocervical Epithelial Cells

J. A. DiPaolo, C. D. Woodworth, P. M. Furbert-Harris, and C. H. Evans

One of the earliest indications that cervical cancer is a sexually transmitted disease comes from the observation of Rigoni-Stern [1] who in 1842 published a statistical paper indicating that "cancer of the uterus" was rare among virgins and nuns and quite common among married women and widows. Although the precise etiology of cervical neoplasia is unknown, the epidemiological profile of the disease makes it almost certain that an infectious agent, as a result of sexual transmission, plays a part in the carcinogenesis [2–4]. In the past few years evidence from various sources, including cytology, histology, and immunohistochemistry, has shown that there is an association between human papillomavirus infection and cervical neoplasia [5–7]. Furthermore, human papillomas have a single host species (*Homo sapiens*) and multiply in mucosal or differentiated cutaneous epithelium at specific anatomical sites. Today there are over 50 various human papillomavirus genotypes; of these a small number, i.e., 6, 11, 16, 18, 31, 33, and 35, and 52b, are found in intraepithelial and invasive cervical cancer [8–11]. Human papillomaviruses 6 and 11 are associated predominantly with benign lesions (condylomas or low-grade dysplasia), whereas the other types occur in invasive cervical carcinomas. Types 16 and 18 occur in most invasive cervical carcinomas; about 95% of women with cervical cancer are found to be positive for a papillomavirus. Similarly, papillomavirus can be detected in men exactly as would be expected in a sexually transmitted disease.

Although experimental data support the hypothesis that human papillomaviruses play a key etiological role in cervical neoplasia, determining the role of papillomaviruses has been epidemiologically difficult. Recently, cervical cells of a large percentage of normal control women have been found positive for the same papillomavirus identified in cervical neoplasia [12–14]. This has cast some doubt on the role of the virus in cervical cancer. One report even suggests that the apparent association between papillomavirus and cervical neoplasia disappears after age-adjustment. Investigation of cervical human papillomavirus infection in normal women using the polymerase chain reaction suggests that the prevalence of human papillomavirus infections in normal women is greater than previously reported when Southern, dot-blot, and filter in situ hybridization techniques were used [15]. Thus, elucidation of the role of human papillomaviruses in the process of carcinogenesis has become imperative.

As yet papillomaviruses have not been grown in vitro. It is accepted that replication requires differentiated keratinocytes and that the virus remains undetectable in the basal layer. It has been suggested that the replication cycle of

Genital Papillomavirus Infections
Eds.: Gross/Jablonska/Pfister/Stegner

the papillomavirus is synchronized to epithelial cell differentiation. As differentiation proceeds from the basal to the spinous and granular stages, virus DNA replication occurs in the cell nuclei. The complete virus with protein coat is found only in fully differentiated squamous cells. Therefore, because the virus cannot be grown in vitro, it is appropriate to use recombinant DNA technology to transfect human papillomavirus (HPV) DNA into human cervical cells to determine their role in carcinogenesis. Whereas transfection of 3T3 cells with HPV 16 produces tumors in nude mice [16], recombinant DNA of HPV types associated with invasive cervical cancer immortalize human epithelial cells but no tumors result when these cells are injected into immunodeficient nude mice [17, 18].

A serum-free medium has been developed which will support the growth of normal exocervical cells (derived from the transformation zone). When coupled with spontaneous squamous cell carcinomas that are grown in culture it becomes possible to examine the entire spectrum from normal to immortal to malignant in terms of a variety of different parameters. The sensitivity of cells containing HPV DNA to the host immune response is a potentially important property in the development of cervical dysplasia and neoplasia. Leukoregulin, a lymphokine identified in our laboratory, has the unique ability to increase the sensitivity of tumor cells to NK and LAK lymphocyte killing. This report summarizes our studies examining leukoregulin im-

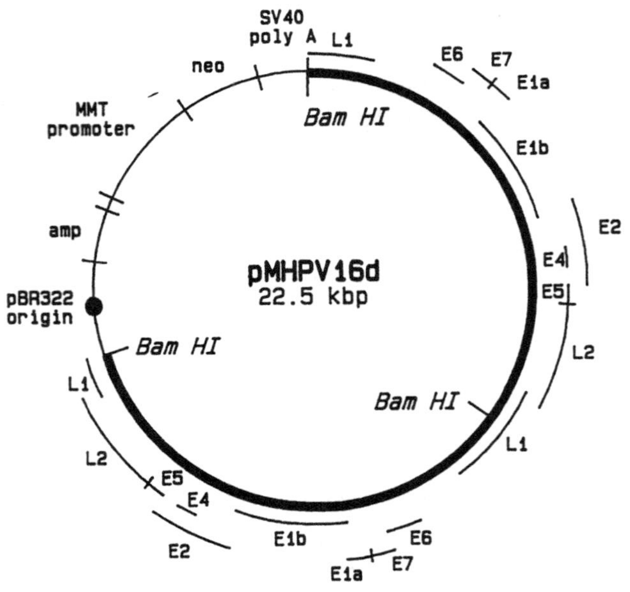

Fig. 1. Structure of recombinant plasmid pMHPV16d. The HPV sequences consist of a head-to-tail dimer indicated by *bold lines*. *Arcs* outside the circle indicate HPV open reading frames. HPV16d was cloned into the *Bam*HI site of the vector. *Neo*, neomycin resistance gene; *amp*, ampicillin resistance gene; *MMT*, metalothiomine; *ori*, bacteria replication origin

munomodulation of HPV 16 DNA immortalized human cervical epithelial cells to NK and LAK lymphocyte killing.

Secondary cultures of normal human exocervical cells in serum free medium were transfected with the recombinant plasmid pMHPV16d (Fig. 1) and selected for resistance to the antibiotic G418. After 10 to 14 days selected colonies appeared and these grew rapidly (population doubling time of 30–40 h) and could be subcultured repeatedly. The individual cells in these colonies were similar in morphology to secondary cultures of normal exocervical cells (Fig. 2). To date nine HPV 16-immortalized lines have been established. Cultures are considered immortalized because they have exceeded the life span of normal cervical cells and continue to proliferate at the same rate as did the original nontransfected controls. Cell lines immortalized by HPV 16 DNA varied in morphology and rate of growth. At the earliest passages (30–50 population doublings) analyzed, cell lines were composed of a mixture of small undifferentiated cells and large flat squamous cells. With increasing passage cell lines grew more rapidly and contained more small undifferentiated cells than large squamous cells. Therefore, higher passage HPV-immortalized cell lines appear to have a more "transformed" phenotype.

To determine whether HPV 16 DNA was present in the HPV 16-immortalized cervical cell lines, high-molecular-weight DNA was obtained from four lines at low passages (50–80 population doublings). All lines contained an average of 2–20 copies per cell of DNA sequences homologous to HPV 16 after *Bam*HI digestion, which releases the HPV 16 sequences from vector DNA (Fig. 3a). In addition to the expected 7.9-kbp complete HPV 16 genome, all lines also contained rearranged HPV 16 DNA sequences, suggesting integration into host cell DNA. When DNA samples from the same HPV 16-immortalized cells were digested with *Bam*HI plus *Eco*RV (a noncut enzyme for pMHPV16d) some rearranged fragments were further digested, sug-

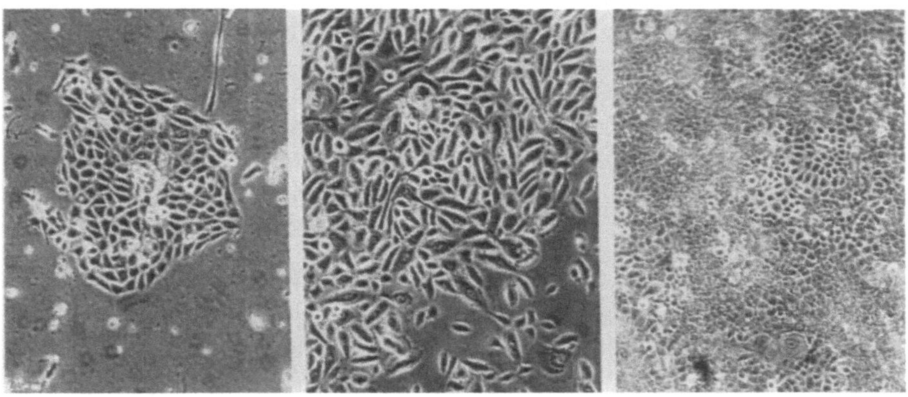

a b c

Fig. 2a–c. Morphological appearance of normal HCX 10 population doublings (**a**) HPV-immortalized HCX at low passage (**b**), and a cervical squamous carcinoma cell line (**c**)

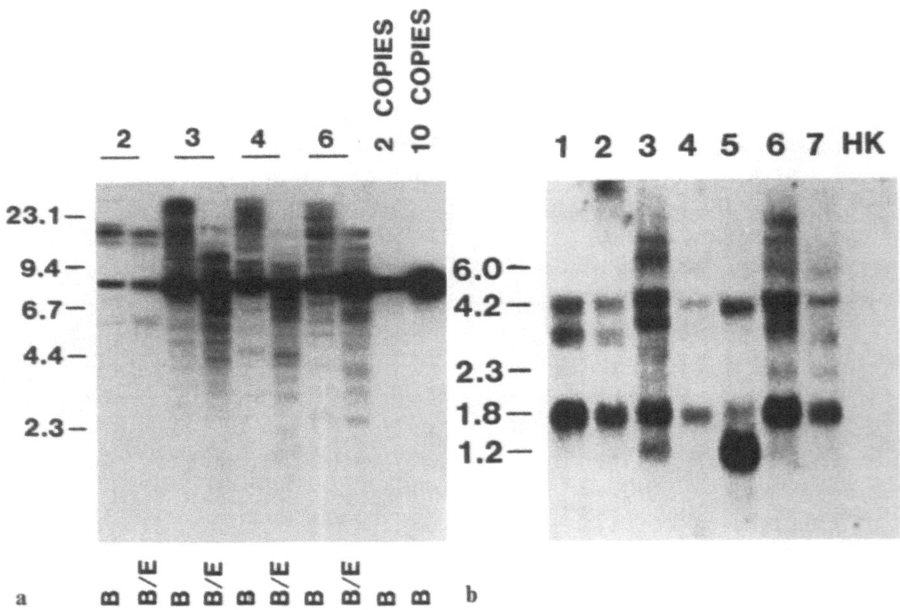

Fig. 3 a, b. Detection of HPV 16 DNA and RNA in immortalized cell lines. **a** Total cellular DNA was extracted from four cell lines; 10 µg of DNA were digested with restriction enzymes, electrophoresed on 0.8% agarose gels, and electroblotted to Gene Screen filters. DNA was digested with *Bam*HI (**B**), which separates vector from HPV 16 sequences, or with *Bam*HI plus *Eco*RV (B/E). *Eco*RV does not digest pMHPV16d. Filters were hybridized to the ^{32}P-labeled *Bam*HI fragment of HPV 16 DNA under stringent conditions. The *last two lanes* represent a reconstruction experiment in which an amount of pMHPV16d DNA equal to two or ten copies per cell was mixed with DNA from nontransfected cells. **b** Northern blot containing 10 µg/lane of polyadenylate-containing RNA isolated from seven HPV 16-immortalized cell lines; the filter was hybridized to the 7.9-kbp *Bam*HI fragment of HPV 16. *Numbers* at the margins indicate molecular weights

gesting that rearranged fragments contained junctions between cellular and HPV 16 DNA. Therefore, a portion of the HPV 16 sequences appear to be integrated into cellular DNA. Expression of HPV 16-specific sequences was measured in seven HPV 16-immortalized cell lines. All lines contained 1.8- and 4.2-kb HPV mRNAs and each cell line also expressed other HPV transcripts (Fig. 3 b).

Keratin synthesis in HPV 16-immortalized cervical cells was compared with both normal cultured cervical cells and three cervical carcinoma cell lines (squamous C4-1 and QG-U and adeno HeLa; Fig. 4). Secondary cultures established from endo- or exocervical epithelium differed markedly in keratin expression. Exocervical cells expressed keratins 5 and 14 at high levels, and low levels of simple epithelial keratins (nos. 7, 8, 18, and 19). In contrast, simple keratins were upregulated in endocervical cells and keratin 14 was absent. Keratin expression in HPV 16-immortalized cell lines was similar to non-

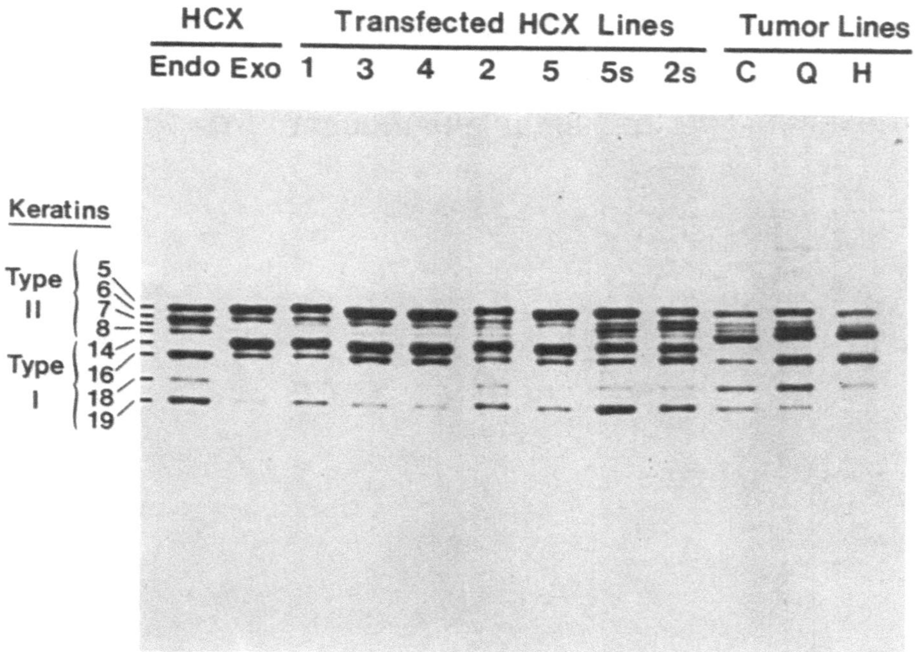

Fig. 4. SDS gradient gel electrophoresis of [^{35}S]methionine-labeled cytoskeletal extracts of HCX. *Lanes 1* and *2*, extracts of secondary cultures of endo- and exocervical cells; *lanes 3* to *7*, extracts of HPV 16-immortalized HCX (cell lines HCX16-1, 2, 3, 2, and 5, respectively); *lanes 8* and *9*, extracts of serum-selected sublines derived from HCX16-5 and HCX16-2, respectively. Tumor cell lines shown in *lanes 10* to *12* (C, C4-1; Q, QG-U; H, HeLa)

transfected exocervical epithelial cells. This indicates that the HPV-immortalized cell lines were probably derived from exocervical epithelium and that HPV 16 had only a minor effect on keratin expression.

The cell-mediated effector arm of the immune response would be expected to play a key role in the host defense against HPV infection and the possible development of a neoplastic state. Support for this is the presence of T cells, and the absence of a significant number of B cells, in cervical metaplasia and in normal cervical epithelium [19]. However, in HPV infections and cervical intraepithelial neoplasia there is a general depletion of intraepithelial T lymphocytes with T4+ helper cells being more depleted than T8+ suppressor cells [19].

Since there is a partial or complete absence of specific T-cell reactivity in patients with HPV infections and cervical intraepithelial neoplasia, and since substantial evidence exists for the role of natural killer (NK) lymphocytes in both viral infections and in neoplasia [20–22], it is important to investigate the presence and function of this form of natural lymphocytotoxicity in these viral-associated diseases. Along these lines, Tay et al. [23] have evaluated the

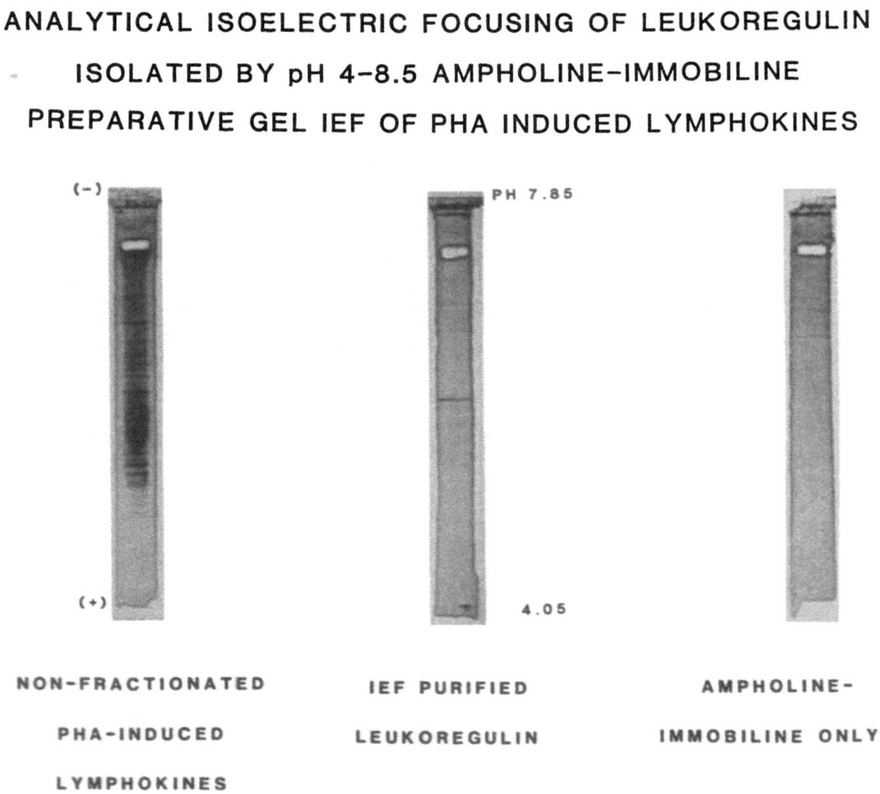

Fig. 5. Leukoregulin was isolated from phytohemagglutinin (PHA)-stimulated normal human peripheral blood lymphocytes [24]. Ficoll-Hypaque-isolated normal mononuclear leukocytes were cultured at 1×10^6 cells/ml in serum-free RPMI 1640 medium with PHA at 10 µg/ml in 2-l plastic cell culture roller bottles at 37° C in a 5% CO_2, 95% air atmosphere for 48 h. The lymphokine containing culture medium was separated from the cells by filtration through a Millipore GVLP 0.4-µm membrane and concentrated 50-fold by ultrafiltration over a YM10 Amicon membrane. The lymphokine concentrates were diafiltered against pH 7.4, 0.01 *M* sodium phosphate buffered saline-0.1% PEG, sterilized by filtration through a 0.22-µm Millex-GV filter unit and stored at $-30°$ C. Leukoregulin with a MW_{HPLC} of 50 kD and an ampholine pI of 5.3 was purified by sequential DEAE ion exchange, isoelectric focusing, and HPLC molecular sizing chromatography, and its activity quantitated by growth inhibition of human K562 erythroleukemia cells [25]. By definition, one leukoregulin cytostatic unit causes 50% growth inhibition of 2×10^4 K562 cells after 72 h incubation

presence of NK lymphocytes in tissue specimens from patients with HPV infection and cervical intraepithelial neoplasia. NK lymphocytes, positive for both the leu 7 (HNK-1) and leu 11 (NK-15) surface antigens, are present in most HPV infections and in one-half of the cervical intraepithelial neoplasia specimens tested. NK lymphocytes and the physiological mechanisms influencing NK lymphocytotoxicity, therefore, may be the components of the effector arm of the immune response important in the host defense against HPV infection and the development of HPV-associated neoplasia.

Leukoregulin is a recently isolated lymphokine (Fig. 5), produced by activated lymphocytes, which increases tumor cell membrane permeability [24, 25] and drug uptake [26], inhibits tumor cell replication [25], and upregulates the sensitivity of tumor cells to killing by NK and lymphokine-activated killer (LAK) lymphocytes [27, 28]. Unlike most lymphokines which activate, or modulate, immunologic cells, leukoregulin has no identified immunoregulatory role for cells in the normal immune system. Instead, it directly induces an anticarcinogenic state in normal cells and directly modulates the integrity and proliferative ability of tumor and other abnormal target cells. In addition to leukoregulin prevention of carcinogenesis and modulation of preneoplastic and neoplastic target cell proliferation, the exposure of these cells to leukoregulin increases their sensivitity to NK and LAK lymphocyte cytotoxicity [25, 27, 29, 30] as illustrated in Fig. 6. The increase in target cell sensitivity to NK and LAK lymphocytotoxicity is frequently two- to fourfold but with some NK and LAK cells it may increase 30-fold. The degree of leukoregulin upregulation of sensitivity to NK or LAK lymphocytotoxicity is not the same for each target cell population. This increase in sensitivity to natural cytotoxicity occurs within 30 min of leukoregulin treatment and is reversible. The upregulation is, moreover, in opposition to the target cell desensitization or protective action of interferon [27, 31]. Leukoregulin is further distinguished from interferon in that it does not directly modulate NK lymphocyte cytotoxic activity and at the target cell negates the desensitization to NK induced by interferon, i.e., even in the presence of interferon, leukoregulin enhances the sensitivity of the target cell to NK lymphocytotoxicity [30].

The ability of leukoregulin to directly induce lysis of some target cells [25] and increase the sensitivity of cells to natural lymphocytotoxicity suggests that this lymphokine directly or indirectly destabilizes the target cell plasma membrane. Flow cytometric measurements confirm this hypothesis by revealing that leukoregulin induces a series of alterations in target cell membrane physiology. Changes in cell surface conformation as indicated by fluorescence depolarization of fluorochrome labeled lectin bound to the cell surface and in membrane fluidity as indicated by fluorescence depolarization of 1,6-diphenylhexatriene [32], and increases in membrane permeability measurable by the influx of propidium iodide and efflux of intracellular fluorescein [25] occur with 30 min and reach a maximum within 2 h after leukoregulin treatment of human K562 erythroleukemia cells. The membrane changes, in addition, correlate well in their time of appearance and in their reversibility with leukoregulin-induced target cell increased sensitivity to NK lymphocyte cytotoxicity [25, 27]. The coordinate changes in membrane permeability and in cell volume in leukoregulin-treated target cells also develop in target cells during a NK lymphocytotoxicity reaction in the absence of exogenous leukoregulin [27]. The similarity of the target cell changes in leukoregulin-

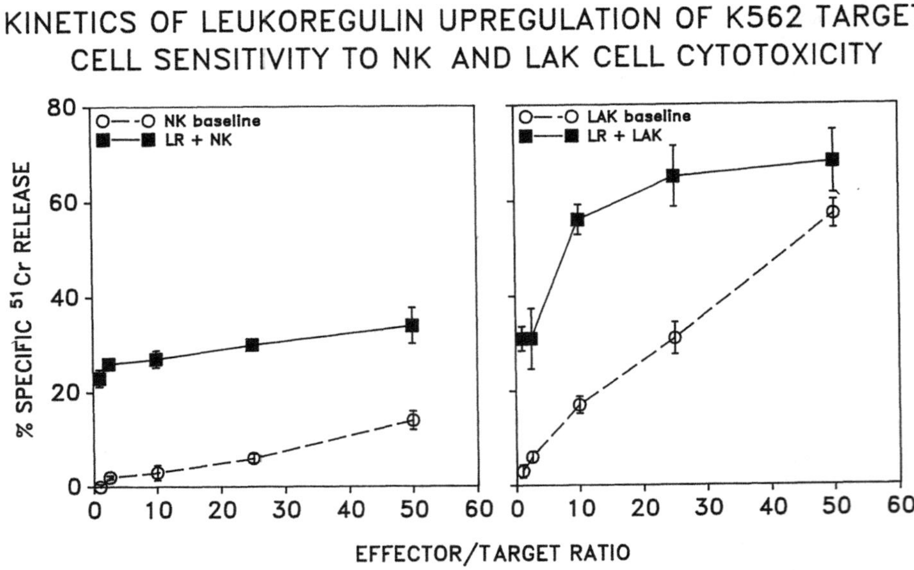

Fig. 6. Leukoregulin enhancement of the sensitivity of human K562 leukemia cells to NK and LAK lymphocytotoxicity. Nylon wool nonadherent mononuclear lymphocytes were isolated from Ficoll-Hypaque-fractionated fresh normal human apheresis preparations and served as NK effector lymphocytes [27]. Ficoll-Hypaque-fractionated mononuclear cells were cultured at 1×10^6 cells/ml in RPMI 1640 medium containing human rIL-2 (100 µg/ml) and 5% human AB serum for 7 days at 37° C in a 5% CO_2, 95% air in water saturated atmosphere and used as LAK effector lymphocytes. K562 cells were labelled with 100 µC ^{51}Cr for 45 min and treated with 2.5 U leukoregulin/ml for 5 or 60 min. Then, 100 µl of effector lymphocytes at 5×10^6, 2.5×10^6, 1×10^6, 2.5×10^5, and 1×10^5/ml was mixed with 100 µl of K562 target cells to yield E:T cell ratios of 50:1, 25:1, 10:1, 2.5:1, and 1:1. Experiments were performed in triplicate in 12 × 75 mm Falcon polystyrene culture tubes. The tubes were centrifuged at 50 *g* for 3 min, and incubated at 37° C in a 5% CO_2, 95% air in water saturated atmosphere for 4 h. The culture supernatants were counted for chromium release in a LKB gamma counter. Spontaneous release (target cells + medium) ranged between 5% and 15%. Percent specific release was calculated as follows:

$$\% \text{ Specific release} = \frac{(\text{Test cpm} - \text{spontaneous cpm})}{(\text{Total cpm} - \text{spontaneous cpm})} \times 100$$

Values reported are the mean ±2 S.E.

treated target cells and in NK cell-target cell mixtures suggests that leukoregulin is one of the mediators in natural lymphocyte cytotoxicity.

Changes in membrane permeability and in membrane fluidity develop within minutes of tumor target cell exposure to leukoregulin, reaching a maximum and returning to constitutive levels within hours. Similar changes in permeability induced by calcium ionophores and other stimulators of increased intracellular ionic calcium [24] suggest that leukoregulin exerts its action through rapid changes in calcium flux. Using the fluorescent calcium chelator indo-1, increases in intracellular ionic calcium are detectable within seconds after tumor target cell exposure to leukoregulin [33]. The rise in intracellular ionic calcium following leukoregulin binding and transmembrane signaling is largely independent of extracellular calcium concentration and may be due primarily to calcium release from intracellular endoplasmic reticulum calcium stores. Following the transient increase in intracellular ionic calcium and several minutes after exposure of the cell to leukoregulin, the plasma membrane exhibits a burst of rapidly opening and closing cation channels [34]. The cation channel activity like the increase in intracellular ionic calcium is transient. Cation ion channel activity is also followed by a period that is refractory to leukoregulin induction reminiscent of the refractory period following stimulation by neurotransmitters. The increase in target cell intracellular ionic calcium and the induction of cation-selective ion channel activity are followed by an increase in membrane-associated protein kinase C activity [35] which occurs concurrently with leukoregulin induced alterations in membrane fluidity, permeability, and cell surface conformation. The latter inturn precede release of ^{51}Cr from the target cells in a NK lymphocytotoxicity reaction and leukoregulin-induced cessation of target cell proliferation. The sequence of events in the molecular pathway of leukoregulin action in preneoplastic and tumor cells is: (a) leukoregulin interaction with the target cell plasma membrane; (b) transmembrane signaling resulting in cation ion channel activation, increased intracellular ionic calcium, increased plasma membrane fluidity, permeability, and cell surface conformational change, activation of membrane-associated protein kinase C; (c) increased sensitivity to natural lymphocytotoxicity, and (d) inhibition of cell division (Table 1).

HPV 16-immortalized human cervical epithelial cell lines at increasing stages of transformation exhibit a changing pattern of sensitivity to NK and LAK lymphocytotoxicity and the modulation of their sensitivity to lymphocytotoxicity by leukoregulin. The steps involved in the establishment of HPV 16-immortalized cervical epithelial cells are illustrated in Fig. 7. The sensitivity of the cervical cells to natural lymphocytotoxicity and modulation of the sensitivity by leukoregulin has been evaluated in relation to the time of establishment of the cells in culture and their transfection with HPV 16 DNA.

Table 1. Molecular events in Leukoregulin action

0 min	Leukoregulin tumor cell interaction
1 min	Increased intracellular ionic calcium
	Increased plasma membrane cation channel activity
5 min	Increased plasma membrane permeability
15 min	Increased drug and macromolecule uptake
30 min	Increase protein kinase C activity
	Increased sensitivity to NK and LAK cytotoxicity
60 min	Inhibition of protein synthesis
6 h	Inhibition of DNA synthesis and cell replication

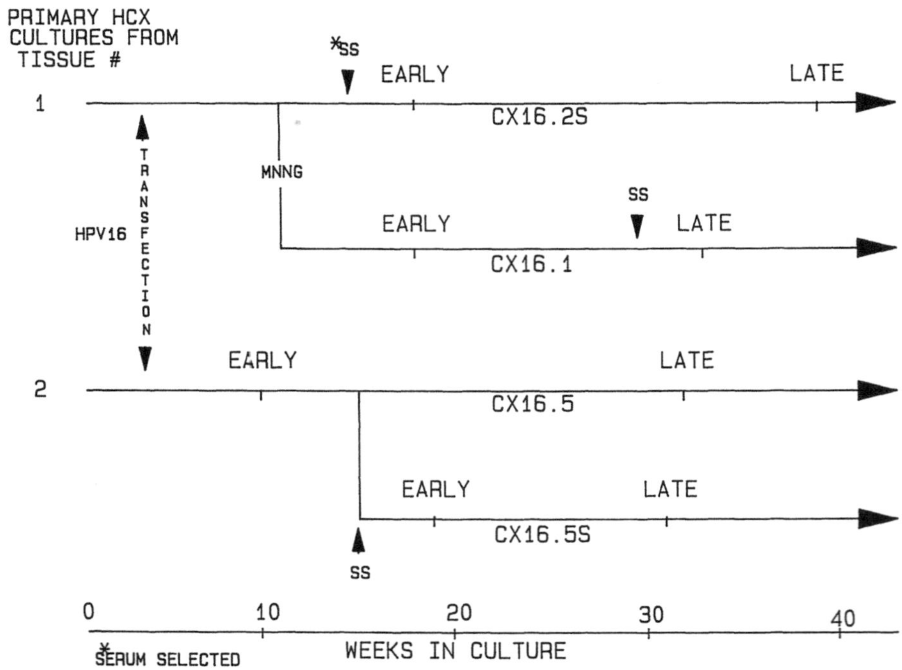

Fig. 7. Several human papillomavirus-immortalized cervical lines established from two different individuals were evaluated in this study [18]. Secondary cultures of normal human exocervical epithelial cells were transfected with a recombinant human papilloma virus DNA which contained the neomycin gene [17] and selected for resistance to the antibiotic G418. Cell lines were derived by pooling cells from several resistant colonies and subculturing at 1:10 in chemically defined medium (MCDB153-LB [17]). Subsequently, sublines resistant to terminal differentiation were selected by culturing cells in MCDB153-LB + 5% fetal bovine serum (FBS). Serum-selected cells were chosen to compare with the HPV 16-positive cervical carcinoma line, QGU, because QGU cells are normally grown in RPMI medium containing 5% FBS. One early passage cervical epithelial cell strain, HCX16-1, was also treated with the carcinogen *N*-methyl-*N'*-nitro-*N*-nitrosoguanidine (MNNG) at passage seven to evaluate the possible effects of chemical and viral cocarcinogen treatment. Normal nontransfected cervical cells, along with HPV 16 DNA positive QGU cervical carcinoma cells, served as nonmalignant and malignant endpoints in the transformation process. All cells were maintained in monolayer cultures in 100-mm plastic tissue culture dishes

Early passage cells were evaluated after culturing for 11–18 weeks in culture, while the later passage cells were examined after 30 weeks in culture. Both early (Fig. 8) and late (Fig. 9) passage HPV 16-immortalized cervical epithelial cells were resistant to NK lymphocytotoxicity. Leukoregulin caused a slight increase ($P = < 0.05$) in susceptibility to NK lymphocytotoxicity in the early passage HCX16-2S cervical cells. Late passage cells were more, if not completely, resistant to NK lymphocytotoxicity and their NK sensitivity was not upregulated by leukoregulin. On the other hand, both early and late passage

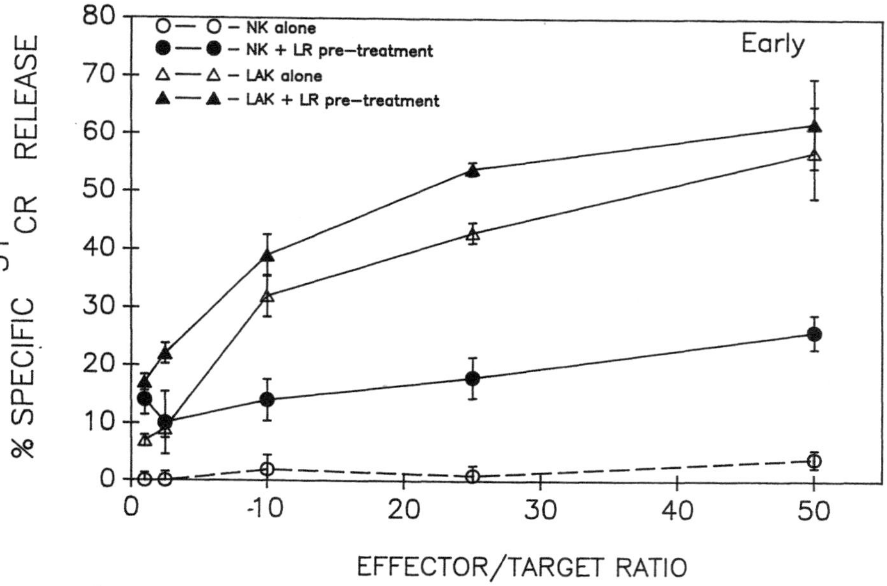

Fig. 8. Leukoregulin enhancement of the sensitivity of early passage HPV 16 DNA immortalized human cervical epithelial cells to NK and LAK lymphocytotoxicity. Confluent monolayer cultures of normal cervical epithelial cells, HPV 16-immortalized cervical epithelial cells, and HPV 16 positive cervical epithelial tumor cells were incubated with 100 μCi sodium chromate for 16–18 h. The cells were washed with 10 ml RPMI 1640 medium-10% FBS and incubated with 5 ml RPMI 1640 medium containing 1% FBS or the same medium containing leukoregulin (2.5 μ/ml) for 1 h. The cells were trypsinized (0.1% trypsin + 25 ml PBS, pH 7.4 + 25 ml EDTA Versene, 1:5000) for 5 min and single cell suspensions washed in 50 ml RPMI 1640 medium-10% FBS and adjusted to 1×10^5 cells/ml. The 4-h chromium release cytotoxicity assay was performed as described in Fig. 7. Values are the mean ± 2 S.E.

immortalized cervical epithelial cells were sensitive to LAK lymphocyte cytotoxicity. Leukoregulin markedly upregulated the sensitivity to LAK lymphocytotoxicity, more so with the late passage cervical epithelial cells at lower lymphocyte/target cell ratios of from 2.5:1 to 25:1.

A differential sensitivity of early and late passage HPV 16-immortalized cervical epithelial cells was seen with each cell line independent of previous chemical carcinogen cotreatment or serum selection of the immortalized cells. Although leukoregulin's slight augmentation of sensitivity to NK lymphocytotoxicity was seen only with the early passage cells, leukoregulin always increased the sensitivity of both the early and late passage HPV-immortalized cervical epithelial cells to LAK lymphocytotoxicity. The increased susceptibility to LAK lymphocytotoxicity was 1.5 to fourfold higher for the late passage compared with the early passage HPV-immortalized cervical epithelial cells. This was observed with HPV 16 DNA immortalized cervical epithelial cells developed in the presence (HCX16-2S and HCX16-5S) or the absence (HCX16-1/MNNG and HCX16-5) of serum.

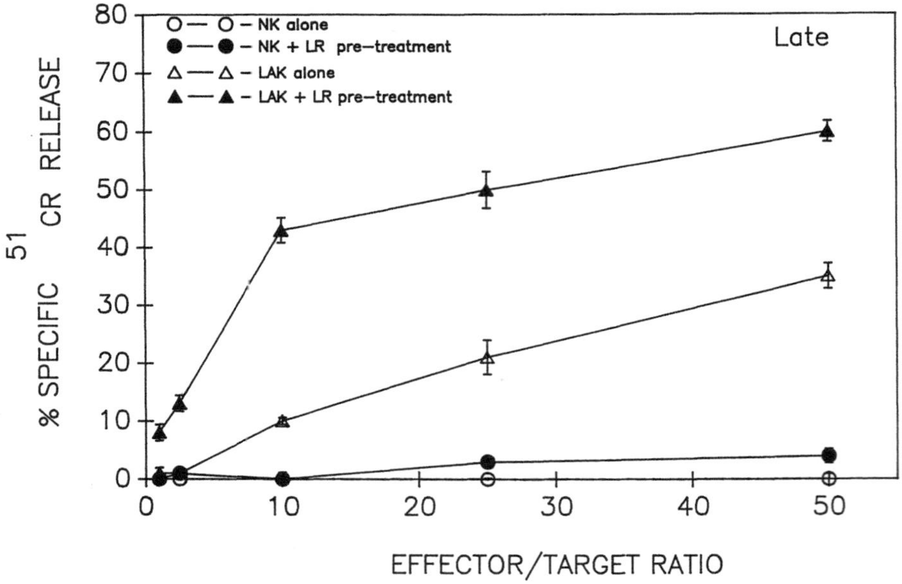

Fig. 9. Leukoregulin enhancement of the sensitivity of late passage HPV 16 DNA immortalized human cervical epithelial cells to NK and LAK lymphocytotoxicity. The assay was performed as described in Fig. 7. Values are the mean ± 2 S.E.

The response of late passage HPV 16-transfected cervical cells was similar to that of HPV 16 positive QGU cervical carcinoma cells (Fig. 10). QGU cervical carcinoma cells were very resistant to NK, and leukoregulin induced only a slight increase in susceptibility to NK lymphocytotoxicity, but like the HPV 16 DNA-immortalized cervical epithelial cells, the carcinoma cells were LAK sensitive and leukoregulin markedly upregulated their sensitivity. Thus, late passage HPV 16-immortalized cervical epithelial cells, exhibited a greater leukoregulin upregulation in sensitivity to LAK than to NK lymphocyte cytotoxicity, mimicking the pattern of leukoregulin-induced upregulation observed with the HPV 16 positive cervical carcinoma cells.

These results demonstrate that HPV 16-immortalized cervical epithelial cells undergo a differential cytokine-controlled increase in their sensitivity to natural lymphocytotoxicity during the period following transfection and immortalization with HPV 16 DNA. Leukoregulin increases the sensitivity of the immortalized cells to NK and to LAK lymphocytotoxicity 18–20 weeks after transfection with HPV 16 DNA, with a greater degree of upregulation by leukoregulin occurring for LAK compared with NK sensitivity. This differential leukoregulin upregulation of LAK lymphocytotoxicity increases as the cervical cells continue to replicate after HPV immortalization and eventually approaches the same sensitivity of HPV 16 DNA positive cervical carcinoma cells. Tay et al. [23] have shown that few NK cells are present in cervical epithelia with HPV infections and in cervical intraepithelial neoplasia. The

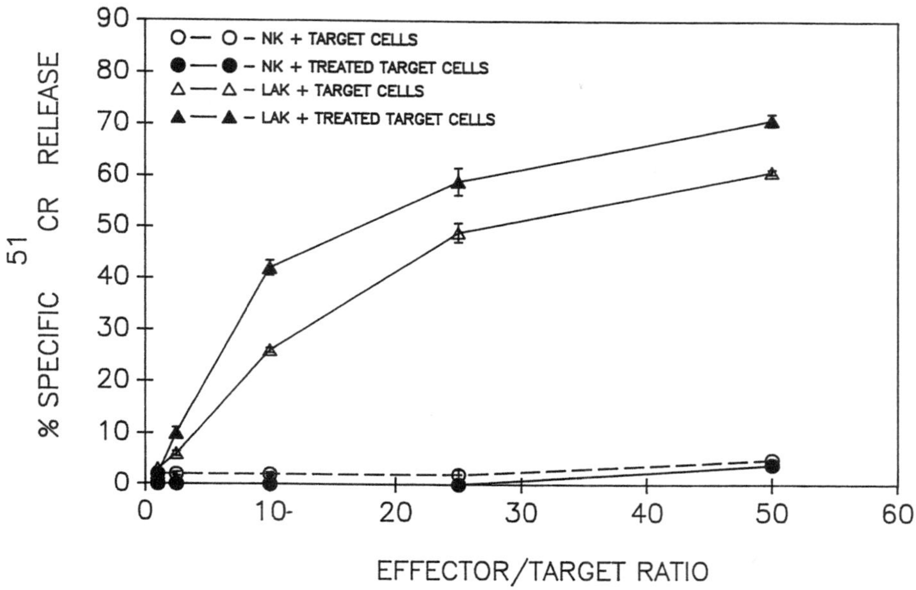

Fig. 10. Leukoregulin enhancement of the sensitivity of HPV 16 DNA positive QGU cervical carcinoma cells to NK and LAK lymphocytotoxicity. Assay was performed as desribed in Fig. 7. Values are the mean ± 2 S.E.

resistance of HPV 16-immortalized cervical epithelial cells, like HPV 16 positive cervical carcinoma cells, to NK lymphocyte killing in vitro suggests that NK cells alone may be ineffective in the destruction of dysplastic and neoplastic cervical cells.

The absence of T-cell specific lymphocytes in HPV infections [19] and the ineffectiveness of NK lymphocytes, even with cytokine upregulation, as shown by the differential upregulation of sensitivity for LAK but not NK lymphocytotoxicity in late passage HPV immortalized cells, indicates that LAK rather than NK lymphocytotoxicity is potentially more important in the control of HPV infection and carcinogenesis. NK lymphocytes can also function in an immunomodulatory capacity, by modulating other effector cells, as well as their own activity. They produce and release an assortment of cytokines, such as IFNγ [36], IL-2 [37], and leukoregulin [25] which act to amplify and regulate the immune response. IFN [38] and IL-2 [39] can act directly on the NK cells to enhance their killer activity. Leukoregulin instead modulates natural lymphocytotoxicity activity by upregulating the sensitivity of tumor target cells to NK [27] and LAK [28] lymphocyte killing. Interferon, particularly IFNα, has been used in the clinical treatment of human genital papillomavirus infections [40, 41]. The response rate of these patients to IFN treatment is dependent on the HPV type with HPV 16/18 infections showing a lower response rate [42]. IFN can be applied topically but is usually administered systemically. In the latter situation, it is difficult to assess whether

IFN, which can both upregulate effector lymphocytes [33] and downregulate target cell sensitivity [27, 43, 44], is acting directly on the abnormal target cell or on the immune effector cell, or both.

The results presented here show that leukoregulin treatment of late passage HPV-immortalized cells has very little effect on the sensitivity of the cells to NK cytotoxicity. On the other hand, both the HPV-immortalized cells and the cervical carcinoma cells are very sensitive to IL-2-activated killer LAK cells and this sensitivity is increased by leukoregulin. Since it has been shown that LAK precursor cells do exist in vivo and that their activity can be enhanced by exogenous IL-2 [45], it is possible that LAK effector cells may play a significant and possibly critical role in vivo in the immune response against cervical epithelial cell infection by HPV and the development of neoplasia. NK cells may function as immunomodulators, releasing substances that activate and amplify LAK effector cells, as evidenced by the in vitro reactivity of LAK cells to the cervical target cells and their corresponding modulation by leukoregulin which upregulates the target cell sensitivity to LAK.

Women with generalized immunodepression are reported to be at risk of developing cervical neoplasia [46, 47]. However, cervical neoplasia is usually found in women who do not present clinically with systemic immunosuppression, hence, if in fact immunosuppression is a significant factor in cervical carcinogenesis the suppression is probably localized. One contributing factor to local immunosuppression may be seminal fluid which contains a number of immunosuppressant agents [48] which exert a suppressive affect on both humoral and cellular immune responses [49, 50], including NK cytotoxicity [51]. The risk of cervical neoplasia increases with sexual promiscuity and with early onset of sexual activity. This results in a greater exposure to seminal fluid and an increased potential exposure to genitally transmitted HPV 16 virus which may itself be immunosuppressive [52]. The dramatic loss of antigen presenting Langerhans' cells in the cervix of women with HPV infection and cervical intraepithelial neoplasia [53] further supports the possibility of a localized immunodeficient state which would facilitate continued growth of the virus and its interaction with other factors in the progression of cervical cells to the neoplastic state.

These observations indicate that leukoregulin alone, or in combination with antiviral or anticancer drugs, may play a significant role in mediating the successful treatment of HPV 16 infections and prevention of cervical dysplasia and neoplasia. The development of HPV-immortalized human cervical epithelial cells to study cervical carcinogenesis provides a valuable model system to define the physiological and biotherapeutic role of leukoregulin and other cytokines in the prevention and control of cervical dysplasia and neoplasia.

References

1. Rigoni-Stern (1842) Fatti statistici relativi alle malattie cancerose che servirono di base alle poche cose dette dal dott. G Servire Progr Pathol Ther 2:507–517
2. Martinez I (1969) Relationship of squamous cell carcinoma of the cervix uteri to squamous cell carcinoma of the penis among Puerto Rican women married to men with penile carcinoma. Cancer 24:777–780
3. Rotkin ID (1973) A comparison review of key epidemiological studies in cervical cancer related to current searches for transmissible agents. Cancer Res 33:1353–1367
4. Singer A, Reid BL, Coppleson M (1976) A hypothesis: the role of a high-risk male in the etiology of cervical carcinoma. Am J Obstet Gynecol 126:110–115
5. Broker TR, Botchan M (1986) Retrospectives and prospectives. In: Botchan M, Grodzicker T, Sharp PA (eds) Cancer cells 4/DNA tumor viruses. Cold Spring Harbor Laboratory, Cold Spring Harbor, pp 17–36
6. Pister H (1987) Relationship of papillomaviruses to anogenital cancer. Obstet Gynecol Clin North Am 14:349–361
7. Peto R, zur Hausen H (eds) (1986) Bradbury report 21: viral etiology of cervical cancer. Cold Spring Harbor Laboratory, Cold Spring Harbor, p 362
8. Campion MJ, McCañce DJ, Cuzick J, Singer A (1986) Progressive potential of mild cervical atypia: prospective cytological and virological study. Lancet 1:237–249
9. Lorincz AL, Temple GF, Kurman RJ, Jenson AB, Lancaster WD (1987) Oncogenic association of specific human papillomavirus types with cervical neoplasia. J Natl Cancer Inst 79:671–677
10. Boshart M, Gissmann L, Ikenberg H, Kleinheinz A, Scheurlen W, zur Hausen H (1984) A new type of papillomavirus DNA, its presence in genital cancer biopsies and in cell lines derived from cervical cancer. EMBO J 3:1151–1157
11. Yajima H, Noda T, de Villiers EM, Yajima A, Yamamoto K, Noda K, Ito Y (1988) Isolation of a new type of human papillomavirus (HPV52b) with a transforming activity from cervical cancer tissue. Cancer Res 48:7164–7172
12. Reeves WC, Caussy D, Brinton LA, Brenes MM, Montalvan P, Gomez B, DeBritton RC, Morice E, Gaitan E, Loo de Lao S, Rawls WE (1987) Case-control study of human papillomaviruses and cervical cancer in Latin America. Int J Cancer 40:450–454
13. de Villiers EM, Wagner D, Schneider A, Wesch H, Miklaw H, Wahrendorf J, Papendick U, zur Hausen H (1987) Human papillomavirus infections in women with and without abnormal cervical cytology. Lancet 2:703–706
14. Meanwell CA, Cox MF, Blackledge G, Maitland NJ (1987) HPV 16 DNA in normal and malignant cervical epithelium: implications for the aetiology and behaviour of cervical neoplasia. Lancet 1:703–707
15. Young LS, Bevan IS, Johnson MA, Blomfield PI, Bromidge T, Maitland NJ, Woodman CBJ (1989) The polymerase chain reaction: a new epidemiological tool for investigating cervical human papillomavirus infection. Br Med J 298:14–18
16. Yasumoto S, Burkhardt AL, Doniger J, DiPaolo JA (1986) Human papillomavirus type 16 DNA-induced malignant transformation of NIH 3T3 cells. J Virol 57:572–577
17. Pirisi L, Yasumoto S, Feller M, Doniger J, DiPaolo JA (1987) Transformation of human fibroblasts and keratinocytes with human papillomavirus type 16 DNA. J Virol 61:1061–1066
18. Woodworth CD, Bowden PE, Doniger J, Pirisi L, Barnes W, Lancaster WD, DiPaolo JA (1988) Characterization of normal human exocervical epithelial cells immortalized in vitro by papillomavirus types 16 and 18 DNA. Cancer Res 48:4620–4628

19. Tay SK, Jenkins D, Maddox P, Singer A (1987) Lymphocyte phenotypes in cervical intraepithelial neoplasia and human papillomavirus infection. Br J Obstet Gynaecol 94:16–21
20. Herberman RB, Ortaldo JR (1981) Natural killer cells: their role in defense against disease. Science 214:24–30
21. Lopez C, Kirkpatrick D, Fitzgerald P (1982) The role of NK (HSV-1) effector cells in resistance to herpesvirus infection in man. In: Herberman RB (ed) NK cells and other natural effector cells. Academic, New York, pp 1445–1450
22. West WH, Cannon GB, Kay HD, Bonnard GD, Herberman RB (1977) Natural cytotoxic reactivity of human lymphocytes against a myeloid cell line: characterization of the effector cells. J Immunol 118:355–361
23. Tay SK, Jenkins D, Singer A (1987) Natural killer cells in cervical intraepithelial neoplasia and human papillomavirus infection. Br J Obstet Gynaecol 94:901–906
24. Barnett SC, Evans CH (1986) Leukoregulin-increased plasma membrane permeability and associated ionic fluxes. Cancer Res 46:2686–2692
25. Ransom JH, Evans CH, McCabe RP, Pomato N, Heinbaugh JA, Chin M, Hanna MG Jr (1985) Leukoregulin, a direct-acting anticancer immunological hormone that is distinct from lymphotoxin and interferon. Cancer Res 45:851–862
26. Evans CH, Baker PD (1988) Tumor-inhibitory antibiotic uptake facilitated by leukoregulin: a new approach to drug delivery. J Natl Cancer Inst 80:861–863
27. Evans CH, Hēinbaugh JA, Ransom JH (1987) Flow cytometric evaluation of leukoregulin as an intrinsic molecular mediator of natural killer lymphocyte cytotoxicity. Lymphokine Res 6:277–297
28. Evans CH, Barnett SC, Gelleri BA, Furbert-Harris P, Sheehy PA, Barker JL, Baker PA, Wilson AC, Farley EK, D'Alessandro F (in press) Biological and molecular characteristics of leukoregulin action. In: Groopman J, Evans C, Golde D (eds) Mechanisms of action and therapeutic applications of biologicals in cancer and immune deficiency disorders. Liss, New York
29. Ransom JH, Evans CH (1982) Lymphotoxin enhances the susceptibility of neoplastic and preneoplastic cells to natural killer cell mediated destruction. Int J Cancer 29:451–458
30. Trinchieri G, Santoli D (1978) Anti-viral activity induced by culturing lymphocytes with tumor-derived or virus-transformed cells. Enhancement of human natural killer cell activity by interferon and antagonistic inhibition of susceptibility of target cells to lysis. J Exp Med 147:1314–1333
31. Evans CH, Ransom JH (1984) The anticancer action of lymphotoxin. In: Goldstein AL (ed) Thymic hormones and lymphokines. Plenum, New York, pp 357–364
32. Barnett SC, Evans CH (1986) Leukoregulin increased plasma membrane permeability and associated ionic fluxes. Cancer Res 46:2686–2692
33. Barnett SC, Evans CH (1986) Calcium dependent membrane destabilization by the antitumor lymphokine leukoregulin. Fed Proc 45:488
34. Sheehy PA, Barnett SC, Evans CH, Barker JL (1988) Activation of ion channels in tumor cells by leukoregulin, a cytostatic lymphokine. JNCI 80:868–871
35. Barnett SC, Evans CH (1988) Leukoregulin induced translocation of protein kinase C activity in K562 cells. Clin Exp Immunol 73:505–509
36. Djeu JY, Stocks N, Zoon K, Stanton GJ, Timonen T, Herberman RB (1982) Positive self regulation of cytotoxicity in human natural killer cells by production of interferon upon exposure to influenza and herpes viruses. J Exp Med 156:1222–1234
37. Kasahara T, Djeu JY, Dougherty SF, Oppenheim JJ (1983) Capacity of human large granular lymphocytes (LGL) to produce multiple lymphokines: interleukin-2, interferon, and colony stimulating factor. J Immunol 131:2379–2385
38. Trinchieri G, Perussia B (1984) Human natural killer cells: biologic and pathologic aspects. Lab Invest 50:489–513
39. Henney CS, JuriBayashi K, Kern DE, Gillis S (1981) Interleukin-2 augments natural killer cell activity. Nature 291:335–338

40. Schneider A, Papendick U, Gissmann L, De Villiers EM (1987) Interferon treatment of human genital papillomavirus infection: importance of viral type. Int J Cancer 40:610–614
41. Weck PK, Whisnant JK (1987) Therapeutic approaches to the treatment of human papillomavirus diseases. Cancer Cells 5:393–402
42. Herberman RB, Ortaldo JR, Bonnard GD (1979) Augmentation by interferon of human natural and antibody-dependent cell-mediated cytotoxicity. Nature 277:221–223
43. Trinchieri G, Granata D, Perussia B (1978) Interferon-induced resistance of fibroblasts to cytolysis mediated by natural killer cells: specificity and mechanism. J Immunol 126:335–340
44. Uchida A, Vanky F, Klein E (1985) Natural cytotoxicity of human blood lymphocytes and monocytes and their cytotoxic factors: effect of interferon on target cell susceptibility. JNCI 75:849–857
45. Anderson TM, Ibayashi Y, Tokuda Y, Colquhoun SD, Holmes EC, Golub SH (1988) Effects of systemic recombinant interleukin-2 on natural killer and lymphokine activated killer activity of human tumor infiltrating lymphocytes. Cancer Res 1180–1183
46. Porreco R, Penn I, Droegemueller W, Greer B, Makowski E (1975) Gynecologic malignancies in immunosuppressed organ homograft recipients. J Obstet Gynecol 45:359–364
47. Obalek S, Glinski W, Hafleck M, Orth G, Jablonska S (1980) Comparative studies on cell-mediated immunity in patients with different warts. Dermatologica 161:73–83
48. James K, Hargreave TB (1984) Immunosuppression by seminal plasma and its possible clinical significance. Immunol Today 5:357–363
49. Lord EM, Sensabaugh GF, Sites DP (1977) Immunosuppressive activity of human seminal plasma. 1. Inhibition of in vitro lymphocyte activation. J Immunol 118:1704–1711
50. Majumddar S, Bapna BC, Mapa MK, Gupta AN, Devi PK, Subrahmanyam D (1982) Effect of seminal plasma and its fractions on in vitro blastogenic response to mitogen. Int J Fertil 27:224–228
51. Rees RC, Valley P, Clegg A, Potter CW (1986) Suppression or natural and activated human antitumour cytotoxicity by human seminal plasma. Clin Exp Immunol 63:687–695
52. Jablonska S, Orth G, Lutzner MA (1982) Immunopathology of papillomavirus-induced tumours in different tissues. Springer Semin Immunopathol 5:33–62
53. Tay SK, Jenkins D, Maddox P, Campion M, Singer A (1987) Subpopulations of langerhans' cells in cervical neoplasia. Br J Obstet Gynaecol 94:10–15

Prospects for Vaccination

S. Schmidt and H. Pfister

Introduction

The available data on the transforming potential of human papillomaviruses (HPV) and their frequent association with anogenital carcinomas are strong indicators of their important role in carcinogenesis and have raised speculations about possible cancer prevention by vaccination against HPV. In the PV field a vaccine may theoretically fulfil two purposes: therapy and prevention.

The spontaneous regression of PV-induced tumours is supposed to be based on immune mechanisms resembling those of allograft rejection (Jablonska and Majewski, this volume). Clinical observations have shown that the immune response may be very slow, however, and PV-induced lesions often persist for many months. One explanation for this fact are the low levels of viral or virus-induced antigens; another is that the antigens, located within the epidermis, cannot be detected by immunocompetent cells in the circulation. It may therefore be promising to vaccinate with appropriate antigen preparations to facilitate the recognition of HPV-specific antigens by the immune system. Such a therapeutic vaccination was successfully practiced by Evans et al. (1962), who achieved an increased rate and frequency of tumour regression by injecting rabbits with papilloma extracts.

On the other hand, it was shown in a number of animal model systems (Spradbrow 1987) that a recovered host appears resistant to reinfection, which provides a basis for preventive vaccination. The immunity of recovered hosts is very likely to be directed against tumour cell surface antigens rather than virus capsid because it was also effective in reinfection experiments, where virus neutralization was avoided by transfection with purified viral nucleic acid (Evans and Ito 1966).

Immunotherapy

Induction of Cell-mediated Immunity

For treating skin warts, condylomata acuminata and laryngeal papillomas of man, autogenous vaccines were used (Holinger et al. 1968; Powell et al. 1970; Abcarian and Sharon 1982; Stephens et al. 1979; Biberstein 1944). The results obtained with homogenized tumour tissues, administered native or formalin-inactived, have been variable. In all these studies, little attention was paid to vaccine safety. Although vaccination accidents were never observed in human

Genital Papillomavirus Infections
Eds.: Gross/Jablonska/Pfister/Stegner

subjects the possible hazard is illustrated by the development of squamous cell carcinomas at the injection sites of a live canine oral PV vaccine in dogs (Bregman et al. 1987).

The safety problem can be overcome nowadays by the use of genetically engineered vaccines. For this purpose, a chosen gene must be cloned into a suitable expression system; this is easily achieved with all HPV types whose DNAs are fully sequenced. The DNAs of the remaining known genotypes have been cloned and are available for further characterization. The difficulties arise when one tries to identify the appropriate gene. Up to now, there is no information about which antigens are able to stimulate rejection.

The role of virus capsid proteins in the immune response is not yet well defined, but it seems to be limited. They are never expressed in proliferating basal cells of warts, which are probable targets for cytotoxic T-lymphocytes and NK cells. In the lower epidermal layers, only products of the "early" open reading frames (ORF) have been detected. If cytotoxic immune cells could be primed to recognize early antigens, it might be possible to treat even advanced lesions, where the expression of virus structural proteins is generally very low. Virus proteins exposed on the surface of infected cells might be focused on. The gene products of bovine papillomavirus (BPV) 1 ORFs E6 and E5 were shown to be associated with non-nuclear membranes of transformed cells (Androphy et al. 1985; Schlegel et al. 1986). In other virus systems, internal antigens have been shown to be also important for the recognition of virus-infected cells by cytotoxic T-lymphocytes (Bennink et al. 1987). There is no way to theoretically predict which viral antigens will trigger cell-mediated immunity best. In vitro assays will have to be established to determine the proteins which are properly processed for being recognized by cytotoxic immune cells. To this end, papillomavirus early proteins must be expressed in eukaryotic cells, for instance via recombinant vaccinia viruses (Brown et al. 1986). Preliminary studies by Orth et al. (unpublished) indicate that vaccination of rabbits with a vaccinia virus expressing ORF E6 of the cottontail rabbit PV (CRPV) accelerates the regression of CRPV-induced tumors.

Identified immunogenic proteins can be offered to the immune system following generally discussed strategies for vaccine development. If a live vaccine virus is used, it must be kept in mind that the proteins encoded by ORFs E6, E7 and E5 of human and/or bovine PVs are able to transform cells and it will be critical to exclude potentially oncogenic functions.

The heterogeneity of PVs poses another problem. A vaccine containing antigens from all relevant types will not be practicable. Therefore, a search for immunogenic epitopes which are at least partially cross-reactive is indispensable.

Inactivation of Blocking Factors

Another target for immunotherapy might be the so-called blocking factors, which were detected in sera of PV-infected patients (Glinski et al. 1976). They were shown to suppress cytotoxic reactions in vitro and this activity is supposed to protect the HPV-induced tumours in vivo. The NK cell-mediated lysis of HPV16-infected keratinocytes turned out to be inhibited by preincubation of active peripheral blood mononuclear cells with sera from patients suffering from HPV 16-associated bowenoid papulosis or anogenital carcinomas (Malejczyk et al. 1989). This inhibition seemed to be target specific, because control cells were still lysed by the preincubated NK cells. It seems possible that an immune response triggered by vaccination with blocking factors themselves could neutralize their inhibiting function so that preexisting NK cell activity could operate and eliminate the tumour cells. There is evidence for such a mechanism in cattle, where PV-associated ocular squamous-cell carcinomas can be cured by vaccination with a phenol-saline extract of an allogenic carcinoma. The active principle of the vaccine was a low-molecular-weight protein which specifically suppressed parameters of cell-mediated immunity against the ocular carcinomas (Spradbrow 1987).

Preventive Vaccination

To date, there are no data on successful prophylactic vaccination except with regard to bovine fibropapillomas. A vaccine consisting of a formalinized homogenate of these fibropapillomas is on the market (Olson et al. 1959, 1962). Herds of cattle revealed a lower wart incidence when immunized by two consecutive intradermal injections. The injection of BPV1 L1 fusion proteins, expressed in *Escherichia coli,* induced antibodies that neutralized the virus in vitro (Pilacinski et al. 1985). Calves were inoculated intramuscularly in field trials with 5–10 mg fusion protein and an $Al(OH)_3$ adjuvant, boostered 1 month later, and challenged intradermally after another month. Only one of 13 vaccinates developed warts, in contrast to nine of nine controls. These trials are highly suggestive of neutralizing antibodies against virions but it is doubtful if neutralization of an infectious virus can account for the protection achieved by the papilloma vaccine under natural, less invasive conditions of infection. There is some possibility of this in the case of BPV1, but it must be questioned whether these findings are transferable to other papillomaviruses. Natural infection by BPV1 starts with a fibroma, and 4–6 weeks later the epithelium starts proliferating. Antibodies might well prevent infection, and thus the transformation of fibroblasts. In contrast to BPV, HPVs induce epithelial lesions only. Therefore, HPVs are more likely to get into a host cell without meeting immune-competent cells and antibodies floating in the circulation. Indeed, there was no success at all in a vaccine trial with bovine papillomas, which are more comparable to human warts, whereas parallel vaccination of the same animals with fibropapillomas protected against these tumours (Barthold et al. 1974). Experiments with CRPV demonstrated the effectiveness of an anti-tumour cell immunity in wart prevention (Kreider 1963; Evans and Ito 1966). These studies indicate that the same mechanisms are operative in both prevention and regression of warts. Thus, if vaccination succeeds in inducing the regression of PV-induced tumours, it should also prevent their development.

The way things are today, the consequence of an anti-papilloma cell immunity on the cancer risk cannot be evaluated. Firstly, there are no data concerning the cell-mediated elimination of latently infected cells. Secondly, it is not known whether malignant conversion takes place in latently infected cells or only in florid lesions. PV-associated cancers arise mostly from precursor lesions and the regression of experimentally induced rabbit papillomas virtually eliminates the cancer risk, so that latently infected cells appear to be at very low risk. A final statement cannot be made, however, before there are results from an appropriate vaccination programme.

Conclusions

A vaccine of therapeutic significance should be able to raise a cell-mediated immunity directed against tumour cells. Such a vaccine could also prevent papilloma growth following primary infections, and PV-associated cancers are not likely to develop in the case of an effective anti-papilloma cell immunity. For developing a vaccine, early gene functions of papillomaviruses should be focused on, because proliferating cells do not produce virus capsid proteins.

There is no formal proof that HPV causes cancer, but a vaccination programme should not depend on this either. A necessary role of HPV in genital cancer development would, in fact, be proved best by prevention following vaccination. Furthermore, morbidity resulting from genital warts and laryngeal papillomatosis is reason enough to set up a vaccine programme against HPVs.

References

Abcarian H, Sharon N (1982) Long-term effectiveness of the immunotherapy of anal condyloma acuminatum. Dis Colon Rectum 25:648–651

Androphy EJ, Schiller JT, Lowy DR (1985) Identification of the protein encoded by the E6 transforming gene of bovine papillomavirus. Science 230:442–445

Barthold SW, Koller LD, Olson C, Studer E, Holtan A (1974) Atypical warts in cattle. J Am Vet Med Assoc 165:276–280

Bennink JR, Yewdell JW, Smith GL, Moss B (1987) Antiinfluenza virus cytotoxic T-lymphocytes recognize the three viral polymerases and a nonstructural protein: responsiveness to individual viral antigens is major histocompatibility complex controlled. J Virol 61:1098–1102

Biberstein H (1944) Immunization therapy of warts. Arch Dermatol Syph 50:12–22

Bregman CL, Hirth RS, Sundberg JP, Christensen EF (1987) Cutaneous neoplasms in dogs associated with canine oral papillomavirus vaccine. Vet Pathol 24:477–487

Brown F, Schild GC, Ada GL (1986) Recombinant vaccinia viruses as vaccines. Nature 319:549–550

Evans CA, Ito Y (1966) Antitumor immunity in the Shope papilloma-carcinoma complex of rabbits. III. Response to reinfection with viral nucleic acid. JNCI 36:1161–1166
Evans CA, Gorman LR, Ito Y, Wieser RS (1962) Antitumor immunity in the Shope papilloma-carcinoma complex of rabbits. I. Papilloma regression induced by homologous and autologous tissue vaccines. JNCI 29:277–285
Glinski W, Jablonska S, Langner A, Obalek S, Haftek M, Proniewsky M (1976) Cell-mediated immunity in epidermodysplasia verruciformis. Dermatologica 153:218–227
Holinger PH, Schild JA, Maurizi DG (1968) Laryngeal papilloma: review of etiology and therapy. Laryngoscope 78:1462–1474
Kreider JW (1963) Studies on the mechanism responsible for the spontaneous regression of the Shope rabbit papilloma. Cancer Res 23:1593–1599
Malejczyk J, Majewsk S, Jablonska S, Rogozinski TT, Orth G (1989) Abrogated NK-cell lysis of human papillomavirus (HPV)-16-bearing keratinocytes in patients with precancerous and cancerous HPV-induced anogenital lesions. Int J Cancer 43:209–214
Olson C, Segre D, Skimore LV (1959) Immunity to bovine cutaneous papillomatosis produced by vaccine homologous to the challenge agent. J Am Vet Assoc 135:499–502
Olson C, Leudke AJ, Brobst DF (1962) Induced immunity of skin, vagina, and urinary bladder to bovine papillomatosis. Cancer Res 22:463–468
Pilacinski WP, Glassman DL, Glassman KF, Reed DE, Lum MA, Marshall RF, Muscoplat CC (1985) Development of a recombinant DNA vaccine against bovine papillomavirus infection in cattles. In: Howley PM, Broker TR (eds) UCLA symposia on molecular and cellular biology, new series, papillomaviruses: molecular and clinical aspects, vol 32. Liss, New York, pp 257–271
Powell LC Jr, Pollard M, Jinkins JL Sr (1970) Treatment of condyloma acuminata by autogenous vaccine. South Med J 63:202–205
Schlegel R, Wade-Glass M, Rabson MS, Yang Y-C (1986) The E5 transforming gene of bovine papillomavirus encodes a small hydrophobic polypeptide. Science 233:464–467
Spradbrow PB (1987) Immune response to papillomavirus infection. In: Syrjänen K, Gissmann L, Koss LG (eds) Papillomaviruses and human disease. Springer, Berlin Heidelberg New York Tokyo, pp 334–370
Stephens CB, Arnold GE, Butchko GM, Hardy CL (1979) Autogenous vaccine treatment of juvenile laryngeal papillomatosis. Laryngoscope 89:1689–1696

Therapy

Patient Applied Podofilox in the Treatment of Genital Warts: a Review

K. R. Beutner

Introduction

Since the early 1940s podophyllum resin has been used for the treatment of genital warts [1]. Physician's use, acceptance, and perceived efficacy of this resin varies greatly. This is in part a reflection of the heterogeneity of podophyllin preparations as well as a lack of a well-defined treatment regimen for the resin. Since Kaplan's original report nearly 50 years ago little has changed in how this resin is used in the treatment of genital warts. The effect of resin source, concentration frequency of application, and vehicle on safety and efficacy have not been studied.

We are currently in a transition from the use of podophyllum resin to the use of podofilox (formerly podophyllotoxin). This transition is analogous to the transition a number of decades ago when digitalis replaced foxglove tea for the treatment of heart failure. This transition has been made possible primarily through the pioneering efforts of Geo von Krogh [2–6]. The topics of podophyllin and podofilox have been recently reviewed in detail [7, 8]. The purpose of this article is to specifically review patient-applied podofilox as a therapy for genital warts.

Terminology

The name for the genus podophyllin is derived from the Greek word meaning "foot-leaf" because of the palmate, foot-like appearance of the plant's leaf. The North American *Podophyllum peltatum* (common names: May apple, mandrake, Indian apple, wild lemon, and duck's foot) and the Indian *Podophyllum emodi* are the two species from which podophyllum is derived. Podophyllum resin is derived from the roots of podophyllum species; podophyllin is an alcoholic extract of podophyllum, and podofilox is one of a group of biologically active lignans contained in and purified from podophyllum.

Podophyllin versus Podofilox

The composition of podophyllin depends upon the species of plant from which it is derived [3, 9]. The lignans podofilox, alpha-peltatin, and beta-

Genital Papillomavirus Infections
Eds.: Gross/Jablonska/Pfister/Stegner

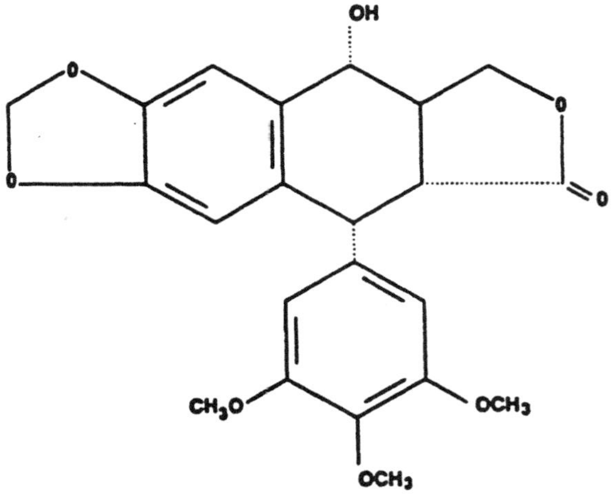

Fig. 1. Chemical structure of podofilox

peltatin are found in *P. peltatum*. While *P. emodi* resin contains primarily podofilox with lower concentrations of 4-demethylpodophyllotoxin. The total lignan content in resin derived from *P. peltatum* is 20% and from *P. emodi* 40%. In addition to these lignans, podophyllin contains other substances some of which have been identified. Quercetin, a known mutagen [23] and carcinogen [24], is present in a number of plants including podophyllum species and the bracken fern (*Pteridium aquilinum*).

The four lignans found in the resin are chemically very similar. Podofilox is the best characterized, the most cytotoxic [6], and the most thoroughly studied for the treatment of genital warts. Podofilox (Fig. 1) is a crystalline white powder. The chemical name of podofilox is 5,8,8a,9-tetrahydro-9-hydroxy-5-(3,4,5-trimethoxyphenyl)furo [3′,4′:6,7]naphtho[2,3,d]-1,3-dioxol-6(5aH)-one. The chemical formula is $C_{22}H_{22}0_8$ with a molecular weight of 414.4 daltons. In addition to the natural lignans a number of chemical derivatives of podofilox, VM 26 (Vumon, Teniposide) and VP 16-213 (Vepesid, Etoposide) are employed as cancer chemotherapeutic agents [10].

Mechanism of Action

The application of podofilox to genital warts results in the relatively acute destruction of the wart tissue over a few days with relative sparing of normal skin followed by healing. Podofilox has a number of biological activities which could explain this observed clinical effect. Contrary to popular misconception, podofilox is not simply a caustic agent but rather is a drug with a number of significant biologic activities.

Podofilox binds to tubulin at a site close to, but not identical with, that of colchicine and prevents the polymerization of tubulin into microtubules [11,

12]. In doing so podofilox can influence cellular events dependent upon intact microtubules. Podofilox has also been demonstrated to block mitosis, inhibit nucleoside transport [12], damage small vessel endothelium [6], stimulate macrophage proliferation and production of interleukin-1 and interleukin-2 [13], suppress immune responses [25], and inhibit mitochondrial function [26]. Exactly which of these activities results in destruction of wart tissue is not known.

Toxicity

There have been no reported cases of severe systemic reactions to podofilox. Approximately 16 severe systemic reactions to podophyllum resin have been reported [14]. Most of these reactions followed ingestion or application of large quantities of the resin. Podofilox is not a mutagen in the Ames bacterial mutagenicity test. In the cell transformation assay podofilox demonstrates no transforming activity. Based on the mouse micronucleous test podofilox is considered a potential clastogen. In laboratory animals topically applied podofilox had no adverse effect on fertility or reproductive performance and was not teratogenic (D. King 1987, personal communication).

There has always been concern that application of podofilox to a genital wart may acutely accelerate the papilloma to carcinoma sequence and/or alter the histology of the wart making it appear to be a cancer. These fears have been dispelled by Wade and Ackerman [15]. In their study genital warts were treated with podophyllum resin and sequentially biopsied prior to therapy and: 24 h, 48 h, 72 h, 1 week, and 6 weeks after therapy or histopathologic evaluation. The histologic changes noted were transient and in no way mimicked squamous cell carcinoma. In this study the resin produced epidermal palor, necrosis of keratinocytes, bizarre mitotic figures, and an increase in the extent of the dermal infiltrate which also changed from lymphohistiocytic to a mixed cell infiltrate with the addition of neutrophils. Despite these changes, orderly maturation persisted and multinucleated and dyskeratotic cells were not seen.

Early studies of podofilox as a cancer chemotherapy agent provide some information on the systemic toxicity of this drug [16, 17]. Patients with a variety of malignancies were treated with intravenous podofilox at a dose of 0.5–1.0 mg/kg/day. Systemic toxic effects of the bone marrow and gastrointestinal system were noted at an average cumulative dose of 6 mg/kg.

Systemic Absorption

Von Krogh has studied the pharmacokinetics of topical 0.5% podofilox [18]. This study demonstrated that most patients can be treated with <100 μl of solution per application. At this dose systemic absorption was rarely detected. A treatment cycle of six applications, twice daily for 3 days of 0.5% podofilox, would represent a total dose of 3.48 mg of podofilox (0.05 mg/kg for a 70 kg patient) given over 3 days. Each 100 µl application contains 0.58 mg of podofilox. In fact, most patients require less then 100 µl per application of 0.5% podofilox to treat their warts. Von Krogh has demonstrated that 41 patients with <15 disseminated condylomata volumes of 5–40 µl were adequate. In 14 patients with >15 disseminated warts 20–70 µl were used. Based on these considerations, von Krogh developed patient applied podofilox for the treatment of genital warts in an effort to increase efficacy and avoid toxicity.

Rationale for Patient-Applied Podofilox

In the late 1940s Sullivan demonstrated that podofilox was the major active component of the resin [19]. Despite this observation purified podofilox was not developed for the treatment of warts until the late 1970s. In his initial studies, von Krogh demonstrated that 8% podofilox applied in the clinic was at least as safe and effective as the resin [3]. To increase efficacy and decrease adverse reactions as well as the dose of biologically active lignans given to patients, von Krogh next proposed the concept of patient-applied low dose podofilox [4].

Patient application not only decreases the frequency of clinic visits and allows patients to participate in their care, but also allows for more frequent application of the drug. Traditionally the resin is applied in a clinic every 1 to 2 weeks. Such a treatment plan would be logical for a physical modality or a caustic agent such as trichloracetic acid but is less than ideal for a drug such as podofilox. Given the known biologic activity of podofilox more frequent application should enhance destruction of wart tissue.

The traditional resin therapy utilizes a high dose of these biologic lignans. A single application of 100 µl of 25% podophyllum resin provides 2.5 to 10.0 mg of podofilox in addition to 0.5–2.5 mg of the other lignans. At this dose one would expect significant systemic absorption [18]. As will be reviewed shortly, the most thoroughly studied patient-applied podofilox therapy is the use of 0.5% podofilox applied twice daily for 3 days.

Clinical Trials

The first clinical trial of patient-applied podofilox for the treatment of genital warts by von Krogh was a dose ranging study [4]. Patients applied 0.5% or 1.0% podofilox twice or thrice daily for 3 days. The efficacy of such therapy was compared with 8% podofilox applied twice in a clinic at 72-h intervals. A total of 214 men with warts predominantly in the preputial cavity of the penis were studied. The observed cure rates were 54% and 48% for the patient applied and clinic applied therapy respectively.

Table 1. Characteristics of 25% podophyllum resin and 0.5% podofilox

	25% Podophyllum resin[a]	0.5% Podofilox
Podofilox (mg/ml)	25–100	5
Other lignans (mg/ml)[b]	5–25	0
Quercetin	+	−
Systemic absorption	++	±
Known stability	?	+
Patient applied	No	Yes

[a] Content varies depending upon species and source
[b] Alpha-peltatin, beta-peltatin, and 4-demethylpodophyllotoxin

Table 2. Reported biologic activities of podofilox

	Reference
Mitotic arrest	9
Prevents polymerization of tubulin	12
Inhibits nucleoside transport	6
Damages endothelium of small vessels	13
Inhibits lymphocyte response to mitogens	13
Induces interleukin-1 and interleukin-2	13
Augments macrophage proliferation	13
Suppresses immune responses	25
Inhibition of mitochondria metabolism	26

Table 3. Summary of clinical trials of patient-applied podofilox for the treatment of genital warts

Author/ Reference	Podofilox concentration[a]	Type of treatment cycle	Number of treatment cycles	Patients cured (%) (n)
Von Krogh [4]	1.0%	TID × 3 days	1	44 (20/46)
	1.0%	BID × 3 days	1	63 (17/27)
	0.5%	TID × 3 days	1	38 (11/29)
	0.5%	BID × 3 days	1	49 (35/71)
Von Krogh [5]	0.5%	QD × 4 days	1	35 (18/51)
	0.5%	BID × 4 days	1	70 (26/37)
	0.5%	QD × 5 days	1	42 (14/33)
	0.5%	QD × 5 days	1	63 (15/26)
Mazurkiewicz [20]	0.5%	BID × 4 days	1–4	68 (17/25)
	0.5%[b]	BID × 4 days	1–4	65 (17/26)
Edwards [21]	0.5%	BID × 3 days	1–6	88 (28/32)
Beutner [22]	0.5%	BID × 3 days	1–4	44 (25/26)

[a] Alcoholic solution except as indicated
[b] Cream for mutation

A wart-free state was induced in 53% and 41% of patients treated twice and thrice daily, respectively. There was no significant difference in response rate with 0.5% podofilox and 1.0% podofilox. These cure rates were recorded at 3 months. Between the end of treatment and the 3-month check, 23% of patients who were lesion free at the end of therapy relapsed.

Variables which influenced efficacy were history of previous successful therapy and anatomical location of the warts. Patients who previously had been treated and cleared were more responsive than patients who had never been treated. This was interpreted as indicating that recurrent warts are more responsive than primary infections. In this study warts in the preputial cavity were most responsive, 68% (86/141) of warts in this anatomical site cleared while only 40% (13/32) of warts at other anatomical sites cleared.

There were no systemic reactions and local reactions occurred in about one-half of the patients in all treatment groups. There was no difference in the local reactions noted in patients treated with 0.5% or 1% podofilox. The cure rates with the two different concentrations were comparable. Von Krogh concluded from these findings that twice daily application of podofilox for 3 days appeared to be the best regimen.

These findings were subsequently expanded upon by evaluating once and twice daily applications for 4 and 5 days [5]. Rate of cure was comparable with the twice a day application for 4 or 5 days. Seventy percent (26/37) of patients treated twice daily for 4 days and 63% (15/24) patients treated twice daily for 5 days were cured. However, local adverse reactions were greater with 4 or 5 days of therapy than previously noted with a 3-day treatment period.

Local adverse reactions were less with once daily application for 4 or 5 days. The once daily application also reduced the cure rate. Once daily application for 4 days resulted in resolution of warts in 35% of cases (18/51) while once daily application for 5 days cleared 42% of patients. Thus, once daily application decreased the cure rate and twice daily for 4 or 5 days increased local adverse reactions. Twice daily application for 3, 4, and 5 days resulted in local adverse reactions in 51%, 81%, and 100% of patients respectively. Based on a balance between efficacy and local adverse reactions, von Krogh concluded that twice daily application of 0.5% podofilox for 3 days is the best therapeutic regimen.

The next major advance was the addition of multiple treatment cycles, the rationale being that sequential treatment cycles may increase efficacy. A treatment cycle is defined as the twice daily application of 0.5% podofilox for 3 consecutive days followed by a 4-day drug-free period. Using such treatment cycles patient-applied podofilox has been compared with traditional resin therapy in single-blind comparative trials.

The first of these reports was by Mazurkiewicz and Jablonska [20]. This was a single-blind, comparative trial of 0.5% podofilox cream vs 0.5% podofilox solution vs 20% podophyllum resin. The self-treated patients applied the test material for up to six treatment cycles. The podophyllin group was treated in the clinic weekly for a maximum of six times. Clinical evaluations were made by an observer who was not aware of the patients treatment group. A wart-free state was noted in 68% (17/25) of the podofilox solution group, 65% (17/26 of the podofilox cream group, and 38% (8/24) of the podophyllin group.

In another single-blind trial, Edwards and associates [21] compared patient applied 0.5% podofilox solution and clinically applied podophyllin. In this study patients received one to six treatment cycles of self-applied 0.5% podofilox or clinically applied 20% podophyllin. Again, patient-applied podofilox produced a greater response than the resin. Complete resolution of warts was achieved in 88% (28/32) of self-treated patients while 63% (12/19) of resin-treated patients were clear at the end of the treatment phase.

In this study there were no systemic adverse reactions, but local reactions of burning, erythema, erosions, and pain were noted. While 87% of clinically treated patients noted no local reactions, 51% of the patient-applied group noted some local reactions. These reactions were predominantly mild, with rarely moderate or severe reactions.

The final step in the development of this therapy was the completion of a double-blind, placebo-controlled trial [22]. In a multicenter trial, patient-applied 0.5% podofilox was compared with placebo in the treatment of genital warts in men. The study population consisted of 109 patients who were randomly allocated to active or placebo groups. Patients were prestratified such that one-half of each group had had their warts for more than 12 months and one-half for less than 12 months.

Patients received a minimum of one and a maximum of four treatment cycles. Each treatment cycle consisted of twice daily application for 3 days followed by a 4-day drug free period. Two weeks after the end of the treatment period 73.6% of podofilox treated warts and 8.3% of placebo treated warts had cleared ($p=0.0001$). The original wart area was reduced by 82% in the podofilox group and 4% in the placebo group ($p=0.0001$).

At some point during the study 25 of 56 podofilox-treated patients but none of the 53 placebo-treated patients were completely clear of all warts. Evaluating warts demonstrated that 82% (206/317) of podofilox-treated warts and 13% (20/158) of placebo-treated warts cleared during the treatment phase of this study. Approximately one-third of the warts that cleared recurred. Recurrence was defined as the appearance of a wart in the anatomic site which was treated. Interestingly, one-third of both the placebo- and podofilox-treated patients developed warts at sites anatomically remote from the treatment site. Duration of infections (<12 months vs >12 months) did not influence efficacy.

Also of note is that contrary to previous studies, most warts (88%) in this study were on penile skin and not in the preputial cavity. It is generally believed that the moist genital warts found in the preputial cavity are more responsive to topical therapy than warts on the more heavily keratinized skin of the penile shaft. In the original report by von Krogh [4], using a single treatment cycle, only 13% (2/16) of warts on the penile skin cleared. A study of warts predominantly on penile skin must be considered a very stringent test of efficacy.

There were no significant systemic reactions. Less than one-half of the podofilox-treated patients experienced local adverse reactions at any time during the study. Previous studies demonstrated that about one-half of the patients treated with 0.5% podofilox solution experienced some local adverse reaction. This is not unexpected for a drug that destroys wart tissue. To what extent the alcoholic vehicle contributed to these reactions is not known. In this last trial local adverse reactions of pain, burning, inflammation, and erosion were evaluated weekly and quantitated as none, mild, moderate, or severe.

No local reactions were noted in 46%–64% of the patients. Inflammation was the most frequent local reaction being reported in 64% of the podofilox-treated patients and 6% of the placebo patients at any time during the study. Destruction of wart tissue was accompanied by superficial erosions in 63% of the podofilox- and 4% of the placebo-treated patients. As might be expected of an alcoholic vehicle, burning was noted in 36% of placebo- and 59% of podofilox-treated patients. Pain at the time of application was noted by 46% and 13% of the podofilox- and placebo-treated patients respectively. All of these local reactions resolved after completion of treatment. These local reactions were judged as mild or moderate in most patients. Only eight occurrences of a severe reaction were noted.

Analysis of local adverse reactions in terms of the number of times an event was noted in relation to the number of observations gives a different picture. The podofilox and placebo patients were seen a total of 182 and 165 times respectively. At each of these visits local adverse reactions were noted. Inflammation, erosion, burning, and pain were noted in 40%, 37%, 39%, and 24% of the visits respectively, in the podofilox group during the treatment period. At 165 visits of the placebo patients inflammation, erosion, burning, and pain were noted at 2%, 1%, 20%, and 5% of the visits respectively.

Summary

Based on large open trials, comparative trials against podophyllum resin, and a multicenter placebo-controlled trial patient-applied 0.5% podofilox has been demonstrated to be safe and effective for the treatment of external genital warts. As judged by the mild local reactions for the treatment of external genital warts. As judged by the mild local reactions noted in these clinical trials as well as the basic knowledge of podofilox toxicity this treatment regimen appears to offer a substantial margin of safety.

The major shortcoming of this treatment appears to be a fairly high recurrence rate. This is not surprising given the latent chronic nature of genital human papillomavirus infection. Recurrence is a reality of all therapies for genital warts.

Von Krogh has noted that patients who previously responded to any therapy had a higher cure rate with podofilox than patients who have never been treated. A next step in the development of this therapy may be evaluating the effect of repeated treatment courses on recurrence. Perhaps with each course of therapy there will be a proportion of patients who remain wart free for a prolonged period of time. As we await the final cure or prevention for genital warts, patient-applied low dose podofilox represents a reasonable first line therapy for many patients.

References

1. Kaplan IW (1941, 1942) Condylomata acuminata. New Orleans Med Surg J 94:388–390
2. von Krogh G, Maibach HI (1983) Cutaneous cytodestructive potency of lignans. II. A comparative evaluation of macroscopic-toxic influence on rabbit skin subsequent to repeated 10-day applications. Dermatologica 167:70–77
3. von Krogh G (1978) Topical treatment of penile condylomata acuminata with podophyllin, podophyllotoxin, and colchicine. Acta Derm Venereol 58:163–168
4. von Krogh G (1981) Penile condylomata acuminata: an experimental model for evaluation of topical self-treatment with 0.5%–1.0% ethanoic preparations of podophyllotoxin for three days. Sex Transm Dis 8:179–186
5. von Krogh G (1987) Topical self-treatment of penile warts with 0.5% podophyllotoxin in ethanol for four or five days. Sex Transm Dis 14(3):135–140
6. von Krogh G, Maibach HI (1982) Cutaneous cytodestructive potency of lignans. I. A comparative evaluation on epidermal and dermal DNA synthesis and on dermal microcirculation in the hairless mouse. Arch Dermatol Res 274:9–20
7. Beutner KR (1987) Podophyllotoxin in the treatment of genital human papillomavirus infection: a review. Semin Dermatol 6:10–18
8. Miller RA (1985) Podophyllin. Int J Dermatol 24:491–498
9. Kelly MG, Hartwell JL (1954) The biological effects and the chemical composition of podophyllin. JNCI 14:967–1010
10. Jardine I (1980) Podophyllotoxins in anti cancer agents based on natural product models. Academic, New York
11. Manso-Martinez R (1982) Podophyllotoxin poisoning of microtubules at steady-state: effect of substoichiometric and superstoichiometric concentrations of drug. Mol Cell Biochem 45:3–11
12. Loike JD, Horowitz SB (1976) Effects of podophyllotoxin and VP-16-213 on microtubule assembly in vitro and nucleoside transport in HeLa cells. Biochemistry 15:5435–5442
13. Zheng Q-Y, Wiranowska M, Sadlik JR, Hadden JW (1987) Purified podophyllotoxin (CPH-86) inhibits lymphocyte proliferation but augments macrophage proliferation. Int J Immunopharmacol 9(5):539–549
14. Cassidy DE, Drewry J, Fanning JP (1982) Podophyllum toxicity: a report of a fatal case and a review of the literature. J Toxicol Clin Toxicol 19(1):35–44
15. Wade TR, Ackerman AB (1984) The effects of resin of podophyllin on condyloma acuminatum. Am J Dermatopathol 6(2):109–122
16. Savel H (1964) Clinical experience with intravenous podophyllotoxin. Proc Am Assoc Cancer Res 5:56
17. Savel H (1966) The metaphase-arresting plant alkaloids and cancer chemotherapy. Prog Exp Tumor Res 8:189–224
18. von Krogh G (1982) Podophyllotoxin in serum: absorption subsequent to three-day repeated application of 0.5% ethanoic preparation on condylomata acuminata. Sex Transm Dis 9:26–33
19. Sullivan M, Friedman M, Hearin JT (1948) Treatment of condylomata acuminata with podophyllotoxin. South Med J 41:336–337
20. Mazurkiewicz W, Jablonska S (1986) Comparison between the therapeutic efficacy of 0.5% podophyllotoxin preparations and 20% podophyllin ethanol solution in condylomata acuminata. Z Hautkr 61:1387–1395

21. Edwards A, Atma-Ram A, Thin RN (1988) Podophyllotoxin 0.5% versus podophyllin 20% to treat penile warts. Genitourin Med 64:263–265
22. Beutner KR, Conant MA, Friedman-Kien AE, Illeman M, Artman NN, Thisted RA, King DH (1989) Patient-applied podofilox for treatment of genital warts. Lancet I:831–834
23. Wang CY, Pamukiu AM, Bryan GT (1976) Braken fern a naturally occurring carcinogen. In: Stock CC, Santamavia L, Mariani P, Govini S (eds) Ecological perspectives on carcinogens and cancer control. Medicine Biologie Environment 4:565–572
24. Pamukcu AM, Yalcinier S, Hatcher JF, Bryan GT (1980) Quercetin a rat intestinal and bladder carcinogen present in bracken fern (*pteridium aquilinum*). Cancer Res 40:3468–3472
25. Brigati C, Sander B (1985) CPH-86, a highly purified podophyllotoxin, efficiently suppresses in vivo and in vitro immune responses. J Immunopharmacol 7:285–304
26. Horrum MA, Jennett RB, Ecklund RF, Tobin RB (1986) Inhibition of respiration in mitochondria and in digitonin treated rat hepatocytes by podophyllotoxin. Mol Cell Biochem 71:79–85

Cryosurgery – Basic Principles and Treatment of Anogenital HPV Lesions

E. Dachów-Siwiec, W. Mazurkiewicz, and E. W. Breitbart

A constant increase in the incidence of condylomata acuminata in the 1970s and in the present decade [1] has created a need for finding the most effective therapy. To date cytotoxic destruction, surgical removal, and immunotherapeutic approaches have been used in the treatment of genital warts, however, none of them has been entirely satisfactory [2]. Podophyllin, which is widely used, is a treatment of low efficacy [3], may have systemic toxic potential [4] and is suspected to have teratogenic effects [5]. Rarely used colchicine has unacceptable adverse local effects such as temporary leukopenia, agranulocytosis, or aplastic anemia [6, 7]. Other chemotherapeutic agents such as thiotepa and bleomycin have only limited usefulness. The main disadvantage of thiotepa is the possibility of systemic absorption and bone marrow depression [8], while that of bleomycin is the requirement of a long period of treatment lasting several months coupled with painful application and a high cost [9]. 5-Fluorouracil (5-FU) according to various authors has a controversial cure rate ranging from 33% to 60% [10, 11]. The treatment is usually connected with painful ulcerations. Other agents, of chemical cautery type such as bi- and trichloracetic acid lack penetrating control and numerous treatments are required [12]. Immunotherapeutic methods involve chiefly interferons and autogenous vaccines. Interferons administered in the form of injections can cause systemic side effects [13], while the same agent in the form of a gel preparation does not [14]. The disadvantages of this method of treatment are the requirement for numerous visits and the fact that mainly small lesions respond to the therapy [15]. There is a report of a cure rate of more than 80% using autogenous vaccine therapy [16]. However, Malison et al. [17] did not confirm these results and found that autogenous vaccine does not induce a higher frequency of wart regression than the use of "placebo" vaccine, prepared from the patient's own normal skin. According to Bunney [18], due to a potential oncogenic effect of viral DNA, this method of treatment should not be used. Some authors advocate scissor excision, however, it is a time-consuming procedure [19, 20] and has a low cure rate. This treatment modality may be indicated during pregnancy, when chemotherapeutic agents are contraindicated. Lately, laser surgery has become more popular as a method of treatment of genital warts. It has several positive aspects such as autosterilization and the possibility of controlling bleeding, however, the cost of treatment is still expensive [22].

During the past decade, liquid nitrogen, due to its physicochemical characteristics (inactive chemically, nontoxic, with a boiling point of −196° C), is

Genital Papillomavirus Infections
Eds.: Gross/Jablonska/Pfister/Stegner

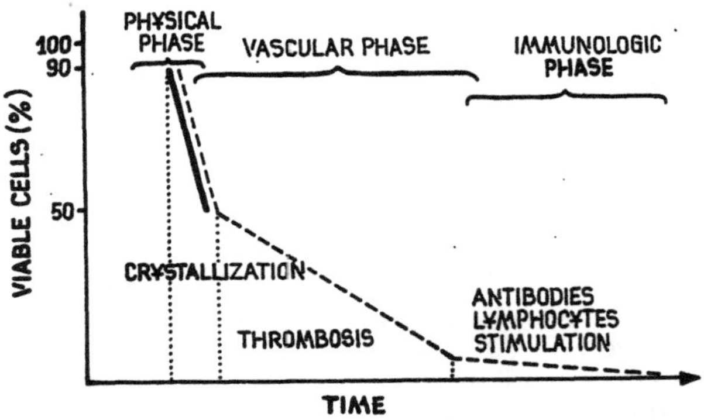

Fig. 1. Schematic representation of the phases of cryodestruction

used widely in the treatment of various skin and mucous membrane lesions [23]. Cryosurgery is based upon biophysical principles, which are presented below.

Basic Considerations of the Cryolesion

A cryolesion is formed by the application of cold (−196° C) to living tissue. This procedure turns pure water in the cell solution into ice crystals through dehydration. The resulting effects are modifications of the solute concentrations in the biological structure and of the membrane systems through the disappearance of structural water [26].

At least three phases of freezing are involved in cryodestruction: the physical, vascular, and immunological (Fig. 1).

The Physical Phase

The sequential changes in a cellular biologic system are [26, 27] as follows (Fig. 2):

1. During freezing:
   a) Down to −5° C, the system remains in a liquid state.
   b) Between −5° and −15° C, extracellular crystals are formed while the intracellular medium remains supercooled and unfrozen.
   c) Below −15° C, an isothermic, transitional state occurs, after which any further increase in solute concentration results in a decrease of the freezing point of the remaining unfrozen liquids.
   d) Finally, the solid state or complete crystallization is attained at the lowest eutectic point (−21° C for NaCl, −51° C for $CaCl_2$, −80° C for LiCl).

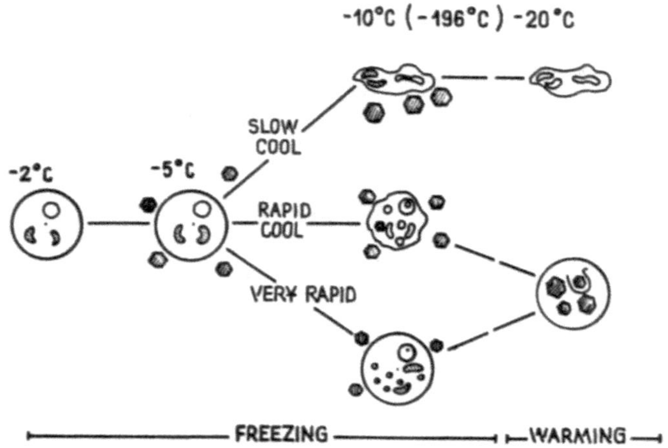

Fig. 2. Schematic representation of physical events in cells during freezing and in frozen cells during warming. (From Mazur [27])

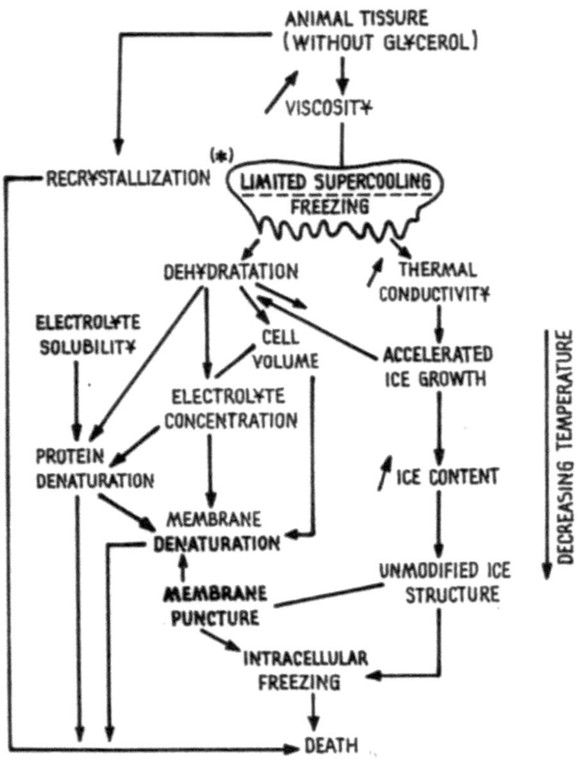

Fig. 3. Schematic representation of the mechanisms of action of freezing damage in animal tissue. (From Baust [28])

2. During thawing:
   a) After a suboptimal freezing, a rapid thawing exposes the cell to a high electrolytic concentration and to an elevated temperature, which could induce a migratory intra- or extracellular recrystallization.
   b) After supraoptimal freezing, a slow thawing induces the migratory recrystallization phenomenon, while a rapid thawing theoretically avoids this recrystallization, thus preventing cellular exposure to both a high and toxic electrolytic concentration and to an elevated temperature.

The elements constituting the determination of cellular cryolesions are (Fig. 3) [26, 28]: extracellular crystallization, intracellular crystallization, cellular dehydration with cellular collapse, increase of intracellular electrolytic concentration, denaturation of membraneous lipoproteins, and thermic shock.

The above factors are closely related and intervene in different degrees in the determination of cellular cryolesions. Mechanical effects due to the formation of ice crystals [29, 30] as well as more complex biochemical effects, secondary to cellular dehydration [31], occur.

The Vascular Phase

Cryoinduced thrombosis completes the cellular destruction produced by the physical phase of the cryocycle. This thrombosis appears to result from the association of the following factors:

1. Vasoconstriction of the arterioles and venules which occurs during slight hypothermia from +11° to +3° C and which may even have a cryoprotective effect [32]
2. Modification of the vascular endothelium
3. Increase in the permeability of the vascular walls with an increase in blood viscosity, a lowering of the intracapillary hydrostatic pressure, and a decrease in blood flow
4. The formation of clots, which are veritable microemboli that coexist with the microthrombi resulting from the slowdown of blood flow.

The Immunological Phase

Cryosurgery (in situ freezing) constitutes an antigenic stimulus capable of generating a specific immunologic response against autologous antigens of the tissue frozen or cryostimulation. The phenomenon of cryostimulation has, in the light of reports of metastatic tumor destruction, suggested that cryosurgery may be applicable not only for ablation of primary tumor but also as a means of induction of host immune response against tumor antigens, i.e., cryoimmunotherapy [30]. We have conducted experiments employing liquid nitrogen cryotherapy in the treatment of condylomata acuminata, especially those resistant to other methods of treatment.

Material and Methods

Patients, who had been treated unsuccessfully with at least six to eight applications of 20% podophyllin solution or a 3- to 4-day series of 0.5% alcohol solution or cream with podophyllotoxin, were selected, including children with condylomata for whom podophyllin is contraindicated.

One hundred and twenty persons were treated, of whom 19 (16%) were children between the ages of 13 months and 14 years (average age 6), 21 (18%) women between the ages 23 and 54 years (average age 36), and 80 (66%) men from 19 to 57 years (average age 30).

In all of the children, condylomata were localized in the anal region, and in 2 girls, aged 4.5 and 12 years, also in the vulva. The number of lesions varied from a few to a large number. They were 0.1 to 1 cm in diameter to coalescent lesions occupying the whole anal region.

The adults were divided according to the location of the lesion into two groups:

1. Those with condylomata exclusively in the anal region. Their number ranged from a few to more than ten or coalescent lesions.
2. Those with genital condylomata. (In some cases, lesions were also perianal.)

In group 2, the lesions in the men occupied mostly the glans penis, the foreskin, and, in a few cases, the meatus urethrae. In the women, condylomata appeared, mostly, on the labia minora and majora; in some cases, single lesions were present in the anal region.

A routine examination for parasites was performed in patients with lesions in the anal region.

In patients with genital lesions examinations were made for *Neisseria gonorrhoea, Chlamydia trachomatis, Candida albicans,* and *Trichomonas vaginalis* in the urethra in both males and females, and in females additionally in the uterine cervix and vagina.

The parents of children were examined for condylomata. Also, mothers were interviewed to determine whether, during pregnancy, they had had con-

Table 1. Results of cryosurgical treatment of patients with condylomata acuminata

Location of lesion	Number of patients		Cured		Without improvement	
			(*n*)	(%)	(*n*)	(%)
Anal region	Children	19[a]	19	100	0	0
	Adults	19	18	95	1	5
Genital region	Adults	70[b]	64	91	6	9
Total		108	101	94	7	6

[a] Two girls, aged 4.5 and 12 years, had condylomata acuminata also on the pudenda
[b] In several patients, single condylomata acuminata were also found in the anal region

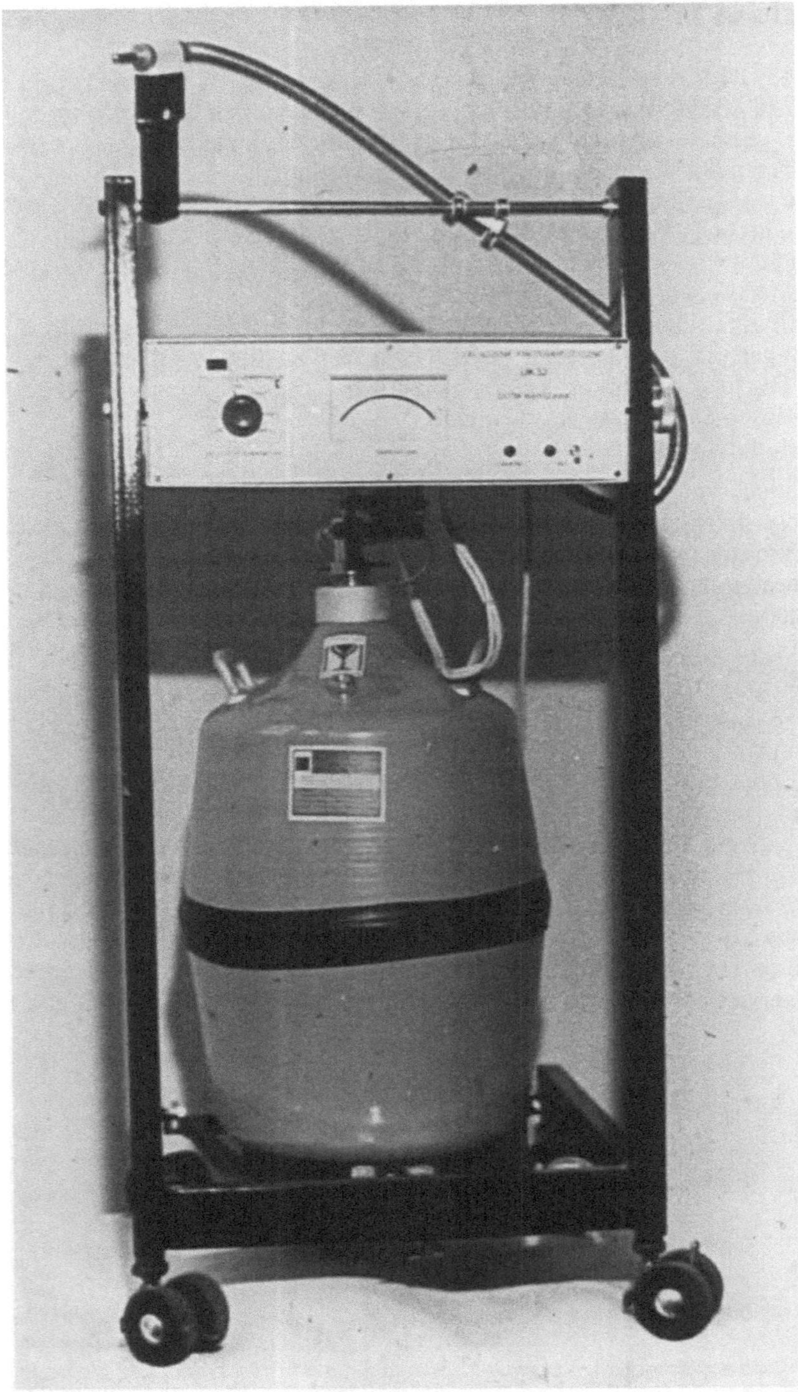

Fig. 4. Apparatus UK 20 used in cryotherapy with liquid nitrogen

dylomata, either external or of the uterine cervix. The adults were also questioned regarding any possible homosexual contacts and the children for possible sexual abuse.

Method of Treatment

Two methods of treatment were used: contact and spray. In the first, wooden applicators with cotton wound around the tips were dipped into a container of liquid nitrogen. Individual lesions were treated using this dipstick method, freezing the entire surface of the lesion and a 5-mm margin of healthy skin. For large and/or grouped lesions the spray method was applied. For this purpose cryosurgical apparatus of Polish construction, UK 20 (Fig. 4) with interchangeable cryoapplicators of various diameters was used. The freezing time ranged from a few to 30 or more s. The entire surface of the lesions with a 10- to 20-mm margin of healthy skin was frozen (Fig. 5a, b, c). It has been suggested that recurrences after removal of clinically manifest HPV-associated lesions may be related to HPV DNA persisting in normal-looking, adjacent skin [33]. Therefore, a treatment that extends 10–20 mm beyond visible lesions may improve the percentage of cures. Treatments were given every 7 to 14 days, depending on the inflammatory reaction. The period of follow-up was at least 3 months.

Widespread condylomata on the surface of the foreskin and the glans penis were frozen in consecutively small areas over a period of time.

To ease the burning pain which accompanied, temporarily, the unfreezing period, the lesions were anesthetized with lignocain spray.

Table 2. Frequency of factors predisposing the development of condylomata in 19 children

Contributing cause		Number of patients	(%)
Helminthiasis of the alimentary tract	Oxyuriasis	4	21.0
	Lamblia intestinalis	1	5.0
	Ascaris lubricoides	1	5.0
Mothers with condylomata acuminata at delivery		2[a]	11.0
Trichomonas vaginalis in vaginal secretions		1	5.0
Passive anal homosexual contacts		1	5.0
Unknown reasons		9	48.0
Total		19	100.0

[a] Enterobius vermicularis eggs also found in one child

Treatment of Children – Course of Healing, Complications, and Results

The group of 19 children consisted of 12 girls, aged 13 months to 14 years (average age 6 years) and 7 boys, aged 2 to 14 years (average age 6 years). The number of treatments ranged from one to six (average three). The treatments resulted in a complete remission of lesions in all of the children.

A few minutes after the freezing reddening, swelling, and subsequently blistering appeared on the surface of the skin. Postoperative care included the application of physiological saline compresses and Clotrimazol cream.

One child developed a secondary bacterial infection which extended the healing period beyond 14 days. No scars were observed (Fig. 6), however, in some cases, discrete hypopigmentation remained.

Factors Predisposing to the Development of Condylomata Acuminata

Out of 19 children, helminthiasis of the alimentary tract was confirmed in six (31%). In four children oxyuriasis eggs were found in the feces; one child had ascaris lumbricoides eggs and one other, lamblia intestinalis cysts. *Trichomonas vaginalis* was found in the vagina of a 14-year-old girl. Two mothers (one with a child who had oxyuriasis) had condylomata during childbirth. A 14-year-old boy had had, since the age of 10, numerous passive homosexual contacts (Table 2).

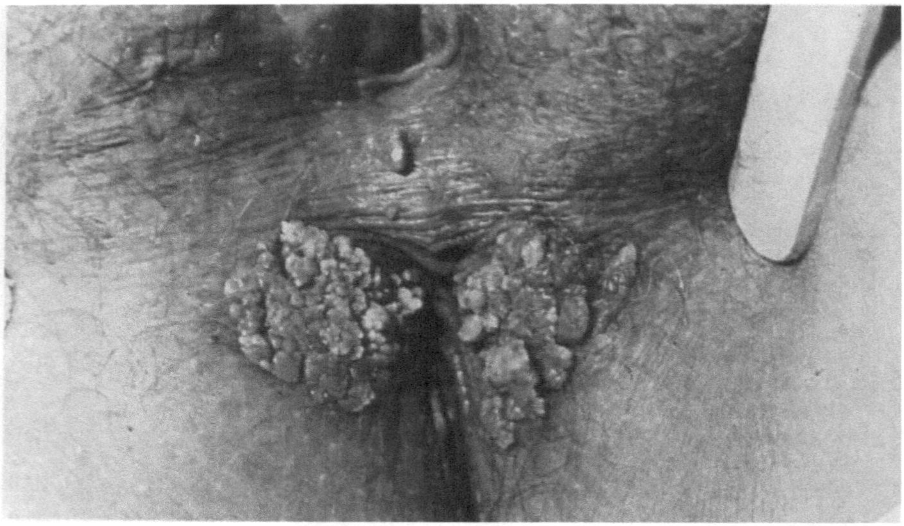

a

Fig. 5 a–c. Entire surface of the lesions with a 10- to 20-mm margin of healthy skin was frozen

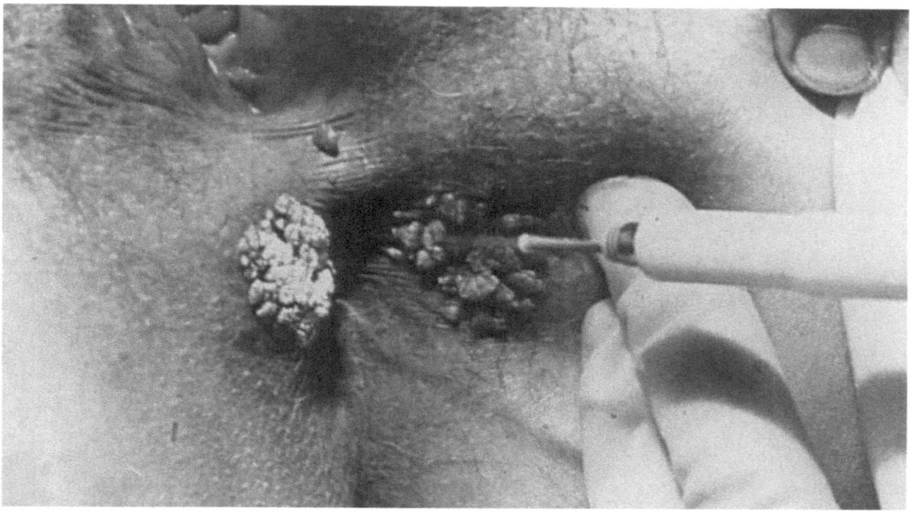

b

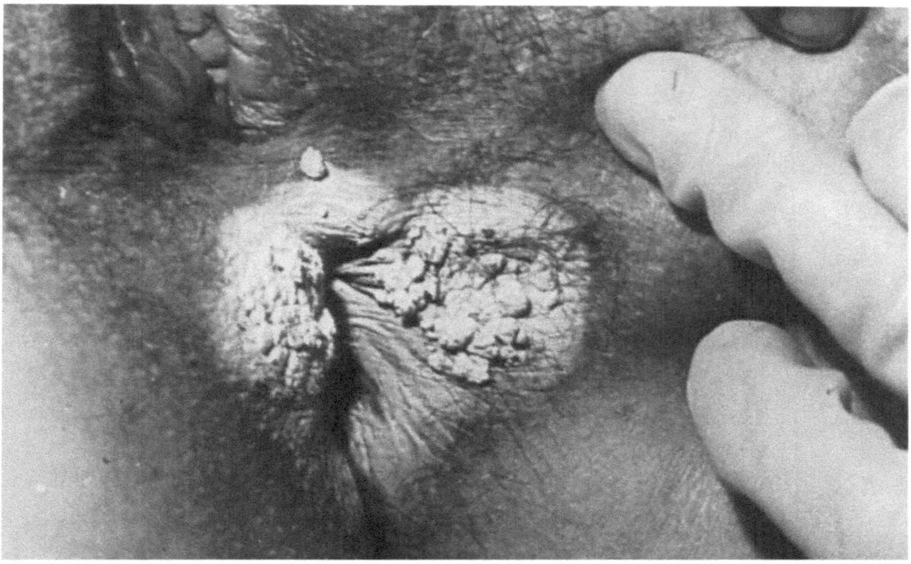

c

Treatment of Adults for Condylomata in the Anal Region – Healing, Complications, and Results

Nineteen persons, ranging in age from 22 to 53 years were treated. Of this group, 18 (95%) were cured. From one to eight treatments (average four) were necessary to achieve cures. In one patient (5%) with very widespread lesions due to lack of improvement after two treatments surgical removal was necessary.

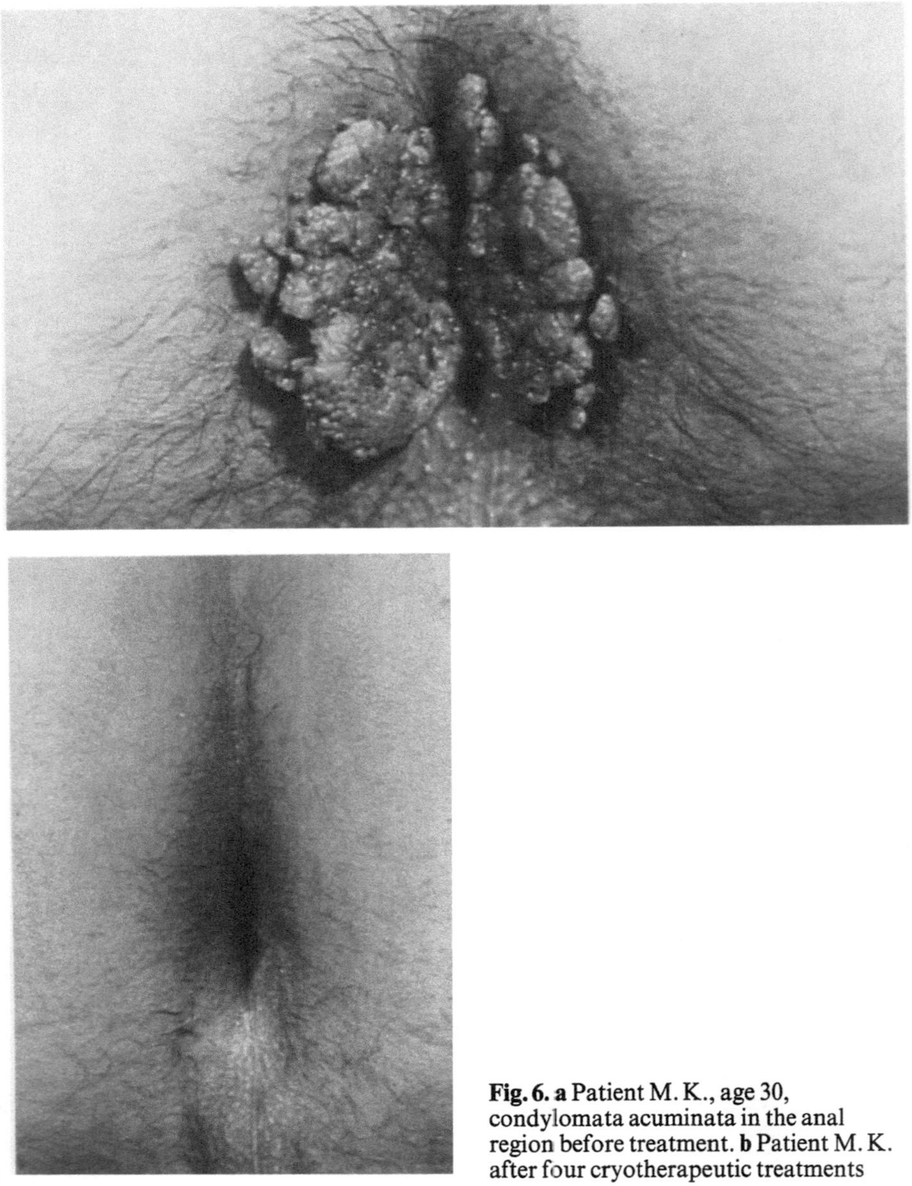

a

b

Fig. 6. a Patient M. K., age 30, condylomata acuminata in the anal region before treatment. **b** Patient M. K. after four cryotherapeutic treatments

In addition to the inflammatory and blistered changes which appeared shortly after treatment, three patients, within 24 h, developed chills and a sense of ill-feeling with fever of over 38° C. These side effects regressed within the next 24 h. One patient was treated with antibiotics because of a secondary infection.

All adult patients were advised to apply physiological saline compresses. The lesions healed without leaving scar. In a few persons with a darker complexion there remained focal depigmentation and in one patient delicate atrophic scars.

Treatment of Adults for Genital Condylomata – Healing, Complications, and Results

Of the 82 patients treated, 12 (15%) did not appear for follow-up control. Of the remainder, 64 (78%) were cured while in 6 (7%) there was no improvement. Since in two of these patients there was an increase in condylomata after two to three treatments, they were recommended for surgery.

The number of treatments ranged from 1 to 14 (average 5), in that for females the range was from 1 to 8 (average 4.5) and for males, 1 to 14 (average 5.2). In three male patients there remained tiny, atrophic scars on the glans penis in the area of the urethra, while in the remaining patients the condylomata regressed without leaving scars. In one patient with numerous, widespread condylomata on the glans penis and the surface of the foreskin developed a phimosis developed which required an operation. Additional examinations revealed slightly positive results of VDRL in two patients previously treated for syphilis; oxyuriasis was detected in one patient with anal condylomata, and nongonococcal urethritis (NGU) was found in 5 (11%) male cases.

Discussion

Cryosurgery in our study was unsuccessful predominantly in patients with depressed cell-mediated immunity. The side effects were phimosis in one patient and secondary infections in two.

Cryosurgical treatment did not generally leave any scars. Only in one case of anal condylomata and in three of condylomata of the urethra tiny scars remained. Slight depigmentations in the anal region were noticed in a few persons with darker complexion. Three persons, with widespread anal condylomata suffered temporary discomfort with a high fever lasting for 24 h.

The high percentage (31%) of parasite infections in the alimentary tracts of children indicates the need for appropriate examination of all children before treatment of condylomata acuminata. There is also a need for thorough interviews concerning sexual abuse in children, as shown by the case of the 14-year-old boy who has been constantly abused since the age of 10. There is also need for examination of all patients for *Trichomonas vaginalis*, in our series present even in a 14-year-old girl. The coexistence of NGU in 11% of adult patients indicates the need for thorough examination in this direction.

After cryotherapy with liquid nitrogen the condylomata regressed in 101 (94%) out of a total of 108 patients treated and followed up for at least 3 months (Table 1).

In seven patients (6%) there were no reactions to the treatment and 12 (10%) did not appear for further observations. The best responses were obtained in condylomata of the anal region, where all of the children and 18 out of 19 adult patients were free of lesions.

The results obtained in treating condylomata in the anogenital region by cryotherapy were better than after other current treatment modalities. Jensen [20] reported results of treatment using 25% podophyllin. There were 76% patients with condylomata cleared at 3 months, but 65% relapsed within 1 year. Other authors who used podophyllin have also obtained a low cure rate, such as 30% [34] and 22% [3]. Interferons have cure rates ranging from 36% [35] to 69% [15]. The results of treatment by scissor excision were 53% [21] and 71% [19].

Better results were reported with the use of CO_2 laser treatment. Badieramonte et al. [36] reported an 80% cure rate in perianal condylomata. However, with the use of cryosurgery for anogenital condylomata results were much better, namely 90% [25] and 91% [24].

References

1. Chuang TY (1987) Condylomata acuminata (genital warts). J Am Acad Dermatol 16:376–384
2. Von Krogh G (1987) Treatment of human papillomavirus induced lesions of the skin and anogenital region. In: Syrjänen K, Gissmann L, Koss LG (eds) Papillomaviruses and human disease. Springer, Berlin Heidelberg New York, pp 297–333
3. Simmons PD (1981) Podophyllin 10% and 25% in the treatment of anogenital warts. Br J Vener Dis 57:208–209
4. Slater GE, Rumack BH, Peterson RG (1978) Podophyllin poisoning. Obstet Gynecol 52:94–96
5. Karol MD, Conner ChS, Watanabe AS, Murphrey KJ (1980) Podophyllum: suspected teratogenicity from topical application. Clin Toxicol 16:283–286
6. Gigay JG, Robison JR (1971) The successful treatment of intraurethral condyloma acuminata with colchicine. J Urol 105:809–811
7. Von Krogh G, Ruden AN (1980) Topical treatment of penile condylomata acuminata with colchicine at 48–72 h intervals. Acta Derm Venereol (Stockh) 60:87–89
8. Halverstadt D, Parry LW (1969) Thiotepa in the management of intraurethral condylomata acuminata. J Urol 101:729–731
9. Mishima Y, Matunaka M (1972) Effect of bleomycin on benign and malignant cutaneous tumours. Acta Derm Venereol (Stockh) 52:211–215
10. Wallin J (1977) 5-Fluorouracil in the treatment of penile and urethral condylomata acuminata. Br J Vener Dis 53:240–243
11. Von Krogh G (1976) 5-Fluorouracil creme in the successful treatment of therapeutically refractory condylomata acuminata of the urinary meatus. Acta Derm Venereol (Stockh) 56:297–301
12. Willcox RR (1977) How suitable are available pharmaceuticals for the treatment of sexually transmitted diseases? (2) Conditions presenting as sores or tumors. Br J Vener Dis 53:340–347
13. Scott GM, Secher DS, Flowers D, Bate J, Cantell K, Tyrrell DAJ (1981) Toxicity of interferon. Br Med J 282:1345–1348

14. Choo YC, Hsu C, Seto WH, Miller DG, Merigan TC, Ng MH, Ma HK (1985) Intravaginal application of leukocyte interferon gel in the treatment of cervical intraepithelial neoplasia (CIN). Arch Gynecol 237:51–54
15. Gall SA, Hughes CE, Trofatter K (1985) Interferon for the therapy of condylomata acuminata. Am J Obstet Gynecol 153:157–163
16. Abcarian H, Sharon N (1977) The effectiveness of immunotherapy in the treatment of anal condylomata acuminatum. J SurgRes 22:231–236
17. Malison MD, Morris R, Jones LW (1982) Autogenous vaccine therapy for condylomata acuminatum. A double blind controlled study. Br J Vener Dis 58:62–67
18. Bunney MM (1986) Viral warts: a new look at an old problem. Br Med J 293:1045–1047
19. Gollock JM, Slatford K, Hunter JM (1982) Scissor excision of anogenital warts. Br J Vener Dis 58:400–401
20. Jensen SL (1985) Comparison of podophyllin application with simple surgical excision in clearance and recurrence of perianal condylomata acuminata. Lancet 23:1146–1148
21. Thomson JPS, Grace RH (1978) The treatment of perianal and anal condylomata acuminata: a new operative technique. JR Soc Med 71:180–185
22. Wheeland RG, Walker NPJ (1986) Lasers – 25 years later. Int J Dermatol 25:209–216
23. Lubritz RR (1977) Cryosurgery of benign and premalignant cutaneous lesions. In: Zacarian SA (ed) Cryosurgical advances in dermatology and tumors of the head and neck. Charles and Thomas, Springfield, pp 55–73
24. Ghosh AK (1977) Cryosurgery of genital warts in cases in which podophyllin treatment failed or was contraindicated. Br J Vener Dis 53:49–53
25. Dodi G, Infantino A, Moretti R, Scalco G, Lise M (1982) Cryotherapy of anorectal warts and condylomata. Cryobiology 19:287–288
26. Le Pivert PJ (1980) Basic considerations of the cryolesion. In: Ablin RJ (ed) Handbook of cryosurgery. Dekker, New York, pp 15–68
27. Mazur P (1977) The role of intracellular freezing in the death of cells cooled at supraoptimal rates. Cryobiology 14:251–272
28. Baust JG (1973) Mechanism of cryoprotection in freezing tolerant animal systems. Cryobiology 10:197–205
29. Reite CB (1966) Mechanical forces as a cause of cellular damage by freezing and thawing. Biol Bull 131:197–203
30. Asahina E (1965) Freezing process and injury in isolated animal cells. Fed Proc 24:5183–5187
31. Lovelock JE (1957) Denaturation of lipid-protein complexes as a cause of damage by freezing. Proc R Soc Med 147:426–430
32. Rothenborg HW (1977) Cryoprotective properties of vasoconstriction. Cryobiology 14:349–361
33. Ferenczy A, Masaru M, Nagai N, Silverstein SJ, Crum ChP (1985) Latent papillomavirus and recurring genital warts. N Engl J Med 313:784–788
34. Khawaja HT (1986) Treatment of condylomata acuminatum. Lancet 25:208–209
35. Eron LJ, Judson F, Tucker S et al. (1986) Interferon therapy for condylomata acuminata. N Engl J Med 315:1059–1064
36. Badieramonte G, Chiesa F, Lupi M, Marchesini R (1987) Laser microsurgery in oncology: indications, techniques and results of 5-year experience. Lasers Surg Med 7:478–486

Laser Therapy in HPV Infections – General Aspects

S. Heinzl

Introduction

Human papillomavirus (HPV) infections, such as condylomata acuminata, can be treated in numerous ways. A variety of drugs, e.g. podophyllin, podophyllotoxin, 5-fluorouracil and trichloroacetic acid, are available for their pharmacological management [5, 8]. Alternatively, these focal lesions can be removed surgically, using a scalpel, sharp curette, electrosurgery, cryosurgery or laser surgery. Recently, interferon has also been recommended for the treatment of HPV infections. Further research is required to establish whether or not this is merely an adjunctive form of treatment.

All these treatment modalities are based on the destruction of the tissue lesions involved. Laser surgery has recently gained an increasing number of adherents because it possesses a Number of advantages, including:

- High precision
- Minimal tissue damage
- No damage to surrounding tissue
- Good haemostasis (vessels up to 0.5 mm are sealed)
- Negligible postoperative edema
- Rapid healing
- Fine scar formation
- Few complications (infections, wound secretion, minimal pain)

Like any procedure, laser therapy also has its disadvantages. These include the cost of the equipment, the stringent safety requirements and the immobility of the instrumentation.

Fundamentally different laser devices can be used in the treatment of HPV infections. The CO_2 laser and the neodymium-YAG laser have gained widespread acceptance. CO_2 lasers are characterised by surface absorption; i.e. their effect takes place at the surface. In this way the entire lesions can be destroyed gradually. The neodymium-YAG laser differs in that it is characterised by volume absorption; i.e. the beam penetrates to a depth of a few millimetres where it causes tissue coagulation (Fig. 1).

In the case of the CO_2 laser, the beam is transmitted to the emission aperture via a rigid system, whereas with the neodymium-YAG laser the beam is transmitted via a flexible light-guiding system. Treatment can be administered using a handpiece, of which several different varieties exist (Fig. 2). Alternatively, the beam can also be transmitted through an optical output device,

Genital Papillomavirus Infections
Eds.: Gross/Jablonska/Pfister/Stegner

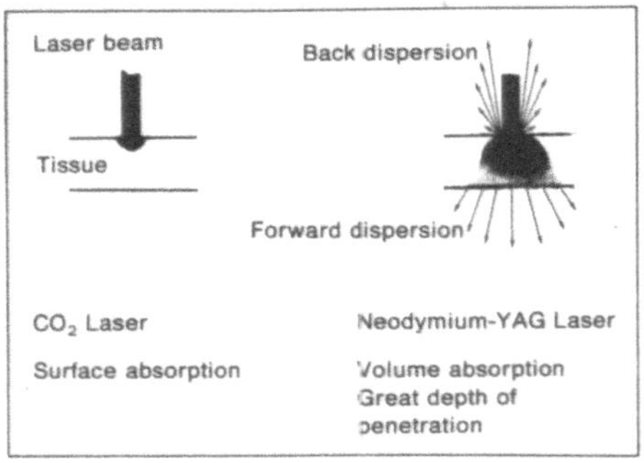

Fig. 1. Tissue effects of the CO_2 and neodymium-YAG laser (modified from Frank and Hofstetter [2])

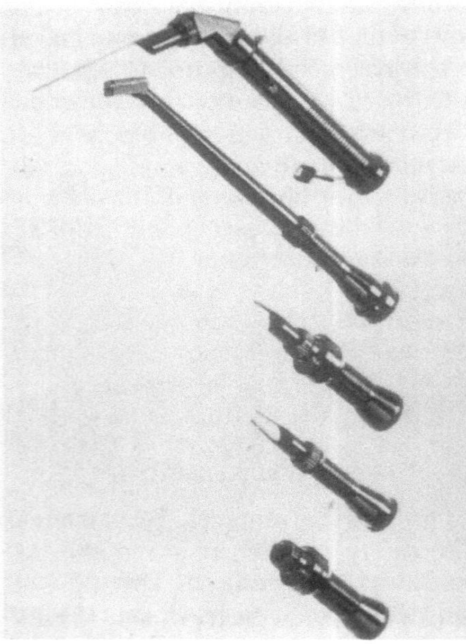

Fig. 2.

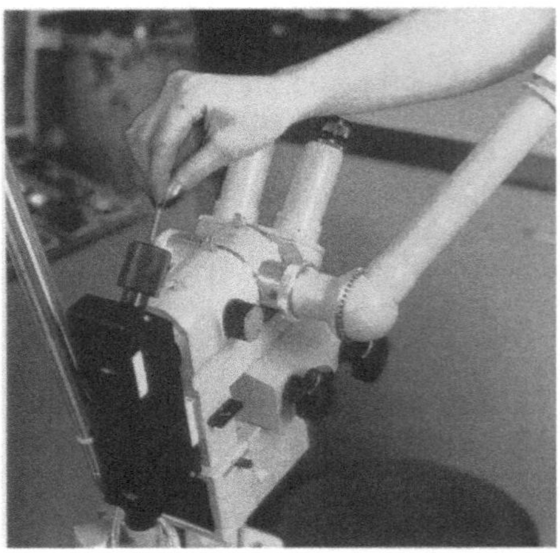

Fig. 3.

Fig. 2. Various laser handpieces

Fig. 3. Micromanipulator (joy-stick) attached to a colposcope

either a colposcope or an operating microscope, and a micromanipulator is then used to direct the beam within the field of view (Fig. 3). This has the advantage that the operative field is seen in high magnification, and consequently even the most minute lesions can be identified.

It is important that a precise diagnosis be established before any treatment is implemented. The entire genital tract should be examined and a search should be made in particular for any intraepithelial neoplasia that may be present. The coexistence of HPV infection and intraepithelial neoplasia has already been confirmed by many investigators. Diagnosis is based on a variety

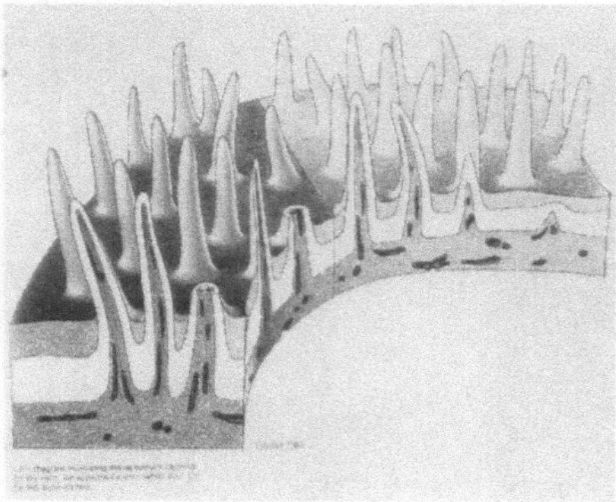

Fig. 4.

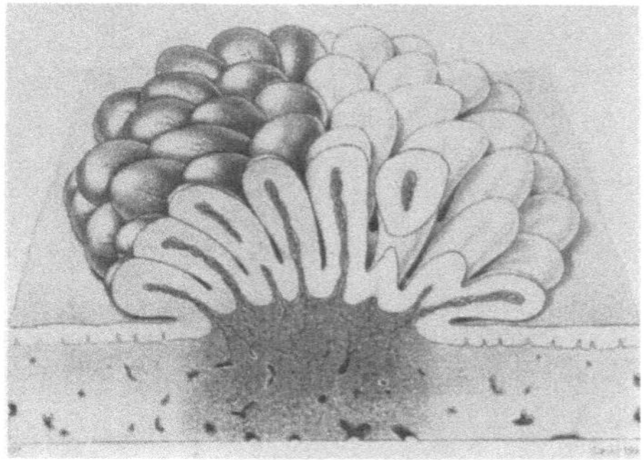

Fig. 5.

Fig. 4. Schematic diagram of a condylomatous lesion (from Cartier [1])

Fig. 5. Schematic diagram of a condyloma acuminatum (from Cartier [1])

of methods. While inspection and colposcopy will indicate the site of the lesions, its nature is determined using the techniques of cytology, histology and virology. Condylomata acuminata of the vulva can usually be diagnosed with the naked eye. The condylomata are generally multiple; they exhibit papillary growth and may develop into cauliflower-like clusters. Flat genital warts of the vulva, or condylomatous vulvitis, are much more difficult to recognise. These lesions have a velvety coat which turns white following application of acetic acid. This is clearly illustrated by the schematic diagrams of Cartier [1] (Figs. 4 and 5).

In the region of the vagina and the cervix, a distinction is made between florid condyloma acuminatum, pointed condyloma, flat condyloma and condylomatous cervicitis or vaginitis. Condyloma acuminatum is usually easy to identify. It, too, is characterised by papillary structures and multiple growth. Pointed condylomata have a dense white epithelium with clearly demarcated edges and a coarse irregular surface. The points of the condylomata appear punctuated and their surface is uneven. Cervical intraepithelial neoplasia must always be considered in the differential diagnosis. Flat condylomata cannot be differentiated colposcopically from intraepithelial neoplasia. A white epithelium is evident, with or without a granulating surface. No vessels are visible. A mosaic pattern is common, with a punctuated appearance being encountered less frequently. Condylomatous cervicitis or vaginitis is characterised by a surface displaying small areas of unevenness or pointed protrusions. The epithelium is red and covered with small white spots. This type of lesion may occur alone or in conjunction with florid or flat condylomata. A Pap smear is not suitable for the diagnosis of vulvar condylomata. In contrast, it is indispensable for lesions in the region of the vagina and cervix. When condylomata are present, three cell types are found, namely the koilocytotic cell, the dyskeratotic cell and the condylomatous parabasal cell.

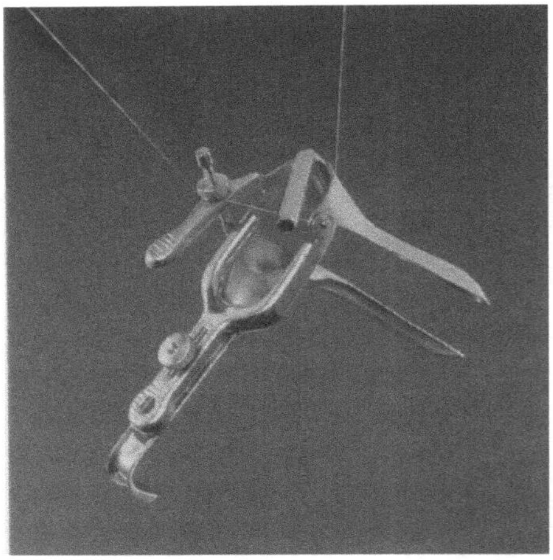

Fig. 6. Speculum with integral extractor tube

Histological diagnosis is always indicated where findings are inconclusive and where there are grounds for suspecting intraepithelial neoplasia. Very liberal use should be made of histopathological diagnosis. Virus typing permits assignment to the individual HPV types. Although definite conclusions cannot yet be drawn from this, it may be expected to have consequences for therapy at some future date.

Laser therapy is relatively straightforward. Aids such as a mirror or swab sticks are sometimes required. The need for anaesthesia depends on the site and extent of the lesions. Anaesthesia may be dispensed with if the lesions are located in the region of the cervix or vagina. Anaesthesia is required for vulvar or perineal lesions. Depending on the extent of these and on the outcome of discussions with the patient, anaesthesia may be local, regional or general. A specially designed speculum with an integral smoke extractor is required for treatment involving the vagina and cervix (Fig. 6). This removes any smoke produced, thus giving a clear view of the operative field during treatment.

Original Observations

Over the past 10 years, 864 women attending our Dysplasia Surgery have undergone CO_2 laser therapy at the University Gynaecology Clinic in Basel. Their average age was 24 years. The youngest patient was 14 and the oldest was 73 years old; 55% of all patients were under 25 years of age. The condylomata were located at various sites (Table 1).

Outpatient population comprised an unfavourable selection: 96 women (11.1%) had already been treated once for HPV infection, and 19 (2.2%) had been treated previously for vulvar or cervical intraepithelial neoplasia. Previous treatment had already been implemented in a total of 13.3% of all patients.

Table 1. Location of condylomatous lesions ($n = 864$)

Location	No.	%
Vulva (perineum)	468[a]	54.2
Vulva, vagina	36[a]	4.2
Vulva, vagina, portio	168[a,b]	19.4
Portio	186[b]	21.5
Vagina	6	0.7

[a] 36 associated with VIN
[b] 145 associated with CIN

Treatment was given on an outpatient basis in cases where the procedure was performed without anaesthesia or with a local anaesthetic. Patients receiving a regional or general anaesthetic were admitted to the same-day surgical unit (morning admission, evening discharge). Great emphasis was placed on scrupulous hygiene as an accompanying postoperative measure. Condom-protected coitus was recommended during the immediate postoperative period. Baths with sea salt, Betadine (povidone-iodine) or Esemtan (0.2% phenylphenol) have also proved effective. Ongoing cleansing with disinfecting agents is recommended in cases of cervical or vaginal involvement.

All patients were followed up in accordance with a predefined timetable. In cases of simple HPV infection, local findings were inspected after 1 week. Cytology, colposcopy and virological tests were repeated after 3 months and 1 year. Where cervical or vulvar intraepithelial neoplasia was also present, initial follow-up took place after 1 week, as in the first group. Cytology, colposcopy and virological tests were then performed after 3, 6, 9 and 12 months.

Primary complications occurred in 55 women (6.4%). Fifty-four women complained of pain requiring treatment, 36 women developed an infection at the ablation site and bleeding occurred in one patient.

Follow-up examination revealed recurrences in 143 women (16.6%). The incidence of recurrence was only 15% among patients without vulvar and cervical intraepithelial lesions compared with 21.9% among patients with vulvar and cervical intraepithelial neoplasia. The incidence of recurrence in our problem group was very high (91.4%). This group consisted of 27 HIV-positive patients, three patients with Hodgkin's lymphoma, three patients with leukaemia and two patients with other malignant diseases.

Patients with recurrences generally underwent repeat treatment. In the group without cervical and vulvar intraepithelial neoplasia (CIN and VIN), recurrence was detected in 4.2% of cases. Recurrences were found in 7.3% of the group with CIN and VIN. Compared with results published in the literature, these findings are in the middle of the range. Statistics collected from 18 papers reporting 2967 cases show an average incidence of persistence or recurrence of 17.6% (5.0%–41.2%). A major problem in identifying recurrences is to differentiate between persistence and genuine recurrence. It is probable that many of the patients will become reinfected after therapy.

Case Reports

Case 1: A 23-year-old woman with a 3-year history of recurring condylomata acuminata. This condition had previously been treated at other hospitals, with podophyllin and by cryosurgery, and in each case the healing phases had been protracted. Condylomata acuminata were diffusely distributed over the posterior commissure of the labia (Fig. 7) and were ablated using a CO_2 laser (Fig. 8). One week later

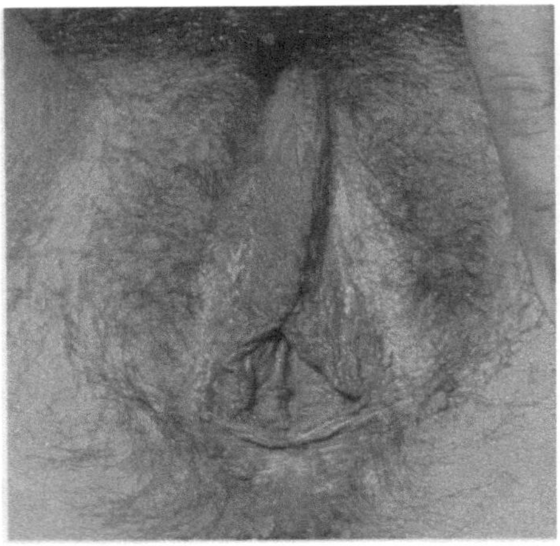

Fig. 7.

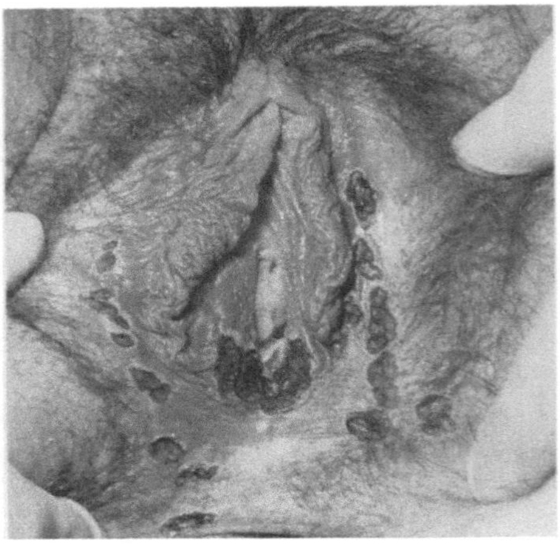

Fig. 8.

Fig. 7. Case 1: 23-year-old woman with diffuse condylomata acuminata

Fig. 8. The same patient shortly after treatment

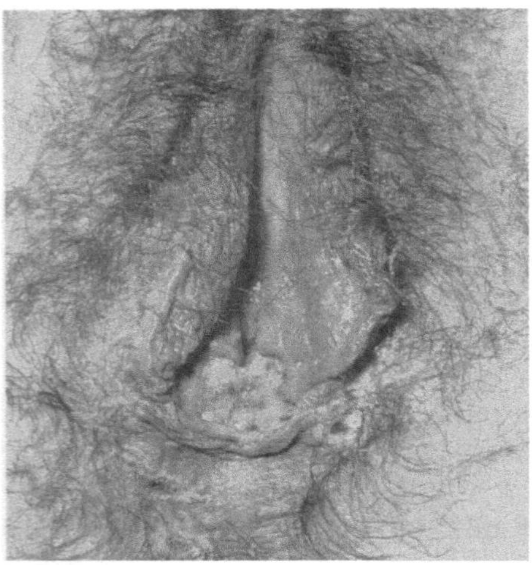

Fig. 9.

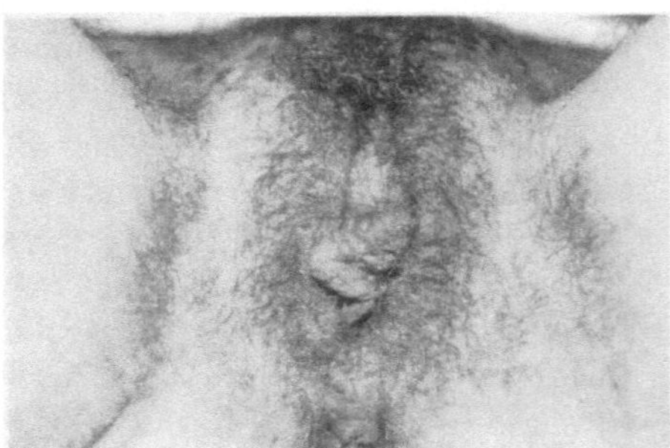

Fig. 10.

Fig. 9. The same patient 1 week after treatment

Fig. 10. The same patient 6 months after treatment

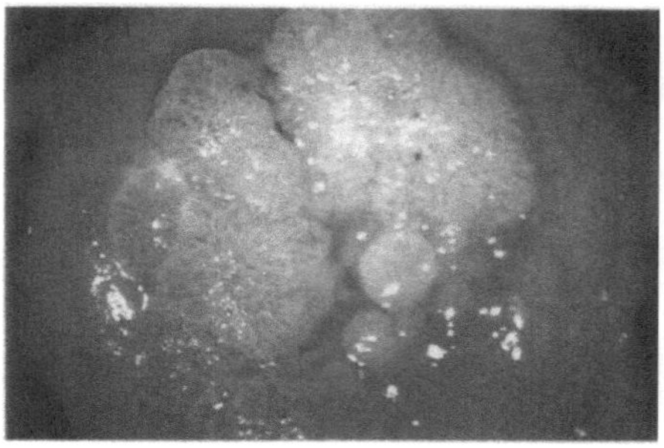

Fig. 11.

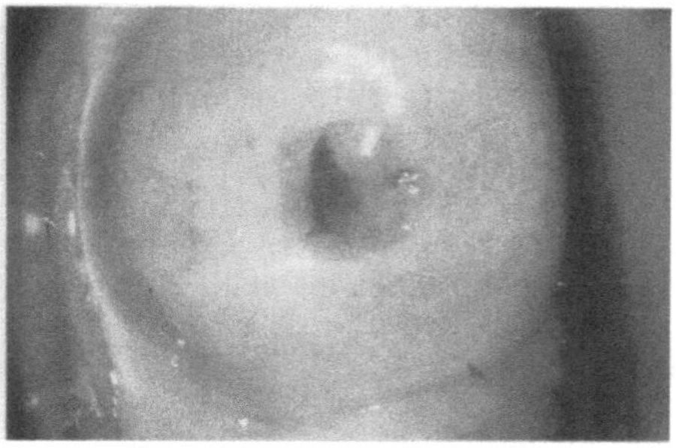

Fig. 12.

Fig. 11. Case 2: 17-year-old girl with a condyloma at the portio vaginalis cervicis

Fig. 12. The same girl 3 months after treatment

the surgical wound was merely covered with a greasy coating (Fig. 9), and after 14 days the wound was fully healed and the patient was symptom free. The patient appeared for regular follow-up, and there have been no recurrences since laser therapy (Fig. 10).

Case 2: A 17-year-old girl was treated for a large condylomata at the portio vaginalis cervicis (Fig. 11). The condylomata was ablated using a CO_2 laser. Three months later the appearance of the portio was normal (Fig. 12). The patient has now been free from recurrence and symptoms for 5 years.

Conclusion

In summary, the laser is a useful treatment modality for HPV infections. Overall, the incidence of recurrence is still relatively high, although our patient population does represent a negative selection. Firstly, they had already been treated previously for HPV infection, and secondly, there was an association with cervical and vulvar intraepithelial neoplasia. Major problems exist in patients with "compromised hustle" and promiscuity. It is probable that the situation can be improved by adjunctive treatments such as interferon or immunostimulants. The initial results from such work are highly promising and require corroboration in further studies.

References

1. Cartier R (1984) Practical Colposcopy. Fischer, Stuttgart New York
2. Hofstetter A, Frank F (1979) Der Neodym-YAG-Laser in der Urologie. Editiones Roche, Basel
3. Heinzl S (1986) Indikationen für die Laser-Technik in der Gynäkologie – Ergebnisse. In: Bender HG, Beck L (eds) Operative Gynäkologie. Springer, Berlin Heidelberg New York Tokyo

4. Heinzl S (1986) Die Anwendung des Lasers bei Erkrankungen des unteren Genitaltraktes. Gynakol Prax 10:523
5. Krebs HB (1989) Genital human papillomavirus infection. Clin Obstet Gynecol 32:107
6. Reid R (1987) Human papillomavirus. Obstet Gynecol Clin North Am 14:329
7. Stegner HE (1981) Zur Klassifikation virusbedingter Dysplasien der Cervix uteri in der zyto-histologischen Routinediagnostik. Gynäkologe 14:252
8. Syrjänen K, Gissmann L, Koss LG (1987) Papillomaviruses and human disease. Springer, Berlin Heidelberg New York Tokyo

Laser Therapy of Anogenital Papillomavirus Infections – The View of the Dermatologist

M. Landthaler, U. Hohenleutner, D. Haina *, and O. Braun-Falco

Biophysical Considerations

Treatment of HPV papillomas is an important indication for the use of lasers in dermatology. There are two lasers especially suitable, the CO_2- and the Nd-YAG laser (Fig. 1).

The CO_2 laser emits in the infrared range at a wavelength of 10600 nm. The energy of this laser is strongly absorbed by all types of tissue, independent of their colour.

Since total absorption of the CO_2 energy occurs in about 0.1 mm of the skin, very high power densities can be attained in small tissue volumes. Therefore, a focused CO_2 laser beam of about 0.1 mm in diameter is suitable for incising and cutting tissue, and the laser is used as a so-called light scalpel. With a defocused beam of about 2 mm spot-size, tissue can be vaporized under visual control, and adjacent to the zone of vaporization there is a zone of 0.2–0.3 mm of coagulation, in which vessels up to a diameter of about 0.5 mm

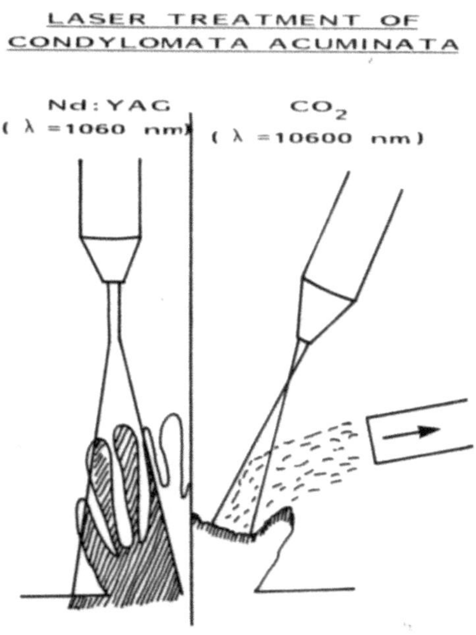

Fig. 1. Exophytic virus papillomas can be coagulated by means of the Nd:YAG laser (*left*) or vaporized by means of the CO_2 laser (*right*). Smoke formation during CO_2-laser application demands effective evacuation

* Gesellschaft für Strahlen- und Umweltforschung, München

Genital Papillomavirus Infections
Eds.: Gross/Jablonska/Pfister/Stegner

are sealed. The advantages of the CO_2 laser, therefore, are perfect control of tissue destruction, reduced risk of bleeding and reduced postoperative swelling [16, 25].

Since there is smoke formation during CO_2 laser surgery smell and smoke have to be eliminated by means of an evacuator. This is especially important because laser smoke presents a definite health hazard for patients and the medical staff. Walker and co-workers found viable bacterial spores in the laser smoke when the power density was less than 500 W/cm^2 [32]. Mullarky et al. [26] detected morphologically intact bacteria in the laser smoke, and Nezhat et al. [27] found particles of a size between 0.1 and 0.8 μm. In an experimental study, Baggish and co-workers demonstrated an interstitial pneumonia in rats inhaling the laser smoke [2]. Garden et al. found intact viral DNA in laser smoke [12], and in 1989 Zeller stated that laser smoke is mutagenic [33].

The Nd-YAG laser emits near infrared radiation at a wavelength of 1060 nm. Radiation at this wavelength is less absorbed and less scattered compared with visible light and therefore penetrates more deeply into the skin. The Nd-YAG laser is therefore suitable for coagulation of tissue. The depth of coagulation can be controlled by laser power, spot size, and exposure time. Cooling the skin surface with water during irradiation prevents superficial vaporization of tissue and increases depth of coagulation.

Clinical Applications

Virus papillomas can be coagulated by means of the Nd-YAG laser [3, 17] or vaporized by means of the CO_2 laser [25]. Since irradiation by both of the lasers is very painful, treatment has to be performed with the patient under local or general anaesthesia.

The literature abounds with reports concerning laser therapy of virus papillomas, and cure rates with the CO_2 laser of up to 98% are reported [1, 4, 5, 7–10, 15, 18–22, 28, 30] (Table 1). However, if one analyses these reports it becomes evident that repeated treatments were necessary in many cases, and often patients afflicted with uncomplicated, only short-lasting condylomata acuminata were included in these studies.

There have been only a few controlled studies comparing CO_2 laser surgery with other therapeutic modalities. Duus et al. determined the cure rates after CO_2 laser and electrosurgery and found no significant difference: 43% of the laser-treated patients were cured by a single treatment, compared with 36% of patients treated by electrosurgery [6]. Ferency compared laser treatment with topical application of 5-fluorouracil and found cure rates of 69% in the laser group and 90% in the 5-FU group [10]. Krebs obtained a 90% cure rate by combining laser treatment with topical 5-FU, compared with 65% by laser treatment alone [22].

In an open, noncontrolled study we have treated 102 patients with condylomata acuminata, 84 men (mean age 34 ± 13 years) and 18 women (mean age 31 ± 12 years). Sixty-six of these 102 patients (65%) were cured by CO_2

Table 1. Cure rates with laser therapy for HPV papillomas as reported in the literature

Reference	Cure rate (%)
Bellina [4]	97
Ferenczy [7]	82
Scott and Castro [30]	98
Ferenczy [9]	69
Ferenczy [8]	92.5
Ferenczy [10]	81
Grundsell et al. [15]	91
Kryger-Baggesen and Falk-Larsen [23]	75–87
Reid [28]	93
Rosemberg [29]	88
Krebs and Wheelock [22]	79
Baggish [1]	43
Duus BR et al. [6]	43
Carpinello VL et al. [5]	44

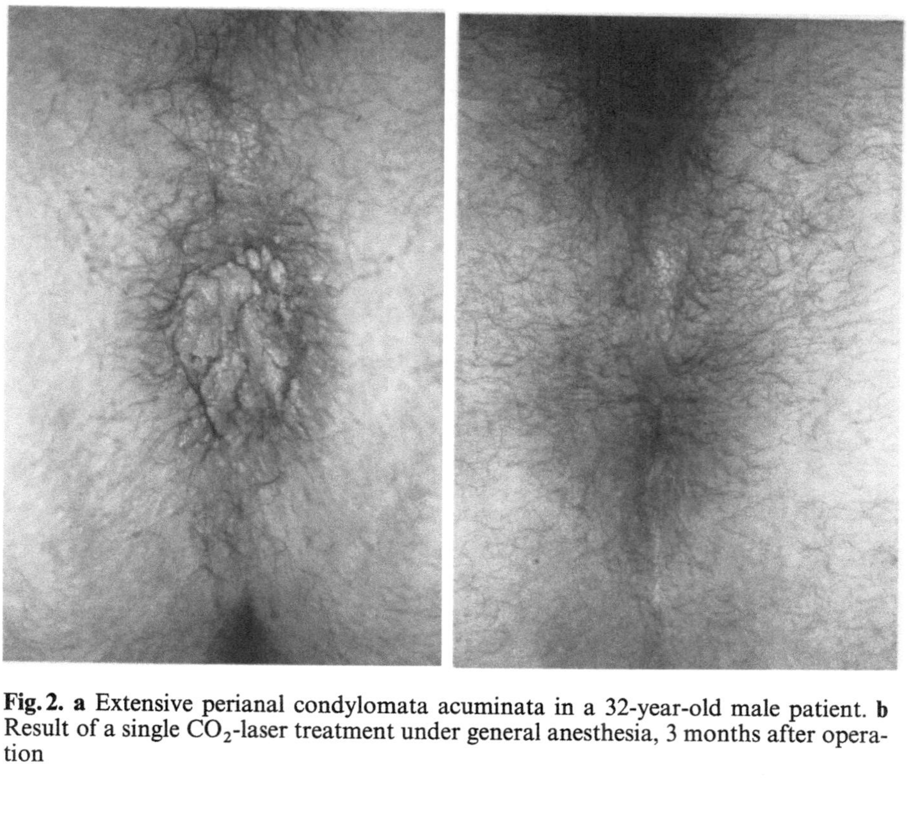

a b

Fig. 2. a Extensive perianal condylomata acuminata in a 32-year-old male patient. **b** Result of a single CO_2-laser treatment under general anesthesia, 3 months after operation

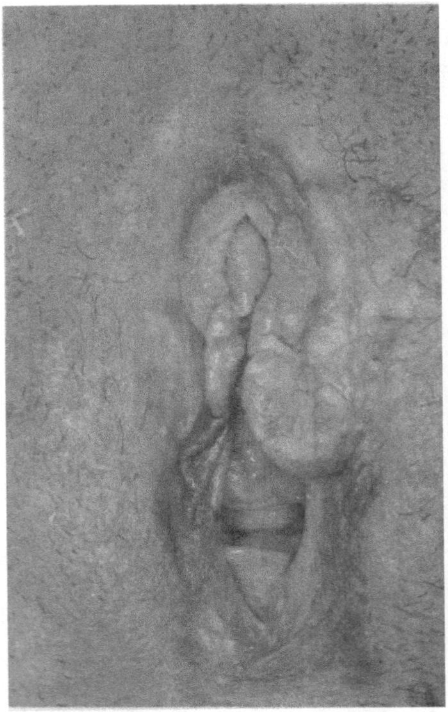

a

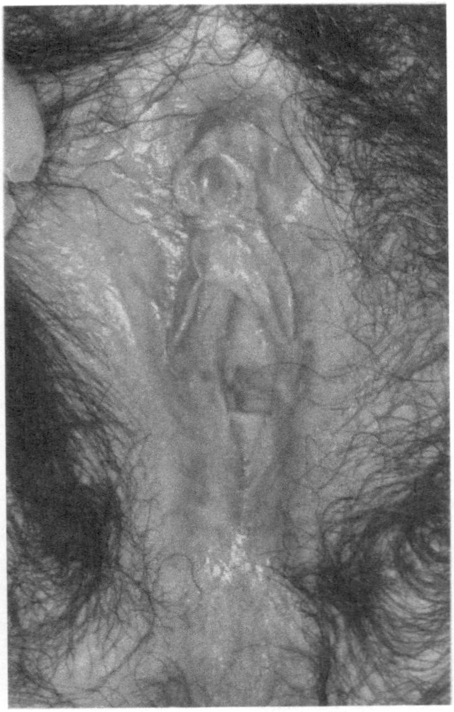

c

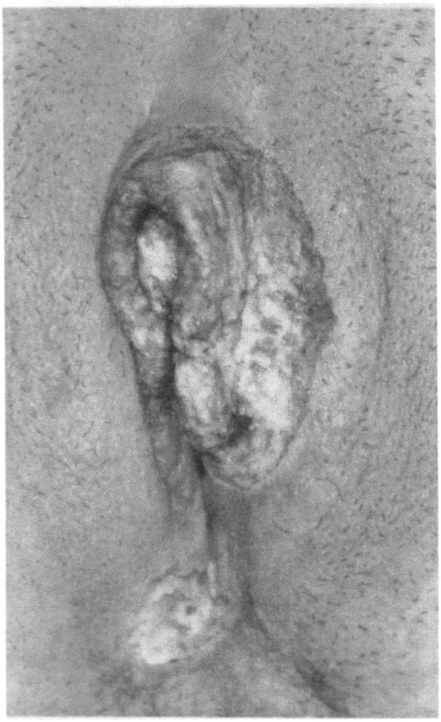

b

Fig. 3. a Plaquelike bowenoid papules with verrucous surface in a 34-year-old female patient. **b** Immediate postoperative view with completely bloodless surgical field. Vaporization was performed to superficial dermis. Healthy dermal tissue has a chagrin leatherlike appearance. **c** Perfect result without recurrence and mutilation of the vulva

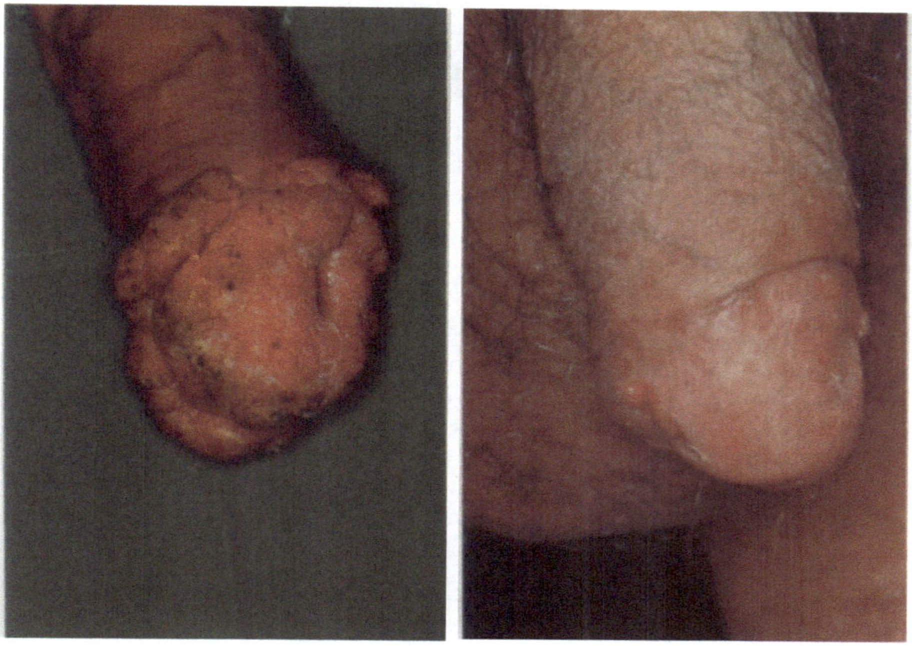

a b

Fig. 4. **a** Condylomata acuminata (Buschke-Löwenstein tumor) in a 41-year-old male patient. **b** Result of six treatments under local anesthesia with the Nd:YAG laser

laser treatment, but in several patients up to three treatments were performed (Fig. 2). In addition to the visible lesions, 5–10 mm of surrounding skin was treated, since it is known that HPV can be found in clinically normal skin [11].

In addition to the patients with condylomata acuminata (HPV 6/11), we have treated 24 patients (13 men, 11 women) with bowenoid papulosis (HPV 16/18), by means of the CO_2 (Fig. 3) and Nd-YAG laser respectively, and a cure rate of 87% was obtained. The advantage of laser therapy was that mutilation of the genital area could be avoided in each of the patients [24].

Lasers are also extremly helpful in patients afflicted with condylomata acuminata gigantea (Buschke-Löwenstein tumours), and mutilation of the genitalia can be avoided (Fig. 4) [3, 13, 18].

Combination of Laser Therapy and Interferons

Since controlled studies in the literature and our own experience revealed a rather high recurrence rate of condylomata acuminata with laser therapy, it was logical to look for an adjuvant postoperative treatment to improve the cure rate.

As interferons have immune-stimulating and antiviral effects we have started a prospective and randomized study of patients afflicted with extensive condylomata acuminata requiring CO_2 laser surgery under general anaesthesia. One group of patients was treated with the CO_2 laser while under general anaesthesia and was followed up every 4 weeks for up to 3 months. The other group of patients was also treated while under general anaesthesia with the CO_2 laser but received in addition low-dose [14] interferon-α (Intron A, Essex) according to the following regimen: 1×10^6 IE s.c. on postoperative days 1–6, 2 weeks' intermission, and again 1×10^6 IE s.c. on postoperative days 21–26.

Although the study is not yet completed, there are already two important findings:

1. The adjuvant postoperative therapy with interferon-α improved the cure rate: Seven of 14 patients (50%) in the laser group had a recurrence within 3 months, compared with two of 14 patients (13%) receiving postoperative interferon.
2. The recurrence rate in patients with extensive, partially long-standing condylomata acuminata who require surgery with general anaesthesia is rather high, even after careful CO_2 laser surgery.

The observation that the recurrence rate can be reduced by postoperative treatment with interferon is in agreement with data from the literature. Tiedemann and Ernst treated condylomata acuminata with electrosurgery alone and combined with postoperative intralesional interferon, and they reported a higher cure rate in interferon-treated patients [31]. Gross et al. successfully treated an immunosuppressed patient with condylomata acuminata gigantea with postoperative application of interferon-α ointment [13].

Conclusions

In our experience, lasers clearly represent progress in the treatment of HPV papillomas. Advantages of CO_2 and Nd-YAG lasers are reduced risk of bleeding, a clear surgical field, precise handling and reduced postoperative swelling. The visual control of tissue destruction is better with the CO_2 laser and wound healing is quicker after CO_2 laser surgery (2–4 weeks vs. 4–6 weeks). On the other hand, there is extensive smoke formation during CO_2 laser surgery, carrying risks for patients and medical staff. Bleeding never occurred during Nd-YAG laser surgery, and we therefore prefer the Nd-YAG laser for treatment of condylomata acuminata in HIV-positive patients.

Whether the recurrence rate following CO_2 and Nd-YAG laser therapy is really significantly lower compared with other surgical procedures has to be established in prospective randomized studies.

Since the first clinical studies concerning the combination of laser surgery and postoperative application of interferon are promising, further studies should be done to determine the optimum treatment regimen (type of interferon, route of application, dosage etc.).

References

1. Baggish MS (1985) Improved laser techniques for the elimination of genital and extragenital warts. Am J Obstet Gynecol 153:545–550
2. Baggish MS, Elbakry M (1987) The effects of laser smoke on the lungs of rats. Am J Obstet Gynecol 156:1260–1265
3. Bahmer FA, Tang DE, Payeur-Kirsch M (1984) Treatment of large condylomata of the penis with the neodymium-YAG laser. Acta Derm Venereol (Stockh) 64:361–363
4. Bellina JH (1983) The use of the carbon dioxide laser in the management of condyloma acuminatum with eight-year follow-up. Am J Obstet Gynecol 147(4):375–378
5. Carpinello VL, Zderic SA, Malloy TR, Sedlacek T (1987) Carbon dioxide laser therapy of subclinical condyloma found by magnified penile surface scanning. Urology 29:608–610
6. Duus BR, Philipsen T, Christensen JD, Lundvall F, Sondergard J (1985) Refractory condylomata acuminata: a controlled clinical trial of carbon dioxide laser versus conventional surgical treatment. Genitourin Med 61:59–61
7. Ferency A (1983) Using the laser to treat vulvar condylomata acuminata and intraepidermal neoplasia. Can Med Assoc J 128:135–137
8. Ferency A (1984) Laser therapy of genital condylomata acuminata. Obstet Gynecol 63:703–707
9. Ferency A (1984) Comparison of 5-fluorouracil and CO_2 laser for treatment of vaginal condylomata. Obstet Gynecol 64:773–778
10. Ferency A (1984) Treating genital condyloma during pregnancy with the carbon dioxide laser. Am Obstet Gynecol 148:9–12
11. Ferency A, Mitao M, Nagai M, Silverstein SJ, Crum CF (1985) Latent papillomavirus and recurring genital warts. N Engl J Med 313:784–788
12. Garden JM, O'Banion MK, Shelnitz LS, Pinski KS, Bakus AD, Reichmann ME, Sundberg JP (1988) Papillomavirus in the vapor of carbon dioxide laser-treated verrucae. JAMA 259:1199–1202
13. Gross G, Roussaki A, Pfister H (1988) Die postoperative Interferon-Hydrogel-Behandlung. Hautarzt 39:684–687
14. Gross G, Roussaki A, Broszka J (1988) Low doses of systematically administered recombinant interferon-gamma in the treatment of genital warts. J Invest Dermatol 90:242 (abstr)
15. Grundsell H, Larsson G, Bekassy Z (1984) Treatment of condylomata acuminata with the carbon dioxide laser. Br J Obstet Gynaecol 91:193–196
16. Haina D, Landthaler M, Waidelich W (1981) Physikalische und biologische Grundlagen der Laseranwendung in der Dermatologie. Hautarzt 32:397–401
17. Hofstetter AG, Keiditsch E, Schmiedt E, Frank F (1984) Der Neodym-YAG-Laser in der Urologie. Derzeitiger Stand der klinischen Erfahrungen. Fortschr Med 102:885–890
18. Hohenleutner U, Landthaler M, Braun-Falco O, Schmoeckel C, Haina D (1988) Condylomata acuminata gigantea (Buschke-Löwenstein-Tumor): Behandlung mit dem CO_2-Laser und Interferon. Dtsch Med Wochenschr 113:985–987
19. Kaplan J, Giler S (1984) CO_2 laser surgery. Springer, Berlin Heidelberg New York Tokyo
20. Knoll LD, Segura JW, Benson RC, Goellner JR (1988) Bowenoid papulosis of the penis: successful management with neodymium-YAG laser. J Urol 139:1307–1309
21. Krebs HB (1988) Combination of laser plus 5-fluorouracil for the treatment of extensive genital condylomata acuminata. Lasers Surg Med 8:135–138
22. Krebs HB, Wheelock JB (1985) The CO_2 laser for recurrent and therapy-resistant condylomata acuminata. J Reprod Med 30:489–492
23. Kryger-Baggesen N, Falck Larsen J, Hjortkjaer Pedersen P (1984) CO_2 laser treatment of condylomata acuminata. Acta Obstet Gynecol Scand 63:341–343

24. Landthaler M, Haina D, Brunner R, Waidelich W, Braun-Falco O (1986) Laser therapy of bowenoid papulosis and Bowen's disease. J Dermatol Surg Oncol 12:1253–1257
25. Landthaler M, Haina D, Hohenleutner U, Seipp W, Waidelich W, Braun-Falco O (1988) Der CO_2-Laser in der Dermatotherapie – Anwendung und Indikation. Hautarzt 39:198–204
26. Mullarky MB, Norris CW, Goldberg ID (1985) The efficacy of the CO_2 laser in the sterilization of skin seeded with bacteria: survival at the skin surface and the plume emissions. Laryngoscope 95:186–187
27. Nezhat C, Winer WK, Nezhat F, Nezhat C, Forrest D, Reeves WG (1987) Smoke from laser surgery: is there a health hazard? Lasers Surg Med 7:376–382
28. Reid R (1985) Superficial laser vulvectomy. I. The efficacy of extended superficial ablation for refractory and very extensive condylomas. Am J Obstet Gynecol 151:1047–1052
29. Rosemberg SK (1985) Carbon dioxide laser treatment of external genital lesions. Urology 25:555–558
30. Scott RS, Castro DJ (1984) Treatment of condyloma acuminate with carbon dioxide laser: a prospective study. Lasers Surg Med 4:157–162
31. Tiedemann KH, Ernst FM (1988) Kombinationstherapie von rezidivierenden Condylomata acuminata mit Elektrokaustik und Alpha-2-Interferon. Aktuel Dermatol 14:200–204
32. Walker NPJ, Matthews J, Newsom SWB (1986) Possible hazards from irradiation with the carbon dioxide laser. Lasers Surg Med 6:84–86
33. Zeller WJ (1989) Kanzerogenität von Rauch bei Laservaporisation und Elektrokoagulation. (Anfrage aus der Praxis). Hautarzt 40:115

Laser Surgery in HPV Infections of the Female Genital Tract

A. Singer

The CO_2 laser is invaluable in treating both benign and premalignant lesions of the genital tract that are associated with HPV. Benign lesions are exclusively condylomatous, while the premalignant are related to cervical, vaginal, vulval and perinatal intraepithelial neoplasia. As most lower genital tract premalignant lesions are in some way associated with HPV, they have been called pre-invasive HPV-associated genital lesions.

It has recently been suggested by some authors (e.g. Reid 1989; Richart 1990) that cervical lesions should be graded into two groups, i.e. high and low grade lesions. High grade lesions would comprise cervical intraepithelial neoplasia grade II and III (CIN II/III) and the low grade lesions would be composed of the CIN I and the subclinical papillomavirus infections (Richart 1990). High grade lesions seem to represent an homologous population of aneuploid lesions most of which are induced by oncogenic human papillomaviruses. It has been shown that 7% of all aneuploid cervical lesions will disappear while 81% remain unchanged and 12% progress to malignancy (Fu et al. 1981).

Low grade lesions are a heterogeneous mixture of genuine precursers and benign human papillomavirus (HPV infections). They are a population of lesions with an unknown malignant potential. A number of studies have shown that there is a progression rate from CIN I to CIN III dependent on the type of HPV virus found.

Diagnosis

Clinical Features of Pre-Invasive HPV-Associated Cervical Lesions

Minor Grade Lesions

Minor grade lesions manifest themselves colposcopically in a number of ways. The usual appearance is that of micropapillae and fine vascular loops analogous to a miniature condyloma. Another common feature is that of flat plaques of epithelium that stain white after the application of acetic acid (Fig. 1). Reid (1989) has devised a scheme for differentiating between minor grade and major grade colposcopic lesions dependent on such features as typography, surface configuration, angioarchitecture, iodine staining, and presence or absence of aceto-white epithelium.

Genital Papillomavirus Infections
Eds.: Gross/Jablonska/Pfister/Stegner

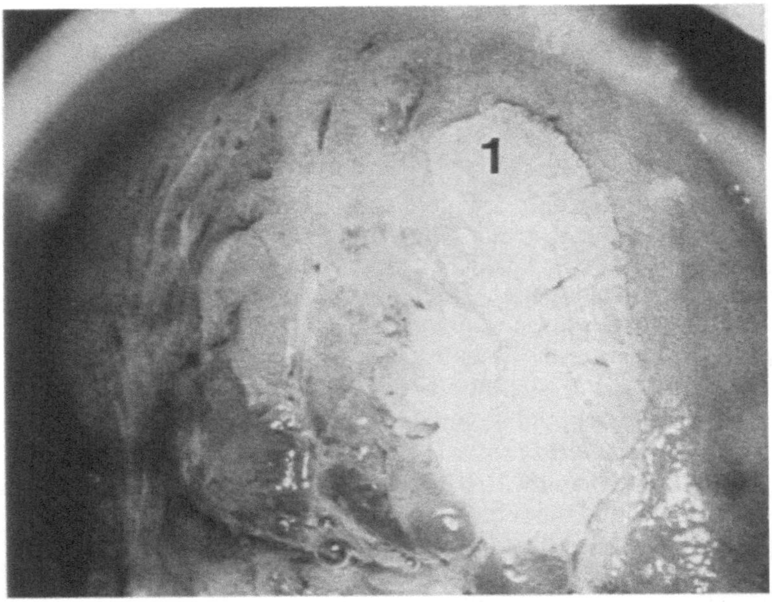

Fig. 1. Copper photograph of a minor grade cervical lesion. Close examination around *area 1* will reveal micropapilliferous surface pattern associated with a fine vascular loop analogous to a miniature condyloma

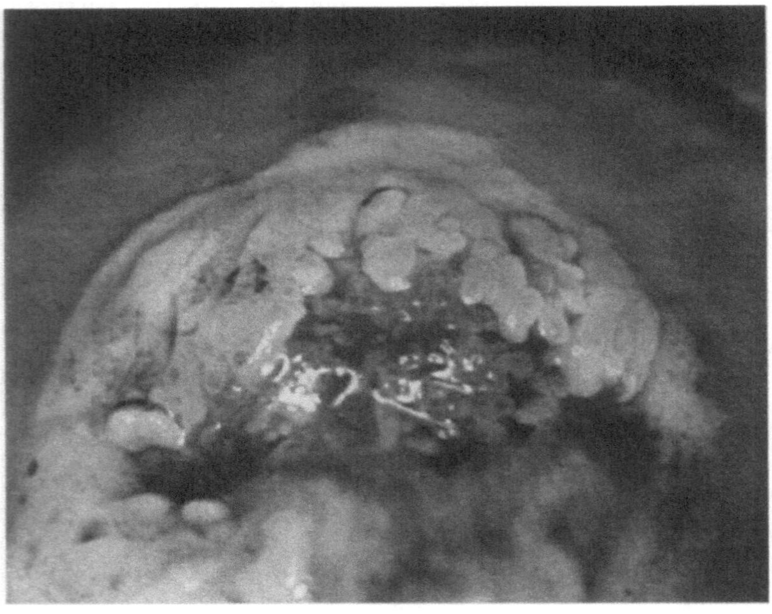

Fig. 2. A higher grade cervical lesion on the anterior lip with a typical dull oyster-white appearance

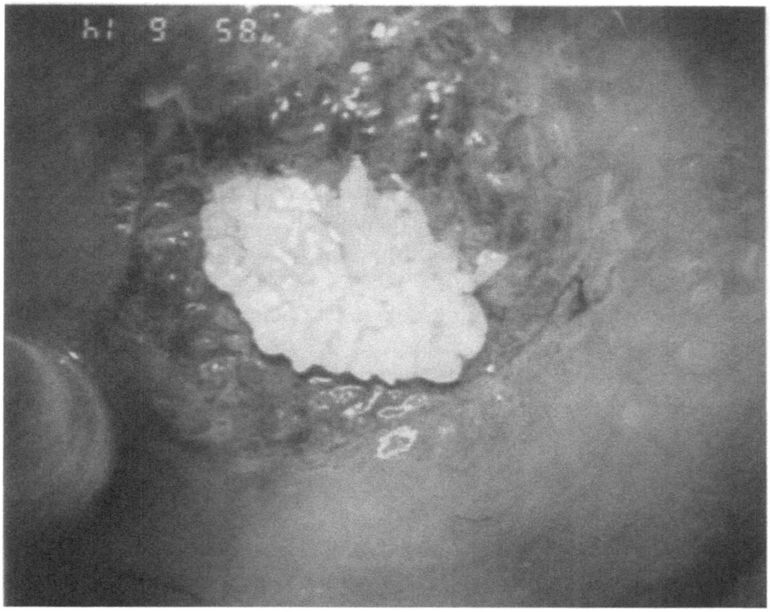

Fig. 3. Benign condylomatous lesions protruding from the endocervical canal

Higher Grade Lesions

These lesions are distinguished by a symmetrical shape, a straight peripheral margin, a flat contour, and a dull, oyster-white colour. Using the Reid colposcopic index (Reid 1989) scheme differentiation can be made in a broad manner between the minor and major grade lesions (Fig. 2). The target biopsy is still needed to produce differentiation.

Benign Condylomatous Lesions

These are uncommon on the cervical epithelium although recently they have been seen with greater frequency. External cervical papillomas are usually caused by HPV types 6 or 11. They may affect either the transformation zone or the original squamous epithelium and be associated with similar lesions in other parts of the lower genital tract. They are easily recognised by the naked eye and their colour ranges from pink to white (Fig. 3).

Clinical HPV infections are extremely common and in the past have been overlooked because the lesions were entirely invisible to the naked eye. Their presence is a potent cause of minor abnormalities in Papanicolaou smears. The natural history of these lesions has as yet not been determined. They most likely represent a low grade neoplastic lesion although this is yet to be confirmed.

Pre-invasive HPV-Associated Vaginal Disease

Benign Disease

The most common vaginal colposcopic sign of HPV infection is the finding of a miriad of tiny aceto-white macules or filaments highlighted against the flat pink vaginal mucosa (Reid 1989). Histologically, each of these white areas correspond to a "pin-point" of parakerototic epithelium capping a prominent intraepithelial papillary. More well-developed papillomavirus infections can be seen as obvious condylomata.

As with the cervix, low and high grade lesions are present (vaginal intraepithelial neoplasia, VAIN). Their features are usually a sharply defined white area located either as a single structure or, more commonly, as multifocal disease throughout the vagina. Iodine staining gives a most characteristic whitish appearance.

Pre-invasive HPV-Associated Vulval Disease

It is now well recognised that HPV DNA can regularly be detected in vulva intraepithelial neoplasia (VIN). Kaufman et al. (1988) found evidence of HPV DNA in 38 of 46 patients (83%) presenting with in situ carcinoma of the vulva (VIN III). Likewise, (Buscema et al. 1988), HPV type 16 DNA was recovered in 81% of VIN III but in only 12% of condyloma. The types 6 and 11 accounted for 70% of condyloma and 0% of VIN. In the invasive cancers type 16 was the predominant virus. Downey et al. (1988) have also shown that of nine cases of condylomatous carcinoma of the vulva a pre-existing condyloma was the antecedent history in 77% with a median of 9 months before the documentation of an invasive lesion. HPV DNA was demonstrated in 55% of these tumours which had a mixture of both HPV 6 and 16 DNA.

Benign Lesions

The condylomata accuminata in the vulval region range from small benign readily treated lesions to an extensive superficial growth that covers large areas of the skin with multiple local recurrence and almost relentless progression. Colposcopic examination of the vulva is sometimes necessary to delineate the extent of these lesions, especially the small ones (Fig. 4); 5% acetic acid is applied for at least 3 min and this permits the identification of the two distinct types of subclinical HPV-associated lesions. They may be single or fused papillary-like lesions or aceto-white epithelium (Fig. 5). These lesions can be extremely symptomatic causing intense pruritus. They may be difficult to distinguish from the physiologic vestibular papillary lesions which are found in this area.

The clinical appearance of VIN lesions can be variable. They may be clinically obvious because of their leukoplakia-like character but may well be diagnosed only after biopsy of seemingly benign condylomatous lesions (Fig. 6).

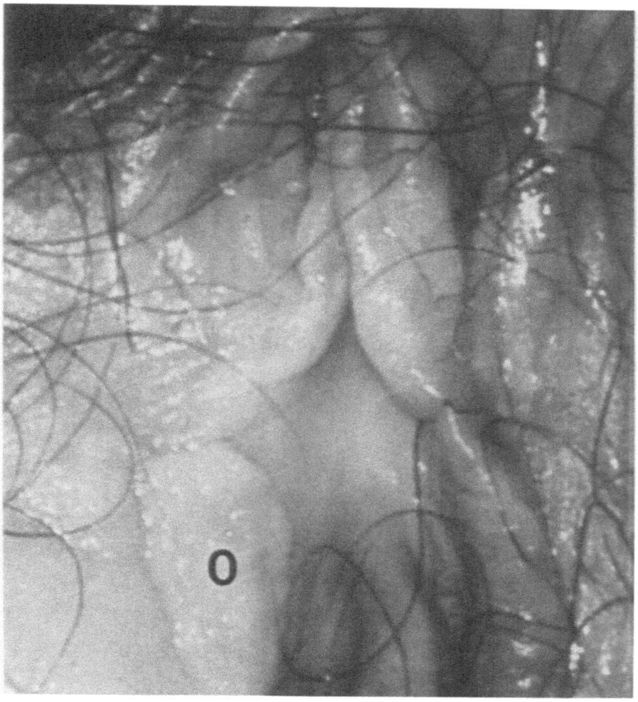

Fig. 4. Benign subclinical HPV vulvar lesion. Subclinical HPV lesion virtually invisible before the application of 5% acetic acid (*0*)

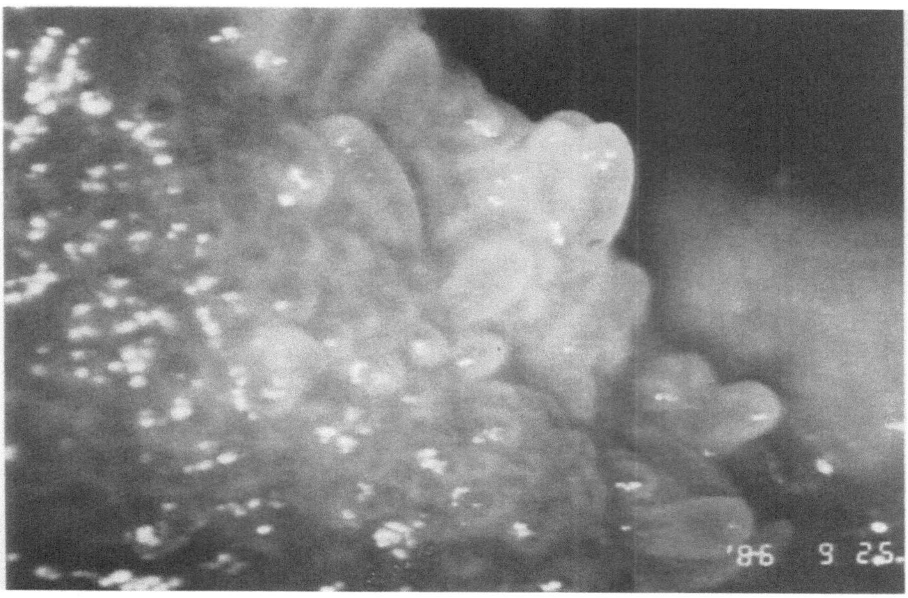

Fig. 5. Benign subclinical HPV vulvar lesion. Profuse papillary-like lesions photographed around the hymenal ring and inner labial surface. Usually associated with intense vulval pruritus

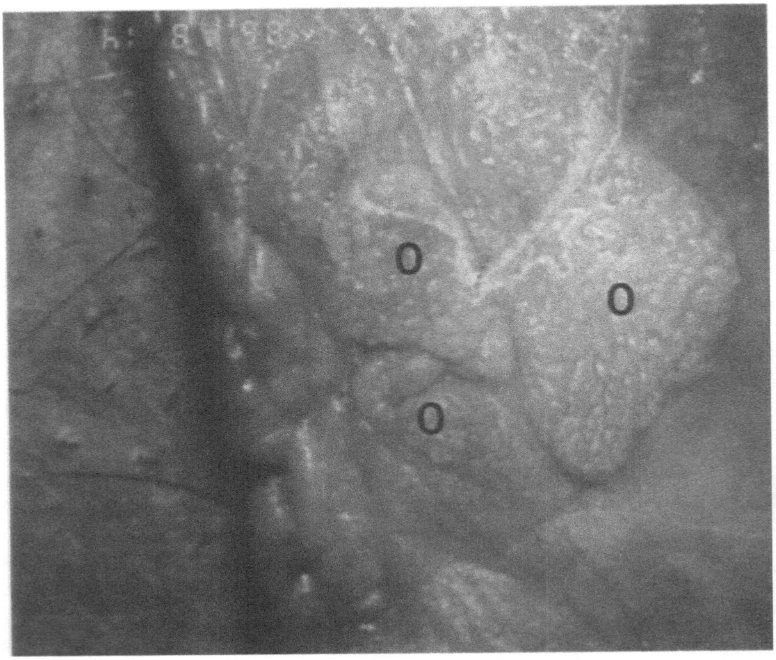

Fig. 6. Vulvar intra-papillary neoplasia. Seemingly benign condylomatous lesions which on biopsy showed CIN (lesion again was virtually invisible until the application of 5% acetic acid; lesion is at *0*)

With the help of colposcopy, the lesion can be circumscribed and a punch biopsy taken. Not only may they present as leukoplakia patches but also as condylomatous lesions within a leukoplakia patch. Condylomatous lesions that do not respond to medical treatment certainly raise suspicion of VIN and they also recur as the same type, be they ulcerative or of a locally destructive nature.

Treatment

Cervix: Pre-invasive HPV-Associated Lesions

Major Grade Lesions (CIN II/III)

It would seem prudent to consider all CIN II and III lesions as potentially malign and thus to submit them to treatment (Kinlen and Spriggs 1978; Buckley et al. 1982; McIndoe et al. 1984). However, the urgency with which these lesions should be treated is difficult to assess. Even more difficult is the pinpointing of the particular lesions that will progress to invasion. Recent evidence on the assessment of DNA ploidy levels by Fu et al. (1981) may be of benefit. These authors have suggested that up to 80% of CIN III lesions are aneuploid and therefore have a higher malignant potential than the seemingly less malign polyploid/diploid lesions. In a later discussion, Fujii et al. (1984) suggested that even the minor CIN I and II lesions that are aneuploid are more likely to progress to CIN III and invasion than are polyploid or diploid lesions. However, these techniques are usually unavailable to the clinician and so are impractical for use in day-to-day practice. It is therefore still recommended that CIN II–III lesions have to be seen and treated as soon as possible.

Minor Grade Lesions

Although significant regression does occur in these lesions, they do not constitute a high-risk group for future progression. There would seem to be two lines of management of the CIN lesions.

1. Anticipatory Management. Here the woman, and usually she would be young, could be followed at regular intervals by cytology and colposcopy examination. This would allow identification of the lesions that have regressed. If the smear or colposcopic impressions indicate that there is a development of a more serious lesion, then obviously treatment would be instituted.

2. Immediate Treatment of Lesions. Both forms of management have advantages and disadvantages. The main advantage of a more conservative approach is that the possibility of unnecessary treatment would be avoided because, theoretically, approximately 30% of these lesions will regress. However, the chances of eventual progression are significant. In a recent study from this unit (Campion et al. 1986) it was shown that 26% of a group of mild dyskaryotic lesions progressed to a severe form over a 3-year observation period. HPV DNA hybridisation revealed that 90%, of the lesions contained HPV 16 DNA compared with only 15% in the nonprogressive lesions.

Against these obvious advantages are a number of pertinent disadvantages. There is a certain amount of anxiety engendered in the regular follow-up and cytological assessment that is necessary with the former method of management. This anxiety can far outweigh any disadvantages of immediate treatment. A regular review of these young women, sometimes at 3- to 6-month intervals, places an enormous strain on not only the patient but also the overworked gynaecological and cytological departments. This department has recently shown (Campion et al. 1987) that 80% of young women are emotionally traumatised by the finding of an abnormal smear. The success and effectiveness of modern treatment, estimated to be in the region of 95% after one treatment, has been well described in a number of centres and it would seem to us to be a major argument for the immediate treatment concept.

The increase in the prevalence of these minor lesions which are associated with HPV is occurring worldwide and will present major logistic problems to gynaecological departments. For example, at many centres there has been a threefold increase in presentation with these lesions over a 5-year period. This has occurred especially in younger women (Elliott et al. 1989).

Why Use Conservative Methods Such as the CO_2 Laser?

There is no denying that the more radical methods of removing cervical premalignancy, that is, in the form of cone biopsy or hysterectomy, are safe techniques. However, this is if safety is defined as conferring a low risk for the development of subsequent malignancy (Bevan et al. 1981). But can the same safety, probably in the region of 90% (Burghardt and Holzer 1980), be achieved by a less radical or traumatic procedure than cone biopsy? We believe that at present the answer is an overwhelming yes. Increasing use of colposcopy with its magnified illumination allows the potentially malign lesions of CIN to be assessed. The gynaecologist can make a definite decision based on colposcopy and directed colposcopic biopsy as to wheter the particular lesion is suitable for the more conservative or locally destructive methods of treatment.

Local treatment is appropriate if the lesion is seen in its entirety and when its upper extension, which usually lies just within the endocervix, can be clearly identified. If this is not the case, or the colposcopist's impression is that there may be an early invasive lesion, or when adequate follow-up cannot be guaranteed because of the patient's noncompliance, then the gynaecologist should not embark on the locally destructive techniques which will be described below. In these cases, a cone biopsy is mandatory. If these criteria are adhered to, then we will not see development of the situation, as has happened in some other countries, of malignant disease arising after locally destructive techniques, particularly after cryotherapy (Townsend et al. 1981). In this clinic, where new patients are being seen at the rate of 1800 annually the cone biopsy rate is about 24% for CIN III lesions. This value is higher than that which many "expert colposcopy clinics practice" (DiSaia et al. 1974; Creasman et al. 1981), but we believe it is a realistic one. This may reflect our unwillingness, and that of other gynaecologists in this country, to place reliance on the findings from endocervical curettage, believing that if the upper extent of a lesion cannot be defined, then a cone biopsy should be performed. This is contrary to American teaching; an endocervical curettage is an integral part of the colposcopic examination (DiSaia and Creasman 1984a).

With the upsurge in the popularity of colposcopy (McGregor 1984) there is a definite trend towards more conservative therapy. After a decade of usage of the various techniques (Chanen and Rome 1983; Duncan 1983; Creasman et al. 1984), there seems to be a consensus that there is a high safety quotient with locally destructive techniques. Success rates of between 80% and 95% are regularly quoted for all these techniques. It would therefore appear that conservative techniques should take precedence over radical methods although the latter certainly have a place as diagnostic and therapeutic methods available to the gynaecologist when treating premalignant disease of the cervix.

Conservative Methods of Management

Selection of Patient; Impact of Colposcopy and Colposcopically Directed Biopsy

The findings of an abnormal cervical smear should prompt an immediate response from the gynaecologist. A mildly dyskaryotic smear (sometimes listed as CIN I) may indicate a precancerous state or be the result of a vaginal infection. It is thus prudent to treat any associated infection, but, if the abnormality is still present on a second smear the patient should be referred for colposcopy. A more severely abnormal cytological smear, usually described as showing moderate to severe dyskaryosis (CIN II/III) or containing malignant cells, warrants immediate referral for colposcopy.

Colposcopy involves the magnified and illuminated viewing of the cervix with a direct view being obtained of the lesion. With an associated colposcopically directed biopsy, the gynaecologist is in a position to advise and undertake the optimum form of therapy. Colposcopically directed biopsy is accurate in nearly 98% of cases (Kirkup and Singer 1980) and this in itself is a significant advance.

Colposcopic examination must be directed towards defining those lesions that are most suitable for either conservative treatment or more radical treatment involving cone biopsy or hysterectomy. Certain criteria must be adhered to – and they have been listed above – if the conservative methods of local destruction are to be employed.

Once the pathology has been assessed, then treatment can be organised. As discussed previously, it is our belief that all CIN lesions should be treated. This being so, it is then only a matter of choosing the technique that is available and most familiar to the gynaecologist.

Techniques for Local Destruction

The conservative treatment methods all depend on the eradication of the localised lesions by physical destruction. It is imperative that any such method is capable of destroying the CIN lesions contained within the cervical glands or, more correctly, crypts. Anderson and Hartley (1980) have shown that such tissue existed to an average depth of 3.80 mm (range 1.242–5.22 mm) in 350 conisation specimens that were studied. Therefore, to be totally destructive these methods must destroy a depth of at least 6 mm, especially if the CIN is situated in crypts. The locally destructive techniques are four in number:

1. Cryotherapy, or freeing the area by application of probes, usually with no anaesthesia (Charles and Savage 1980; Coney et al. 1983; Creasman et al. 1984
2. Electrodiathermy, usually under general anaesthesia; or loop diathermy excision (Chanen and Rome 1983)
3. Cold coagulation, usually with no or only local anaesthesia (Duncan 1983)
4. Carbon dioxide (CO_2) laser destruction, usually with no or only local anaesthesia (Baggish 1986). However, the CO_2 laser is the only method to be discussed. It allows versatility and accuracy to be employed in treating these lesions.

Carbon Dioxide Laser Therapy Method

The use of CO_2 laser therapy is now widely accepted as one of the most effective treatments of CIN (Singer and Walker 1982). The word "laser" is an acronym derived from the words "light amplification by stimulated emission of radiation." The laser converts energy such as heart, light, or electricity into radiant energy at a specific wave length determined by the type of laser. For example, the carbon dioxide laser, the one most widely used in gynaecology, produces energy at a wave length of 10.6 *m which is the infrared portion of the spectrum where it is invisible to the human eye. This energy, by a system of mirrors and lenses, can be focused to a specific spot 1.5 to 2 mm in diameter and at its focal point it releases an enormous amount of energy. Any tissue at the focal point of the laser is vaporised at the speed of light. The laser itself is attached to a colposcope and at all times the area to be destroyed is under the direct vision of the person performing the laser surgery. Manipulation of the beam is extremely simple.

This technique offers considerable advantages despite the high cost of the initial capital outlay for equipment. The healing of tissues subjected to laser evaporation is unusually rapid and far superior to that produced by cryosurgery or electrocoagulation diathermy, in which marked necrosis occurs in healthy adjacent tissue. The laser produces no destruction of normal surrounding tissue and a remodelled cervix is generated in which the endocervical canal is readily accessible and any recurrence can easily be seen colpo-

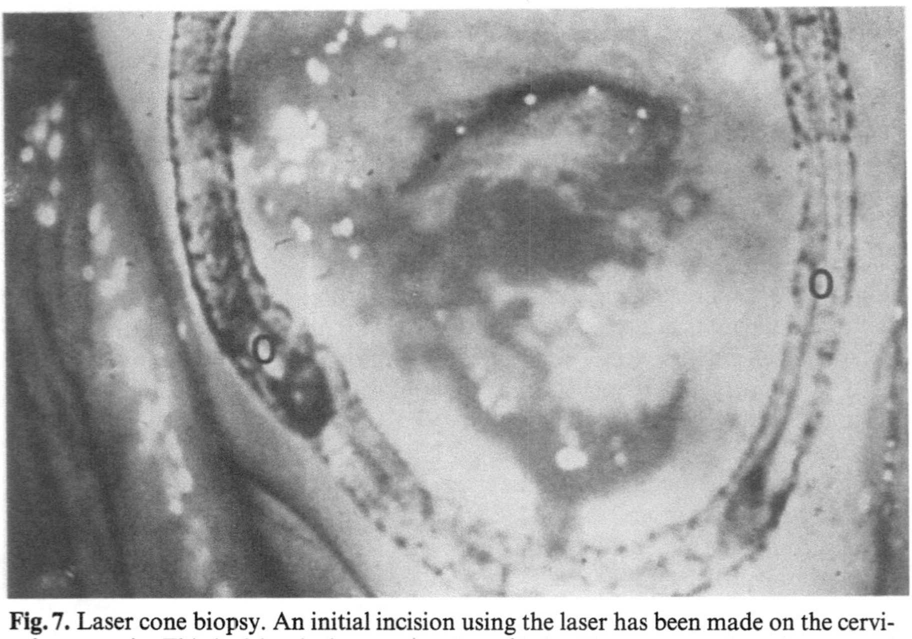

Fig. 7. Laser cone biopsy. An initial incision using the laser has been made on the cervical ectocervix. This incision is deepened and a cylinder-like segment of cervix removed for examination. (Incision line is at *0,0*)

scopically. There appear to be no complications associated with subsequent pregnancy in treated patients (Baggish 1986). Patient acceptance of the procedure performed under local anaesthesia is high, although 10% experience mild to moderate discomfort during the 5 to 10-min procedure. Blood loss during the procedure is minimal since haemostasis of smaller vessels is instantaneous. However, minimal bleeding in the postoperative period, occurring up to 2 weeks after the procedure, complicates about 10% of cases. Should this require treatment, application of silver nitrate to the bleeding site rapidly produces coagulation.

Cone biopsy can also be performed with the CO_2 laser as an outpatient procedure (Partington et al. 1987; McIndoe et al. 1989) (Fig. 7). Local anaesthetic and vasoconstrictive agents are injected into the cervix at the 2, 4, 8, and 10 o'clock positions. A claw instrument is then used to retract the incised tissue medially and a cone or cylinder-type block of tissue is removed.

A high primary cure rate is obtained, in the region of 90% after one application. The follow-up time is still short (10 years) but it is difficult to believe that a significant recurrence rate will develop, especially considering the depth and completeness of destruction associated with laser evaporation. The carbon dioxide laser has the significant advantage of versatility, being successfully used for associated premalignant lesions in the vagina and vulva.

Cervix: Benign Lesions

The CO_2 laser is the ideal method to use for destruction of these benign lesions which are either clinical condyloma accuminata or specific HPV lesions. Those that are to be destroyed can be outlined colposcopically and destruction to a depth of 7 mm undertaken. The surrounding normal-appearing cervical tissue may well contain HPV DNA in a latent form. Certainly when treating such lesions in the vagina and vulva it is necessary to ablate normal tissue as it has been shown (Ferenczy et al. 1985) that HPV DNA exists in these sites.

Vagina: Pre-Invasive HPV-Associated Lesions

Vaginal Intraepithelial Neoplasia

As with similar cervical lesions, VIN lesions are asymptomatic and only detectable cytologically. Many of the women will already have had CIN and therefore a hysterectomy but a number will still have their uterus. If colposcopy is performed before hysterectomy then many of these lesions will have been noted and dealt with. However, a number will occur up to many years after the hysterectomy; they present problems in management because of their association with the scar tissue of the vaginal vault.

Careful colposcopic examination is essential and the exact dimensions and limits of the lesions must be outlined. This is particularly so in those lesions extending into the vaginal vault or particularly into the angles of the vagina.

The use of a long-handled iris hook or a three-pronged claw, as used for cone biopsy is essential in defining the lesions in the site. Biopsy must be undertaken before any laser treatment is proposed. Many postmenopausal patients, especially those who have had previous invasive disease of the cervix, will benefit from the intravaginal application of oestrogen cream for at least 3 weeks before examination. The entire examination must be evaluated as the multifocal nature of the disease has been described above.

Laser Management

The technique of laser therapy for VAIN is different from that for CIN (Stafl et al. 1977). Generally much lower power densities are utilised and local analgesics are injected into the tissue to serve as both a protective barrier from vaporising too deeply as well as providing pain relief. Many small lesions, especially those that are localised, can be treated in the outpatient situation. However, large areas and those particularly in the vault must be treated with the patient having a general anaesthetic.

Two types of laser therapy are used on VAIN lesions. These areas can either be removed by vaporisation (Fig. 8) or excised using the laser to perform a type of "skinning" surface excision (Baggish et al. 1989) (Fig. 9).

Vaporisation is carried out usually employing a spot of between 1.5 to 2 mm. The larger spot reduces the chance of deep penetration and provides better haemostasis. Initially a 2 mm border of normal tissue around the lesion is vaporised down to a depth of approximately 2 mm and during the vaporisation of the vagina the contrast between the mucosa and submucosal tissue is easy to recognise. In nearly all cases a 1 mm depth will remove all of the mucosa. However, a 2 mm depth is recommended to complete the vaporising of the abnormal tissue. Glands in the vagina are seldomly found and they only penetrate to a few millimeters. Haemostasis can be achieved by defocusing the beam or the application of an astringent such as a solution of feric subsulphate (Monsel's Solution).

The second technique involves those lesions that are large in area or where there is some doubt, even though a punch biopsy is taken, of the exact pathology of the area. If the facility is available then a superpulse mode is used with a wattage of 10 and a spot size of about 1.5 mm. The lesion is outlined with a 2 mm area of normal tissue surrounding. A claw or hook is then used to retract the edge of the incised area inwards after which the lesion can be undercut and removed (Fig. 9). The area is then left open and healing occurs very quickly. Excised lesions towards the vault of the vagina may produce adhesions and so the area must be checked at 10-day intervals for at least 6 weeks.

Lesions that extend into the angles of the vagina and where the complete margins cannot be seen will need to be treated by other techniques such as radiotherapy (Woodman et al. 1988). The risk of an early invasive cancer within the "hidden area" in the vaginal angles of those women who have had a hysterectomy is not uncommon and only those areas that can be completely visualised to their full extent, should be removed.

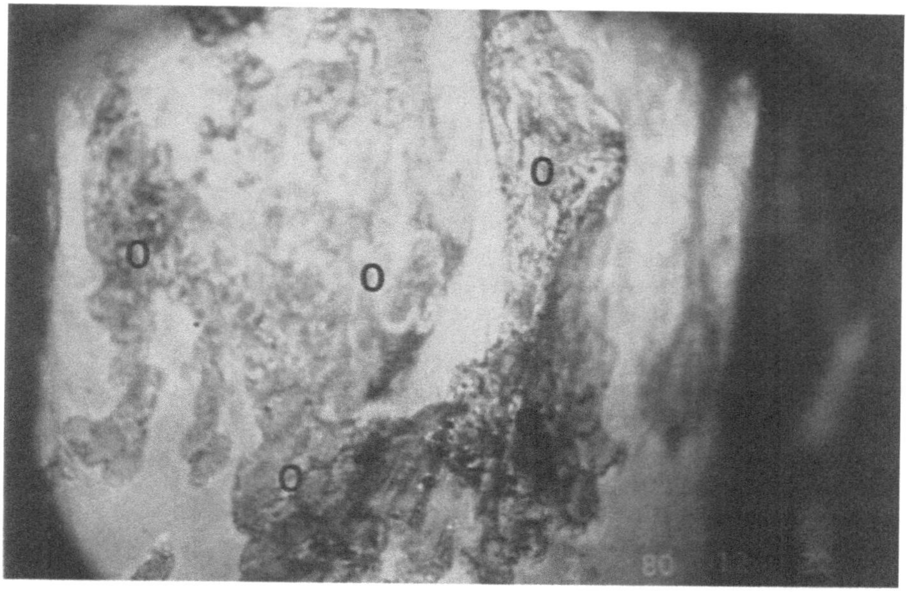

Fig. 8. Vaporization of VAIN vaginal vault. Multifocal VAIN disease has been vaporized to a depth of 1.5 mm by laser. These areas are marked *0* (an area of normal tissue has been treated at the same time so as to prevent recurrence)

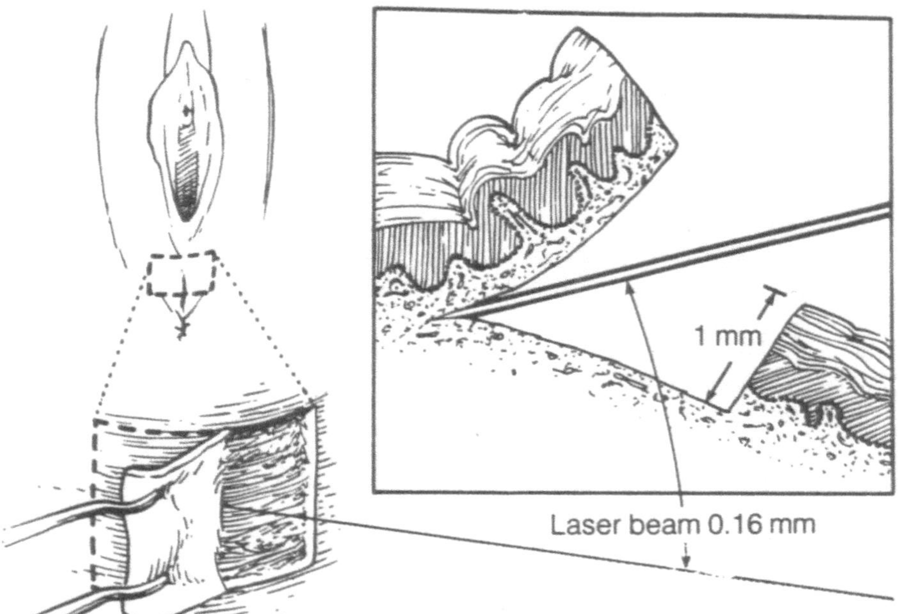

Fig. 9. Surface excision biopsy using the "skinning technique". This technique can be used not only for VAIN but also for VIN (as illustrated). A small laser beam was used and tissue is undermined at a depth 1 mm from the surface. (From Baggish et al. 1989)

Vagina: Benign Disease

The method of laser vaporisation for condyloma accuminata is different from that described above for premalignant lesions. The depth of vaporisation is less and it is important to treat several millimeters of surrounding normal tissue. The central area of proliferative tissue is initially vaporised to a depth of 1 mm and then shallow brushing of less than 1 mm in depth to a distance of 3 mm to 5 mm surrounding the lesions follows. When there are many vaginal lesions and a high recurrence is common, adjuvant treatment with 5 fluorouracil cream may also be given in these instances.

Vulva: Pre-Invasive HPV-Associated Lesions

Vulval Intraepithelial Neoplasia

There is no agreement as to the premalignant potential of VIN lesions. Indeed Friedrich (1984) has said that "the disease is as likely to regress as progress". He defines those patient who may be followed conservatively as those who are pregnant, or temporarily immuno-suppressed or relatively asymptomatic, while women who need treatment are those with detectable symptoms, usually of pruritus, and who are elderly and who lack the facility for follow-up, or in those women in whom the lesion fails to regress. Recent evidence suggesting a malignant potential to these lesions has been the finding of a malignant type of karyotypic pattern (aneuploid) in some of them (Friedrich 1984), with in our own studies the isolation of HPV 16 DNA in the majority of them. This latter human papillomavirus type is found in about 90% of invasive lesions of the cervix and vulva.

In many cases there is another neoplasm present in the vagina or cervix. In some series the figure is as high as 60%. Schlaerth et al. (1984) described 20 women with extension of intraepithelial disease of the vulva to the posterior perineum and anal areas. These women ranged in age from the very young to the very old with an average age of 34.7 years. In 17 of them, there were associated cervical intraepithelial lesions; in two there was the presence of invasive cancer of the vulva.

Which Method of Treatment: Conservative or Radical?

Whatever method is used, be it radical in the form of surgery or conservative with the usage of laser evaporation, there is a recurrence rate dependent on the freedom of the excised or destroyed margins of the lesions (Baggish 1985). In the case of surgery it is 50% if the margins are involved and 10% if they are free (Friedrich 1984). Comparison with recurrence rates for conservative methods, namely laser, is not possible because standardisation between patients is difficult. Data from two large studies suggest success rates in excess of 90%. Choosing surgery or laser treatment depends on many factors which include:

1. The skill of the operator.
2. The position and extent of the lesion. Perianal and anal disease, with or without extension into the natal cleft, seems to be more suitable for surgery. Similarly, disease of the labia minora or clitoris may be more conducive to laser therapy.
3. The age of the patient, with younger women more likely to be concerned about the potentially mutilating effect of surgery with the resulting bias towards laser therapy. However, recent reports by DiSaia and Creasman (1984b) and Schlaerth et al. (1984) describe limited scarring and constriction of the vulva after surgery, especially where a split thickness skin graft has been employed.

It seems as though surgery in the form of skinning vulvectomy may have a significant place in treatment, but it is still a major operative procedure requiring general anaesthesia and a prolonged convalescence. Local surgical excision of the common solitary lesion has proved effective but it is difficult to use in patients with multiple lesions.

Conservative Methods: The CO_2-Laser. Under colposcopic guidance, it is possible to control the extent of tissue destruction with great precision and the depth of penetration by visually monitoring the change from epithelium to dermis (Baggish 1985). The minimal heat damage and rapid postoperative healing phase produces a clean wound that is less likely to be affected or to be associated with significant morbidity (Fig. 10).

The Technique. The colposcope allows for the definition and extent of the lesion to be ascertained and thus enables a decision to be made as to the use of either local or general anaesthesia. Many authors employ local anaesthesia and undertake the procedure in the outpatient department or office when discrete local lesions, measuring up to about 4 cm in diameter are found. However, coalescent or any extensive progress of the lesion requires general anaesthesia.

The power setting of the CO_2 laser ranges from 10 to 31 W with a spot size of between 1.5 and 1.8 mm and power density settings of 300 W/cm^2 to 1100 W/cm^2. The author prefers lower power densities in the region of 300 to 400 W/cm^2 in view of the delicate nature of the vulvar and perianal tissues. Interrupted or continuous beam therapy is employed under a colposcopic magnification of six times. The stratified squamous epithelium on the vulva is less than 1 mm in thickness and the deepest rete peg rarely plunges more than 1 mm into the underlying dermis. Skin appendages are not commonly involved in the neoplastic process and they rarely exceed 2 to 3 mm into the stroma (Baggish et al. 1989; Shatz et al. 1989). However, treatment to a depth of 3 mm should eliminate any intraepithelial disease and at the same time will not result in scar formation (Baggish et al. 1989; Shatz et al. 1989).

a
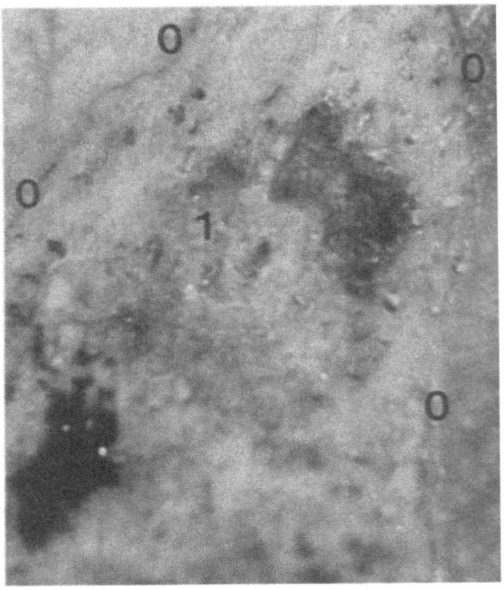

b
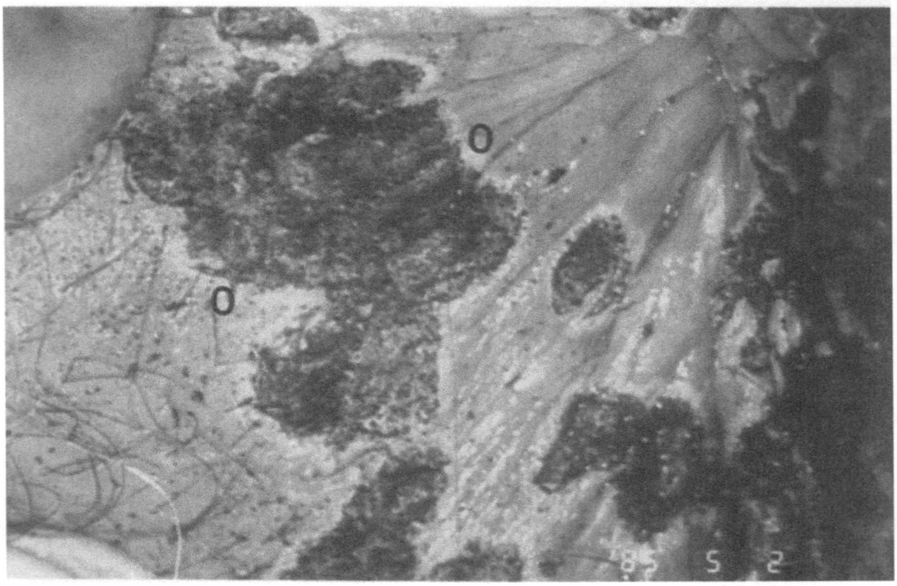

Fig. 10 a, b. Laser vaporization technique showing extent of tissue destruction. **a** An area of vulval skin has been treated with the laser. The surface epithelium has been removed and the dermal papillae can be seen within the treated lesion. There is minimal damage to the surrounding tissue. This can be achieved by using a high powered density with rapid movement of the beam across the tissue. The edge of destruction is seen at *0* and *0*. **b** Similar laser destruction has occurred in relation to perianal warts. However, a low power density with slow beam movement over the tissue has been used with resultant heat damage to the surrounding tissue. Blanching of the tissue at the edge of the wound can be seen at *0* and *0*

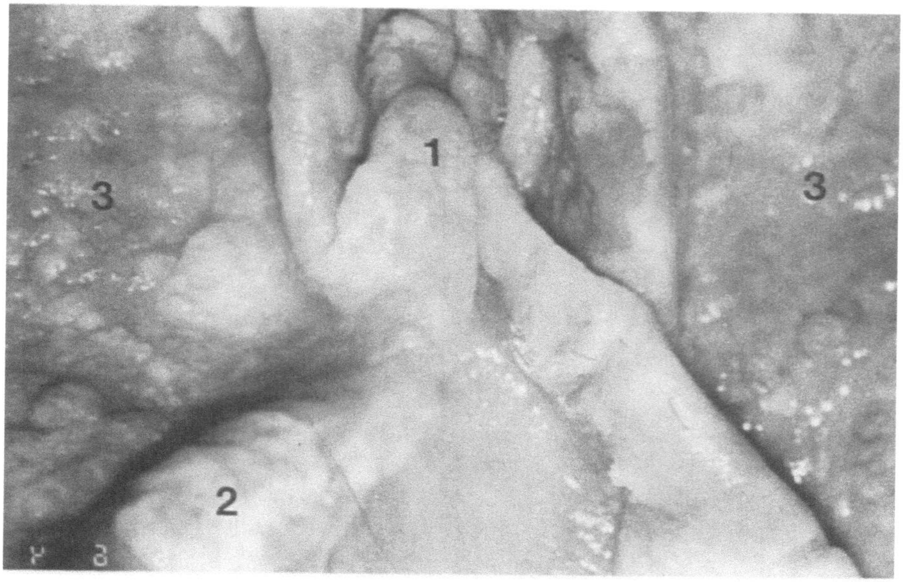

Fig. 11. Premalignant vulval lesions involving hair-bearing areas. An intraepithelial neoplastic lesion is seen involving the clitoral area: *1*, with extension onto the labia majora; *2*, labia majora, and *3* inner thigh. This presents a problem in treatment

The lesion is initially outlined by the laser beam and then "painted" by a series of vertical and horizontal strips until the entire epithelium is removed. The endpoint for removal is ascertained under visual control and occurs when the underlying dermis, which has a tanned, shiny appearance, comes into view. Reid (1984; Reid et al. 1984) advocates a technique in which the initial thermal damage produces minimal epithelial destruction, allowing the overlying skin and vaporised tissue to be gently wiped off. This reveals a characteristic vascular dermal papillary layer. These papillae are then carefully ablated till the grey straight fibres of the reticular dermis come into view.

Lesions involving the hair-bearing areas (Fig. 11) or sebaceous gland regions of the vulva present a problem in management. Once cannot be sure that the penetration of the intraepithelial disease into these structures has been adequately cleared by the superficial nature of the laser treatment (Baggish et al. 1989; Schatz et al. 1989). However, to date, the development of malignancy after CO_2-laser treatment of vulval lesions has not been reported but it is the author's view that extensive disease in these lateral areas of the vulva should be assessed very carefully prior to the undertaking of laser treatment.

However, there are a number of authors who advocate the surgical excision of any VIN present in hair-bearing areas. This can either be performed with the knife or using the laser (Fig. 12). After excision the area would need to be sutured (Fig. 13).

The postoperative regime, as advocated by Baggish and Dorsey (1981) is most important and comprises the use of four times daily soaking of the vulva

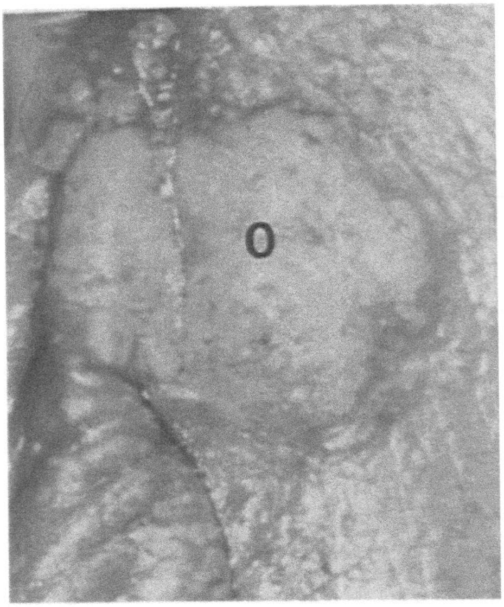

Fig. 12. A small focal lesion is seen at *0*. Its lateral path is involving hair-bearing areas

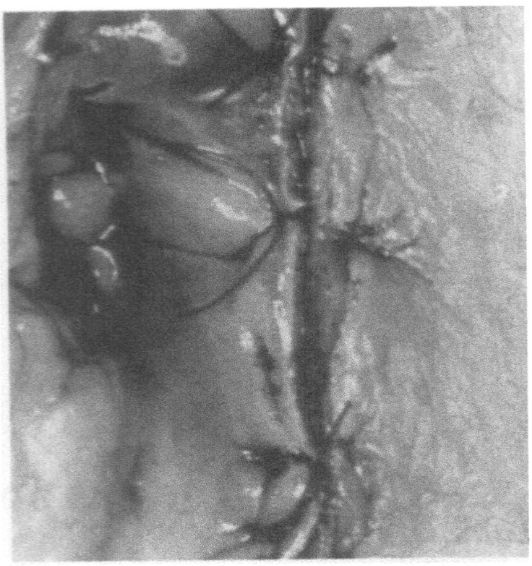

Fig. 13. Knife excision of VIN involving hair-bearing areas. The VIN area pictured in Fig. 12 has now been removed by an excision biopsy. The "skinning technique" (Fig. 9) could also have been employed in this case

in warm seawater baths followed by electric fan drying of the area. This regime is gradually modified as the healing progress with local anaesthetic or testosterone propionate 25% in petroleum base for pain or pruritic symptoms respectively. With this regime, postoperative infection is rare.

Results. A high rate of cure has been reported in some series. Townsend et al. (1982) treated 33 patients by laser and were successful in 31 (94%). In 14 cases two or more treatments were needed while in 2, five applications of laser ablation were applied.

Baggish and Dorsey (1981) reported 32 of 35 intraepithelial lesions of the vulva to be cured but 26 of their patients required three or more treatments and 2 needed six. Pain was a significant complication in 11 of them, occurring at about 4 days, while bleeding continued in six up to 7 days.

The laser area is covered by mature epithelium by 3 weeks and complete healing is present at 6 weeks. In the author's experience, the laser area in the fourchette may take up to 8 to 10 weeks to be completely healed. Moderate pain and discomfort sometimes lasts for this length of time.

Vulva: Benign Lesions

The CO_2 laser eliminates condylomatous lesions to a depth of 1 mm or less. As the virus replicates in the basal layers of the epithelium it is not necessary to vaporise deeper than this point. The heat created by the laser beam increases the temperature within the cell to over 100° C. Higher temperatures are also produced by the subsequent carbonisation and this further results in the destruction of the virus. If superficial destruction like this is performed then scarring is not produced. Even extensive laser vaporisation will not lead to scar formation on the vulva.

Not only are the condylomatous lesions destroyed with the laser but also the microwart or subclinical lesions. The technique necessitates the use of the colposcope for illuminated and magnified vision.

Condylomata accuminata especially in pregnancy are highly vascular and so the CO_2 laser must be used on low power density so as the seal vascular channels as vaporisation occurs. Since treatment is superficial only small vessels, e.g. less than 0.5 mm in diameter, are encountered and are easily sealed off.

The actual technique involves the vaporisation of the warts from outside inwards and the reduction of the lesions to the level of the surrounding normal skin surface (Fig. 14). There is no advantage in treating deeply since the virus does not penetrate into these deeper tissues, as explained above. A number of authors use the brush and skin technique which allows the epithelium to be destroyed to a distance surrounding the actual condylomatous lesion. This technique is particularly applicable on the mucous membrane surfaces of the labia minora. The hymenal ring and vestibule are favourite sites for proliferating condyloma and may not be apparent until the application of acetic acid. Also the urethra may contain some warts and is exposed by opening up the external meatus with either a moist cotton-tip applicator or some small Dejardin's forceps.

Excision of large warts or clusters of warts, especially if they have a pedicle, is frequently undertaken. A small spot size in the region of 0.5 mm and a higher power density (1000 W/cm^2) is employed. Brisk bleeding may be encountered when an artery is cut and rather than damaging the skin excessively by attempting to control the bleeding by laser coagulation a placement stitch may be indicated.

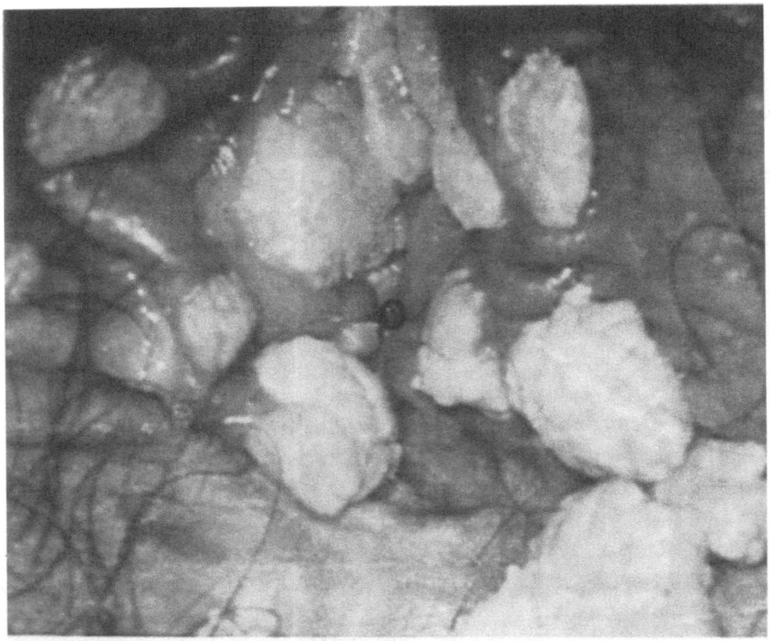

a

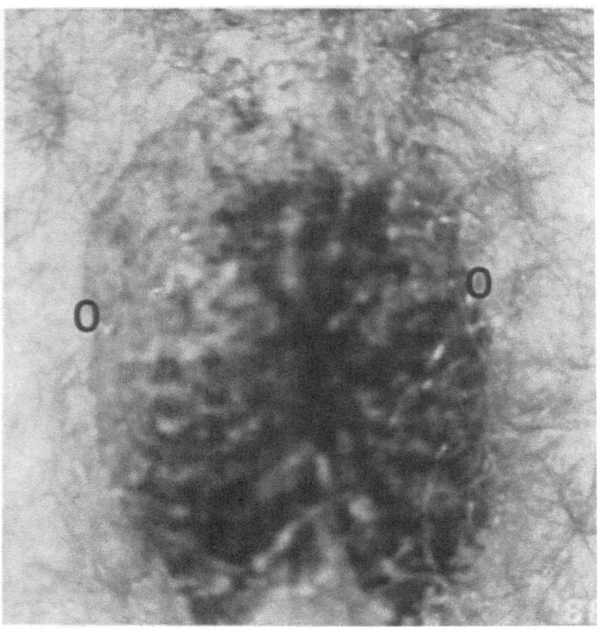

b

Fig. 14 a, b. Perianal wart destruction by vaporization. **a** Typical crop of perianal warts is seen. The anal canal is at *0*. **b** A similar set of warts have been removed and it can be seen that level of viral tissue destruction is at the level of the surrounding normal skin surface. The area of normal skin surrounding these lesions must be treated as otherwise a high recurrence rate can be expected. There is minimal damage to the surrounding normal tissue outside the treated area (*0* and *0*)

Postoperative pain may sometimes occur but is easily relieved using an analgesic and Sitz baths and the technique described above for the postoperative treatment of VIN.

References

Anderson MC, Hartley RB (1980) Cervical crypt involvement by intraepithelial neoplasia. Obstet Gynecol 55:546–550

Baggish MS (1985) Laser for the treatment of vulvar intraepithelial neoplasia. In: Baggish MS (ed) Basic and advanced laser surgery in gynaecology. Appleton Century Crofts, Norwalk, Connecticut, pp 195–205

Baggish MS (1986) A comparison between laser excisional biopsy and laser vaporisation for the treatment of cervical intraepithelial neoplasia. Am J Obstet Gynecol 155:39–44

Baggish MS, Dorsey JH (1981) CO_2 laser for the treatment of vulvar carcinoma in situ. Obstet Gynecol 57:371–375

Baggish MS, Sie E, Adelson M, Oates R (1989) Quantitative evaluation of the skin and accessory appendages in vulvar intraepithelial neoplasia. Obstet Gynecol 74(2):169–174

Bevan JR, Attwood ME, Jordan JA, Lucas A, Newton JR (1981) Treatment of preinvasive disease of the cervix by cone biopsy. Br J Obstet Gynaecol 88:1140–1144

Buckley CH, Butler EB, Fox H (1982) Cervical intraepithelial neoplasia. J Clin Pathol 35:1–13

Burghardt E, Holzer E (1980) Treatment of carcinoma in situ: evaluation of 1609 cases. Obstet Gynecol 55:539–545

Buscema J, Naghashfar Z, Sawad AE (1988) The predominance of human papillomavirus type 16 in vulval neoplasia. Obstet Gynecol 71:4 601–606

Campion A, McCance D, Singer A (1986) Progressive potential of mild cervical atypia: prospective psychological colposcopic and biological study. Lancet ii:237–240

Campion A, Brown JR, McCance DJ et al. (1987) Psychosexual trauma of abnormal cervical smear. Br J Obstet Gynaecol 95:175–181

Chanen W, Rome RM (1983) Electrocoagulation diathermy for cervical dysplasia and carcinoma in situ: a 15 year survey. Obstet Gynecol 61:673–679

Charles EH, Savage EW (1980) Cryosurgical treatment of cervical intraepithelial neoplasia. Obstet Gynecol Surv 35:539–541

Coney P, Walton LA, Edelman DA, Fowler WC Jr (1983) Cryosurgical treatment of early cervical intraepithelial neoplasia. Obstet Gynecol 62:463–466

Creasman WT, Ckarke-Pearson DL, Week JC Jr (1981) Results of outpatient therapy of cervical intraepithelial neoplasia. Obstet Gynecol 63:145–149

Creasman WT, Hinshaw WM, Clearke-Pearson DL (1984) Cryosurgery in the management of cervical intraepithelial neoplasia. Obstet Gynecol 63:145–149

DiSaia PJ, Creasman WT (eds) (1984a) Clinical gynaecologic oncology. Mosby, St Louis, pp 17–18

DiSaia PJ, Creasman WT (eds) (1984b) Clinical gynaecologic oncology. Mosby, St Louis, pp 22–25 and 37–41

DiSaia PJ, Townsend DE, Morrow LP (1974) The rationale for less than radical treatment for gynaecologic malignancy in early reproductive years. Obstet Gynecol Surv 29:581–593

Downey GO, Okagaki T, Ostrow R et al. (1988) Condylomatous carcinoma of the vulva with special reference to the human papillomavirus DNA. Obstet Gynecol 72(1):68–74

Duncan ID (1983) The Semm cold coagulator in the management of cervical intraepithelial neoplasia. Clin Obstet Gynecol 26:996–1006

Elliott PM, Tattersall MH, Coppleson M et al. (1989) Changing patterns of cervical cancer in young women. Br Med J 298:288–290
Ferenczy A, Mitho O, Nagai N et al. (1985) Latent papillomavirus and recurrent genital warts. N Engl J Med 313:784–788
Friedrich E (ed) (1984) Vulvar disease, 2nd edn. Saunders, Philadelphia, pp 90–99
Fu Y, Reagan W, Richart RM (1981) Definition of presursors. Gynecol Oncol 12:220–232
Fujii T, Crum CP, Winkler B, Fu YS, Richard RM (1984) Human papillomavirus infection and cervical intraepithelial neoplasia. Histopathology and DNA content. Obstet Gynecol 63:99–104
Jordan JA, Mylotte MJ (1982) Treatment of CIN by destruction laser. In: Jordan JA, Sharp F, Singer A (eds) Preclinical neoplasia of the cervix. Proceedings of study group. RCOG, London
Kaufman R, Bornstein J, Adam E et al. (1988) Human papillomavirus and herpes simplex virus in vulval squamous cell carcinoma in situ. Am J Obstet Gynecol 158:854–862
Kinlen LJ, Spriggs AI (1978) Women with positive cervical smears but without surgical intervention. Lancet ii:463–465
Kirkup W, Singer A (1980) Colposcopy in the management of the pregnant patient with abnormal cervical cytology. Br J Obstet Gynaecol 87:322–325
McGregor JE (1984) Colposcopy and ablative therapy. Br Med J 289:1204–1205
McIndoe WA, McLean MR, Jones RW, Mullins PR (1984) The invasive potential of carcinoma in situ of the cervix. Obstet Gynecol 64:451–458
McIndoe AG, Robson M, Tidy J, Anderson M (1989) Laser excision rather than vaporisation, the treatment of choice in cervical intraepithelial neoplasia. Obstet Gynecol 74(2):165–168
Partington CK, Soutter WP, Turner MJ, Hill AS (1987) Laser excisional biopsy under local anaesthesia. J Obstet Gynaecol 8:48–52
Reid R (1984) Superficial laser vulvectomy; a new surgical technique for CO_2 laser ablation. Am J Obstet Gynecol
Reid R (1989) HPV-associated lesions of the cervix: biology and colposcopic features. Clin Obstet Gynecol 32(1):157–179
Reid R, Elfont EA, Zirkin RM, Fuller TA (1984) Superficial laser vulvectomy II. The anatomic and biophysical principles permitting accurate control over depth of dermal destruction with CO_2 laser. Am J Obstet Gynecol
Richart R (1990) A modified terminology for cervical intraepithelial neoplasia. Obstet Gynecol 75(2):131–132
Schlaerth JB, Morrow CP, Nalick R, Gaddis O (1984) Anal involvement by carcinoma in situ of the perineum in women. Obstet Gynecol 64(3):406–411
Shatz P, Bergeron C, Wilkinson E, Arseneau J, Ferenczy A (1989) Vulvar epithelial neoplasia and skin appendage involvement. Obstet Gynaecol 74(5):769–774
Singer A, Walker P (1982) Commentary: what is the optimum treatment of cervical premalignancy? Br J Obstet Gynaecol 89:335–337
Stafl A, Wilkinson EJ, Mattingly RF (1977) Laser treatment of cervical and vaginal neoplasia. Am J Obstet Gynecol 2:128–136
Townsend DE, Richard RM, Marks E, Neilsen J (1981) Invasive cancer following outpatient evaluation and therapy for cervical disease. Obstet Gynecol 57:145–149
Townsend DE, Levine RV, Richard MR, Crum CP, Petrilli ES (1982) Management of vulval intraepithelial neoplasia. Obstet Gynecol 60(1):49–52
Woodman CB, Mould JJ, Jordan JA (1988) Radiotherapy in the management of vaginal intraepithelial neoplasia with hysterectomy. Br J Obstet Gynaecol 95(10):976–980

Epidermal Cells as Target for Immunological Reactions

D. Niederwieser, J. Auböck, P. Fritsch, and C. Huber

Introduction

Recently it was shown that keratinocytes (Ks) commonly express class II major histocompatibility complex (MHC) antigens (Ag) adjacent to T-lymphocyte infiltrates in numerous immunologically mediated skin diseases [1, 2]. Subsequently it has been demonstrated that de novo Ks synthesize class II MHC Ag and increase class I MHC Ag upon exposure to interferon-gamma (IFN-γ) in vitro and in vivo [3–6]. Although it is generally presumed that these IFN-induced alterations in MHC Ag expression have a major impact on immune function, there are only few data available on this question [7].

In an attempt to clarify this point, we exposed Ks cultures to IFN-γ or IFN-α2, prepared single cell suspensions, and then asked whether IFN-mediated changes of cell surface MHC Ag would affect Ks immunogenicity in alloimmune response. We demonstrated that class II MHC Ag-positive Ks are incapable of stimulating resting lymphocytes, but can maintain proliferation of activated antigen specific lymphocytes (i.e., T blasts). When used as targets, Ks are not susceptible to lysis by classical cytotoxic T lymphocytes (CTLs) unless class I MHC Ag expression on Ks is increased by pretreatment with IFN-γ [8].

Materials and Methods

For experimental details see [8].

Peripheral blood mononuclear cells (PBMCs) were separated from heparinized blood of healthy donors and skin donors by centrifugation on Ficoll Hypaque (Lymphoprep). *Fresh epidermal cells* (ECs) were prepared by trypsinization of normal human skin obtained from patients undergoing plastic surgery. *Ks* were expanded from fresh ECs using a 3T3 feeder layer technique according to Green et al. [9]. Secondary K cultures were used for all experiments.

IFN treatment of Ks cultures was carried out with human rIFN-γ (Genentech) and human rIFN-α2 (Boehringer-Ingelheim) for 72 h with concentrations ranging from 1 to 3000 U/ml. In most experiments 500 U/ml were used. *Class I and class II MHC Ag expression* on PBMCs, ECs, and Ks was measured by fluorescence activated cell sorter (FACS) analysis using FITC-

Genital Papillomavirus Infections
Eds.: Gross/Jablonska/Pfister/Stegner

conjugated monoclonal antibody (mAb) specific for HLA-DR (clone L234; Becton Dickinson) or HLA-A,B molecules (MAS 12532, Seralab).

Lymphocyte proliferation assays. PBMCs, ECs, and IFN-treated or untreated Ks were used as stimulators. PBMCs, memory cells or interleukin-2 dependent T cell lines (i.e. T blasts) enriched for CD4-positive cells served as responder.

Cytotoxic assays. PBMCs, ECs, and IFN-treated or untreated Ks served as target cells. Effector cells were alloreactive cytotoxic T lymphocytes which had been generated in primary mixed leukocyte reaction (MLR) by using PBMCs from skin donors as stimulators.

Blocking experiments. We carried out blocking experiments on the in vitro cellular immune responses with mAb recognizing CD3 and CD4 (VIT-3 and VIT-4, respectively, a kind gift of Dr. W. Knapp) and common determinants on class I MHC Ag (MAS 1532, Seralab) and class II MHC Ag (anti-HLA-DR, clone L243, Becton Dickinson).

Results

IFN-α2 and IFN-γ alter the expression of MHC Ag on Ks. In previous studies the physiologic concentrations of IFN produced by immune cells upon stimulation with Ag or virus were found to range between 500 and 3000 U/ml at the peak of the immune response [10]. Therefore these concentrations were used to investigate the role of IFN in lymphocyte-Ks interactions. Exposure of Ks to IFN-γ with 500 (Fig. 1) to 3000 U/ml led to a marked, dose-dependent increase of class I MHC Ag. In contrast, IFN-α2 even at 3000 U/ml

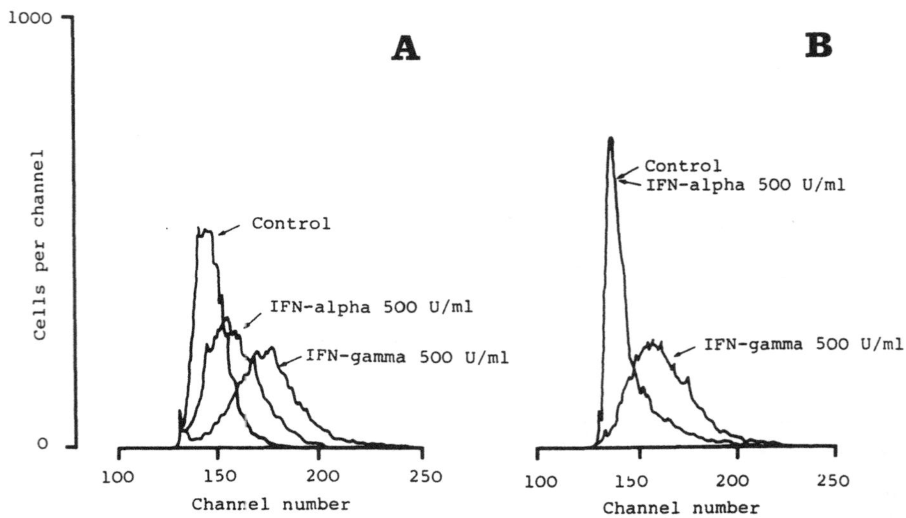

Fig. 1 A, B. Effect of IFN-α2 and IFN-γ on class I MHC Ag (A) and class II MHC Ag (B) expression on cultured keratinocytes (*Ks*)

Table 1. IFN-γ pretreated keratinocytes (Ks) express class II MHC products on their surface, but do not stimulate unsensitized allogeneic lymphocytes and suppress primary MLR [a]

Responder cells	Allogeneic stimulator cells	(x 10^{-4})	^{3}H-Thymidine incorporation (cpm) (mean $\pm$ SEM)
Unsensitized $PBMC_A$ 5×10^4	$PBMC_B$	5	38856 $\pm$ 8803
	EC_B	5	46154 $\pm$ 8322
	K_B	5	860 $\pm$ 183
	K_B [b]	5	781 $\pm$ 234
	K_B [c]	5	549 $\pm$ 79
	Medium		1038 $\pm$ 173
	$PBMC_A$	5	419 $\pm$ 68
	$PBMC_C$	5	24311 $\pm$ 2031
	$+K_A$	5	6543 $\pm$ 999
		2.5	12130 $\pm$ 649
		1.25	13625 $\pm$ 822
	$+K_A$ [d]	5	5178 $\pm$ 544
		2.5	12234 $\pm$ 853
		1.25	11439 $\pm$ 762
None	$PBMC_B$	5	539 $\pm$ 37
	$PBMC_C$	5	341 $\pm$ 61
	EC_B	5	632 $\pm$ 25
	K_B	5	311 $\pm$ 32

[a] The proliferative response of unsensitized peripheral blood mononuclear cells (PBMCs) to graded doses of mitomycin-C-treated stimulator cells was determined on days 3, 5, and 7. Stimulator cells were: (a) autologous PBMCs ($PBMC_A$), (b) allogeneic PBMCs ($PBMC_B$, $PBMC_C$), fresh ECS (EC_B), and cultured Ks (K_B) either untreated or pretreated with IFN-γ or IFN-α2, (c) allogeneic PBMCs ($PBMC_C$) with graded doses of autologous Ks (K_A) added. Only peak responses obtained on day 5 are shown
[b] K_B pretreated with IFN-γ (500 U/ml, 72 h)
[c] K_B pretreated with IFN-α2 (500 U/ml, 72 h)
[d] K_A pretreated with IFN-γ (500 U/ml, 72 h)

induced only a small increase of class I MHC Ag, which approximated that seen with 10 U/ml IFN-γ. However, expression of class II MHC molecules on Ks was induced exclusively by IFN-γ (Fig. 1).

HLA-DR positive Ks do not induce proliferation of resting unprimed or memory lymphocytes, but stimulate T blasts. To investigate the functional role of IFN-γ-induced class II MHC Ag on Ks, we compared the stimulatory capacity of PBMCs, ECs, and class II MHC Ag-negative and -positive Ks – all from the same donor – for three different allogeneic responder cell populations: unsensitized PBMCs, memory cells, and an alloreactive T cell line enriched for CD4-positive lymphoblasts.

PBMCs and ECs (containing about 3% Langerhans cells [11]) stimulated proliferation of unsensitized PBMCs (Table 1) and memory cells (not shown). In contrast both HLA-DR-negative and HLA-DR-positive Ks were not stimulatory at all. When Ks untreated or pretreated with IFN-γ (Table 1) or

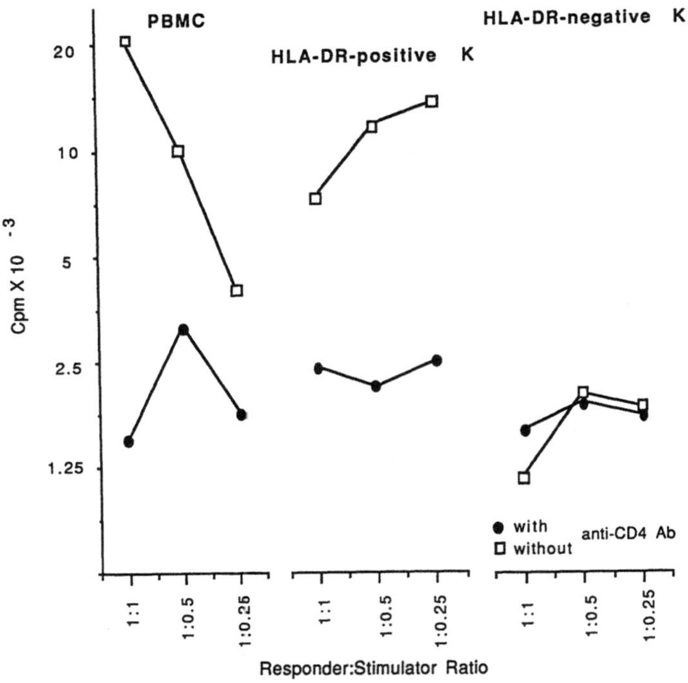

Fig. 2. Capacity of IFN-γ treated, HLA-DR-positive Ks to induce proliferation of alloreactive CD4-positive T blasts. T blasts (5×10^4) were stimulated with graded doses of peripheral blood mononuclear cells (PBMCs), IFN-γ treated (HLA-DR-positive) or untreated (HLA-DR-negtive) Ks, all derived from the original stimulator. Note that Ks are inhibitory at higher doses and that addition of anti-CD4 Ab blocks proliferation

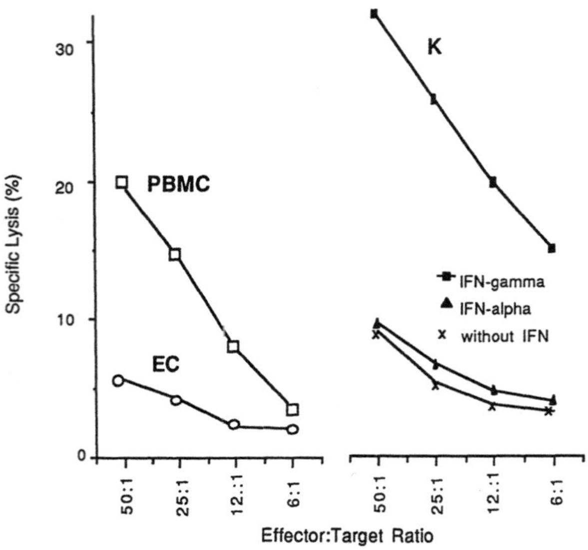

Fig. 3. Comparison of the lytic susceptibility of PBMCs, fresh epidermal cells, and untreated or IFN-treated Ks to lysis by CTL. Ks become susceptible to lysis after pretreatment with IFN-γ (500 U/ml, 72 h) but not after pretreatment with IFN-α2 (500 U/ml, 72 h)

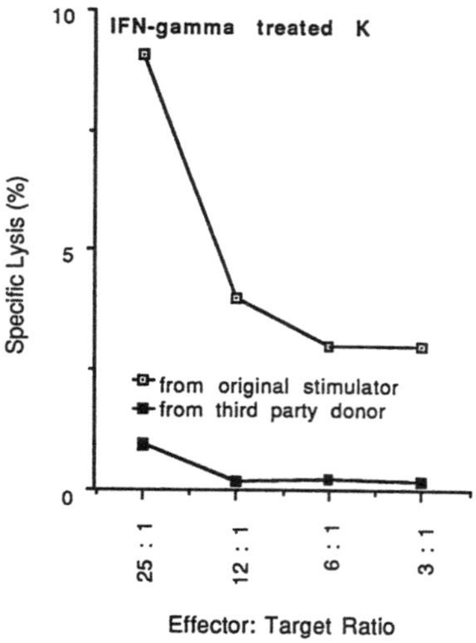

Fig. 4.

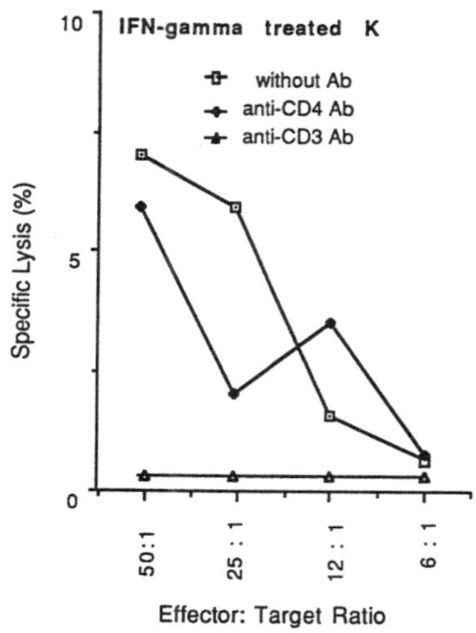

Fig. 5.

Fig. 4. Lysis of IFN-γ pretreated Ks by CTL is specific for the original stimulator cell. IFN-γ treated Ks from an unrelated third party donor are not subject to lysis

Fig. 5. Lysis of IFN-γ pretreated Ks by CTL can be blocked at the effector cell by anti-CD3 but not by anti-CD4 Ab

Ks conditioned media (not shown) were added to MLR a pronounced dose-dependent inhibition occurred.

Although HLA-DR-positive Ks were unable to induce proliferation of small, resting lymphocytes they stimulated CD4-enriched T blasts. Our findings that anti-CD4 MAb could block proliferation (Fig. 2) and that HLA-DR-negative Ks completely failed to induce proliferation indicate that class II MHC molecules represent the crucial stimulating antigenic structures on Ks.

Ks become susceptible to immune destruction by CTL after IFN-γ pretreatment. The lytic susceptibility of PBMCs, ECs, and IFN-treated or untreated Ks to alloreactive CTL was very different (Fig. 3). In comparison to PBMCs, ECs and Ks of the same donor were rather poor targets. However, when Ks were pretreated with IFN-γ concentrations (>100 U/ml) sufficient to induce a pronounced increase in MHC Ag expression, Ks became susceptible to lysis by alloreactive CTL in a dose-dependent manner. Strikingly, Ks treated with identical antiviral concentrations of IFN-α2 were not more susceptible to lysis than untreated Ks (Fig. 3). Even at high IFN-α2 doses suceptibility remained minimal.

Further experiments suggested that immune destruction of Ks is mediated by classical CTL recognizing class I MHC Ag rather than by NK cells or by anomalous killer cells. Evidence was based on the following findings: first, lysis of IFN-γ treated Ks by effector cells is specific for the donor of the original stimulator cell (Fig. 4); only IFN-γ treated Ks derived from the donor

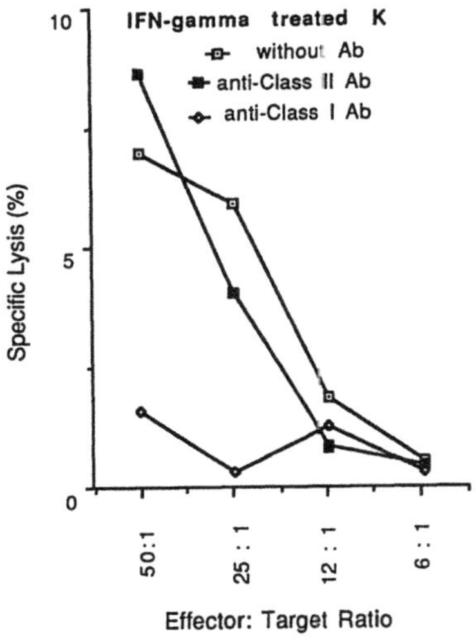

Fig. 6. Lysis of IFN-γ pretreated Ks by CTL can be blocked at the target cell by anticlass I but not by anticlass II MHC Ab

of the original stimulator cell, but not from a different unrelated skin donor were effectively killed. Second, lysis of IFN-γ treated Ks could be prevented by pretreating effector cells with mAb recognizing CD3 determinants, whereas mAb against CD4 Ag was ineffective (Fig. 5). Third, mAb against common determinants on class I MHC Ag but not against class II MHC Ag were capable of blocking lysis of IFN-γ treated Ks at the level of the target cell (Fig. 6).

Discussion

It has been shown that IFN-γ treated Ks express class II MHC Ag [4–6], but are unable to stimulate a primary MLR [7]. This is in agreement with our findings, that Ks exposed to physiologic concentrations of IFN-γ expressed significant levels of cell surface class II MHC molecules, but stimulated neither unprimed nor memory lymphocytes. Furthermore, mixing experiments demonstrated that Ks significantly inhibited MLR, an effect which has already been observed by others [7]. Therefore, we were unable to establish whether the failure of Ks to stimulate resting lymphocytes is merely a consequence of inhibition or also due to absence of stimulatory capacity. In contrast class II MHC Ag-bearing Ks were able to maintain clonal proliferation of T blasts. This finding is not surprising, since it has been reported that primed and unprimed T cells are distinguished by different accessory/antigen presenting cell (APC) requirements [12]. Several cell types including nonleukocytes [13–15] can effectively present Ag to T blasts. Resting T lymphocytes are stimulated almost exclusively by Ag in association with dendritic cells, which seem specialized for initiation of immune responses [16]. Activated B cells and IFN-γ treated endothelial cells have also been reported to stimulate resting T cells [17, 18], but are much less active than dendritic cells.

When untreated Ks were used as targets in alloimmune reaction, they were not destroyed by CTL generated in primary MLR. However, pretreatment with IFN-γ concentrations (> 100 U/ml), which caused a strong increase in class I MHC Ag expression, rendered Ks susceptible to lysis in a dose-dependent fashion. Such IFN-γ doses are within the physiologic range and are demonstrable during immune reactions in vitro and in vivo [10]. The same concentrations of IFN-α2 failed to induce significant susceptibility of Ks to killing. Lysis of Ks was mediated by classical CTL since it was genetically restricted and prevented by anti-CD3 mAb pretreatment at the effector level and by anticlass I mAb pretreatment at the target level. Lysis of Ks was not induced by increased fragility due to a possible toxic effect of high IFN-γ concentrations, because treated as well as untreated Ks were equally susceptible to natural killing.

In summary, our results suggest that a critical threshold amount of class I MHC molecules must be present on the cell surface to make K targets susceptible to immune destruction by CTL. Apparently such a high density of class I MHC Ag is only achieved by IFN-γ but not by IFN-α2 at the concentrations studied.

If our in vitro findings are applicable in vivo and not only for alloantigens but also for Ag recognized in the context of self-MHC Ag, e.g., virus protein, the implications are several-fold. IFN-γ induced class II MHC Ag of course would not initiate but amplify immune reactions in loco. Ks displaying basic, constitutive levels of class I MHC Ag would be resistant to lysis by alloreactive or self-MHC-restricted CTL, even when bearing a given target Ag. This refractory state, however, would be abolished, as soon as IFN-γ were endogenously released or exogenously applied for therapeutic reasons, such as for the treatment of viral papillomas.

References

1. Volc-Platzer B, Majdic O, Knapp W, Wolff K, Hinterberger W, Lechner K, Stingl G (1984) Evidence of HLA-DR antigen biosynthesis by human keratinocytes in disease. J Exp Med 159:1784
2. Auböck J, Romani N, Grubauer G, Fritsch P (1986) Expression of HLA-DR antigens on keratinocytes is a common feature of diseased skin. Br J Dermatol 114:465
3. Skoskiewicz MJ, Colvin RB, Schneeberger EE, Russell PS (1985) Widespread and selective induction of major histocompatibility complex-determined antigens in vivo by gamma-interferon. J Exp Med 162:1645
4. Basham TY, Nickoloff BJ, Merigan TC, Morhenn VB (1984) Recombinant gamma-interferon induces HLA-DR expression on cultured human keratinocytes. J Invest Dermatol 83:88
5. Volc-Platzer B, Leibl H, Luger T, Zahn G, Stingl G (1985) Human epidermal cells synthesize HLA-DR alloantigens in vitro upon stimulation with gamma-interferon. J Invest Dermatol 85:16
6. Auböck J, Niederwieser D, Romani N, Fritsch P, Huber C (1985) Human interferon-gamma induces expression of HLA-DR on keratinocytes and melanocytes. Arch Dermatol Res 277:270

7. Nickoloff BJ, Basham TY, Merigan TC, Torseth JW, Morhenn VB (1986) Human keratinocyte-lymphocyte reactions in vitro. J Invest Dermatol 87:86
8. Niederwieser D, Auböck J, Troppmair J, Herold M, Schuler G, Boeck G, Lotz J, Fritsch P, Huber C (1988) IFN-mediated induction of MHC antigen expression on human keratinocytes and its influence on in vitro alloimmune responses. J Immunol 140:2556
9. Green H, Kehinde O, Thomas J (1979) Growth of cultured human epidermal cells into multiple epithelia suitable for grafting. Proc Natl Acad Sci USA 76:5665
10. Woloszczuk W, Troppmair J, Leiter E, Flener R, Schwarz M, Kovarik J, Pohanka E, Margreiter R, Huber C (1986) Relationship of interferon-gamma and neopterin levels during stimulation with alloantigens in vivo and in vitro. Transplantation 41:716
11. Wolff K, Stingl G (1983) The Langerhans cells. J Invest Dermatol 80:17s
12. Inaba K, Steinman RM (1984) Resting and sensitized T lymphocytes exhibit distinct stimulatory (antigen-presenting cell) requirements for growth and lymphokine release. J Exp Med 160:1717
13. Umetsu DT, Katzen KD, Jabara HH, Geha RS (1986) Antigen presentation by human dermal fibroblasts: activation of resting T lymphocytes. J Immunol 136:440
14. Lipscomb MF, Lyons CR, Nunez G, Ball EJ, Stasny P, Vial W, Lem V, Weissler J, Miller LM, Toews GB (1986) Human alveolar macrophages: HLA-DR-positive macrophages that are poor stimulators of a primary mixed leukocyte reaction. J Immunol 136:497
15. Rubinstein D, Roska AK, Lipsky PE (1986) Liver sinusoidal lining cells express class II major histocompatibility antigens but are poor stimulators of fresh allogeneic T lymphocytes. J Immunol 137:1803
16. Steinman RM, Inaba K (1985) Stimulation of the primary mixed leukocyte reaction. CRC Crit Rev Immunol 5:331
17. Krieger JF, Chesnut RW, Grey HM (1986) Capacity of B cells to function as stimulators of a primary mixed leukocyte reaction. J Immunol 137:3117
18. Geppert TD, Lipsky P (1985) Antigen presentation by interferon-treated endothelial cells and fibroblasts: differential ability to function as antigen-presenting cells despite comparable Ia expression. J Immunol 135:3750

Immunomodulating Effects of Interferons: Conclusions for Therapy

J. Brzoska and H.-J. Obert

Properties of Interferons and Treatment Strategies

Interferons (IFN) possess antiviral and antiproliferative properties and modulate cell differentiation and immune functions (Stanton et al. 1987; Gastl and Huber 1988). The antiviral, antiproliferative, and cell-differentiating effectiveness increases with increasing dose. Above a certain level, which depends on the type of IFN and the cells being studied, the effectiveness cannot be increased any further. The phenomenon may be described graphically as a saturation curve. The maximum effect is usually achieved with a relatively high dose (Munoz and Carrasco 1984; Domke-Opitz et al. 1986; Mayer-Eichberger et al. 1981; Schiller et al. 1986; Kuebler et al. 1987; Fisher and Grant 1985; McGlave et al. 1987; Stella et al. 1988). The immunomodulating properties of IFN, however, especially in vivo, are usually characterized by a bell-shaped curve (Shalaby et al. 1984; Edwards et al. 1985; Kleinerman et al. 1986; Talmadge et al. 1987; Weiner et al. 1988; Siegel 1988). As a rule, the best effect is achieved with a relatively low dose. With higher doses, the effect may vanish, or the medication may even have a negative effect (Fig. 1).

The various properties of IFN lead to a variety of strategies for treatment of papillomavirus-induced warts. High doses of IFN must be applied if their direct antiviral and antiproliferative effects are to be exploited. For example, a daily dose of 25×10^6 IU IFN-β has been in use for the treatment of severe, acute viral infections such as disseminated herpes zoster or virus encephalitis (Heidemann and Obert 1985; Prange and Weber 1986). The drawback with high-dose therapies is that they produce considerable side effects (Quesada et al. 1986; Jones and Itri 1986). In cases of strictly localized virus infection, this disadvantage of systemic application can be overcome by using localized treatment, which leads to high local concentration and requires only low doses (0.1–6×10^6 IU). Due to its high tissue affinity, IFN-β seems to be the most appropriate type of IFN for such applications (Ishihara et al. 1983; Horiuchi et al. 1985; Bocci 1985; Hündgen 1988). Low-dose local therapy is already in use successfully in cases of labial and genital herpes (Glezerman et al. 1988), common warts (Niimura 1983; Horiuchi et al. 1985; Remy et al. 1987), genital warts (Vance et al. 1986; Remy et al. 1987; Friedman-Kien et al. 1988; Reichman et al. 1988), epidermodysplasia verruciformis (Androphy et al. 1984), cervical intraepithelial neoplasia (De Palo et al. 1986; Choo et al. 1986), and other benign or malignant tumors (Ishihara et al. 1983; Edwards et al. 1986; Greenway et al. 1986; Vonderheid et al. 1987; von Wussow et al. 1988; Remy and Demmler 1988; Baba et al. 1988; Schmoll and Schöpf 1988; Wick-

Genital Papillomavirus Infections
Eds.: Gross/Jablonska/Pfister/Stegner

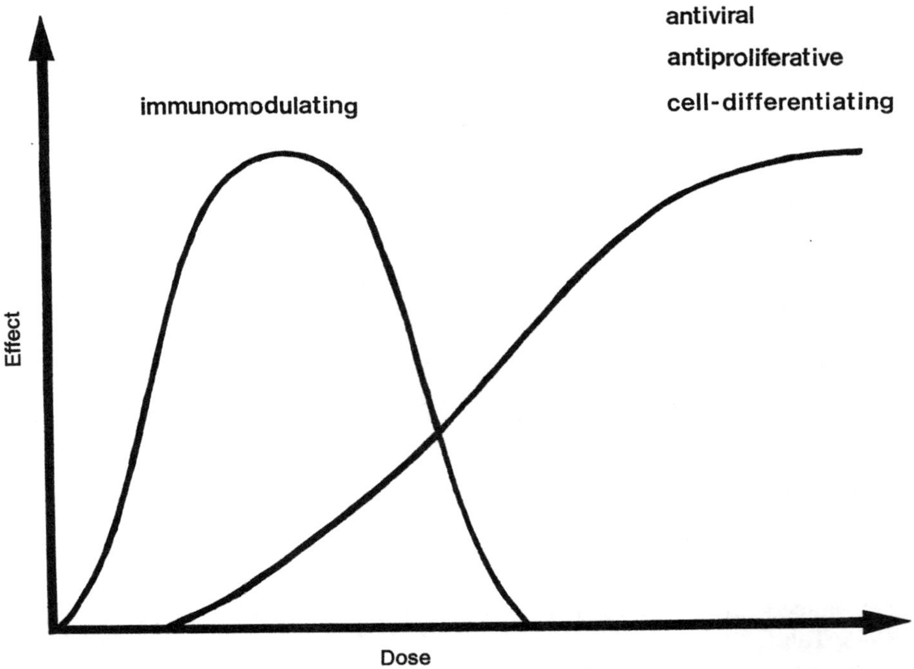

Fig. 1. Schematic representation of the dose-response curve for various effects of interferons

ramasinghe et al. 1989). However, the disadvantage of local therapy becomes clear when it is applied to patients with multiple lesions. Apart from the time required and the increased systemic side effects associated with the accumulated dose, not all patients tolerate this form of treatment because of the considerable pain incurred. In addition, clinically inconspicuous, infected areas may be neglected or insufficiently affected by the local therapy.

Another treatment strategy makes use of the modulating effect of the IFN on immune functions. In cases of papillomavirus infection, this seems sensible because patients with persistent or recurring multiple warts often have a defect in cellular immunity, and because cell-mediated immune reactions with local signs of inflammation are observed during healing of warts (Kirchner 1986; Gross et al. 1986a; Cauda et al. 1987; Fierlbeck et al. 1989). However, the real problem with the immunomodulating effects of IFN is that they can function not only as activators but also as inhibitors. Several examples of the dual effects of IFN are presented here.

Effect of Interferons on the Activity of Monocytes/Macrophages

Among the types of IFN, IFN-γ is the most potent macrophage-activating factor (Vilcek et al. 1985; Russel and Pace 1987, Murray 1988). In fact, if IFN-α or IFN-β are applied together with IFN-γ in vitro, they can even have inhibiting effects (Yoshida et al. 1988). In monocytes and macrophages, IFN-γ can stimulate spontaneous and antibody-mediated cytotoxicity, phagocytosis, metabolic production of oxygen radicals and lymphokines such as interleukin 1 (IL-1), antimicrobial activity, and expression of cell surface structures such as Fc receptors and HLA class I and II antigens (Vilcek et al. 1985; Russel and Pace 1987; Bonnem and Oldham 1987; Murray 1988; Johnston 1988). The lowest effective subcutaneous or intramuscular dose which produces activation of monocytes is about 10 $\mu g/m^2$ body surface (Sechler et al. 1988), the highest about 500 $\mu g/m^2$ (Kleinerman et al. 1986). The optimum dose lies between the two at about 100 $\mu g/m^2$ (Maluish et al. 1988).

It has been shown in in vitro studies that treatment of monocytes and macrophages with IFN-γ alone only moderately stimulates their production of IL-1, a key mediator of inflammation (Oppenheim et al. 1986; Dinarello 1988). However, IFN-γ potentiates the reaction of the cells to mitogens if the incubation with IFN-γ precedes that with the mitogens (Virelizier and Arenzana-Seisdedos 1985; Eden and Turino 1986; Gerrard et al. 1987). With already activated monocytes and macrophages or with cells that have been incubated first with mitogens and then with IFN-γ, the production of IL-1 is not increased but inhibited (Brandwein 1986; Eden and Turino 1986; Ghezzi and Dinarello 1988; Ruschen et al. 1989). The same effect has been demonstrated in vivo as well (Zabel et al. 1988). In patients with lung cancer who are treated for 5 days with 50 μg IFN-γ/day, the monocytes from peripheral blood produce distinctly more IL-1 than before; the alveolar macrophages, however, which are already producing more IL-1 spontaneously, are not affected (Table 1). This difference is not dependent on the mononuclear cell type, as shown by studies on peripheral blood monocytes from patients with rheumatoid arthritis who are treated with IFN-γ (Lemmel et al. 1987). Monocytes which have been shown by chemiluminescence testing to be

Table 1. Spontaneous IL-1 production (cpm) in a patient with lung cancer after five daily subcutaneous doses of 50 μg IFN-γ. (From Zabel et al. 1988)

	Before therapy	During therapy
Peripheral blood monocytes	317	4529
Alveolar macrophages	1285	993

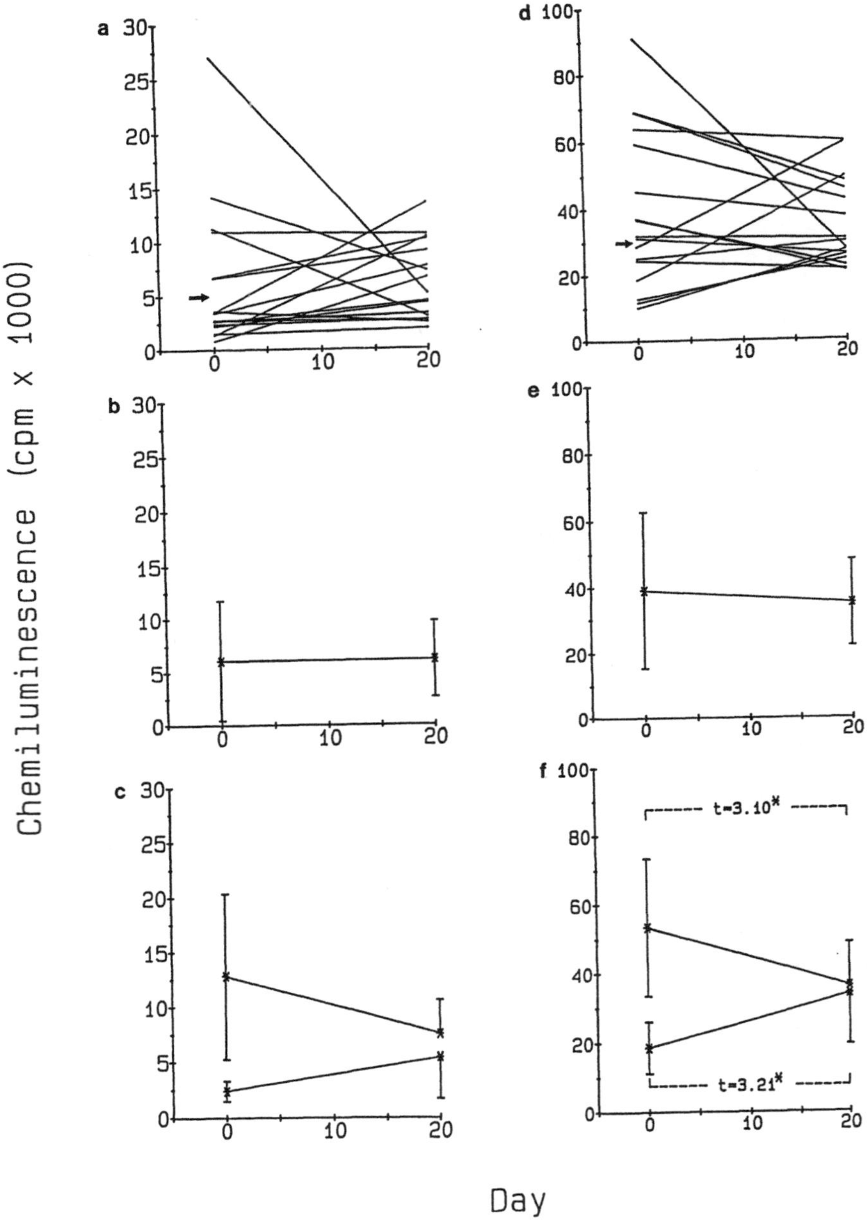

Fig. 2 a–f. Activity (chemiluminescence) of peripheral blood monocytes (**a–c**) and neutrophil granulocytes (**d–f**) from patients with rheumatoid arthritis after 20 days of treatment with 100 μg IFN-γ s.c. daily. **a, d** Individual patients' values. **b, e** Mean and standard deviation for all values. **c, f** Mean and standard deviation after division into low and high baseline values. *Arrow* (**a, d**) shows the dividing line. Level of significance, $*p<0.05$ (*t* test for paired data)

moderately activated are reduced in activity as a result of the therapy, whereas an increase in activity is seen in those monocytes which are more or less inactive at the beginning (Fig. 2a–c). With daily administration, the stimulating effect of IFN-γ vanishes gradually (Maluish et al. 1988). The loss of effectiveness is found even with only one injection per week (Aulitzky et al. 1987).

Effects of Interferons on the Activity of Neutrophil Granulocytes

In vitro, IFN-α and IFN-β have only little effect on the function of neutrophil granulocytes (Schuff-Werner et al. 1984; Saito et al. 1987; Lappegard et al. 1988). In vivo, however, an effect can be demonstrated. In tumor patients who are treated for 1 year with daily intramuscular doses of 3×10^6 IU natural IFN-α, the oxidative metabolism of the granulocytes is affected (Einhorn and Jarstrand 1984). During the 1st week of therapy the granulocytes are activated; later, however, they are found to be inhibited. IFN-γ has a much greater effect on the functions of neutrophil granulocytes than IFN-α or IFN-β (Schuff-Werner et al. 1984; Kapp et al. 1986; Lappegard et al. 1988). In patients with chronic granulomatous disease, cell activation may be induced with subcutaneous application of as little as 10 μg IFN-γ/m^2 body surface (Sechler et al. 1988). As with the monocytes, IFN-γ can affect the neutrophil granulocytes in two different ways, depending on the basal activity of the cells. Thus, in patients with rheumatoid arthritis, IFN-γ therapy results in stimulation of the initially inactive neutrophil granulocytes, whereas the activity of the cells which are already activated is reduced by the treatment (Fig. 2d–f).

Effect of Interferons on the Activity of Natural Killer Cells

Natural killer (NK) cells are able to kill certain tumor cells and virus-infected cells spontaneously, i.e., without a previous encounter or immunization (Welsh 1984). Patients with recurring genital warts have NK cell activities which are reduced in comparison with controls (Cauda et al. 1987). Patients who respond to IFN-α therapy with partial or complete remission react, unlike nonresponders, with a significant and lasting increase in the activity of their NK cells (Tyring et al. 1988; Table 2). The optimum single intramuscular dose for stimulation of the NK cell activity is 3×10^6 IU IFN-α; doses below 1×10^6 IU or above 10×10^6 IU have no effect. Unlike the situation in vivo, the in vitro dose response is not characterized by a bell-shaped curve but by a saturation curve (Edwards et al. 1985). Intramuscular or intravenous injections of IFN-β at doses of about 3×10^6 IU also lead to activation of the NK cells (Ogawa et al. 1982; Schönfeld et al. 1984; Tentori et al. 1987). IFN-γ can also stimulate NK cell activity (Ernstoff et al. 1985; Thompson et al. 1987).

Table 2. Effect of treatment with IFN-α_{2b} on NK cell activity in patients with genital warts. (From Tyring et al. 1988)

Patients	n	Before treatment	During treatment	After treatment
Nonresponders	19	24.3 ± 3.7	28.0 ± 5.4^{a}	29.9 ± 6.0^{a}
Responders	16	26.1 ± 2.6	55.8 ± 4.9^{b}	51.7 ± 6.9^{b}

Responders, patients with partial or complete remission. 1×10^6 IU was administered three times a week for 3 weeks. The NK cell activity was determined before initiation of treatment, at the end of the 2nd week, and 4 weeks after the end of the therapy.
[a] NS
[b] Significant at $p < 0.01$

The optimum dose by the subcutaneous route is about 200 μg (G. Fierlbeck, personal communication).

After continuous, repeated administration of IFN, the stimulating effect on the immune system gradually vanishes; later, it can even become inhibitory (Ogawa et al. 1982; Ernstoff et al. 1983; Hirsch and Johnson 1985; Cuellar et al. 1985). Inhibition of the activity of the NK cells can also arise after the very first administration of IFN-α, whereby the basal activity of the cells plays a decisive role. The NK cell activity is inhibited in two-thirds of patients in whom levels are already high before the treatment (Lotzova et al. 1983).

Effect of Interferons on the Erythrocyte Sedimentation Rate

The erythrocyte sedimentation rate (ESR) is an important parameter of general, humoral-systemic inflammation. Most patients with genital warts have normal ESR values (≦10 mm/h). In these patients, IFN-γ treatment increases the ESR significantly. In contrast, in patients whose ESR is increased at the beginning of the therapy, IFN-γ treatment leads to a decrease of the values (G. Fierlbeck, personal communication; Table 3). Patients with

Table 3. Changes in the ESR in patients with genital warts during systemic treatment with IFN-γ. (From G. Fierlbeck, personal communication)

ESR	n	Increase	Decrease	No change
≦10 mm/h	36	29	3	4
>10 mm/h	5	1	4	0

A dose of 100–200 μg IFN-γ was given subcutaneously daily for 1 week. The ESR values (1-h, according to Westergren) were determined before initiation and 1 day after termination of therapy rheumatoid arthritis in whom the ESR is elevated also respond to IFN-γ therapy with a marked reduction in this value (Lemmel et al. 1988; Klein 1988; Sprekeler 1988).

Conclusions for Therapy with Interferons

The optimum modulating effect on the immune system with systemically administered IFN is obtained with a dose of $1–10 \times 10^6$ IU IFN-α or IFN-β and of 20–200 μg (ca. $0.4–4 \times 10^6$ IU) IFN-γ. The IFN must be administered only for a short time if it is intended to have a stimulating effect. The therapy can then be repeated after a break of several weeks. Therapeutic schemes using such intermittent applications are already in use (Ernstoff et al. 1983; Medenica and Slack 1985; Gross et al. 1986b; Gross 1987). Besides this active treatment, passive (adoptive) immunotherapy consisting of transfer of autologous in vitro stimulated cytotoxic cells may prove to be successful (Lacerna et al. 1988). Long-term continuous administration of IFN for several months can result in reduced immune responses. The nature of the modulating effect on the immune system (activation or inhibition) seems to be determined by the basal activity of the NK cells. In addition, dose and type of IFN, route, site and time of administration, and the presence of other immunomodulating factors can be important. Concerning the effects of IFN-γ on inflammatory processes, we have formulated the following hypothesis (Brzoska and Obert 1987; Fig. 3). If the inflammatory response is of low

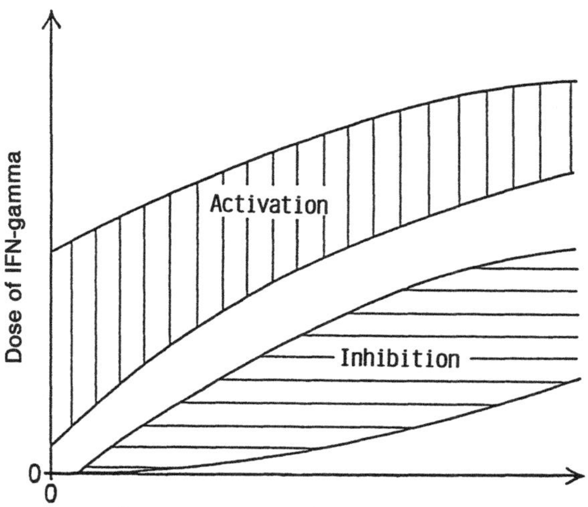

Fig. 3. Effect of exogenously applied IFN-γ on the inflammatory response as a function of the level of activity before each application. If the inflammatory activity is low, IFN-γ stimulates the reaction; if the activity is high, IFN-γ has an inhibitory effect. For more details see text

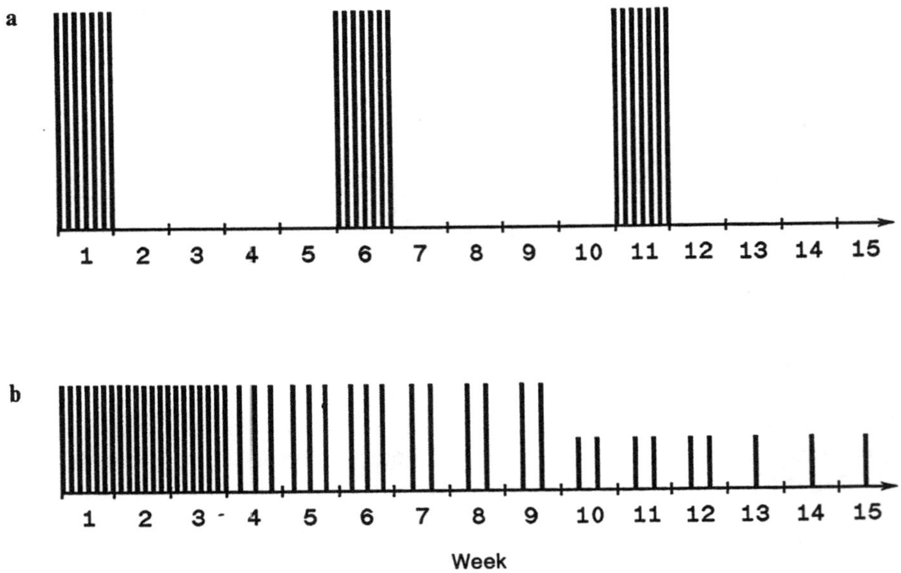

Fig. 4 a, b. Therapeutic regimen for the use of IFN-γ. Each column represents one injection; the height of the column shows the dose. **a** Treatment of genital warts. **b** Treatment of rheumatoid arthritis

grade, it may be stimulated by a particular dose of IFN-γ. If there is active inflammation, the dose which was previously stimulatory now inhibits the inflammatory response. If IFN-γ induces an inflammatory response, its continued administration at the same dose leads eventually to diminution of the response and later even to complete inhibition of the inflammatory process. In consequence, it is not expedient to carry out continuous therapy since this initiates a negative feedback mechanism. On the other hand, if one wishes to achieve lasting inhibition of an existing inflammation, the dose must be successively reduced, to avoid losing effectiveness or even producing a renewed stimulation of the inflammatory process.

Accordingly, the following therapeutic regimens are currently in use for IFN-γ (Fig. 4). For patients with multiple genital warts in whom an inflammatory response is to be induced, 50–200 μg is administered systemically every day for 1 week followed by 4 weeks without treatment. If the patient does not respond to the therapy within this time, the cycle may be repeated several times (Gross 1987; Gross et al. 1988). In patients with a chronic inflammatory disease, such as rheumatoid arthritis, the treatment is initially continuous with daily doses of 50 μg. When the symptoms improve, the dose and frequency of administration must be reduced in order to maintain the effectiveness (Obert and Brzoska 1986). Further studies will demonstrate whether these therapeutic regimens can be optimized and also whether they can be applied to other diseases as well.

Acknowledgements. We would like to thank Prof. Dr. E.-M. Lemmel, Baden-Baden, and Dr. G. Fierlbeck, Tübingen, for their kind permission to include some of their unpublished data.

References

Androphy EJ, Dvoretzky I, Maluish AE, Wallace HJ, Lowy DR (1984) Response of warts in epidermodysplasia verruciformis to treatment with systemic and intralesional alpha interferon. J Am Acad Dermatol 11:197–202

Aulitzky W, Gastl G, Aulitzky WE, Nachbaur K, Lanske B, Kemmler G, Flener R, Frick J, Huber C (1987) Interferon-γ for the treatment of metastatic renal cancer: dose-dependent stimulation and downregulation of beta-2 microglobulin and neopterin responses. Immunobiology 176:85–95

Baba T, Hoshino M, Uyeno K (1988) Resolution of cutaneous lesions of granuloma annulare by intralesional injection of human fibroblast interferon. Arch Dermatol 124:1015–1016

Bocci V (1985) Distribution, catabolism and pharmacokinetics of interferons. In: Finter NB, Oldham RK (eds) Interferon, vol 4. Elsevier, Amsterdam, pp 47–72

Bonnem EM, Oldham RK (1987) Gamma-interferon: physiology and speculation on its role in medicine. J Biol Response Mod 6:275–301

Brandwein SR (1986) Regulation of interleukin 1 production by mouse peritoneal macrophages. J Biol Chem 261:8624–8632

Brzoska J, Obert HJ (1987) Interferon gamma: ein janusköpfiger Mediator bei Entzündungen. Arzneimittelforschung 37(2):1410–1416

Cauda R, Tyring SK, Grossi CE, Tilden AB, Hatch KD, Sams WM, Baron S, Whitley RJ (1987) Patients with condyloma acuminatum exhibit decreased interleukin-2 and interferon gamma production and depressed natural killer activity. J Clin Immunol 7:304–311

Choo YC, Seto WH, Hsu C, Merigan TC, Tan YH, Ma HK, Ng MH (1986) Cervical intraepithelial neoplasia treated by perilesional injection of interferon. Br J Obstet Gynaecol 93:372–379

Cuellar B, Ruiz M, Orbach-Arbouys S (1985) Decreased natural killer cell activity after prolonged administration of interferon in cancer patients. Immunol Lett 10:137–139

De Palo G, Stefanon B, Del Vecchio M (1986) Clinical patterns, follow-up and treatment of papilloma virus lesions of the uterine cervix. In: De Palo G, Rilke F, zur Hausen H (eds) Herpes and papilloma viruses. Raven, New York, pp 305–327

Dinarello CA (1988) Biology of interleukin 1. FASEB J 2:108–115

Domke-Opitz I, Straub P, Kirchner H (1986) Effect of interferon on replication of herpes simplex virus types 1 and 2 in human macrophages. J Virol 60:37–42

Eden E, Turino GM (1986) Interleukin-1 secretion by human alveolar macrophages stimulated with endotoxin is augmented by recombinant immune (gamma) interferon. Am Rev Respir Dis 133:455–460

Edwards BS, Merritt JA, Fuhlbrigge RC, Borden EC (1985) Low doses of interferon alpha result in more effective clinical natural killer cell activation. J Clin Invest 75:1908–1913

Edwards L, Levine N, Weidner M, Piepkorn M, Smiles R (1986) Effect of intralesional α_2-interferon on actinic keratoses. Arch Dermatol 122:779–782

Einhorn S, Jarstrand C (1984) Functions of human neutrophilic granulocytes after in vivo exposure to interferon alpha. Infect Immun 43:1054–1057

Ernstoff MS, Fusi S, Kirkwood JM (1983) Parameters of interferon action. Immunological effects of whole cell leukocyte interferon (IFN-α) in phase I–II trials. J Biol Response Mod 2:528–539

Ernstoff MS, Reich S, Nishoda Y, Kirkwood JM (1985) Immunological assessment of melanoma patients treated with recombinant interferon gamma (rIFN-γ, Biogen, Inc., Cambridge, MA) in a phase I/II trial. Proc Am Assoc Cancer Res 26:280
Fierlbeck G, Schiebel U, Müller (1989) Immunohistology of genital warts in different stages of regression after therapy with interferon gamma. Dermatologica 179:191–195
Fisher PB, Grant S (1985) Effects of interferon on differentiation of normal and tumor cells. Pharmacol Ther 27:143–166
Friedman-Kien AE, Eron LJ, Conant M, Growdon W, Badiak H, Bradstreet PW, Fedorczyk D, Trout JR, Plasse TF (1988) Natural interferon alpha for treatment of condylomata acuminata. JAMA 259:533–538
Gastl G, Huber C (1988) The biology of interferon actions. Blut 56:193–199
Gerrard TL, Siegel JP, Dyer DR, Zoon KC (1987) Differential effects of interferon-α and interferon-γ on interleukin 1 secretion by monocytes. J Immunol 138:2535–2540
Ghezzi P, Dinarello CA (1988) IL-1 induces IL-1. III. Specific inhibition of IL-1 production by IFN-γ. J Immunol 140:4238–4244
Glezerman M, Lunenfeld E, Cohen V, Sarov I, Movshovitz M, Doerner T, Shoham J, Revel M (1988) Placebo-controlled trial of topical interferon in labial and genital herpes. Lancet I:150–152
Greenway HT, Cornell RC, Tanner DJ, Peets E, Bordin GM, Nagi C (1986) Treatment of basal cell carcinoma with intralesional interferon. J Am Acad Dermatol 15:438–443
Gross G (1987) Interferone zur Behandlung von Condylomata acuminata. Dtsch Med Wochenschr 112:571
Gross G, Roussaki A, Ikenberg H, Drees N (1986a) Interferon-α bei Condylomata acuminata and juvenilem Diabetes mellitus. Dtsch Med Wochenschr 111:1351–1355
Gross G, Ikenberg H, Roussaki A, Drees N, Schöpf E (1986b) Systemic treatment of condylomata acuminata with recombinant interferon-alpha-2α: low-dose superior to the high-dose regimen. Chemotherapie 32:537–541
Gross G, Roussaki A, Brzoska J (1988) Low doses of systemically administered recombinant interferon-gamma effective in the treatment of genital warts. J Invest Dermatol 90:242
Heidemann E, Obert HJ (1985) Clinical trials and pilot studies with natural interferons in Germany. In: Kirchner H, Schellekens H (eds) The biology of the interferon system 1984. Elsevier, Amsterdam, pp 557–567
Hirsch RL, Johnson KP (1985) Natural killer cell activity in multiple sclerosis patients treated with recombinant interferon-α_2. Clin Immunol Immunopathol 37:236–244
Horiuchi S, Naito S, Baba T, Uyeno K (1985) Treatment of common warts with intralesional injection of human fibroblast interferon. In: Kono R (ed) Herpes virus and virus chemotherapy. Elsevier, Amsterdam, pp 321–324
Hündgen M (1988) Pharmakologie der Interferone-alpha, -beta und -gamma. In: Schmoll HJ, Schöpf E (eds) Lokale und systemische Tumortherapie mit Interferonen. Zuckschwerdt, Munich, pp 5–16 (Aktuelle Immunologie, vol 5)
Ishihara K, Hayasaka K, Yamamoto A, Hasegawa F (1983) Clinical responses of patients with malignant skin neoplasia to intralesional treatment with three types of interferons (α and β). In: Kishida T (ed) Interferons. ISIFN, Kyoto, pp 222–227
Jones GJ, Itri LM (1986) Safety and tolerance of recombinant interferon alpha-2α (Roferon®-A) in cancer patients. Cancer 57:1709–1715
Johnston RB (1988) Monocytes and macrophages. N Engl J Med 318:747–752
Kapp A, Kirchner H, Wokalek H, Schöpf E (1986) Modulation of granulocyte oxidative response by recombinant interferon-α2 and γ. Arch Dermatol Res 278:274–276
Kirchner H (1986) Immunobiology of human papillomavirus infection. Prog Med Virol 33:1–41
Klein HO (1988) Therapie der rheumatoiden Arthritis mit rekombinantem Interferon-gamma. Fortschr Med 106:721–725

Kleinerman ES, Kurzrock R, Wyatt D, Quesada JR, Gutterman JU, Fidler IJ (1986) Activation or suppression of the tumoricidal properties of monocytes from cancer patients following treatment with human recombinant γ-interferon. Cancer Res 46:5401–5405

Kuebler JP, Oberley TD, Meisner LF, Sidky YA, Reznikoff CA, Borden EC, Cummings KB, Bryan GT (1987) Effect of interferon alpha, interferon beta, and interferon gamma on the in vitro growth of human renal adenocarcinoma cells. Invest New Drugs 5:21–29

Lacerna LV, Stevenson GW, Stevenson HC (1988) Adoptive cancer immunotherapy utilizing lymphokine activated killer cells and gamma interferon activated killer monocytes. Pharmacol Ther 38:453–465

Lappegard KT, Benestad HB, Rollag H (1988) Interferons affect oxygen metabolism in human neutrophil granulocytes. J Interferon Res 8:665–677

Lemmel EM, Franke M, Gaus W, Hartl PW, Hofschneider PH, Miehlke K, Machalke K, Obert HJ (1987) Results of a phase-II clinical trial on treatment of rheumatoid arthritis with recombinant interferon-gamma. Rheumatol Int 7:127–132

Lemmel EM, Brackertz D, Franke M, Gaus W, Hartl PW, Machalke K, Mielke H, Obert HJ, Peter HH, Sieper J, Sprekeler R, Stierle H (1988) Results of a multicenter placebo-controlled double-blind randomized phase III clinical study of treatment of rheumatoid arthritis with recombinant interferon-gamma. Rheumatol Int 8:87–93

Lotzova E, Savary CA, Quesada JR, Gutterman JU, Hersh EM (1983) Analysis of natural killer cell cytotoxicity of cancer patients treated with recombinant interferon. JNCI 71:903–910

Maluish AE, Urba WJ, Longo DL, Overton WR, Coggin D, Crisp ER, Williams R, Sherwin SA, Gordon K, Steis RG (1988) The determination of an immunologically active dose of interferon-gamma in patients with melanoma. J Clin Oncol 6:434–445

Mayer-Eichberger S, Treuner J, Joester KE, Niethammer D (1981) Effects of IFN-α and IFN-β on the proliferation of human cells in vitro. In: De Maeyer E, Galasso G, Schellekens H (eds) The biology of the interferon system. Elsevier, Amsterdam, pp 169–172

McGlave P, Mamus S, Vilen B, Dewald G (1987) Effect of recombinant gamma interferon on chronic myelogenous leukemia bone marrow progenitors. Exp Hematol 15:331–335

Medenica RD, Slack N (1985) Immunomodulatory activity of human leukocyte interferon in cancer patients: results obtained during pulse therapy schedule. Cancer Drug Deliv 2:91–118

Munoz A, Carrasco L (1984) Comparison of the antiviral action of different human interferons against DNA and RNA viruses. FEMS Microbiol Lett 21:105–111

Murray HW (1988) Interferon-gamma, the activated macrophage, and host defense against microbial challenge. Ann Intern Med 108:595–608

Niimura M (1983) Intralesional human fibroblast in common warts. In: Kishida T (ed) Interferons. ISIFN, Kyoto, pp 324–327

Obert HJ, Brzoska J (1986) Interferon-gamma in der Therapie der chronischen Polyarthritis. Arzneimittelforschung 36(2):1557–1560

Ogawa M, Ezaki K, Okabe K (1982) Clinical and immunologic studies of human fibroblast interferon. In: Kono R, Vilcek J (eds) The clinical potential of interferons. University of Tokyo Press, Tokyo, pp 199–212

Oppenheim JJ, Kovacs EJ, Matsushima K, Durum SK (1986) There is more than one interleukin 1. Immunol Today 7:45–56

Prange HW, Weber T (1986) Zur Behandlung der schweren Virusenzephalitis mit Interferon und Virustatika. Med Klin 81:657–662

Quesada JR, Talpaz M, Rios A, Kurzrock R, Gutterman JU (1986) Clinical toxicity of interferons in cancer patients: a review. J Clin Oncol 4:234–243

Reichman RC, Oakes D, Bonnez W, Greisberger C, Tyring S, Miller L, Whitley R, Carveth H, Weidner M, Krueger G, Yorkey L, Roberts NJ, Dolin R (1988) Treatment of condyloma acuminatum with three different interferons administered intralesionally. Ann Intern Med 108:675–679

Remy W, Demmler M (1988) Örtliche/intratumorale Interferon-Behandlung von Basaliomen. In: Hofschneider PH (ed) Ergebnisse der Beta-Interferon-Therapie bei chronisch-aktiver Hepatitis B, Multipler Sklerose und Krebserkrankungen. Zuckschwerdt, Munich, pp 83–86 (Aktuelle Immunologie, vol 3)

Remy W, Demmler M, Mayerhausen W (1987) Interferone, Anwendung in der Dermatologie. In: Braun-Falco O, Schill WB (eds) Fortschritte der praktischen Dermatologie und Venerologie. Springer, Berlin Heidelberg New York, pp 391–397

Ruschen S, Lemm G, Warnatz H (1989) Spontaneous and LPS stimulated production of intracellular IL-1β by synovial macrophages in rheumatoid arthritis is inhibited by IFN-γ. Clin Exp Immunol 76:246–251

Russell SW, Pace JL (1987) The effects of interferons on macrophages and their precursors. Vet Immunol Immunopathol 15:129–165

Saito H, Hayakawa T, Yui Y, Shida T (1987) Effect of human interferon on different functions of human neutrophils and eosinophils. Int Arch Allergy Appl Immunol 82:133–140

Schiller JH, Groveman DS, Schmid SM, Willson JKV, Cummings KB, Borden EC (1986) Synergistic antiproliferative effects of human recombinant α54- or β_{ser}-interferon with γ-interferon on human cell lines of various histogenesis. Cancer Res 46:483–488

Schmoll HJ, Schöpf E (eds) (1988) Lokale und systemische Tumortherapie mit Interferonen. Zuckschwerdt, Munich (Aktuelle Immunologie, vol 5)

Schönfeld A, Schattner A, Crespi M, Levavi H, Shoham J, Nitke S, Wallach D, Hahn T, Yarden O, Doerner T (1984) Intramuscular human interferon-β injections in treatment of condylomata acuminata. Lancet I:1038–1042

Schuff-Werner P, Wurl M, Gottsmann K, Obert HJ, Nagel GA (1984) Amplification of the granulocyte derived chemiluminescence by interferons. Antiviral Res Abstr 1 No 3:127

Sechler JMG, Malech HL, White CJ, Gallin JI (1988) Recombinant human interferon-γ reconstitutes defective phagocyte function in patients with chronic granulomatous disease of childhood. Proc Natl Acad Sci USA 85:4874–4878

Shalaby MR, Weck PK, Rinderknecht E, Harkins RN, Frane JW, Ross MJ (1984) Effects of bacteria-produced human alpha, beta, and gamma interferons on in vitro immune functions. Cell Immunol 84:380–392

Siegel JP (1988) Effects of interferon-γ on the activation of human T lymphocytes. Cell Immunol 111:461–472

Sprekeler R (1988) Die Behandlung der rheumatoiden Arthritis (RA) mit rekombinantem Gamma-Interferon. In: Heidemann E (ed) Klinische Erfahrungen mit Interferon beta und gamma. Zuckschwerdt, Munich, pp 35–41 (Aktuelle Immunologie, vol 4)

Stanton GJ, Weigent DA, Fleischmann WR, Dianzani F, Baron S (1987) Interferon review. Invest Radiol 22:259–273

Stella CC, Cazzola M, Ganser A, Bergamaschi G, Meloni F, Pedrazzoli P, Bernasconi P, Invernizzi R, Hoelzer D, Ascari E (1988) Recombinant gamma-interferon induces in vitro monocytic differentiation of blast cells from patients with acute nonlymphocytic leukemia and myelodysplastic syndromes. Leukemia 2:55–59

Talmadge JE, Tribble HR, Pennington RW, Phillips H, Wiltrout RH (1987) Immunomodulatory and immunotherapeutic properties of recombinant γ-interferon and recombinant tumor necrosis factor in mice. Cancer Res 47:2563–2570

Tentori L, Fuggetta MP, d'Atri S, Aquino A, Nunziata C, Roselli M, Ballatore P, Bonmassar E, de Vecchis L (1987) Influence of low-dose beta-interferon on natural killer cell activity in breast cancer patients subjected to chemotherapy. Cancer Immunol Immunother 24:86–91

Thompson JA, Cox WW, Lindgren CG, Collins C, Neraas KA, Bonnem EM, Fefer A (1987) Subcutaneous recombinant gamma interferon in cancer patients: toxicity, pharmacokinetics, and immunomodulatory effects. Cancer Immunol Immunother 25:47–53

Tyring SK, Cauda R, Ghanta V, Hiramoto R (1988) Activation of natural killer cell function during interferon-α treatment of patients with condyloma acuminatum is predictive of clinical response. J Biol Regul Homeost Agents 2:63–66

Vance JC, Bart BJ, Hansen RC, Reichman RC, McEwen C, Hatch KD, Berman B, Tanner DJ (1986) Intralesional recombinant alpha-2 interferon for the treatment of patients with condyloma acuminatum or verruca plantaris. Arch Dermatol 122:272–277

Vilcek J, Kelker HC, Le J, Yip YK (1985) Structure and function of human interferon-gamma. In: Ford RJ, Maizel AL (eds) Mediators in cell growth and differentiation. Raven, New York, pp 299–313

Virelizier JL, Arenzana-Seisdedos F (1985) Immunological functions of macrophages and their regulation by interferons. Med Biol 63:149–159

Vonderheid EC, Thompson R, Smiles KA, Lattanand A (1987) Recombinant interferon alfa-2b in plaque-phase mycosis fungoides. Arch Dermatol 123:757–763

Von Wussow P, Block B, Hartmann F, Deicher H (1988) Intralesional interferon-alpha therapy in advanced malignant melanoma. Cancer 61:1071–1074

Weiner LM, Steplewski Z, Koprowski H, Litwin S, Comis RL (1988) Divergent dose-related effects of γ-interferon therapy on in vitro antibody-dependent cellular and nonspecific cytotoxicity by human pheripheral blood monocytes. Cancer Res 48:1042–1046

Welsh RM (1984) Natural killer cells and interferon. CRC Crit Rev Immunol 5:55–93

Wickramasinghe L, Hindson TC, Wacks H (1989) Treatment of neoplastic skin lesions with intralesional interferon. J Am Acad Dermatol 20:71–74

Yoshida R, Murray W, Nathan F (1988) Agonist and antagonist effects of interferon α and β on activation of human macrophages. J Exp Med 167:1171–1185

Zabel P, Kreiker C, Schlaak M (1988) Kombinationsbehandlung von Bronchialkarzinomen mit Polychemotherapie und Interferonen. In: Hofschneider PH (ed) Ergebnisse der Beta-Interferon-Therapie bei chronisch-aktiver Hepatitis B, Multipler Sklerose und Krebserkrankungen. Zuckschwerdt, Munich, pp 76–82 (Aktuelle Immunologie, vol 3)

Interferons in Genital HPV Disease

G. Gross

Introduction

Human papillomaviruses (HPV) are recognized to be a group of small DNA viruses which induce an increasing spectrum of squamous epithelial tumors (warts and papillomas), and of flat, partially subclinical lesions. Furthermore, papillomavirus DNA can be present in clinically and histologically asymptomatic epithelial tissue in the form of latent infections.

Papillomaviruses are thought to infect cells of the basal cell layer of the skin where they induce proliferation of the target cells by as yet unknown mechanisms. Productive virus infection has been demonstrated only in benign lesions and is confined to the stratum granulosum and corneum, indicating that only epithelial cells at a certain stage of differentiation are able to support virus replication (Fig. 1). Thus, proliferating malignant cells represent a nonproductive system for papillomavirus infection [49].

There is evidence to support a correlation between certain phenotypes of HPV and specific morphological and histological patterns of disease [17, 20, 36]. HPV lesions may be single or multiple and the majority of cutaneous lesions resolve within a year, either spontaneously or following the use of simple local therapy. Spontaneous involution of genitally localized HPV lesions such as condylomata acuminata is well known but poorly understood. About one third of primary genital warts regress spontaneously within 6 months. There remain two thirds which do not regress [39]. Without treatment, the warts may grow in size and spread to other anatomic sites. That the lesions persist, are severe, or multiply is probably due to a failure of cell-mediated immunity (Jablonska and Majewski, this volume). Persistent lesions cause many patients considerable discomfort, personal trauma, and also difficulties in hygiene. Recurrences are known to occur, even though lesions are no longer detectable either clinically or histologically. Because of the persistent and recurrent nature of this disease, repeated therapy is often required. Prolonged infections may be a serious risk for the development of cervical intraepithelial neoplasia and carcinoma. Genital HPV infections are increasingly frequent. Today about 10% of the population of the United States between the ages of 15 and 50 years is infected [40]. One cause might be the lack of a safe and efficient therapy. Due to the association with the development of genital carcinoma, it has become important to focus new therapies on the eradication of HPV diseases.

Commonly used genital wart treatments include topical applications of caustic agents such as podophyllin, 5-fluorouracil and idoxuridin [39], the lat-

Genital Papillomavirus Infections
Eds.: Gross/Jablonska/Pfister/Stegner

Table 1. Modes of treatment in HPV-associated diseases reported in the literature. *Methods still in use*

Chemotherapy
Podophyllin/podophyllotoxin
Trichloracetic/bichloracetic acids
5-Fluorouracil
Bleomycin
Methotrexate
Thiotepa
Colchizin
Surgery
Curettage, scissors
Cryotherapy
Electrocautery
CO_2-laser, (neodym-yag-laser)
Antiviral substances
Idoxuridin
Aciclovir
Adenin-arabinosid
Interferon
Immunotherapy
Autogenous vaccine
Levamisol
Transfer factor
Viral antigens
Inosiplex
Interferons
Miscellaneous approaches
Infrared coagulation
Antibiotics
Salicylic acid
Hormones
Corticosteroids
X-rays
Vitamin A, aromatic retinoids
Hypnosis

ter being largely ineffective. Podophillin is known to induce local complications such as pain, itching, or discharge after treatment or to have systemic toxicity [39]. Today, with the use of podophyllotoxin, side effects are less commonly seen [39; Beutner, this volume; von Krogh, this volume]. Only limited or mixed success has been reported from all aggressive chemotherapy with substances such as bleomycin and methotrexate (Table 1). Physical removal of visible lesions is usually employed when topical agents have failed or disease is extensive. Ablative methods include CO_2 laser, cryotherapy with liquid nitrogen, curettage, electrocautery or electrodessication, or a combination of these. All of these therapies may be painful for the patient and may leave

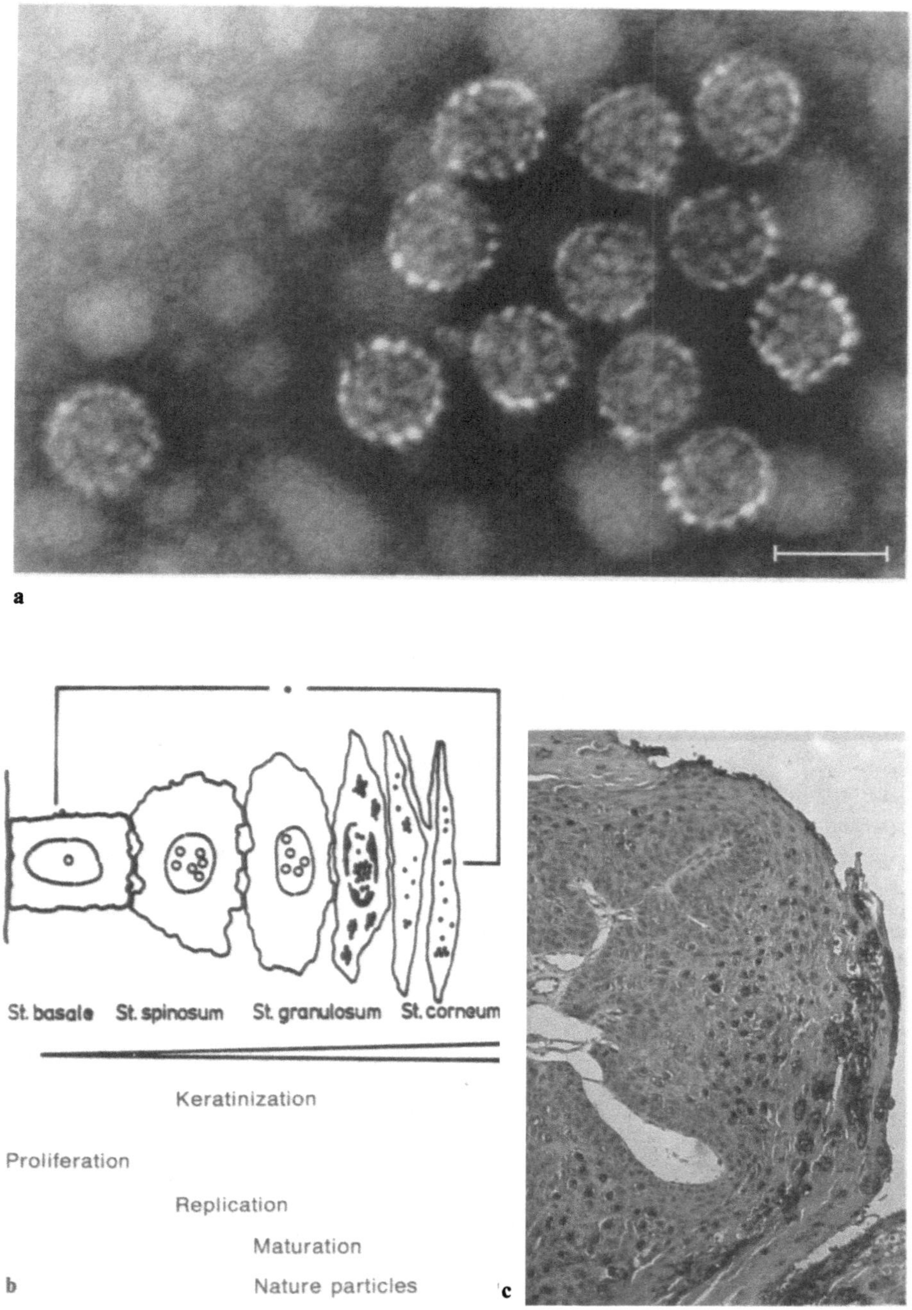

Fig. 1 a–c. **a** Mature papillomavirus particles, diameter of one particle 50 nm. **b** Hypothesis on the pathogenesis of genital condyloma acuminatum (●, virusparticle). **c** Peroxydase-antiperoxydase-assay using polyclonal antibody to group-specific antigen of BPV to detect structural antigens of papillomavirus in a viral wart scars. Recurrence of disease is a major problem with ablative therapy, and recurrence rates range from 7.5% to 80%. This is probably due to the presence of HPV DNA sequences in apparently normal tissue, as for instance shown in surrounding tissue following CO_2 laser removal of genital warts [11]. Apparently, surgical removal of proliferative lesions can leave the virus to reinfect, replicate, and transform surrounding epithelium. Further causes of recurrent disease may be incomplete removal of warty tissue and the multifocality of HPV infection, which is commonly seen in the anogenital region.

Recently, attention has been increasingly focused on the interferons, which possess antiviral, antiproliferative, and immunomodulatory properties [1, 50] and, further, appear to represent a rational choice as a therapy for severe, recalcitrant warts which prove to be resistant to rigorous conventional treatments. Theoretically, in therapeutic strategies the complex nature of HPV diseases must be taken into consideration. Conventional therapies such as surgery, chemotherapy, or others cannot be curative since they do not take into account the pathogenesis of HPV infection (Fig. 1), the role of the immune system, the nature and the course of the disease, the clinical picture with its multifocality in the genital tract, nor the fact that the genital HPV disease is a sexually transmitted condition.

To give an overview of the treatment of genital HPV disease with interferons, it is important to bear in mind the difficulties imposed by disease heterogeneity in terms of morphological and histological features and the HPV genotype, and of duration and severity of disease. In addition, the type of interferon and the treatment regimen need to be considered. It is evident that the immune status of the patients may influence treatment outcome, and there is also evidence to suggest a link between the immune status and the presence of certain HPV genotypes. Additionally, recent studies indicate that the duration of disease, at least in the case of condylomata acuminata, may effect response to therapy with interferons [11].

The terms leukocyte, fibroblast, and immune interferon have been eliminated. Interferons are now classified as alpha, beta, and gamma, according to their antigenic and biological properties (Table 2). More recently highly purified interferons have been produced by recombinant DNA technology.

Table 2. Terminology of available interferons

New	Old
IFN-alpha-N1	Lymphoblastoid IFN
IFN-alpha-2a	Recombinant alpha-A-IFN
IFN-alpha-2b	Recombinant alpha-2-IFN
IFN-alpha-2c	Alpha-2arg-IFN
IFN-beta	Fibroblast-IFN
IFN-gamma	Immune-IFN

In Vitro Experiments with Interferons in Papillomavirus Infections

Interferon was discovered by Isaaks and Lindenmann in 1957 [34]. These investigators identified substances of a group of glycoproteins that were released into the medium surrounding virus-infected cells and seemed to inhibit growth of other viruses. This antiviral activity is not directed against specific viruses, but both DNA and RNA viruses are inhibited. The antiviral effect is indirect. On binding to specific receptors on the cell membrane, interferons induce a multitude of effects which depend on the class of interferon and on the type of cells to which they are bound [50, 59]. Of main importance with regard to the establishment of the antiviral state is the induction of three cellular proteins by the interferons. These include a protein kinase, a (2-5-)-oligoadenylate synthetase (OAS) and an endoribonuclease. The activation of these enzymes may result in both breakdown of viral replication, including bovine papillomavirus, and of viral RNA and inhibition of viral protein synthesis. A large number of other interferon-induced effects may be important in virus inhibition. For instance, interferons may block receptor-mediated endocytosis of viral particles. Thus, a virally infected cell may be stimulated to secrete interferon which binds to uninfected cells and blocks virus penetration. Furthermore, interferons also affect plasma membranes, cellular cytoskeletons, cell growth, and differentiation. These may be important influences, in particular on papillomavirus expression. In addition to induction of antiviral activity, alpha-, beta-, and gamma-interferons all modulate immune responses particularly of the cellular immune system. For example, interferons may stimulate natural killer cell activities and the activity of macrophages. Furthermore, they appear to block humoral immunity, including antibody production [42].

Utilizing in vitro cell systems consisting in mouse embryo fibroblasts Turek et al. showed that mouse interferon may prevent the acute transformation of bovine papillomavirus [63]. Furthermore, the presence of bovine papillomavirus genomes directly correlated with the transformed phenotype, and, finally, interferon inhibited bovine papillomavirus DNA replication in vitro. From these studies one may postulate that interferons prevent the expression of episomal papillomavirus DNA. Studies on simian virus 40 (SV 40) suggest that virus genomes integrated into host chromosomal DNA may remain protected from the antiviral actions induced by interferon [16]. This would be of eminent importance since most papillomavirus-associated malignancies contain integrated genomes [49]. Recently, however, antiproliferative effects of recombinant interferon-alpha and of recombinant interferon-gamma were examined against cell lines derived from a human epidermoid carcinoma of the uterine cervix. These lines were Caski and SiHa, containing the genome of HPV type 16 and SW 756, containing HPV 18 DNA. Interferon-gamma suppressed the growth of all three lines, whereas interferon-alpha was effective in this regard only against SW 756. In this line the two interferons together inhibited cell replication to a greater extent than either interferon alone [37].

Clinical Experiences in Genital HPV Disease

Historical Aspects

Since the 1970s, years before DNA hybridization confirmed virus involvement, clinical experience with interferon in virus-induced papillomas has been collected. The first observations were reported by Krusic et al. in 1972 [41] and by Ikic et al. [32, 33], who treated condylomata acuminata cervical intraepithelial neoplasia (CIN) and cervical carcinoma. In another historical therapeutic approach with natural leukocyte interferon (IFN alpha N1), Strander and Cantell observed spontaneous regression of bilaterally sited plantar warts [61]. This led to the postulation that interferons might be useful as a specific and noninvasive treatment for skin warts. Later on, again the group of Strander was the first to treat severe cases of juvenile laryngeal papillomatosis with interferon given parenterally [29]. Today the systemic interferon treatment of juvenile laryngeal papillomatosis has been confirmed in a number of studies [29, 52, 67].

With the improvements in tissue culture techniques, broad application of DNA recombinant technology, and subsequent development of highly purified products, large quantities of interferon for clinical trials are now available. Different interferons administered by different routes have been evaluated in the treatment of HPV infections, particularly in genital warts. In the meanwhile a variety of clinical trials of interferons have been performed. Mostly interferon-alpha from leukocytes or lymphoblastoid cells (IFN-alpha N1) or recombinant interferon-alpha (interferon-alpha 2a, interferon-alpha-2b or interferon-alpha-2c) were used. Only recently interferon-beta, primarily used for skin warts, and interferon-gamma derived from *Escherichia coli* were also utilized for clinical investigations in genital warts and other genitally located papillomavirus-associated diseases.

The very nature of HPV-induced genital lesions makes interferon a logical therapeutic approach (Fig. 1): The antiviral effects of interferons should limit viral replication in the new or regenerating epithelium. The antiproliferative or cell cycle inhibitory effect should slow the rapid growth of transformed epidermal cells. These actions, in combination with the immunomodulatory effects of interferons should clear up visible lesions and possibly eliminate the viral DNA responsible for the transformation events. Interferons can be given topically as gel or cream, intralesionally, or parenterally in the form of intramuscular, subcutaneous, or intravenous injections. Furthermore, interferon therapy can be applied as an adjuvant to conventional treatment modalities such as ablative treatments, podophyllin or 5-fluorouracil therapy. Finally, interferon can be combined with other treatment modalities such as retinoids [46]. So far it has not been clear which is the optimal treatment regimen for each type of interferon in the different HPV-related diseases, which is the ideal route, the dose, the frequency of administration, and the duration of therapy producing maximal benefit. The evaluation of these issues is complicated by the unpredictable nature of HPV disease, including the role of the host's immune status in disease resolution.

Topical Treatment

Initial studies suggested that interferon-alpha and interferon-beta may be effective when administered topically [33, 44, 65, 66]. However, in a recently conducted randomized, controlled study of topically administered interferon-alpha in women with condylomata acuminata, no beneficial effects of interferon with or without nonoxynol-9 were demonstrated as compared to placebo [38]. In a study with recombinant interferon-alpha 2c gel containing 1 million IU interferon-alpha/1 g gel, lesions disappeared completely in five of six patients with condylomata acuminata [54]. In another open study using interferon-beta gel more primary condylomata acuminata regressed completely in 15 of 18 cases (83%) [43]. In a placebo-controlled double-blind study using interferon-beta at a dose of 0.5×10^6 IU/1 g gel, 15 of 24 patients showed complete or partial remission in contrast to 2 of 11 patients receiving placebo [66]. Discrepancies in the results of reported studies using topically administered interferons for treatment of condyloma acuminatum may result from differences of study design or of the patient populations that have been evaluated. These preparations may not deliver adequate quantities of interferon to the papillomavirus infected tissues, especially in cornified squamous cell lesions of the outer genitalia (Table 3).

Table 3. Topical IFN-treatment of condyloma acuminatum of external genitalia

Study	IFN	Dose (MU/g)	Route	Complete remission	
				IFN	Placebo
Ikic et al. [33]	Alpha-N1	0.04	Ointment	4/10	2/11
De Virgiliis et al. [66]	Beta	0.5	Gel	15/24	2/11
Keay et al. [38]	Alpha-N1	1.0	Gel	10/30	8/28

Intralesional Treatment

In contrast to topical administration of interferons, the intralesional route was shown to be effective in a number of controlled and uncontrolled studies [10, 11, 14, 51, 64, 67]. In these trials in which up to five warts per patient were injected with interferon or placebo twice or three times weekly for periods of up to 8 weeks, rates of complete resolution of interferon-injected warts pertain only to the injected warts and are similar among the different studies, ranging

Table 4. Intralesional IFN-treatment of condyloma acuminatum (dose 1.2–2.0 MU/injection), controlled studies

Study	IFN	Schedule	Complete remission	
			IFN injection	Placebo/control injection
Scott and Csonka [57]	Beta	3 × /week for 3 weeks	0/11[a]	0/11
Eron et al. [11]	Alpha	3 × /week for 3 weeks	42/116	20/116
Vance et al. [64]	Alpha	3 × /week for 3 weeks	16/30	4/29
	Alpha	3 × /week for 3 weeks	6/32	
Friedman-Kien et al. [14]	Alpha-N1	2 × /week for 8 weeks	41/66	14/66
Reichmann et al. [51]	Alpha-N1	3 × /week for 4 weeks	7/15	4/18
	Alpha		11/23	
	Beta		10/20	

[a] 11/11 partial remission

from 36% to 62%. Complete regression was seen in about 20% of placebo-treated lesions. Side effects were similar to those seen in patients receiving interferon parenterally and consisted of influenza-like symptoms or of reversible decrease in white blood cell counts and of slight elevation of liver enzymes. These adverse reactions have frequently lead to dose reductions (Table 4).

The intralesional administration of interferons has only limited practical value since beneficial effects on noninjected lesions have not been observed regularly. Thus, this therapy has no marked advantage in comparison to conventional therapies. Invisible lesions, for instance papillomas in the urethra or in the anal canal, do not respond, and subclinical papillomavirus infections also appeared to be unresponsive. Nevertheless, intralesionally administered recombinant interferon-alpha 2b has recently been licensed in the USA by the Food and Drug Administration for the treatment of anogenital warts.

Parenteral Treatment

Parenteral administration of interferons should theoretically treat all HPV-infected epithelial cells. According to experiments in small animals such as mice or rats, after injection of interferon there is an initial bolus in the circulation, which rapidly falls. Some of this is bound by receptors of different cells and organs and some is excreted through the kidneys. In addition to the antiviral effects, while there is a direct effect of interferon to the HPV-positive cells, there is an indirect effect via the cellular immune system [42, 59]. Also, a range of toxic side effects of interferon have been measured. These were ini-

Table 5. Parenteral IFN-Treatment of condyloma acuminatum in large, controlled studies

Study	IFN	Dose (MU)	Route/ Schedule	CR of patients treated	
				With IFN	With placebo
Schonfeld et al. [56]	Beta	2.0	IM daily × 10 days	9/11	2/11
Gall et al. [15]	Alpha-N1	5.0	IM daily × 14–28 days	11/16	Not done
	Alpha-N1 Alpha-N1	3.0 2.0			Not done
Costa et al. [9]	Beta	2.0	IM daily	6/6 vagina 5/5 vulva 3/8 cervix	0/4 0/1 0/6
Gross et al. [25]	Gamma	1.0	SC daily × 7 days, 4 weeks pause (1 cycle) up to 3 cycles	12/20[a]	7/21[a]

[a] Complete and partial remission

tially thought to be due to contamination of the interferon preparations, but when purified interferon preparations became available, the toxic effects were still seen. These include fever and local inflammation and at higher doses effects on the central nervous system. Most of the studies with parenteral administration of interferon for the treatment of condyloma acuminatum have utilized interferon-alpha preparations in open uncontrolled studies (Table 5) [52, 67]. An exception was one study that employed interferon-beta compared to placebo [56]. In this trial 9 of 11 interferon recipients experienced complete resolution of all lesions, compared to 2 of 11 patients in the placebo group. All patients were women with disease of relatively short duration. Diagnosis was not confirmed by histologic examination in all cases and a cross-over design was used prohibiting accurate determination of recurrence rates. The more recent studies using parenterally administered interferon-alpha for the treatment of condyloma acuminatum have focused on patient populations with disease of long duration which has not responded to conventional modes of treatment. In 1986, an investigation using a cross-over design compared interferon-alpha 2a in a low-dose regimen (1.5×10^6 IU/day) and a high-dose interferon-alpha 2a regimen (18×10^6 IU/day) given subcutaneously into the lateral abdominal skin during 1 week followed by a 4-week therapy-free interval [23]. The low-dose group experienced a significantly lower rate of side effects and laboratorial abnormalities and the efficacy rate was superior by 60% after three treatment cycles (average therapy duration 15 weeks). Comparable results were obtained recently with low doses of recombinant interferon-gamma (50 μg and 100 μg) using the same treatment schedule. Higher daily doses (200 μg and 400 μg) led, however, to worse results or the genital warts

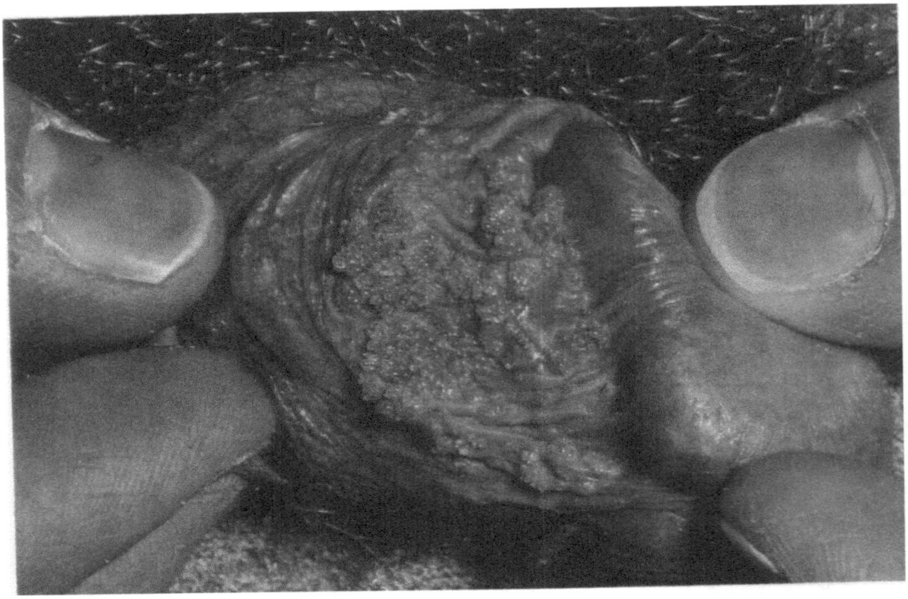

a

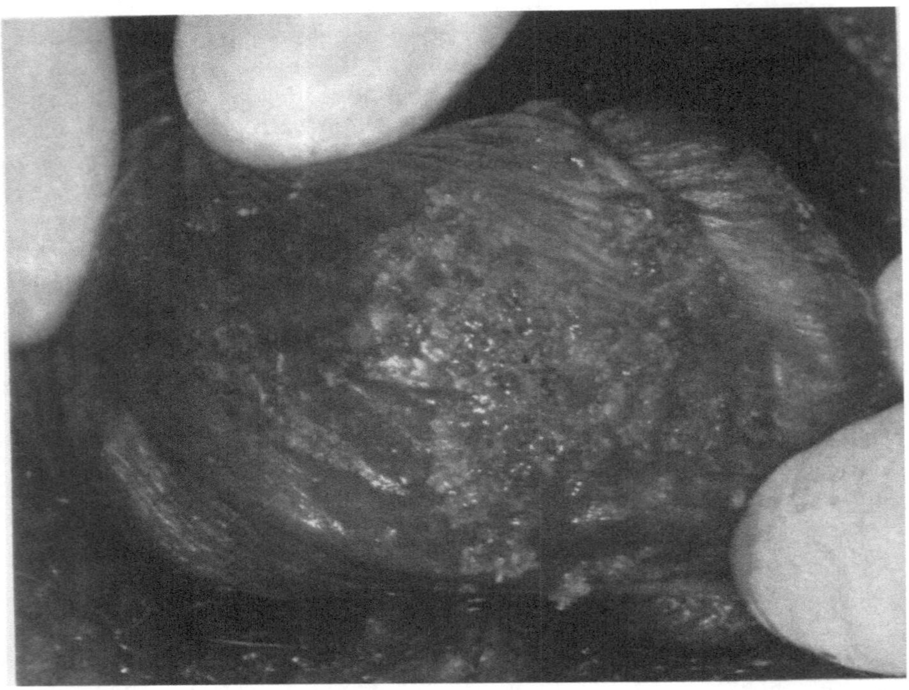

b

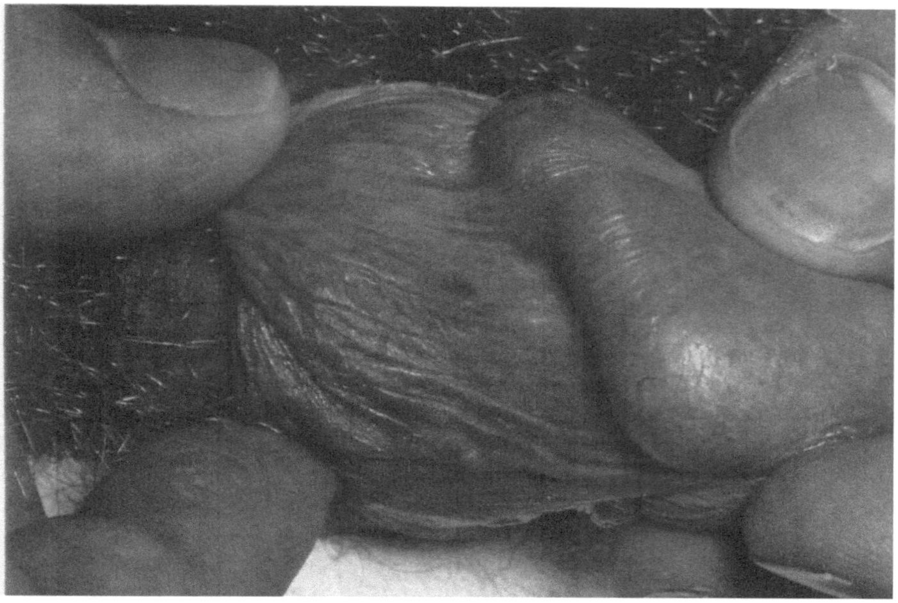

c

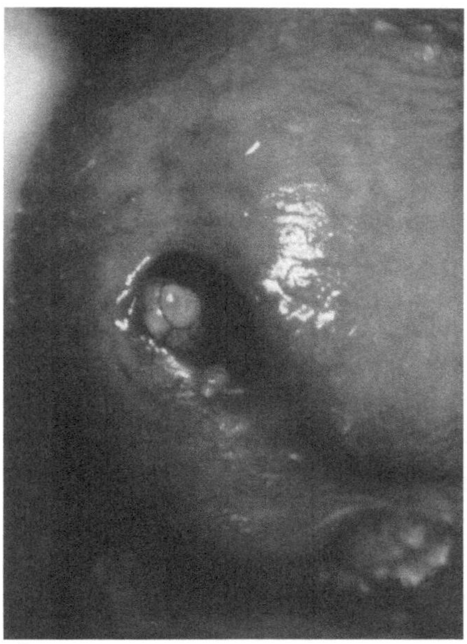

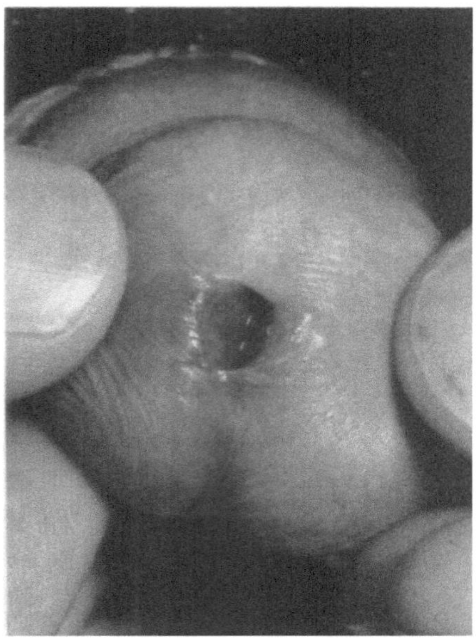

d e

Fig. 2 a–e. Regression of preputial condylomata acuminata (HPV 6 DNA positive, Southern blot) during parenteral low-dose interferon alpha2a-interval therapy. **a** Before therapy. **b** Inflammation after the second treatment cycle (week 10). **c** Complete remission without scars. **d** Condyloma present in the fossa navicularis urethrae of the same patient (prior to therapy). **e** Complete remission together with involution of preputial condylomata did not respond at all (Gross et al., in preparation). These observations stimulated placebo-controlled multicenter studies (Table 5). Forty-one patients with recalcitrant condylomata acuminata were treated with subcutaneous interferon-gamma at a dose of 50 μg or with placebo. Twelve out of 20 (60%) treated with interferon-gamma showed a clinical response (complete or partial response) in contrast to 7 out of 21 patients (33%) treated with placebo [25]. In an ongoing, larger second study 100 patients with chronic recalcitrant genital warts are being treated with either 100 μg interferon-gamma daily or with placebo.

There are a number of advantages to parenteral therapy. Multifocal lesions located at external sites, in the urethra, or in the anal canal do respond (Fig. 2) and recurrences are rarely seen (less than 10%, follow-up 14 months (Gross et al., in preparation). In contrast to continuous therapy (three injections per week during 4–8 weeks) the systemic side effects are less severe and are seen less frequently, during low-dose systemic interval treatment. Flu-like symptoms regularly disappear within the 3rd or 4th day of therapy and, furthermore, serum-neutralizing factors are less frequently detectable in patients receiving interferon at low-dose and at intervals than in patients groups treated with high-dose interferon. Disadvantages of parenteral low-dose interval therapy are the long duration of therapy and the lack of efficacy in immunocompromised and HIV-positive patients (Gross, unpublished data), which has been also reported after intralesional interferon therapy [10]. The same is true for drug addicts [28] and for patients suffering from Hodgkin's disease [26].

Adjuvant Treatment

Interferon can be given as an adjuvant both parenterally and topically after removal of the lesions.

Parenteral Adjuvant Treatment

Parenteral interferon treatment can be safely combined with CO_2-laser therapy. Using low-dose interferon-alpha interval therapy there were complete remissions at follow-up after 8 months in about 80% of patients suffering from therapy-resistant anogenital condylomata acuminata [12, 30]. Similar results were observed by Tiedemann and Ernst [62] in patients with genital warts treated by electrocautery and by intramuscular injections of interferon-alpha (internal therapy: daily dosage 5×10^6 IU during 5 consecutive days, followed by a 3- to 4-week pause). Side effects seen during this therapy were identical with those seen in parenteral monotherapy using interferon-alpha. These results are now being tested for significance in a number of ongoing, prospective, randomized studies. Adjuvant parenteral interferon therapy can also be given after removal of the genital warts by other methods such as cryosurgery, podophyllin, or 5-fluorouracil. Again, so far immunodefective patients and HIV-positive patients did not show comparable benefit from this treatment.

Local Adjuvant Treatment

In contrast to the disappointing effects of topical interferon on genital warts [44, 65–67], application of interferon-alpha hydrogel after destruction of the epithelial barrier by CO_2-laser, electrocautery, or even cryosurgery leads to high concentrations of interferon at the target basal cells of HPV infection (Fig. 1). This was proven in clinical studies not only in immunocompetent patients [2], but early findings suggest activity also in immunosuppressed and in HIV-positive individuals [24, 26]. It is suggested that the mechanism of the adjuvant local treatment is independent of the systemic cellular immunity. The colposcopy-guided CO_2 laser is of special value in removing genital warts and even subclinical flat lesions. This treatment in combination with local interferon gel therapy appears to be highly effective, as shown in a placebo-controlled double-blind study using interferon-alpha 2b gel at a daily dose of 1 million IU/g gel. Twenty of 23 patients, in comparison to 9 of 18 patients under placebo, showed a recurrence-free complete remission [53]. According to a small placebo-controlled study, interferon-beta is likely to be superior to interferon-alpha in the local treatment, which might be due to the higher tissue affinity of interferon-beta [66]. Further controlled studies are necessary for final evaluation. The main attraction of the local therapy is the ease of administration and the lack of side effects. Possibly, efficacy of topical interferon may even improve by using available methods which increase the transport of interferon deeper into the epithelial tissues and lead to an increased availability of interferon at the target cells of HPV infection.

Interferons in Combination with Other Treatments

Interferon also can be given both parenterally or topically together with other primary therapies such as retinoids, which are vitamin A derivatives that also have an antiproliferative effect on various epithelial tissues and some effect on HPV-infected tissue. This was shown earlier in uncontrolled studies in patients with common warts [3, 18], epidermodysplasia verruciformis, or laryngeal papillomatosis [67]. In a recent trial genital warts did not respond to the vitamin A derivative isotretinoin, but only to the combination of it with interferon-alpha N1 [46].

Interferons in HPV-Associated Neoplasia of the Cervix Uteri, Vulva, Vagina, and Penis

As early as 1972 Ikic et al. reported on the response of cervical intraepithelial neoplasia (CIN) to local application of interferon-alpha [32] (Table 6). Similar results were seen by Möller et al. [44]. In three of six cases of CIN II and CIN III complete responses were noted. In the meanwhile a number of reports have been communicated on successful treatment of CIN vulvar intraepithelial neoplasia (VIN) and vaginal intraepithelial neoplasia (VAIN) (Table 6).

Table 6. Interferon treatment of CIN, VIN, VAIN, and PIN

Diagnosis	Type of interferon	Route	Number of evaluable patients	Success of therapy			Authors
				CR	PR	NC	
CIN II–III	Alpha	Gel	15	3	–	12	Ikic et al. [33]
CIN II–III	Alpha	Gel	6	3	2	–	Möller et al. [44]
PIN I[a]	Beta	Intralesional	1	1	–	–	Gross et al. [19]
CIN I–III	Beta	Intralesional/gel	11	6	2	3	de Palo et al. [47]
VIN III	Beta	Intralesional/gel	2	1	–	1	de Palo et al. [47]
VAIN I–II	Alpha	Gel	8	3	2	3	Vesterinen et al. [65]
CIN II–III	Leuko	Gel	7	2	3	2	Choo et al. [7]
CIN II–III	Beta	Perilesional	16	8	2	6	de Palo et al. [48]
CIN III	Alpha	Perilesional	7	6	n.e.	n.e.	Choo et al. [8]
CIN III	Beta	Perilesional	5	2	n.e.	n.e.	Choo et al. [8]
CIN II, III	Leuko	-Gel-kappe	13	3	5	5	Byrne et al. [4]
PIN III[b]	Alpha	SC	3	1	2	–	Gross et al. [22]
CIN I–III	Alpha	SC/gel	6	3	n.e.	n.e.	Schneider et al. [54]
VIN II[b]	Alpha	SC	1	1	–	–	Slotman et al. [60]
CIN II–III	Beta	Perilesional/gel	24	14	3	7	Neis et al. [45]
CIN II–III	Gamma	Gel	24	10	9	5	Schneider et al. [55]
CIN II–III	Gamma	Perilesional	8	4	n.e.	n.e.	Iwasaka et al. [35]
CIN I–II	Alpha	Vaginal cream	9	4	n.e.	n.e.	Yliskoski et al. [68]

CR, complete remission; PR, partial remission; NC, no change; n.e., not evaluated; CIN, cervical intraepithelial neoplasia; PIN, intraepithelial neoplasia of the penis; VIN, intraepithelial neoplasia of the vulva; VAIN, vaginal intraepithelial neoplasia

[a] Initially CR, later recurrence of flat penile lesions (histology: PIN III, virustype HPV 16 (Southern blot)

[b] Low-dose interval therapy

Today there is evidence that interferon in gel form or administered by the intra- or perilesional route are also active in these conditions. Due to very small studies, the results of parenteral administration of adjuvant interferon therapy in CIN are not yet definitely evaluable. Nevertheless, a placebo-controlled study showed that in contrast to earlier reports, leukocyte interferon containing gel has no significant effect on CIN [4]. The cases treated in this study preferentially were CIN III. In contrast to remissions of HPV 6-positive lesions, HPV 16-positive and HPV 6 and HPV 16 double-infected CIN lesions did not respond to the local interferon treatment. Slotmann et al. showed that CIN II, VIN II, and VAIN II respond to parenteral interferon interval therapy in the same way as genital warts do [23, 25, 60], and recently there was a report suggesting that intralesional interferon-gamma treatment is also effective in CIN (Table 6) [35]. Bowenoid papulosis of the male genital skin and of the vulva was effectively treated by low doses of interferons given at intervals only after treatment of long duration [22]. Bowenoid papulosis is a severe intraepithelial neoplasia of the vulva (VIN III) and of the penis (PIN III) consisting of multiple and multifocal flat and papular lesions, the biological course of which is not identical with that of Bowen's disease, though both diseases are HPV associated, especially HPV 16 [20, 21]. Preliminary results of an open randomized trial of recombinant interferon-gamma in patients suffering from this disease suggest that, in contrast to condylomata acuminata, bowenoid papulosis lesions respond better to continuous than to intermittent subcutaneous interferon-gamma injections [27]. This probably indicates that VIN, PIN, and the other intraepithelial neoplasias associated with HPV respond due to the antiproliferative and less to the immunomodulatory effects of interferon. An alternative therapeutic approach to treating bowenoid papulosis, CIN, VIN, and VAIN consists in removal of the neoplastic tissue by CO_2-laser or by electrocautery and adjuvant topical treatment with interferon-alpha or interferon-beta gel.

Side Effects of Interferon Therapy

Most patients treated intralesionally or systemically with interferon exhibit side effects. It is generally assumed that these side effects are dose related. The most common complaint of intralesionally treated patients is pain at the injection site, especially in individuals generally receiving interferon at doses of more than 1.0 MU per injection. The most frequent side effects are fever and flu-like symptoms which often lead to tolerance after the second or third injection. Myalgias, chills, headache, and backache, gastrointestinal discomfort, and emotional lability are seen less frequently. Laboratory abnormalities reported are leukopenia, thrombocytopenia, which in general is reversible after the end of therapy [1]. Interferon may lead to rejection of kidney transplants. Thus, interferons are generally prohibited in renal transplant recipients, who are very often suffering from HPV disease. Whether locally applied interferons can also lead to systemic side effects, especially to a risk of transplant, is not yet clear.

Factors Influencing the Efficacy of Interferons in Genital HPV Disease

The efficacy of interferon is difficult to evaluate because about 22%–25% of condylomata acuminata and about 40% of CIN I and II regress spontaneously or after injections of placebo. In contrast to genital warts less than 3 months old, older warts do not respond to placebo [11].

The importance of the virus type with regard to the efficacy of interferon in genital warts is not definitely clear. HPV 16- or 18-associated lesions seem to have a lower response rate to interferon than those induced by HPV 6 or 11. In general HPV infections such as genital warts with the presence of episomal HPV genome respond better than lesions with moderate or severe dysplasia which are more closely associated with HPV 16 and 18. Considering a case report of HPV 11 and 16-related penile warts in a young man whose condition improved during intralesional interferon-beta therapy and then underwent neoplastic change following cessation of therapy, one might argue that interferon may influence the physical state of HPV DNA [19].

Immunosuppressed and immunodeficient patients such as HIV positive individuals and patients suffering from Hodgkin's disease do not respond in the same way as normal patients to systemic and local interferon treatment [10, 24, 26]. The same is true for drug addicts, also for individuals patients using cannabis [28].

An accurate prediction of future responders to interferon therapy would be useful in order to select appropriate therapy. Although there are conflicting reports of immune defects in otherwise healthy patients with condyloma acuminatum [5], the level of NK-cell activity seems to be one helpful factor [6, 42, 58] (see also Jablonska and Majewski, this volume). Although no clear-cut differences in NK activity were detected, patients who eventually responded to intralesional interferon-alpha 2b had a significantly greater increase of NK activity than did nonresponders. This difference in NK activity was maintained following completion of therapy. Cauda and coworkers observed that lymphocytes from otherwise healthy condyloma acuminatum patients exhibited decreased interleukin-2 and interferon-gamma production, depressed NK-cell activity, and a reduced T-cell helper/supressor ratio, associated with a marked increase in supressor T cells [6].

Conclusions

Today there is no more doubt that papillomaviruses are susceptible to interferon-induced activity in vitro and in vivo. There is evidence that alpha-, beta-, and gamma-interferons are clinically useful for treating genital HPV infections. Interferons have been applied to all forms of genital disease caused by papillomaviruses. So far it is uncertain which of the known human interferons could offer the greatest therapeutic advantage. Interferons have antiviral, antitumoral, and immunomodulating properties. It has become clear that lower doses of interferons are more effective than higher doses in treating genital warts. Undesirable side effects at these low doses are generally tolerable. Common toxic reactions include reversible fever, chills, and mild leukopenia. Furthermore, induction of anti-interferon antibodies is less frequently seen in patients treated at low doses especially when given at intervals. In contrast to the intralesional route, which seems of only limited practical value in the genital HPV disease, the parenteral approach indicates significantly greater therapeutic advantages. The future of interferons in the treatment of genital papillomavirus infections seems to be, at least, as adjuvant treatment to other modalities such as surgical removal. Given parenterally after removal of genital warts interferons lead to prevention of recurrence, as shown in ongoing studies. The same seems to be true for interferon incorporated into in a hydrogel for topical application after surgery. The effect of this treatment seems to depend upon the dosage. This is in contrast to the lack of effect of topical interferons on the surface of genital and skin warts. Local adjuvant treatment using interferon-alpha or interferon-beta hydrogel has a special advantage since it is likely to be also useful in patients suffering from both genital wart disease and from immunosupression or immunodeficiency. In such patients parenteral and intralesional treatments have been shown to be less effective or even ineffective [10, 24, 26].

Further studies should examine the efficacy and toxicity of combinations of interferons such as interferon-gamma with interferon-alpha or interferon-beta. Furthermore, the role of interferon as an adjuvant with conventional therapies should be clarified. Also, intraepithelial neoplasias of the lower genital tract especially of the cervix (CIN) may be treated by local application of interferon gel either alone or in combination with superficial surgery such as CO_2-laser.

In summary, genital HPV disease management has profited greatly from the introduction of interferons since so far neither a specific antiviral therapy such as acyclovir in herpesvirus infection nor a vaccine against HPV is available. Understanding the biology of papillomaviruses, including the replication and dependence on the epithelia for their life cycle and also host immune responses to papillomavirus infections, may help us to develop better protocols for using interferons and other lympho- and cytokines to treat these recurrent infections. Interferons must be viewed as prototype cytokines of recombinant DNA technology since this provides new biologic agents – protein products such as interleukin-2 and tumor necrosis factor – which might also be used to treat papillomavirus-associated disease in the future.

References

1. Androphy EJ (1986) Papillomaviruses and interferon. In: Ciba Foundation. Symposium 120: Papillomaviruses. Wiley, New York, pp 221–228
2. Berthold J, Schöpf E, Völckers W et al. (1989) Topical application of interferon beta as adjuvant in the treatment of recurrent genital warts: results of a placebo-controlled double-blind study. J Invest Dermatol 93:541
3. Boyle J, Dick DC, Mackie RM (1982) Treatment of extensive virus warts with etretinate (Tigason) in a patient with sarcoidosis. Clin Exp Dermatol 8:33–36
4. Byrne MA, Möller BR, Taylor-Robinson D et al. (1986) The effect of interferon on human papillomavirus associated with cervical intraepithelial neoplasia (CIN). Br J Obstet Gynecol 93:1136–1144
5. Carson LF, Twiggs LB, Fukushima U (1986) Human genital papillomavirus infections: an evaluation of immunologic competece in the genital neoplasia-papilloma syndrome. Am J Obstet Gynecol 155:784–789
6. Cauda R, Tyring SK, Grossi CE (1978) Patients with condyloma acuminatum exhibit decreased interleukin-2 and interferon-gamma production and depressed natural killer cell activity. J Clin Immunol 7:304–311
7. Choo YC, Hsu S, Seto W et al. (1985) Intravaginal application of leukocyte interferon gel in the treatment of cervical intraepithelial neoplasia (CIN). Arch Gynecol 237:51–54
8. Choo YC, Seto WH, Hsu S et al. (1986) Cervical intraepithelial neoplasia treated by perilesional injection of interferon. Br J Obstet Gynecol 93:372–379
9. Costa S, Poggi UG, Palmisano L et al. (1988) Intramuscular β-interferon treatment of human papillomavirus lesions in the lower female genital tract. The Cervix and l.f.g.t. 6:203–212

10. Douglas JM, Rogers U, Judson FN (1986) The effect of asymptomatic infection with HTLV-3 on the response of anogenital warts to intralesional treatment with recombinant alpha-2 interferon. J Infect Dis 154:331–334
11. Eron LJ, Harvey L, Toy C et al. (1986) Interferon therapy of condylomata acuminata. N Engl J Med 315:1059–1069
12. Erpenbach K, Derschum W, Wiese H et al. (1989) Results of the combined laser and adjuvant interferon alpha 2b therapy for patients with therapy resistant anogenital condylomata acuminata. In: Gross G, Jablonska P, Pfister H, Stegner HE (eds) Genital papillomavirus infections. Advances in modern diagnosis and therapy, Hamburg (abstracts)
13. Ferency A, Mitao M, Nagai et al. (1985) Latent papillomavirus and recurring genital warts. N Engl J Med 313:784–788
14. Friedman-Kien AE, Eron LJ, Conant U et al. (1988) Natural interferon alpha for treatment of condylomata acuminata. JAMA 259:533–538
15. Gall SA, Hudges CE, Mounts P et al. (1986) Efficacy of human lymphoblastoid interferon in the therapy of resistant condylomata acuminata. Obstet Gynecol 67:643–651
16. Garcia-Blanco MA, Ghosh PK, Jayaram BM et al. (1985) Selectivity of interferon action in simian virus 40-transformed cells superinfected with simianvirus 40. J Virol 53:893–898
17. Gross G, Pfister H, Hagedorn M et al. (1982) Correlation between human papillomavirus (HPV) type and histology of warts. J Invest Dermatol 78:160–164
18. Gross G, Pfister H, Hagedorn M et al. (1983) Effect of oral aromatic retinoid (Ro10-9359) on human papillomavirus-2-induced common warts. Dermatologica 166:48–53
19. Gross G, Ikenberg H, Gissmann L (1984) Bowenoid dysplasia in human papillomavirus-16 DNA positive flat condylomas during interferon beta treatment. Lancet I:1467–1468
20. Gross G, Ikenberg H, Gissmann L et al. (1985) Papillomavirus infection of the anogenital tract: correlation between histology, clinical picture and virus type. Proposal of a new nomenclature. J Invest Dermatol 85:147–152
21. Gross G, Hagedorn M, Ikenberg H et al. (1985) Bowenoid papulosis. Presence of human papillomavirus (HPV) structural antigens and of HPV-16-related DNA sequences. Arch Dermatol 121:858–863
22. Gross G, Roussaki A, Schöpf E et al. (1986) Successful treatment of condylomata acuminata and bowenoid papulosis with subcutaneous injections of low-dose recombinant interferon-alpha. Arch Dermatol 122:749–750
23. Gross G, Ikenberg H, Roussaki A et al. (1986) Systemic treatment of condylomata acuminata with recombinant interferon alpha 2a: low-dose superior to the high-dose regimen. Chemotherapy 32:537–541
24. Gross G, Papendick U (1989) Recombinant interferon alpha 2c gel given as adjuvant to surgery: a method which leads to complete cure without relapse of genital papillomavirus infection even in immunocompromised patients. Arch Dermatol Res 281:142–143
25. Gross G, Degen W, Hilgarth M et al. (1989) Recombinant interferon gamma in genital warts: results of a multicenter placebo-controlled clinical trial. J Invest Dermatol 93:553
26. Gross G, Roussaki A, Pfister H (1989) Recurrent vulvar Buschke-Löwenstein's tumor-like condylomata acuminata and Hodgkin's disease effectively treated with recombinant interferon alpha-2c gel as adjuvant to electrosurgery. In: Fritsch P, Schuler G, Hintner H (eds) Immunodeficiency and skin. Curr Probl Dermatol 18:178–184
27. Gross G, Roussaki A, Papendick U (1990) Efficacy of interferons on bowenoid papulosis and other precancerous lesions. J Invest Dermatol (in press)
28. Gross G, Roussaki A, Ikenberg H et al. (1990) Genital warts do not respond to systemic recombinant interferon alpha-2a treatment during cannabis consumption. Dermatologica (in press)

29. Haglund S, Lundquist PG, Cantell K (1981) Interferon therapy in juvenile laryngeal papillomatosis. Arch Otolaryngol 107:327–332
30. Hohenleuthner U, Landthaler M, Braun-Falco O (1990) Postoperative adjuvante Therapie mit Interferon alpha-2b nach Laserchirurgie von Condylomata acuminata. Hautarzt (in press)
31. Ikenberg H, Gissmann L, Gross G et al. (1983) Human papillomavirus type 16-related DNA in bowenoid papulosis and genital Bowen's disease. Int J Cancer 32:563–565
32. Ikic D, Krusic J, Cupak S et al. (1972) The use of human leukocyte interferon in patients with cervical cancer and basocellular cancer in the skin. Proc Symp Clinical Use of Interferon, Zagreb, pp 167–177
33. Ikic D, Bosnic N, Smerdel S (1975) Double-blind clinical study with human leukocyte interferon in the therapy of condylomata acuminata. Proc Symp Clinical Use of Interferon, Zagreb, pp 239–243
34. Isaacs A, Lindenmann J (1957) Virus interference. I. The interferon. Proc Roy Soc London Biol Sci 147:258
35. Iwasaka I, Hayashi Y, Yokoyama M et al. (1990) Interferon gamma treatment for cervical intraepithelial neoplasia. Gynecol Oncol 37:96–102
36. Jablonska S, Orth G, Glinski W (1981) Morphology and immunology of human warts. In: Bachmann PA (ed) Leukemias, lymphomas and papillomas: comparative aspects. Taylor and Francis, London, pp 107–131
37. Jacobs AJ, Dawond L, Kowacs Z et al. (1989) Inhibition of proliferation of lines derived from human cervical carcinomas by cytotoxic drugs and by recombinant interferons. Gynecol Oncol 32:31–36
38. Keay S, Teng N, Eisenberg M et al. (1988) Topical interferon for treating condylomata acuminata in women. J Infect Dis 185:934–939
39. Von Krogh G (1989) Treatment of human papillomavirus-induced lesions of the skin and anogenital region. In: Syrjänen K, Gissmann L, Koss LG (eds) Papillomaviruses and human disease. Springer, Berlin Heidelberg New York, pp 296–333
40. Koutsky LA, Galloway DA, Holmes KK (1988) Epidemiology of genital human papillomavirus infection. Epidemiol Rev 10:122–162
41. Krusic J, Ikic D, Knezevic M et al. (1972) Clinical and histological findings after local application of human leukocyte interferon in patients with cervical cancer. Proc Symp Clinical Use of Interferon, Zagreb, pp 167–177
42. Maluish AE, Ortaldo JR, Sherwin SA et al. (1983) Changes in immune function in patients receiving natural leukocyte interferon. J Biol Response Modif 2:418–427
43. Marcovici R, Peretz BA, Paldi E (1983) Human fibroblast interferon therapy in patients with condylomata acuminata. Isr J Med Sci 19:104
44. Möller BR, Johannesen P, Osther K et al. (1983) Treatment of dysplasia of the cervical epithelium with an interferon gel. J Obstet Gynecol 62:625–629
45. Neis KJ, Tesseraux M, Claußen C et al. (1989) Lokale Therapie cervikaler intraepithelialer Neoplasien mit natürlichem beta-Interferon. Arch Gynecol 245:550
46. Olsen EA, Kelly FF, Vollmer RT et al. (1989) Comparative study of systemic interferon alpha-n1 and isotretinoin in the treatment of resistant condylomata acuminata. J Am Acad Dermatol 20:1023–1030
47. De Palo G, Stefanon B, Rilke F et al. (1984) Human fibroblast interferon in cervical and vulvar intraepithelial neoplasia associated with papillomavirus infection. Int J Tissue React 6:523–527
48. De Palo G, Stefanon B, Rilke F et al. (1985) Human fibroblast interferon in cervical and vulvar intraepithelial neoplasia associated with viral cytopathic effects. J Reprod Med 30:404–408
49. Pfister H (1984) Biology and biochemistry of papillomaviruses. Rev Physiol Biochem Pharmacol 99:111–181
50. Preble OT, Friedman RM (1983) Biology of disease: interferon-induced alterations in cells: relevance to viral and nonviral diseases. Lab Invest 49:4–8

51. Reichmann RC, Oakes D, Bonnez E et al. (1988) Treatment of condyloma acuminatum with three different interferons administered intralesionally. A double-blind, placebo-controlled trial. Ann Intern Med 108:675–679
52. Reichmann RC (1990) Human papilloma-viruses and interferon therapy. In: Galasso GJ, Whitley RJ, Merigan TC (eds) Antiviral agents and viral diseases of man. Raven, New York, pp 301–325
53. Saastamoinen J (1989) Interferon ointment in genital warts of males. In: Gross G, Jablonska S, Pfister H, Stegner HE (eds) Advances in modern diagnosis and therapy. International symposium on genital papillomavirus infections. Hamburg (abstracts)
54. Schneider A, Papendick U, Gissmann L et al. (1987) Interferon treatment of human genital papillomavirus infections: importance of viral-type. Int J Cancer 40:610–614
55. Schneider A, Kirchmayer R, Wagner D (1989) Efficacy trial of topically applied gamma-interferon in cervical intraepithelial neoplasia. In: Gross G, Jablonska S, Pfister H, Stegner HE (eds) Advances in modern diagnosis and therapy. International symposium on genital papillomavirus infections. Hamburg (abstracts)
56. Schonfeld A, Schattner A, Crespi M et al. (1984) Intramuscular human interferon-β injections in treatment of condylomata acuminata. Lancet I:1038–1041
57. Scott GM, Csonka GW (1979) Effect of injections of small doses of human fibroblast interferon into genital warts. A pilot study. Br J Ven Dis 55:442–444
58. Seltzer A, Doyle A, Kadish AS (1983) Natural cytotoxicity to malignant and premalignant neoplasias and enhancement of cytotoxicity with interferon. Gynecol Oncol 15:340–349
59. Sen GC (1984) Biochemical pathways in interferon-action. Pharmacol and Ther 24:235–257
60. Slotman BJ, Helmerhorst TJM, Wijermans PW et al. (1988) Interferon-alpha in treatment of intraepithelial neoplasia of the lower genital tract: a case-report. Eur J Obstet Gynecol Reprod Biol 27:327–333
61. Strander H, Cantell K (1974) Studies on antiviral and antitumor effects of human leukocyte interferon in vitro and in vivo. In: Waymouth C (ed) The production and use of interferon for the treatment and prevention of human virus infections. Tissue Culture Association, Gaithersburg, pp 49–56
62. Tiedemann KH, Ernst TM (1988) Kombinationstherapie von rezidivierenden Condylomata acuminata mit Elektrokaustik und Alpha-2 Interferon. Akt Dermatol14:200–204
63. Turek LP, Byrne JC, Lowy DR et al. (1982) Interferon induces morphologic reversion with elimination of extrachromosomal viral genomes in bovine papillomavirus transformed mouse cells. Proc Natl Acad Sci USA 79:7914–7918
64. Vance JC, Bart BJ, Hansen RC et al. (1986) Intralesional recombinant alpha-2 interferon for the treatment of patients with condyloma acuminatum or verruca plantaris. Arch Dermatol 122:272–277
65. Vesterinen E, Meyer B, Cantell K et al. (1984) Topical treatment of flat vaginal condyloma with human leukocyte interferon. Obstet Gynecol 64:535–538
66. De Virgilis C, Crippa L, Leopardi O et al. (1987) The role of beta-interferon in the therapy of female genital viral diseases. Int J Immunol 3:147–150
67. Weck PK, Brandsma JL, Whisnant JK (1986) Interferons in the treatment of human papillomavirus diseases. Cancer Metastasis Rev 5:139–165
68. Yliskoski M, Cantell K, Syrjänen K (1990) Topical treatment with human leukocyte interferon of HPV 16 interferons associated with cervical and vaginal intraepithelial neoplasia. Gynecol Oncol 36:353–357

Genital Warts and Intralesional Injections of Interferon

J. C. Vance, B. J. Bart, and O. J. Rustad

Introduction

The incidence of condylomata acuminata has been found to be rapidly increasing over the past several decades. An increase of five times was documented by the Centers for Disease Control in the United States for consultations for condylomata between the years of 1966 and 1981 [1]. That warts may be sexually transmitted has been recognized by physicians for centuries, but the discovery of a possible role of wart virus in oncogenesis is much more recent, and the evidence is of necessity indirect. It is well demonstrated that human papillomavirus (HPV) can be detected in vulvar and cervical carcinoma in situ, bowenoid papulosis, verrucous carcinoma, and squamous cell carcinomas both on the genital areas and elsewhere [2–5]. Warts have been clinically observed to convert into squamous cell carcinomas in patients with epidermodysplasia verruciformis, and HPV has been detected in those malignancies [6, 7]. The conversion of viral warts into carcinomas has been documented in animal models [8].

The combination of a sexually transmitted disease with a potential for cancer development has resulted in a clinical problem having great emotional overlay. The demand by the affected individuals for successful treatment has at the same time revealed to us the weaknesses of our current modalities of therapy. Since there is no HPV-specific vaccine or therapy, we must rely on nonspecific destructive modalities such as trichloracetic acid (TCA), podophyllum or podophyllotoxin, cryotherapy, surgical excision, and laser destruction; chemotherapy with agents such as 5-fluorouracil and bleomycin; or the use of the immune stimulators such as dinitrochlorobenzine (DNCB), BCG, or levamisole [9, 10]. Interferon was described in 1957 and was first tested in patients with warts in the 1970s [11, 12], but it did not become available for large, well-controlled studies until the early 1980s when recombinant DNA technology made large scale production possible [13]. The treatment of warts has been studied using a variety of interferon types and differing grades of purity, using variable dosages and schedules, and administered in forms varying from intravenous, intramuscular, subcutaneous, intralesional, and topically in ointments and creams. The types of warts and types of patients treated and their follow-up periods have also been variable, so comparisons among the various reports are difficult. Our studies at the Hennepin County Medical Center and the University of Minnesota in Minneapolis have been confined to intralesional injections of interferon, which will be reviewed here.

Genital Papillomavirus Infections
Eds.: Gross/Jablonska/Pfister/Stegner

Interferon as a Single Agent

One-Wart Study

Beginning in 1983, we conducted a multicenter trial in which we demonstrated that intralesional injections of alfa-2b interferon (IFN) are safe and effective for the treatment of condylomata [14].

Methods

In that study, a single genital or perianal wart was treated, giving intralesional injections of IFN at a dose of 1 million units, 100000 units, or placebo into the substance of the wart in a 0.1 ml volume. The schedule was three injections per week for 3 consecutive weeks. The response to therapy was followed for 12 weeks.

Results – Efficacy

Ninety-one of 114 patients completed the full 12-week study and qualified for analysis. By week 5 there was a difference among the groups, and by week 12 the high-dose group was significantly improved ($P<0.01$). Complete clearing of the treated wart is the most important clinical criterion, and that occurred in 16 (53%) of 30 patients in the high-dose group as opposed to 6 (19%) of 32 receiving the lower dose of 100000 IU per injection and 4 (14%) of 29 receiving placebo injections. The percent of treated condylomata that cleared over time is presented in Fig. 1.

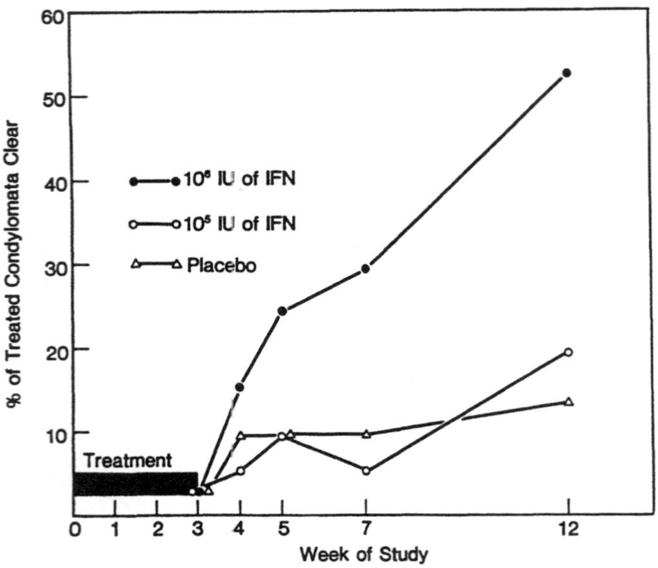

Fig. 1. Percent of treated condylomata that cleared over time (week of study), one-wart study

Results – Safety

The IFN injections were well tolerated. Among those in the high-dose IFN group 25 (68%) of 37 reported some type of side effects, compared with 19 (50%) of 38 in the low-dose IFN group and 17 (44%) of 39 receiving placebo. Only one patient receiving the high-dose IFN reported side effects judged to be severe (nausea), and that patient completed the study. Severe side effects were reported by three subjects in the placebo group (headache, genital pruritus, and urticaria). The most common mild symptom was the expected generalized flulike syndrome, consisting of a variable combination of fever, chills, myalgia, arthralgia, malaise, headache, and/or nausea. It occurred in 57% of patients receiving high-dose IFN, 29% receiving low-dose IFN, and 23% receiving placebo. Stinging, pain, or inflammation occurred at the injection site in 11% of patients receiving high-dose IFN, 18% receiving low-dose IFN, and 23% receiving placebo. Those local reactions did not present a therapeutic problem, since the patients felt they were mild enough to continue the study. No patient was dropped from the study because of laboratory abnormalities, although minor changes were commonly seen. Depression of the white blood cell count was seen in 15 (21%) of 70 patients receiving 1 million units, compared with none in the low-dose group and 4 (5%) of 74 in the placebo group. In the high-dose group, the depression was of approximately 1500 cells/ml and it returned to normal as soon as the therapy was completed. The group receiving 100000 units showed a reduction of approximately 500 cells/ml. The hematocrit level was slightly reduced in the IFN-treated patients and the aspartate aminotransferase level was minimally elevated, but all other laboratory values showed infrequent and minor alterations not thought to be related to the IFN. The injections were thus safe as well as effective and well tolerated.

One- to Three-Wart Study (Eron et al.)

Our observations were confirmed and extended by Eron et al. [15] who injected 1 million units of IFN into one to three warts, and followed the warts for 16 weeks. Their randomized, double-blind trial enrolled 296 patients, of whom only 13 (4%) discontinued because of side effects.

Results – Efficacy

At 1 week after the completion of the 3 weeks of intralesional injections, IFN had produced a large reduction in mean wart area (62% decrease), as compared with placebo (1% increase; $p=0.001$). At the completion of the 16-week follow-up, there was a 39.9% decrease in mean wart area in the IFN-treated group compared with a 46% increase in the placebo group. All treated warts had completely cleared in 36% of the IFN recipients and in 17% of the placebo recipients.

Results – Safety

Despite the higher dose of interferon given (up to 3 million units per treatment session), the injections were still well-tolerated clinically and only mild laboratory abnormalities were detected.

Five-Wart Study

In a double-blind, placebo-controlled study conducted in conjunction with Charles Welander, M.D. of Bowman Gray School of Medicine in Winston-Salem, North Carolina [16] we increased the number of warts treated to five per session.

Methods

The same dose of IFN per wart (1 million units or placebo) and schedule (three injections per week for 3 weeks) were used, but the patients were now followed for a total of 19 weeks. Eighty-two subjects were enrolled and evaluated for safety. Twelve were excluded from the efficacy analysis, primarily due to failure to return or to receive the complete course of injections, leaving 70 to be evaluated for efficacy; 32 in the IFN group and 38 in the placebo group. Dr. Welander enrolled 42 at Bowman Gray and 40 were enrolled at Minnesota. The patients at Bowman Gray differed in being primarily women,

Table 1. Demographic data on all efficacy subjects both centers, five-wart study

Characteristic	IFN	Placebo	Probability
Sex (male, female)	16/16	16/22	NS
Race (white/other)	28/4	34/4	NS
Age in years			
(mean)	29.7	28.4	NS
(range)	18–61	18–60	
Sexual preference (heterosexual/homo-/bi-/NR)	30/0/1/1	32/1/0/5	
Lesion volume index in mm^3			
(mean)	79.9	53.5	NS
(range)	2.4–384	0.6–302	
Site, each lesion (penis/vulva/perianal/other)	47/69/40/4	57/95/30/8	
Previous therapy (none/podophyllin/other)	8/20/4	11/23/4	NS

NR, not recorded; NS, not significant

in being about 7 years younger, and in having had more previous therapy than the patients at Minnesota, but the demographic factors of the two treatment groups pooled together showed no significant differences (see Table 1). Thirty-one subjects dropped out of the study for treatment failure, of whom 26 were in the placebo group and 5 in the IFN group. The greater number of dropouts in the placebo group produces a consistent bias when evaluating the data at each evaluation point, so the data has been calculated either as "endpoint" which is the last valid visit in the study, regardless of the time, or "best response" which measures the best result while in the evaluation phase, regardless of the particular visit at which it occurred.

Results – Efficacy

At the endpoint the IFN-treated condylomata had decreased in mean volume by 40% while the placebo-treated warts had increased in mean volume by 56%, a significant difference ($P<0.01$). The percentage change in patient target lesion volume index is shown in Table 2. As can be seen, only 9 (28%) of 32 patients were clear of all five lesions at the endpoint. In the placebo group, 22 (58%) of 38 patients had exacerbation of their wart volume index by endpoint. Table 3 shows the percentage change in the individual target lesions; 40% of all the 160 IFN-treated warts were clear at the endpoint while 23% of 190 warts treated with placebo were clear.

The response to IFN therapy was related to the initial size of the condylomata. This effect is shown in Table 4 which relates the pretreatment volume index of the warts to the number and percentage clear at the time of best response. It is a common finding with other treatment modalities that larger warts are more difficult to clear.

The time required for a wart to clear varied with whether IFN was used or placebo, as can be seen on Table 5.

Table 2. Percentage change in patient target lesion volume index at endpoint and best response, five-wart study

Patients with Degree of change	Endpoint		Best response	
	IFN	Placebo	IFN	Placebo
Increase	9 (28%)	22 (58%)	1 (3%)	8 (21%)
No Change	0	1	0	3
1%–25%	1	3	0	10
26%–50%	2	1	4	4
51%–75%	3	2	4	1
76%–99%	8	2	12	5
100% clear	9 (28%)	7 (18%)	11 (34%)	7 (18%)
No. of patients	32	38	32	38
Mean % change in volume	−40%	+56%	−80%	−28%
P value	<0.01		<0.01	

Table 3. Percentage change in individual wart volume index, at endpoint and best response, five-wart study

Warts with degree of change	Endpoint		Best response	
	IFN	Placebo	IFN	Placebo
Increase	27(17%)	84(44%)	2(1%)	23(12%)
No change	9	22	4	39
1%–25%	11	15	11	21
26%–50%	7	11	10	28
51%–75%	19	7	22	15
76%–99%	23	7	29	13
Clear	64(40%)	44(23%)	82(51%)	51(27%)
No. of warts	160	190	160	190

Table 4. Pretreatment size range (volume index) related to clearing by IFN at time of best response, five-wart study

Pretreatment, size volume index (mm^3)	No. clear	Percentage clear	No. of warts
< 1–5	23	77	30
6–10	16	64	25
11–20	11	41	27
21–30	6	40	15
31–50	10	55	18
51–100	8	44	18
>100 mm^3	8	29	27
Total	82	51	160

Table 5. Time of clearing of condylomata, five-wart study

Visit	IFN	Placebo
Week 1	0	1
Week 2	7	1
Week 3	13	4
Week 4	32	13
Week 7	15	5
Week 11	6	11
Week 15	4	10
Week 19	5	6
Total	82	51

Table 6. Response of untreated lesions to treatment, five-wart study

Improvement	IFN	Placebo
Exacerbation	0	1
No change	2	10
<50%	6	4
>50%–75%	1	0
>75%–99%	3	0
Clear	7	6
Total	19	21

Nontreated condylomata were also measured in these subjects in order to determine if there was a systemic effect of IFN. As can be seen in Table 6, the effect of IFN on nontarget condylomata at the time of best response was modest. Clearing was seen in 7 (37%) of 19 warts in IFN-treated subjects, and at least moderate improvement (>50%) in 4 (21%). Placebo-treated subjects had clearing of 6 (29%) of 21 warts but none had moderate or marked improvement. No change in or exacerbation of nontarget lesions was seen in only 2 (11%) of 19 IFN-treated subjects but in 11 (52%) of 21 placebo-treated patients.

Results – Safety

Only one IFN patient and no placebo patients dropped out of the study due to side effects, again showing that the drug is well tolerated even at 5 million units per treatment session. Forty (100%) of the IFN-treated patients developed at least one treatment-related adverse experience, and 28 (67%) of 42 placebo-treated patients also noted side effects, showing that the patients were being closely asked about such effects. The same flulike symptoms

Table 7. Percentages of IFN-treated patients reporting treatment-related adverse experiences, five-wart study

Symptom	Overall	Treatment period			Posttreatment
		Week 1	Week 2	Week 3	Week 4
Chills	75	70	28	11	3
Fever	70	60	31	17	6
Myalgia	70	50	36	36	6
Headache	55	50	23	11	0
Nausea	30	25	8	8	3
Fatigue	18	2	13	14	3

Table 8. Reports of intralesional IFN in condylomata

Study	Patients/warts	IFN	Dose/schedule	Response	Comments
Scott and Csonka 1979 [17]	11/11	Fibroblast	300 U/once	1 CR 9% 4 PR 33%	$P=0.045$ over controls
	11/?	Placebo			
Schonfeld et al. 1984 [18]	5/?	Beta IFN (Frone)	3 million U/4×, q.i.d.	5 CR 100% OR	Painful
	5/?	Placebo			
Geffen et al. 1984 [19]	10/32	Alfa IFN (IFN Sciences)	0.8–35 million U/9–28 inj	21 CR 66% 7 PR 22%	
Gross et al. 1984 [20]	1/?	Beta IFN (Frone)	3 million U/4×, q.i.d.	1 PR	Recurred Ca in situ
Gall et al. 1985 [21]	4/?	Lymphoblastoid (Wellferon)	1 million U/2× weekly, ×4	3 CR 75%	After IM failed
Vance et al. 1986 [14]	30/30	Alfa-2b (Intron A)	1 million U/3× weekly, ×3	16 CR 53%	
	32/32	Alfa-2b (Intron A)	0.1 million U/3× weekly, ×3	6 CR 19%	
	29/29	Placebo		4 CR 14%	
Eron et al. 1986 [15]	124/327	Alfa-2b (Intron A)	1–3 million U/3× weekly, ×3	CR 36%	1–3 warts
	132/?	Placebo		CR 17%	1–3 warts
Vance et al. 1986 [16]	32/160	Alfa-2b (Intron A)	5 million U/3× weekly, ×3	64 CR 40%	
	38/190	Placebo		44 CR 23%	
Friedman-Kien et al. 1988 [22]	66/?	Alfa (Alferon N)	0.3–02.1 million U/2× weekly, ×8	CR 73%	
	66/?	Placebo		CR 36%	
Reichman et al. 1988 [23]	23/23	Alfa-2b (Intron A)	1 million U/3× weekly, ×4	11 CR 48%	
	15/15	Alfa-nl (Wellferon)	1 million U/3× weekly, ×4	6 CR 40%	
	20/20	Beta (Roswell Pk)	1 million U/3× weekly, ×4	10 CR 50%	
	18/18	Placebo		4 CR 22%	

CR, Complete Remission; PR, Partial Remission; IM, Ifosfamide predominated, were mild or moderate, were of limited duration, and responded to oral acetomenophen. There was distinct tachyphylaxis of the side effects, with the incidence decreasing over the treatment period, and all were much improved by the fourth week. The most common treatment-related adverse experiences related to the week of the study are shown in Table 7.

Other Published Single-Agent Studies

Table 8 reviews the published reports of intralesional IFN use in condylomata. Several limited studies were done between 1979 and 1985 [17–21] which demonstrated that there was definate potential for intralesional IFN. Recently two well-controlled studies have been reported [22, 23] again confirming the usefulness of IFN given in this manner.

Friedman-Kien and his associates in Virginia and California [22] used natural (leukocyte) IFN (Alferon N) in a randomized, double-blind, placebo-controlled multicenter trial. The dose of IFN depended on the "wart area index," but ranged from 0.3 to 2.1 million units (median, 1.2). The schedule was to give the IFN or placebo twice weekly for up to 8 weeks or until clearing occurred. The warts were completely eliminated in 62% of patients getting IFN and 21% of those getting placebo. Considering the warts individually, 73% were cleared by IFN and 36% by placebo. The patients were followed for up to 1½ years to evaluate for possible relapse, and 9 (25%) of 36 did so in the IFN group while 3 (23%) did so in the placebo group. The mean time until relapse was approximately 4 months for IFN-treated patients and 2 months for placebo-treated patients. Thus while relapse, which includes both recurrence and reinfection, was the same, IFN delayed the time to relapse.

Reichman et al. [23] carried out a double-blind, placebo-controlled, multicenter trial comparing the efficacy of three separate types of IFN; alfa-2b, alfa-n1, and beta. One wart was treated on each of the 76 patients, and the dose given was 1 million units or placebo at a schedule of three injections per week for 4 weeks. The three IFNs were approximately equal in efficacy, with clearing occurring in 48%, 40%, and 50% respectively. Placebo-treated subjects were cleared in 22%. Rates of relapse were similar among recipients of the three IFN preparations, being about one-third, and no significant difference was detected in comparison with placebo.

IFN Combined with Other Agents

Liquid Nitrogen VS Liquid Nitrogen and Interferon

In the clinical situation, everything possible is done to optimize the cure rate. In wart treatments that usually means combining two or more treatment modalities, and in Minneapolis we sometimes combine liquid nitrogen cryotherapy to remove the bulk of the warts with podophyllin resin to help prevent recurrences.

We have recently completed a combination study designed to first remove the bulk of the virally infected tissue employing a widely used physically destructive method, liquid nitrogen cryotherapy [24]. Liquid nitrogen has the advantage of being inexpensive, and quick and easy to perform while producing minimal scarring [9]. Incomplete removal of the warts and recurrences are common, however. To that therapy we have added a course of interferon injections given into three selected warts to prevent recurrences. An equal number of subjects received the single liquid nitrogen treatment but did not receive the interferon injections.

Methods

The study was conducted in a randomized, parallel group, third-party blind fashion in which an investigator not acquainted with the treatment group of the patients did the wart measurements. Patients with the clinical diagnosis of genital or perianal condyloma acuminatum having three or more individual lesions greater than 3 mm but less than 16 mm in their largest diameter were enrolled.

Treatment

Human recombinant alfa-2b IFN (Intron-A) was used, giving 1 million units of IFN in 0.1 ml, the injection volume for each treated wart. One-half of the patients received a single treatment to all warts using liquid nitrogen, the other half were given the same treatment followed by thrice weekly injections of interferon at a dose of 1 million units per wart into each of three warts for 3 weeks. No placebo injections were given.

Results – Efficacy

Eighty-seven subjects were enrolled in the study, of whom 43 were randomized to receive liquid nitrogen alone (LN) and 44 to receive liquid nitrogen plus interferon injections (LN and IFN). Seventy-seven subjects were evaluated for efficacy, 38 in the LN group and 39 in the LN and IFN group. There were no statistically significant differences between the two treatment groups with respect to any of the demographic details (see Table 9). Looking first at the 3-week treatment period, 16 (42%) of the 38 patients receiving LN were cleared of all test site warts at some time-point. In the LN and IFN group, 19 (49%) of 29 subjects were cleared. The difference is not statistically different ($P = 0.65$); see Table 10. The time to clearing was the same in both groups; see Fig. 2. This demonstrates that liquid nitrogen acts more quickly than IFN and so determines time to clearing.

By week 4, a significant difference between the groups was present. All 19 (49%) of the 39 subjects who cleared in the LN and IFN group were still clear, while only 9 (24%) of 38 subjects in the LN group were still clear ($P = 0.03$). Subjects began dropping out of the study beginning at week 8, with only 10

Table 9. Demographic data on all randomized subjects: LN versus LN and IFN study

Characteristic	LN	LN and IFN	Probability
Sex (male/female)	31/12	32/12	>0.99+
Race (white/other)	39/4	41/3	0.71+
Age in years			
(mean)	30.3	28.1	0.14*
(range)	18–61	18–50	
Weight in kg (mean)	73.7	73.4	0.87*
Sexual preference (heterosexual/ homo/bi-/NR)	32/2/1/8	35/4/1/4	0.62+
Total no. of warts			
(mean)	12.3	12.1	0.17*
(range)	3–40	3–60	
Wart volume index in mm^3			
(mean)	49.7	44.6	0.25*
(range)	6.7–407	4.5–296	
Duration of warts in years			
(mean + STD)	2.2+3.8	1.9+3.2	0.43*
(range)	0.2–20	0.1–15	
Location of warts (vulva/perianal/ penile/other)	8/9/20/6	10/10/22/2	0.53+

* Wilcoxon's rank sum test
\+ Fisher's exact test

subjects remaining in the LN group at week 24, while 23 remained in the LN and IFN group (see Table 11). A large number of patients discontinued the 6-month long study because of treatment failure, and the majority were in the LN group. Looking at the last visit or "endpoint" of each subject (independent of the study day) helps negate the effect of patient drop-out; 21 (54%) of 39 were clear at endpoint in the LN and IFN group while 9 (24%) of 38 were clear in the LN group, a significant difference ($P=0.01$).

The percentage of clearing of all treated sites was related to pretreatment mean wart size (volume index), as is shown on Table 12; 64% of the smallest warts (0 to 10 mm^3) cleared under combination therapy, while 20% cleared with liquid nitrogen alone. Of the warts greater than 100 mm^3, only 33% cleared with LN and IFN and 0% with LN.

Other demographic characteristics also correlated with clearing of all test sites (see Table 13). Gender, location of warts, lesion type and size all showed differences, which were the same for both treatment groups. Sexual preference was a factor primarily in the LN and IFN group. It is known that asymptomatic infection with human immunodeficiency virus (HIV) causes a marked reduction in the efficacy of IFN in the treatment of anogenital warts [25]. The duration of the warts did not appear to be an important factor in either group.

Table 10. Subjects having total clearing of all test sites: LN versus LN and IFN study

Visit	Treatment	Clear	Not clear	*n*	*P* value
Week 1	LN and IFN	0	39	39	
	LN	0	38	38	>0.99
Week 2	LN and IFN	6	33	39	
	LN	9	29	38	0.43
Week 3	LN and IFN	12	27	39	
	LN	9	27	36	0.62
Treatment phase best[a]	LN and IFN	19	20	39	
	LN	16	22	38	0.65
Week 4	LN and IFN	19	20	39	
	LN	9	29	38	0.03
Week 24	LN and IFN	18	5	23	
	LN	5	5	10	0.21
Last visit[b]	LN and IFN	21	18	39	
	LN	9	29	38	0.01

[a] Treatment phase visit that had the best response (i.e., the lowest number of lesions present)
[b] Independent of study day

Table 11. Subjects evaluable in study at each visit: LN versus LN and IFN study

Visit	LN	LN and IFN
Week 1	38	39
2	38	39
3	38	39
4	38	39
8	32	37
12	21	32
16	17	28
20	9	23
24	10	23

Table 12. Clearing of all treated warts at last visit as related to pretreatment wart size, LN vs LN and IFN study

Mean wart volume index (mm^3)	Liquid nitrogen + IFN			Liquid nitrogen alone		
	n	No. clear	Percentage clear	*n*	No. clear	Percentage clear
0–10	11	7	64	5	1	20
11–20	12	8	67	13	4	31
21–30	3	2	67	2	0	0
31–40	2	0	0	6	2	33
41–50	3	1	33	3	0	0
51–100	5	2	40	6	2	33
<100	3	1	33	3	0	0
Total	39	21	54	38	9	24

Table 13. Summary of demographic and baseline disease characteristics in subjects clearing all test site lesions at last visit, LN vs LN and IFN study

Risk factor		*n*	LN and IFN	(%)	*n*	LN	(%)
Gender	male	27	13	48	28	6	21
	female	12	8	67	10	3	30
Location	vulva	10	7	70	7	2	29
	perianal	8	2	25	7	1	14
	penis	19	11	58	18	5	28
	other	2	1	50	6	1	17
Sexual preference	heterosexual	32	18	56	29	7	24
	other	7	3	43	9	2	22
Mean target lesion size	$\leqq 20\ mm^3$	23	15	65	18	5	28
	$> 20\ mm^3$	16	6	38	20	4	20
Duration of warts	≦ 1year	27	15	56	23	6	26
	>1 year	12	6	50	15	3	20
Skin site	not on mucosa	26	12	46	26	5	19
	On mucosa	13	9	69	12	4	33

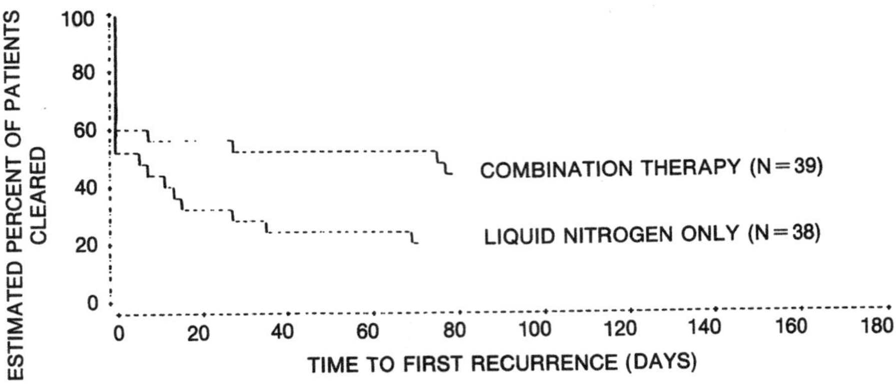

Fig. 2. Time to first recurrence, including all patients evaluated for efficacy, product limit survival estimate, LN vs LN and IFN study. Combination therapy means interferon alpha-2b and liquid nitrogen

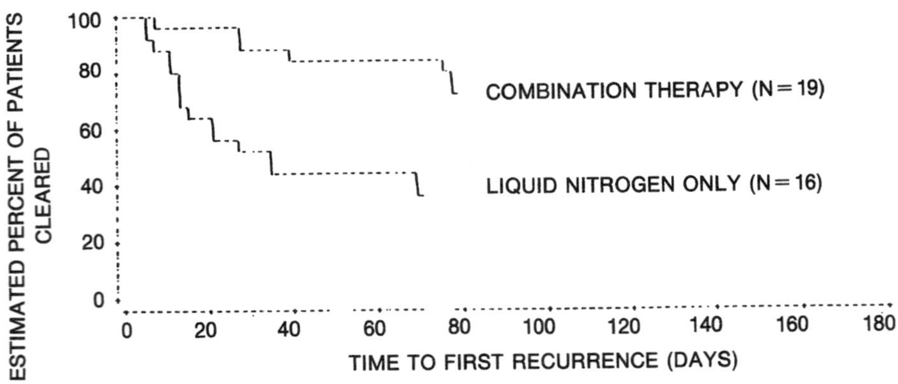

Fig. 3. Time to first recurrence, including only those who cleared by week 4, product limit survival estimate, LN vs LN and IFN study. Combination therapy means interferon alpha-2b and liquid nitrogen

An analysis of the time to first recurrence is demonstrated in Fig. 2, including all patients evaluated for efficacy. Thirty days after clearing of their lesions, approximately 54% of the LN and IFN and 28% of the LN patients remained clear of their lesions. At 60 days, the figures were 51% and 25% respectively. Limiting the analysis to those patients who cleared by week 4 is shown in Fig. 3; 89% of those in the LN and IFN group remained clear for 30 days, and 84% for 60 days. In the LN group, approximately 50% remained clear for 30 days, and 44% for 60 days.

Results – Safety

The adverse experiences encountered were similar to those previously reported for intralesional alfa interferon injections at similar dosages [14, 16].

40 (91%) of 44 subjects in the LN and IFN group reported at least one side effect. Flulike symptoms predominated. No treatment-related adverse experiences were encountered during the 6-month follow-up period, and no patients were dropped from the study for adverse experiences. The laboratory test results showed the treatments to be safe. The white blood cell counts temporarily decreased below normal values in 5 (11%) of the 44 in the LN and IFN group. Two (4%) of them had mild evevation of SGOT.

Conclusions

This study demonstrates that IFN given intralesionally can be of benefit as an adjunct to liquid nitrogen cryotherapy. It increases the number of condylomata which clear and also reduces the rate of recurrence. A weakness of this combination lies with the cryotherapy since it did not clear a high enough percentage of the warts. A useful combination which remains to be tested is repeated cryotherapy, as it is often given in practice, combined with IFN. That would increase the baseline rate of clearing before the IFN treatments. Destructive treatments which can clear 100% of the treated sites, such as electrocautery or laser vaporization, would also be effective since their cure rates and then prevention of recurrence should resemble those in Fig. 3.

Electrocautery Combined with IFN

Tiedemann and Ernst in Berlin have reported the results of their study using IFN as an adjuvant to electrocautery [26]. They enrolled patients with large condylomata, whose warts were of long duration, and with multiple recurrences, i.e., those who presented the most difficult clinical problems. Twenty-two were treated with electrocautery and then after allowing 5 to 10 days for healing, a course of IFN injections was given using alfa-2b (Intron A), at a dose of 5 million units, given intralesionally every 2 to 3 days to a total dose of 25 million units. They were compared with 11 patients who recieved electrocautery alone. There was a recurrence rate of 45% (5 of 11) within 6 weeks in the electrocautery group. In the combination group, only two patients (9%) developed a recurrence in a 6-month follow-up. Three others had transient recurrences which spontaneously resolved. Of importance is the observation that the injections of IFN caused delay in healing of the surgical sites if they were still open at the time the injections were started. It is probable that IFN delays wound healing.

Gerd Gross from Hamburg (F.R.G.) has also recorded the combination therapy of condylomata [27, 28]. He reported a case of a Buschke-Loewenstein-like condyloma in a women with Hodgkin's disease who responded favorably to electrocautery followed by the topical application of recombinant alpha-2c in a hydrogel. It was applied four times daily for 8 weeks. The patient remains clear after 14 months follow-up.

Carbon Dioxide Laser Combined with IFN

Currently in Minnesota we are conducting a trial testing the combination of carbon dioxide laser vaporization of recalcitrant or large anogenital condylomata in conjunction with intralesional injections of alfa-2b interferon. The combination is being compared with laser treatment alone. The injections begin at the time of the surgery, and a mild delay of healing is noted (2 to 4 weeks to complete healing of erosions) but excessive pain or scarring have not been encountered. The IFN is started at 1 million units, and then is moved up to 5 million units per treatment as soon as it is tolerated, taking advantage of the tachyphylaxis phenomenon seen in IFN injections. The injections are given three times weekly for three weeks, and are placed adjacent to the treated sites, rotating the injections so that all adjacent areas receive some injections. To date 23 subjects have been enrolled in the combination group with as much as 18 months follow-up, and only two (9%) have experienced a recurrence. Both have elected to be re-treated with the combination. It is our preliminary impression that IFN injections add to a reduction in the recurrence rate of difficult cases of condylomata treated first with laser vaporization.

Discussion

Interferon given intralesionally is a safe and effective modality for the treatment of condylomata. It is similar to all other modalities of wart treatment in that not all warts clear up. The response rates for IFN-treated warts vary from 73% [22] to 36% [15]. The large differences probably relate to variations in the dose, duration of treatment, number of injections, type of IFN used, demographics of the treated individuals, types of warts treated, and the length of the follow-up period [14–23].

The mechanism by which IFN is effective in the treatment of warts remains unknown. Interferon does produce an antiviral state which limits the further spread of many viruses [29] but it also cause a reduction in cellular proliferation at the affected site, and changes in the host immune responses [30–33]. Since IFN has been shown to be capable of clearing cultured cells of their content of papilloma virus [34], it is hoped that in vivo treatment of affected sites using intralesional IFN may be capable of eradicating occult HPV in the skin thereby preventing recurrences of the condylomata and at the same time reducing risk of malignant proliferation.

Combinations of IFN with other treatment modalities will probably be the major way in which it is used for treating condylomata. Combinations with liquid nitrogen, electrocautery, laser vaporization, and others are currently being investigated and have been found to be useful adjuncts in the effort to control this difficult clinical problem.

References

1. Center of Disease Control (1983) Condyloma acuminatum – United States 1966–1981. MMWR 32:306–308
2. Ikenberg H, Gissmann L, Gross G et al. (1983) Human papillomavirus type 16 related to DNA in genital Bowen's disease and bowenoid papulosis. Int J Cancer 32:563–565
3. Grunebaum AN, Seldis A, Sillman F et al. (1983) Association of human papillomavirus infection with cervical intraepithelial neoplasia. Obstet Gynecol 62:448–455
4. Crum CP, Ikenberg H, Richart RM et al. (1984) Presence of human papilloma virus in genital tumors. N Engl J Med 310:880–883
5. Gross G, Ikenberg H, Gissmann L et al. (1985) Papillomavirus infection of the anogenital region: correlation between histology, clinical picture, and virus type: proposal of a new nomenclature. J Invest Dermatol 85:147–152
6. Jablonska S, Orth G, Lutzner M (1982) Immunopathology of papillomavirus-induced tumors in different tissues. Semin Immunopathol 5:33–62
7. Ruiter M, Van Mullem PJ (1970) Behavior of virus in malignant degeneration of skin lesions in epidermodysplasia verruciformis. J Invest Dermatol 54:324
8. Syverton JT (1952) The pathogenesis of the rabbit papilloma to carcinoma sequence. Ann NY Acad Sci 54:1126
9. Bunney MH (1983) Viral warts. Their biology and treatment. Oxford University Press, London
10. Clark DP (1987) Condyloma acuminatum. Dermatol Clin 5(4):779–788
11. Strander H, Cantell K (1974) Studies on antiviral and antitumor effects of human-leukocyte interferon in vitro and in vivo. In Vitro 3:49–56
12. Ikic D, Brnobic A, Jukovic-Vukclic V et al. (1975) Therapeutic effect of human leukocyte interferon incorporated into ointment and cream on condylomata acuminata. In: Ikic D (ed) Proceedings – symposium on clinical use of interferon. Yugoslav Academy of Sciences and Arts, Zagreb
13. Spiegel RJ (1986) Intron A (interferon alfa-2b): Clinical overview and future directions. Semin Oncol 13(3):89–101
14. Vance JC, Bart BJ, Hansen RC et al. (1986) Intralesional recombinant alpha-2 interferon for the treatment of patients with condylomata acuminatum or verruca plantaris. Arch Dermatol 122:272–277
15. Eron LJ, Judson F, Tucker S et al. (1986) Interferon theapy for condylomata acuminata. N Engl J Med 315:1059–1064
16. Vance JC, Bart BJ, Fish F et al. (1986) Effectiveness of intralesional human recombinant alfa-2b interferon (intron-A) for the treatment of patients with condyloma acuminatum. Clin Res 34(4):993A
17. Scott GM, Csonka GW (1979) Effect of injections of small doses of human fibroblast interferon into genital warts – a pilot study. Br J Vener Dis 55:442–445
18. Schonfeld A, Nitke S, Schattner A et al. (1984) Intramuscular human interferon-beta injections in treatment of condylomata acuminata. Lancet I (8385):1038–1042
19. Geffen JR, Klein RJ, Friedman-Kien AE (1984) Intralesional administration of large doses of human leukocyte interferon for the treatment of condylomata acuminata. J Infect Dis 150(4):612–615
20. Gross G, Ikenberg H, Gissmann L (1984) Bowenoid dysplasia in human papillomavirus-16 DNA positive flat condylomas during interferon-beta treatment. Lancet I (8392):1467–1468
21. Gall SA, Hughes CE, Trofatter K (1985) Interferon for the therapy of condyloma acuminatum. Am J Obstet Gynecol 153:157–163
22. Friedman-Kien AE, Eron LJ, Conant M et al. (1988) Natural interferon alfa for treatment of condylomata acuminata. JAMA 259(4):533–538
23. Reichman RC, Oakes D, Bonnez W et al. (1988) Treatment of condylomata acuminatum with three different interferons administered intralesionally. Ann Intern Med 108:675–679

24. Vance JC, Bart BJ, Krywonis N et al. (1989) Treatment of condylomata: the use of intralesional alfa-2b interferon injections as an adjuvant to liquid nitrogen therapy. J Am Acad Dermatol (submitted)
25. Douglas JM Jr, Rogers M, Judson FN (1986) The effect of asymptomatic infection with HTLV-III on the response of anogenital warts to intralesional treatment with recombinant alfa-2 interferon. J Infect Dis 154:331–334
26. Tiedemann K-H, Ernst T-M (1988) Kombinationstherapie von rezidivierenden Condylomata acuminata mit Elektrokaustik und Alpha-2-Interferon. Aktuel Dermatol 14(7):189–228
27. Gross G (1988) Interferon and genital warts. JAMA 260(14):2066
28. Gross G, Pfister H (1988) Recurrent vulvar Buschke-Loewenstein's tumor-like condylomata acuminata and Hodgkin's disease effectively treated with recombinant interferon alpha 2c gel as an adjuvant to electrosurgery (abstract). J Cancer Res Clin Oncol 114S:147
29. Radke KL, Colby C, Kates JR et al. (1974) Establishment and maintenance of the interferon-induced antiviral state: studies in enucleated cells. J Virol 13:623–630
30. Ortega JA, Ma A, Shore NA et al. (1979) Suppressive effect of interferon on erythroid cell proliferation. Exp Hematol 7:145–150
31. Johnson HM, Baron S (1976) The nature of the suppressive effect of interferon and interferon inducers on the in vitro immune response. Cell Immunol 25:106–115
32. Gresser I (1977) On the varied biological effects of interferon. Cell Immunol 29:406–415
33. Stiehm ER, Kronenberg LH, Rosenblatt HM et al. (1982) Interferon: immunobiology and clinical significance. Ann Intern Med 96:80–93
34. Turek LP, Byrne JC, Lowy DR et al. (1982) Interferon induces morphologic reversion with elimination of extrachromosomal viral genomes in bovine papillomavirus-transformed mouse cells. Proc Natl Acad Sci USA 79:7914–7918

Interferon Ointment in Genital Warts

J. Saastamoinen, K. Syrjänen, S. Syrjänen, and R. Mäntyjärvi

Our work on the treatment of papillomavirus in men is based on the observation of a high recurrence rate of cervical intraepithelial neoplasia in women after treatment. After a cervix operation women often develop a new infection which they obviously contract from their partners. In colposcopic examination there is a nearly 100% incidence of condyloma in these women's partners.

Indications to treat male papillomavirus infections were as follows: there was some degree of PIN (penile intraepithelial neoplasia), the patient had some disturbing symptom of the infection, or his partner had to be treated.

The diagnosis of male genital condyloma was based on both colposcopic and histologic examinations. Patients were informed about the treatment study and were willing to participate. Prior to this study several efforts had been made in treating male condyloma patients. In the present study all patients were first treated with CO_2 laser. After the laser therapy interferon cream was applied once a day on affected areas. Assignment of patients to the group receiving interferon after laser treatment or to the control group was carried out by a double-blind method. The interferon used was Introna (Essex Pharmaceuticals), and the amount of interferon was 50 MU/50 g cream base. The ointment was stored at $-20°$ C, and during therapy it was stored in a refrigerator.

The histologic grading from punch biopsies in both groups is presented in Table 1. In all cases condylomalata was diagnosed; in only one case was there condyloma acuminatum and in one case pigmented papulosis. There was one case of PIN I in both groups and two cases of PIN II in the interferon group. The localization of condyloma in male genitalia varies greatly. As is shown in Table 2, the urethral meatus is almost always affected. Preputial mucosa, preputial skin, and the skin of the penis shaft are the most commonly affected

Table 1. Histologic grading from punch biopsies

	Condyloma latum	Condyloma acuminatum	Papulosis	PIN I	PIN II
Controls ($n=20$)	20	0	1	1	0
Interferon ($n=23$)	23	1	0	1	2

Genital Papillomavirus Infections
Eds.: Gross/Jablonska/Pfister/Stegner

Table 2. Localization of condyloma lesions

Lesion site	Control group	Interferon group
Meatus	19	20
Glans	0	1
Preputium skin	5	2
Preputium mucosa	7	8
Frenulum	4	3
Shaft	4	6
Scrotum	0	1
Anus	0	2

Table 3. Results of treatment

	Control group[a]	Interferon group
Total regression	9 (50%)	20 (87%)
Over 75% regression	0	3 (13%)
Residual disease	9 (50%)	0

[a] Two patients were lost to follow-up

sites. Diagnosis of condyloma sites was based on colposcopic examination after staining with 5% acetic acid. Condylomata sites showed different intensities of white staining with or without red spots. The findings were always confirmed histologically by punch biopsy.

Prior to therapy the affected areas were anesthetized locally with 2% lidocaine. Colposcope-guided laser therapy was given, and great attention was paid to diagnose and treating all condylomatous lesions. The laser evaporization was performed with a margin of 2–3 mm to surrounding tissue. After laser treatment patients received polyvinyliodine ointment locally to avoid infections, and the treated areas were showered with water two to six times a day. The interferon cream was applied in the evening after washing. The interferon cream therapy generally lastet 10–14 days.

The duration of control colposcopy in the control group was 7.8 ± 6.4 weeks and in the interferon group 12.8 ± 2.7 weeks after the laser treatment. In both groups healing of the laser-evaporized areas lasted about 2 weeks. No scaring or other complications of the treatment were seen.

Table 3 presents the final results of treatment. Total regression of the lesions was observed in 50% of patients in the control group and in 87% of patients in the interferon group. In addition, over 75% regression of disease was seen in 13% of cases in the interferon group. Residual disease was found in 50% of cases in the control group and in none in the interferon group.

The results of this study confirm that a 2-week course with topical interferon cream (50 MU/50 g cream base) improves the cure rate ($p < 0.025$) of male genital condyloma lesions over that achieved by CO_2 laser evaporization alone. Recovery of the treated tissue was quite rapid, and no scarring or other complications were found.

Adjuvant Interferon Treatment of Condylomata Acuminata

K.-H. Tiedemann

Introduction

Condylomata acuminata caused by human papillomavirus infection constitute one of the most common sexually transmitted diseases which is showing a worldwide increase of incidence [8, 10, 14]. Numerous therapeutic regimens are available such as conservative methods including application of podophyllin, fluorouracil, colchicine, or bleomycin solutions. The most widely employed surgical methods are electrocaustic resection of the warts, cryotherapy, and Neodym Yag or carbon dioxide laser therapy. Altogether these methods are not satisfying. The therapeutic procedures in most cases lead to a complete removal of the warts, but there is a high recurrence rate with all these methods. As Jensen [12] reported, the recurrence rate after 3 months is 43% after podophyllin application and 18% after electrocautery. Three months later more than 50% of the patients treated by electrocautery will again register the appearance of genital warts. Newer operative procedures such as laser resection may be easier to handle and lead to fewer side effects, but the rate of recurrence will be the same as after electrocaustic resection [15]. According to our observations in large (more than 7 mm diameter) and numerous (>20) warts and in warts which have been recalcitrant for more than 2 years the recurrence rate is higher still than mentioned in the study of Jensen [4, 15]. Thus, it is not uncommon for patients to consult their physician several times a year for more than 5 or even 10 years seeking relief for their recurring genital or perianal warts.

Interferon Therapy of Condylomata Acuminata

Taking into account that human papillomavirus infections are associated with genital cancer [2, 8], it must be said that to date management of a disease with as high an incidence rate as that of condylomata acuminata has been unsatisfactory. New therapeutic regimens are therefore necessary especially to reduce the high recurrence rate. To this end interferons have been introduced in the therapy of condylomata acuminata. Interferons are glycoproteins belonging to the group of immune response modifiers. They are the first members of this group that, due to recombinant technology, are available in large amounts for several years now. There are alpha, beta, and gamma interferons, each with distinct biological activities. They have antiproliferative and im-

Genital Papillomavirus Infections
Eds.: Gross/Jablonska/Pfister/Stegner

mune modifying effects, and the alpha and beta interferons also have strong antiviral effects [13].

Several studies have been conducted to investigate the therapeutic use of interferons in the treatment of condylomata acuminata [4, 7, 9, 17, 18]. Two therapeutic regimens have been employed. One uses sublesional injections to obtain a more local effect of interferon [4, 17, 18] and the other uses subcutaneous or intramuscular injections to achieve a systemic effect of interferon [13]. To date response rates of from 30% to more than 60% have been reported in different studies [7, 9, 17, 18]. However, little overall effect of interferon injections has been seen in problem cases of condylomata acuminata [4].

Adjuvant Interferon Therapy after Operative Resection of Condylomata

An alternative therapeutic program was developed in which interferons were employed especially to stave off recurrences after conservative or surgical treatment. Compared with studies investigating a curative effect of interferon injections alone, the number of studies handling interferon as a recurrence-preventing agent are rare. It is more difficult to carry out an interferon study to reveal a recurrence-reducing effect than it is to show a curative effect. The observation time has to be much longer in such studies. The recurrence-free intervals after operative or podophyllin therapy can involve several months. The observation intervals after these therapies combined with interferon in order to prevent recurrences, have therefore to be ½ year or longer to get significant results. Several other questions arise regarding to the optimal dosage, optimal mode of application, optimal duration of interferon therapy, and whether interferon therapy should be practised before or after operative therapy.

Four studies are presented in Table 1 [5, 11, 16, 19]. Preferably alpha interferon was used in dosages from 1 MU to 5 MU per day. In these studies the therapy interval with interferon was short, only 7 to 8 days. In three studies interferon was given after surgical treatment and in one study it was given before and after surgical treatment. Although the observation time was not very long in most of the studies to facilitate a final conclusion, the results showed a reduction of the recurrence rate to 10% to 20% after combined therapy. This means a reduction of recurrences of 50% compared with operative treatment alone.

Adjuvant Interferon Therapy after Podophyllin Application

Weck et al. [19] investigated the effect of a combination therapy with podophyllin and interferon. Their study showed that interferon therapy alone

Table 1. Studies with adjuvant interferon treatment after operative resection of condylomata

Author	Year	Surgical method	IFN dose application	No. of patients	Observation period	Recurrences
Hohenleutner et al. [11]	1988	CO_2 laser	1 MU, 7× postoperatively	1	6 Months	–
Tiedemann and Ernst [16]	1988	Electrocauterization	5 MU, 5× postoperatively	22	2 Years	2
Weck et al. [19]	1988	Laser	1 MU, 7× preoperatively	25	6 Months	3
Erpenbach et al. [5]	1989	Laser	1 MU, every 8 h for 7 days postoperatively	13	12 Weeks	2

Table 2. Adjuvant interferon treatment after podophyllin application

Author	Year	Comparison of therapy	No. of patients	Response after 6 weeks	
				Clear	<50% Clear
Weck et al. [19]	1988	Interferon	54	11	44
		Podophyllin	54	28	22
		Combination	61	31	15

was less effective in the treatment of condylomata than podophyllin alone. Combined therapy with podophyllin and interferon was a little more effective than each therapy alone (Table 2). In all three cases the therapy was given for 6 weeks. The criterion of successful therapy was a clearance of all warts. More than 40% of the patients did not have complete remission of all warts. Thus, a considerable number of patients still suffered from condylomata after 6 weeks of therapy.

The side effects after combined therapy were the same as after interferon therapy alone. Primarily there were flulike symptoms which were mitigated successfully by administering acetaminophen. Life-threatening effects were not seen, but because high fever may develop special attention is necessary to patients with cardiovascular diseases. Where the interferon injections were given perilesionally after operative treatment before a complete reepithelization of the wounds, we observed a delayed healing period [16]. The patients felt uncomfortable because of their long-lasting and badly healing wounds; this might be due to the antiproliferative effect of the high interferon concentrations in the wound area. In general, the side effects do not lead to a disproportional impairment of the patients well-being in relation to the nature of their chronic disease.

Effect of Interferon in Adjuvant Therapy

The immunological and biochemical effects of interferon in preventing recurrences after operative resection of condylomata acuminata are unknown. Resection of the warts removes most of the tumor mass. But in the majority of cases there is a good chance of autoreinfection from small condylomata (not visible to the surgeon's eye) in cavities such as the upper urethra or parts of the rectum [2]. In such patients antiviral medication may be helpful in preventing reinfections. But the antiviral activity is probably not the only effect of interferon in preventing recurrences. We observed that some warts reappeared after combinant therapy, but disappeared spontaneously without any additional therapy. This may indicate that there is another effect which may be due to the immune modifying effect of interferon. This is in accordance with the fact that interferon is of little effect in the therapy of condylomata acuminata in patients with compromised cellular immune systems, especially in AIDS patients [3]. It seems not unlikely that a cell-dependent immune effect is involved in the prevention of recurrences after adjuvant therapy with interferon alpha. This problem needs further investigation.

Conclusion

Interferons are a new therapeutic tool available since 6–7 years. In the meantime interferons have been used as therapeutic agents with several dose regimens in several diseases in a considerable number of patients. To date the in vivo mechanisms of interferons are not fully understood. It could be argued that unexpected long-term side effects are still possible. On the other hand, it could be argued that recombinant interferons are one of the most thoroughly examined drugs. Tenfold higher doses than in condylomata acuminata have been used for long time in the therapy of malignant diseases [1, 6]. In all cases the side effects were sufficiently manageable under careful clinical control. It also should be taken into consideration that the long-term effect of long-standing condylomata acuminata plays a role in the development of genital cancer and that adequate therapy contributes to the prevention of this. Conventional therapeutic methods to date have not been entirely satisfying. Thus adjuvant interferon therapy was developed to prevent recurrences after surgical dissection and this has certainly been an improvement in the therapy of condylomata acuminata.

References

1. Creagan ET, Ahmann DL, Frytak S, Long HJ, Chang MN, Itri LM (1986) Phase II tials of recombinant leucocyte A interferon in disseminated malignant melanoma. Cancer Treat Rep 70:619–626
2. De Villiers E-M, Wagner D, Schneider A, Wesch H, Miklaw H, Wahrendorf J, Papendick U, zur Hausen H (1987) Human papillomaviruses in women with and without abnormal cervical cytology. Lancet II (8561):703–705
3. Douglas JM, Royers M, Judson FN (1986) The effect of asymptomatic infection with HTLV III on the response of anogenital warts to intralesional treatment with recombinant alpha-2 interferon. J Infect Dis 154(2):331–334
4. Eron LJ, Judson F, Tucker S, Prawer S, Mills J, Murphy K, Hickey M, Rogers M, Flanningan S, Hien N, Katz HI, Gottlieb A, Adams K, Burton P, Tanner D, Taylor E, Peets E (1986) Interferon therapy for condylomata acuminata. N Engl J Med 315(17):1059–1064
5. Erpenbach K, Derschum W, Wiese H, v. Vietsch H (1989) Results of the combined laser and adjuvant interferon alpha 2B therapy for patients with therapy resistant anogenital condylomata acuminata. International symposium of advances in diagnosis and therapy of genital papillomavirus infections, Hamburg, Feb 3–5, 1989 (abstracted)
6. Esposito R, Orlando G,Lazzarin A, Chianura Castagna A, Verani P, Noroni M (1986) Recombinant alpha interferon treatment of AIDS-related Kaposi sarcoma. J Interferon Res 6:46
7. Gall SA, Hughes CE, Trofatter K (1985) Interferon for the therapy of condylomata acuminata. Am J Obstet Gynecol 153(2):157–163
8. Gissmann L, Gross G (1985) Association of human papilloma viruses with human genital tumors. Clin Invest Dermatol 3:124–126
9. Gross G, de Villiers EM, Roussaki A, Papendick U, Schöpf E (1986) Successful treatment of condylomata acuminata and bowenoid papulosis with subcutaneous injections of low dose recombinant interferon-alpha. Arch Dermatol 122:749–750
10. Gross G, Pfister H, Gissmann L, Hagedorn M (1982) Correlation between human papillomavirus (HPV) type and histology of warts. J Invest Dermatol 78:160–164
11. Hohenleutner U, Landthaler M, Braun-Falco O, Schmoeckel C, Hans D (1988) Condylomata acuminata gigantea; Behandlung mit dem CO_2-Laser und Interferon. Dtsch Med Wochenschr 113:985–987
12. Jensen SL (1985) Comparison of podophyllin application with simple surgical excision in clearance of perianal condylomata acuminata. Lancet II:1146–1148
13. Kirchner H (1984) Interferons. A group of multiple lymphokines. Immunopathology 2:347–374
14. Lowy DR, Androphy EJ (1987) Warts. In: Fitzpatric TB, Eisen AZ, Wolff K, Freedberg IM, Austen KF (eds) Dermatology in general Medicine, 3rd edn. McGraw-Hill, New York, pp 2355–2372
15. Duus BR, Philipsen T, Christensen JD, Lundvall F, Søndergaard J (1985) Refractory condylomata acuminata: a controlled clinical trial of carbon dioxide laser versus conventional surgical treatment. Genitourin Med 61:59–61
16. Tiedemann KH, Ernst ThM (1988) Kombinationstherapie von rezidivierenden Condylomata acuminata mit Elektrokaustik und Alpha-2-Interferon. Aktuel Dermatol 14:200–204

17. Trofatter KFJ, English PC, Hughes CE, Gall SA (1986) Human lymphoblastoid interferon in primary therapy of two children with condylomata acuminata. Obstet Gynecol 67(1):137–140
18. Vance JB, Bart BJ, Hansen RC, Reichman RC, Mcewen C, Hatch KD, Berman B, Tanner DJ (1986) Intralesional recombinant alpha-2 interferon for the treatment of patients with condylomata acuminata or verruca plantaris. Arch Dermatol 122:272–277
19. Weck PhD, Debra A, Buddin MSN, Wisnant JK (1988) Interferons in the treatment of genital human papillomavirus infections. Am J Med 85 [Suppl 2a]:159–164

Subject Index

SPRINGER NATURE

GPSR Compliance

The European Union's (EU) General Product Safety Regulation (GPSR) is a set of rules that requires consumer products to be safe and our obligations to ensure this.

If you have any concerns about our products, you can contact us on ProductSafety@springernature.com

In case Publisher is established outside the EU, the EU authorized representative is:

Springer Nature Customer Service Center GmbH
Europaplatz 3
69115 Heidelberg, Germany

FSC
www.fsc.org
MIX
Papier | Fördert gute Waldnutzung
FSC® C083411

Kolibri 360 GmbH
Ferdinand-Jühlke-Straße 7
99095 Erfurt, Deutschland
produktsicherheit@kolibri360.de